Methods in Enzymology

Volume 322
APOPTOSIS

METHODS IN ENZYMOLOGY

EDITORS-IN-CHIEF

John N. Abelson   Melvin I. Simon

DIVISION OF BIOLOGY
CALIFORNIA INSTITUTE OF TECHNOLOGY
PASADENA, CALIFORNIA

FOUNDING EDITORS

Sidney P. Colowick and Nathan O. Kaplan

Methods in Enzymology

Volume 322

Apoptosis

EDITED BY

John C. Reed

THE BURNHAM INSTITUTE
LA JOLLA, CALIFORNIA

ACADEMIC PRESS

San Diego   London   Boston   New York   Sydney   Tokyo   Toronto

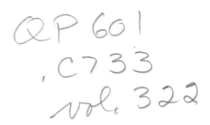

This book is printed on acid-free paper.

Academic Press
A Harcourt Science and Technology Company
525 B Street, Suite 1900, San Diego, California 92101-4495, USA
http://www.academicpress.com

Academic Press Limited
Harcourt Place, 32 Jamestown Road, London NW1 7BY, UK

International Standard Book Number: 0-12-182223-0

PRINTED IN THE UNITED STATES OF AMERICA
00 01 02 03 04 05 06 MM 9 8 7 6 5 4 3 2 1

Table of Contents

Section I. Measuring Apoptosis and Apoptosis-Induced Endonucleases

Section II. Measuring Apoptosis in Lower Organisms

Section III. Proteases Involved in Apoptosis and Their Inhibitors

Section IV. Cell-Free Systems for Monitoring Steps in Apoptosis Pathways

Section V. Mitochondria and Apoptosis

Section VI. Bcl-2 Family Proteins

Section VII. Studying Receptors and Signal Transduction Events Implicated in Cell Survival and Cell Death

Section VIII. Other Methods

Contributors to Volume 322

Article numbers are in parentheses following the names of contributors.
Affiliations listed are current.

JOHN M. ABRAMS (6), *Department of Cell Biology, University of Texas Southwestern Medical Center, Dallas, Texas 75390-9039*

SOUICHI ADACHI (20), *Department of Pediatrics, Kyoto University, Kyoto, Japan 606-8057*

ALEXANDER Y. ANDREYEV (21), *Mitokor, San Diego, California 92121*

MICHELE BARRY (4), *Department of Biochemistry, University of Alberta, Edmonton, Alberta, Canada, T6G 2H7*

ELZBIETA BEDNER (3), *Department of Pathology, Pomeranian School of Medicine, Szczecin, Poland*

R. CHRIS BLEACKLEY (4), *Department of Biochemistry, University of Alberta, Edmonton, Alberta, Canada, T6G 2H7*

CARL D. BORTNER (39), *Laboratory of Signal Transduction, National Institute of Environmental Health Sciences, National Institutes of Health, Research Triangle Park, North Carolina 27709*

RON BOSE (32, 33), *School of Medicine at Washington, University Medical Center, Washington University, St. Louis, Missouri 63110*

ELLA BOSSY-WETZEL (2, 22), *Division of Cellular Immunology, La Jolla Institute for Allergy and Immunology, San Diego, California 92121*

DALE E. BREDESEN (38), *Program on Aging, Buck Center for Research in Aging, Novato, California 94945*

CATHERINE BRENNER (23), *Centre National de la Recherche Scientifique, Unité Propre de Recherche 420, F-94801 Villejuif, France*

ROBIN BROWN (28), *Glaxo Wellcome Medicines Research Centre, Cell Biology Unit, Stevenage, Hertfordshire, SG1 2NY United Kingdom*

KRISTINE BUTROVICH (30), *Division of Molecular Immunology, La Jolla Institute for Allergy and Immunology, San Diego, California 92121*

ELEONORA CANDI (40), *Department of Experimental Medicine and Biochemical Sciences, University of Rome Tor Vergata, 00133 Rome, Italy*

KEVIN T. CHAPMAN (9), *Department of Enzymology, Merck Research Laboratories, Rahway, New Jersey*

PO CHEN (6), *Department of Cell Biology, University of Texas Southwestern Medical Center, Dallas, Texas 75390-9039*

JOHN A. CIDLOWSKI (5, 39), *Laboratory of Signal Transduction, National Institute of Environmental Health Sciences, National Institutes of Health, Research Triangle Park, North Carolina 27709*

ZBIGNIEW DARZYNKIEWICZ (3), *Brander Cancer Research Institute, New York Medical College, Valhalla, New York 10532*

HELENA L. DE ARAUJO VIEIRA (23), *Centre National de la Recherche Scientifique, Unité Propre de Recherche 420, F-94801 Villejuif, France*

QUINN L. DEVERAUX (13), *The Burnham Institute, La Jolla, California 92137*

JOSÉ-LUIS DIAZ (24), *Almirall Prodesfarma, 08024 Barcelona, Spain*

JOSEPH A. DIDONATO (36), *Department of Medicine, University of California San Diego, La Jolla, California 92093-0636*

STEVEN F. DOWDY (44), *Howard Hughes Medical Institute and Division of Molecular Oncology, Departments of Pathology and Medicine, Washington University School of Medicine, St. Louis, Missouri 06311*

MONICA DRISCOLL (7), *Department of Molecular Biology and Biochemistry, Rutgers University Center for Advanced Biotechnology and Medicine, Piscataway, New Jersey 08855*

WILLIAM C. EARNSHAW (1), *Institute of Cell and Molecular Biology, University of Edinburgh, Edinburgh, Scotland*

LISA M. ELLERBY (38), *Program on Aging, Buck Center for Research in Aging, Novato, California 94945*

GARY FISKUM (21), *Department of Anesthesiology, University of Maryland, Baltimore, Maryland 21201*

THOMAS F. FRANKE (37), *Department of Pharmacology, College of Physicians and Surgeons, Columbia University, New York, New York 10032*

STEVEN M. FRISCH (41), *The Burnham Institute, La Jolla, California 92037*

LAWRENCE C. FRITZ (24), *Idun Pharmaceuticals, Inc., La Jolla, California 92037*

CHRISTOPHER J. FROELICH (11), *Department of Medicine, Northwest University Medical School, Evanston, Illinois 60201-1797*

MAURICE GEUSKENS (18), *Department of Molecular Biology, Université Libre de Bruxelles, 1640 Rhode-Saint-Genèse, Bruxelles, Belgium*

ROBERTA A. GOTTLIEB (20), *Division of Hematology, Department of Molecular & Experimental Medicine, The Scripps Research Institute, La Jolla, California 92037*

DOUGLAS R. GREEN (2, 22), *Division of Cellular Immunology, La Jolla Institute for Allergy and Immunology, San Diego, California 92121*

ARNOLD H. GREENBURG (11), *Manitoba Institute of Cell Biology, University of Manitoba, Winnepeg, MB, Canada R3E 0V9*

ERICH GULBINS (32), *Division of Cell Biology, Institute of Physiology, University of Tuebingen, 72076 Tuebingen, Germany*

J. MARIE HARDWICK (43), *Departments of Molecular Microbiology and Immunology, Neurology, and Pharmacology and Molecular Sciences, Johns Hopkins Schools of Public Health and Medicine, Baltimore, Maryland 21205*

CHRISTINE J. HAWKINS (14), *Department of Haematology and Oncology, Royal Children's Hospital, Parkville, VIC 3052, Australia*

BRUCE A. HAY (14), *Division of Biology, California Institute of Technology, Pasadena, California 91125*

JEFFREY HEIBEIN (4), *Department of Biochemistry, University of Alberta, Edmonton, Alberta, Canada, T6G 2H7*

FRANCIS M. HUGHES, JR. (5), *University of North Carolina at Charlotte, Charlotte, North Carolina 28223*

MICHAEL KARIN (35), *Department of Pharmacology, Laboratory of Gene Regulation and Signal Transduction, University of California San Diego, La Jolla, California 92093-0636*

SCOTT H. KAUFMANN (1), *Division of Oncology Research, Mayo Clinic, and Department of Molecular Pharmacology, Mayo Graduate School, Rochester, Minnesota 55905*

NING KE (27), *The Burnham Institute, La Jolla, California 92037*

FRANK C. KISCHKEL (31), *Department of Molecular Oncology, Genentech Inc., San Francisco, California 96305*

RICHARD KOLESNICK (32, 33, 34), *Department of Molecular Pharmacology, Memorial Sloan-Kettering Cancer Center, New York, New York 10021*

ALICIA J. KOWALTOWSKI (21), *Department of Biochemistry and Molecular Biology, Oregon Graduate Institute of Science and Technology, Portland, Oregon 97291-1000*

PETER H. KRAMMER (31), *Tumor Immunology Program, German Cancer Research Center, 69120 Heidelberg, Germany*

GUIDO KROEMER (17, 18, 19, 23), *Centre National de la Recherche Scientifique, Unité Propre de Recherche 420, F-94801 Villejuif, France*

YULIA E. KUSHNAREVA (21), *Department of Anesthesiology, University of Maryland, Baltimore, Maryland 21201*

NATHANAEL LAROCHETTE (18), *Centre National de la Recherche Scientifique, Unité Propre de Recherche 420, F-94801 Villejuif, France*

DUNCAN LEDWICH (7), *Department of Molecular, Cellular, and Developmental Biology, University of Colorado at Boulder, Boulder, Colorado 80309-0437*

BETH LEVINE (43), *Department of Medicine, Columbia University College of Physicians and Surgeons, New York, New York 10032*

NATALIE A. LISSY (44), *Howard Hughes Medical Institute and Division of Molecular Oncology, Departments of Pathology and Medicine, Washington University School of Medicine, St. Louis, Missouri 06311*

XUESONG LIU (15), *Howard Hughes Medical Institute and Department of Biochemistry, University of Texas Southwestern Medical Center, Dallas, Texas 75235*

HANS K. LORENZO (17), *Unité de Biochemie Structurale, Institut Pasteur, F-75724 Paris, France*

ISABEL MARZO (23), *Centre National de la Recherche Scientifique, Unité Propre de Recherche 420, F-94801 Villejuif, France*

SHIGEMI MATSUYAMA (27), *The Burnham Institute, La Jolla, California 92037*

ERIC MELDRUM (28), *Glaxo Wellcome Medicines Research Centre, Cell Biology Unit, Stevenage, Hertfordshire, SG1 2NY United Kingdom*

GERRY MELINO (40), *IDI-IRCCS Biochemistry Lab, University of Rome Tor Vergata, 00133 Rome, Italy*

PETER W. MESNER, JR. (1), *Division of Oncology Research, Mayo Clinic, Rochester, Minnesota*

DIDIER MÉTIVIER (19), *Centre National de la Recherche Scientifique, Unité Propre de Recherche 420, F-94801 Villejuif, France*

MASAYUKI MIURA (42), *Department of Neuroanatomy, Osaka University Medical School, Osaka 565-0871, Japan*

DONALD D. NEWMEYER (16), *La Jolla Institute for Allergy and Immunology, San Diego, California 92121*

DONALD W. NICHOLSON (9, 10), *Departments of Pharmacology, Biochemistry and Molecular Biology, Merck Frosst Centre for Therapeutic Research, Montreal, Quebec, Canada H9R 4P8*

TILMAN OLTERSDORF (24), *Idun Pharmaceuticals, Inc., La Jolla, California 92037*

MARCUS E. PETER (31), *University of Chicago, Ben May Institute for Cancer Research, Chicago, Illinois 60637*

MICHAEL PINKOSKI (4), *Department of Biochemistry, University of Alberta, Edmonton, Alberta, Canada, T6G 2H7*

JOHN C. REED (13, 25, 26, 27), *The Burnham Institute, La Jolla, California 92037*

ISABELLE ROONEY (30), *Division of Molecular Immunology, La Jolla Institute for Allergy and Immunology, San Diego, California 92121*

SOPHIE ROY (10), *Department of Biochemistry and Molecular Biology, Merck Frosst Centre for Therapeutic Research, Montreal, Quebec, Canada H9R 4P8*

GUY S. SALVESEN (8, 12), *The Burnham Institute, La Jolla, California 92037*

KUMIKO SAMEJIMA (1), *Institute of Cell and Molecular Biology, University of Edinburgh, Edinburgh, Scotland*

CARSTEN SCAFFIDI (31), *Laboratory of Immunogenetics, NIAID, National Institutes of Health, Rockville, Maryland 20852*

PASCAL SCHNEIDER (29), *Institute of Biochemistry, University of Lausanne, CH-1066 Epalinges, Switzerland*

SHARON L. SCHENDEL (26), *The Burnham Institute, La Jolla, California 92037*

LIANFA SHI (11), *Manitoba Institute of Cell Biology, University of Manitoba, Winnepeg, MB, Canada R3E 0V9*

ANATOLY A. STARKOV (21), *Department of Anesthesiology, University of Maryland, Baltimore, Maryland 21201*

PETER M. STEINERT (40), *Laboratory of Skin Biology, NIAMS, National Institutes of Health, Bethesda, Maryland 20892-2752*

HENNING R. STENNICKE (8), *The Burnham Institute, La Jolla, California 92037*

TATSUHIKO SUDO (35), *Biodesign Group, The Institute of Physical and Chemical Research, Hiroshima, Saitama 35101, Japan*

SANTOS A. SUSIN (17, 18), *Centre National de la Recherche Scientifique, Unité Propre de Recherche 420, F-94801 Villejuif, France*

NANCY A. THORNBERRY (9), *Department of Enzymology, Merck Research Laboratories, Rahway, New Jersey*

SHIGENOBU TONÉ (1), *Institute of Cell and Molecular Biology, University of Edinburgh, Edinburgh, Scotland*

CATHERINE N. TORGLER (28), *Glaxo Wellcome Medicines Research Centre, Cell Biology Unit, Stevenage, Hertfordshire, SG1 2NY United Kingdom*

ADAMINA VOCERO-AKBANI (44), *Howard Hughes Medical Institute and Division of Molecular Oncology, Departments of Pathology and Medicine, Washington University School of Medicine, St. Louis, Missouri 06311*

OLIVER VON AHSEN (16), *La Jolla Institute for Allergy and Immunology, San Diego, California 92121*

SUSAN L. WANG (14), *Division of Biology, California Institute of Technology, Pasadena, California 91125*

XIAODONG WANG (15), *Howard Hughes Medical Institute and Department of Biochemistry, University of Texas Southwestern Medical Center, Dallas, Texas 75235*

CARL F. WARE (30), *Division of Molecular Immunology, La Jolla Institute for Allergy and Immunology, San Diego, California 92121*

KATE WELSH (13), *The Burnham Institute, La Jolla, California 92137*

YI-CHUN WU (7), *Zoology Department, National Taiwan University, Taipei, 10617, Taiwan, Republic of China*

ZHIHUA XIE (25), *The Burnham Institute, La Jolla, California 92037*

QUNLI XU (27), *The Burnham Institute, La Jolla, California 92037*

DING XUE (7), *Department of Molecular, Cellular and Developmental Biology, University of Colorado at Boulder, Boulder, Colorado 80309-0437*

XIAOHE YANG (11), *Department of Medicine, Northwest University Medical School, Evanston, Illinois 60201-1797*

JUNYING YUAN (42), *Department of Cell Biology, Harvard Medical School, Boston, Massachusetts 02115*

NAOUFAL ZAMZAMI (19), *Centre National de la Recherche Scientifique, Unité Propre de Recherche 420, F-94801 Villejuif, France*

QIAO ZHOU (12), *The Burnham Institute, La Jolla, California 92037*

Preface

At present, apoptosis research represents the fastest growing area of scientific inquiry in all of the life sciences. The importance of programmed cell death in a broad range of normal physiological processes has attracted thousands of researchers to the area. Moreover, the clear relevance of apoptosis to many diseases has also encouraged a veritable stampede of biomedical researchers to the field. It is entirely appropriate, therefore, that this volume of *"Methods in Enzymology"* be devoted to apoptosis.

Like any field of biomedical research, and perhaps more than many, apoptosis and cell death research is rapidly evolving and so too are the methods used in its study. Indeed, the list of topics directly relevant to apoptosis research grows yearly as additional connections are made between known components of cell death pathways and new, sometimes unexpected, molecules or systems, which occasionally can represent entire disciplines of biochemistry and molecular biology by themselves. Thus we offer this contribution with the proviso that it is a work in progress, which hopefully will evolve along with the field in later volumes of *Methods in Enzymology*. It is in part because of the rapidly evolving nature of the field that this book contains what at first glance may appear to be redundancies. In such cases, it was our intention to offer different perspectives and approaches, given that our knowledge is too limited at present to dogmatically promulgate one methodological approach over another. Despite these caveats, it is my hope that this volume will serve the many scientists presently involved in apoptosis research and those who will come anew to the field in the years ahead as graduate students, postdoctoral fellows, or seasoned investigators, making their work just a bit easier and progress in the field more rapid.

It has been a privilege to serve as the editor of this *Methods in Enzymology* volume devoted to apoptosis. The knowledge and experience embodied in the 44 chapters of this book represent the collective wisdom and hard-earned empirical observations of 101 of the world's most talented contributors to the field of apoptosis and cell death research. It has been an honor to work with each of the authors in assembling this compilation of protocols and procedures, and my only regret is that I was unable to either invite or coerce other top scientists in the field to add their experiences and know-how to the text. Any failings of the volume reflect my own inadequacies and not those of the authors to whom we all owe a debt of gratitude for sharing their gifts of knowledge with us. I also owe thanks to Susan Farrar, Eric Smith, and Tara Brown for helping assemble the chapters and to the staff of Academic Press for their patience and steadfast dedication to the volume.

JOHN C. REED

METHODS IN ENZYMOLOGY

Section I

Measuring Apoptosis and Apoptosis-Induced Endonucleases

[1] Detection of DNA Cleavage in Apoptotic Cells

By Scott H. Kaufmann, Peter W. Mesner, Jr., Kumiko Samejima,
Shigenobu Toné, and William C. Earnshaw

At least two discrete deoxyribonuclease activities can be detected during apoptotic death, one that generates 30- to 500-kilobase pair (kbp) domain-sized fragments and another that mediates internucleosomal DNA degradation. The latter nuclease has been identified as the caspase-activated deoxyribonuclease (CAD)/CPAN, a unique enzyme that is normally inhibited by the regulatory subunit ICAD (inhibitor of CAD)/DFF45 (DNA fragmentation factor). In this chapter, techniques widely used to detect DNA cleavage in apoptotic cells, including pulsed-field gel electrophoresis, conventional agarose gel electrophoresis, and terminal transferase-mediated dUTP nick end-labeling (TUNEL), are briefly reviewed. In addition, the use of ICAD to inhibit apoptosis-associated nuclease activity is illustrated. When properly applied, these techniques are widely applicable to the characterization of apoptotic cells.

Introduction

DNA cleavage is widely observed in dying cells that display apoptotic morphological changes. In contrast to the random DNA degradation that occurs in necrotic cells, DNA degradation in apoptotic cells most often involves generation of large DNA fragments—so-called domain-sized 50- to 300-kilobase pair fragments[1-3]—followed by the appearance of a ladder of low molecular weight fragments that are multiples of ~180 bp[4] and reflect double-strand cleavage of linker DNA between nucleosomes.[5-7] It should be kept in mind, however, that other DNA alterations have also been reported in apoptotic cells, including DNA modifications that intro-

[1] P. R. Walker, C. Smith, T. Youdale, J. Leblanc, J. F. Whitfield, and M. Sikorska, *Cancer Res.* **51,** 1078 (1991).
[2] C. E. Canman, H. Tang, D. P. Normolle, T. S. Lawrence, and J. Maybaum, *Proc. Natl. Acad. Sci. U.S.A.* **89,** 10474 (1992).
[3] F. Oberhammer, J. W. Wilson, C. Dive, I. D. Morris, J. A. Hickman, A. E. Wakeling, P. R. Walker, and M. Sikorska, *EMBO J.* **12,** 3679 (1993).
[4] A. H. Wyllie, *Nature (London)* **284,** 555 (1980).
[5] D. R. Hewish and L. A. Burgoyne, *Biochem. Biophys. Res. Commun.* **52,** 504 (1973).
[6] M. Noll, *Nature (London)* **251,** 249 (1974).
[7] L. A. Burgoyne, D. R. Hewish, and J. Mobbs, *Biochem J.* **143,** 67 (1974).

duce alkali-sensitive sites at regular intervals[8] and result in increased levels of single-stranded DNA after heat treatment.[9]

The nuclease(s) responsible for apoptotic DNA cleavage have been the subject of considerable investigation. Although earlier studies implicated deoxyribonuclease I,[10] deoxyribonuclease II,[11] DNase γ,[12,13] or cyclophilins[14] in apoptotic DNA degradation, experiments demonstrating that internucleosomal cleavage occurs downstream of caspase activation[15,16] have focused attention on a magnesium-dependent caspase-activated deoxyribonuclease called CAD[17] or CPAN.[18] Forced overexpression of the endogenous inhibitor of CAD [ICAD, also known as DFF45[16]] prevents internucleosomal DNA degradation in apoptotic cells,[19] strongly implicating CAD in apoptotic internucleosomal DNA cleavage. Although the effect of ICAD on the generation of the high molecular weight DNA fragments has not been reported, the observation that these fragments can occur in the absence of internucleosomal DNA degradation[3,20] suggests the action of a separate domain endonuclease that remains to be identified.

In the following sections we describe several biochemical and histochemical assays for the nuclease activities that accompany apoptosis. In addition, we describe the use of ICAD/DFF45 to characterize nucleases present in cellular extracts capable of inducing apoptotic changes in exogenous nuclei *ex vivo.* Flow cytometry-based assays for DNA degradation are described in [3] in this volume.[21]

[8] L. D. Tomei, J. P. Shapiro, and F. O. Cope, *Proc. Natl. Acad. Sci. U.S.A.* **90**, 853 (1993).

[9] O. S. Frankfurt, J. A. Robb, E. V. Sugarbaker, and L. Villa, *Exp. Cell Res.* **226**, 387 (1996).

[10] M. C. Peitsch, B. Polzar, H. Stephan, T. Crompton, H. R. MacDonald, H. G. Mannherz, and J. Tschopp, *EMBO J.* **12**, 371 (1993).

[11] M. A. Barry and A. Eastman, *Arch. Biochem. Biophys.* **300**, 440 (1993).

[12] D. Shiokawa, H. Ohyama, T. Yamada, K. Takahashi, and S.-I. Tanuma, *Eur. J. Biochem.* **226**, 23 (1994).

[13] D. Shiokawa, H. Ohyama, T. Yamada, and S. Tanuma, *Biochem. J.* **326**, 675 (1997).

[14] J. W. Montague, M. L. Gaido, C. Frye, and J. A. Cidlowski, *J. Biol. Chem.* **269**, 18877 (1994).

[15] Y. A. Lazebnik, S. H. Kaufmann, S. Desnoyers, G. G. Poirier, and W. C. Earnshaw, *Nature (London)* **371**, 346 (1994).

[16] X. Liu, H. Zou, C. Slaughter, and X. Wang, *Cell* **89**, 175 (1997).

[17] M. Enari, H. Sakahira, H. Yokoyama, K. Okawa, A. Iwamatsu, and S. Nagata, *Nature (London)* **391**, 43 (1998).

[18] R. Halenbeck, H. MacDonald, A. Roulston, T. T. Chen, L. Conroy, and L. T. Williams, *Curr. Biol.* **8**, 537 (1998).

[19] H. Sakahira, M. Enari, and S. Nagata, *Nature (London)* **391**, 96 (1998).

[20] X. M. Sun and G. M. Cohen, *J. Biol. Chem.* **269**, 14857 (1994).

[21] Z. Darzynkiewicz and E. Bedner, *Methods Enzymol.* **322**, [3], (2000) (this volume).

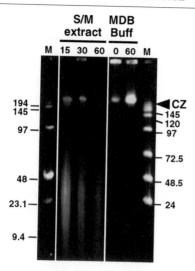

FIG. 1. Pulsed-field gel electrophoresis. Isolated HeLa nuclei were incubated in apoptosis-inducing S/M extracts or in MDB buffer[53] for 0–60 min as indicated, embedded in agarose, and subjected to CHEF gel electrophoresis as described in text. M, Molecular weight markers. Note that DNA above the resolution limit of the gel runs as a sharp band in the compression zone (CZ), which can be mistaken for a discrete fragment.

Pulsed-Field Gel Electrophoresis

Pulsed-field gel electrophoresis can be utilized to detect domain-sized fragments that are generated from genomic DNA during the course of apoptosis (Fig. 1). To examine DNA by this technique, cells or nuclei are encapsulated in agarose to protect the DNA from shearing during subsequent manipulation, lysed in a deproteinizing detergent such as sodium dodecyl sulfate (SDS) or sarkosyl, treated with proteinase K, embedded in the wells of an agarose gel, and subjected to electrophoresis under conditions in which the direction and, if desired, the strength of the electric field are varied during the course of the run. This pulsed alteration in the electrical field causes long DNA molecules to reorient during the course of electrophoresis.[22] Because smaller molecules reorient more quickly than larger molecules, fragments up to 6 megabases (Mb) can be resolved by this approach if gels are run for a sufficient length of time (up to 1 week). The three pulsed-field gel techniques that have been used most widely are orthogonal field-alternating gel electrophoresis (OFAGE[23]), field inversion

[22] D. C. Schwartz and M. Koval, *Nature (London)* **338**, 520 (1989).
[23] D. C. Schwartz and C. R. Cantor, *Cell* **37**, 67 (1984).

gel electrophoresis (FIGE[24]), and contour-clamped homogeneous field gel electrophoresis (CHEF[25]). These are all "tuned" methods, i.e., each method resolves DNA in differing size ranges depending on the pulse amplitudes and frequencies. In each case, DNA above the limit of resolution runs as a relatively sharp compressed band (indicated by CZ in Figs. 1 and 4A). This is an artifact, not a single discrete species. It is also important to note that certain techniques, FIGE in particular, can give different results depending on the pulse program used.[26]

Materials

Suitable molecular weight standards, e.g., oligomers of large viral genomes (λ phage) as well as chromosomes of either the budding *Saccharomyces cerevisiae* (0.26–3 Mb) or fission yeast *Schizosaccharomyces pombe* (3.5–5.5 Mb)

Low melting point agarose (Sea Plaque GTG agarose; FMC BioProducts, Rockland, ME)

Lysis buffer [0.5 M EDTA, 10 mM Tris-HCl (pH 9.5), proteinase K (1 mg/ml), 1% sarkosyl]

TE buffer: 10 mM Tris-HCl (pH 8.0) containing 1 mM EDTA

Contour-clamped homogeneous (CHEF) electrophoresis system (e.g., Bio-Rad, Hercules, CA)

Pulse controller (e.g., Bio-Rad Pulsewave 760 switcher)

Procedure

1. Cells are gently sedimented, resuspended in a small volume of suitable tissue culture medium, diluted with an equal volume of low melting point agar [3% (w/v) GTG agarose dissolved in medium and cooled to 37° prior to use], poured into a suitable mold, and allowed to harden at 4–21°.

2. Gel blocks are soaked in lysis buffer for 20 hr at 50° and then stored in TE buffer at 4°.

3. Blocks are transferred into the wells of a 1% (w/v) agarose gel and sealed in place by the addition of a small volume of agarose.

4. Electrophoresis is performed with a horizontal gel electrophoresis apparatus and a suitable pulse controller to vary current direction. We run gels under thermostatically controlled conditions (14°) in a Bio-Rad CHEF system, using the following parameters: 200 V for 20 hr with a pulse ramp of 4 to 40 sec. Alternatively, samples can be subjected to FIGE, using a

---

[24] H. J. Dawkins, D. J. Ferrier, and T. L. Spencer, *Nucleic Acids Res.* **15**, 3634 (1987).

[25] G. Chu, D. Vollrath, and R. W. Davis, *Science* **234**, 1582 (1986).

[26] P. V. Gejman, N. Sitaram, W. T. Hsieh, J. Gelernter, and E. S. Gershon, *Appl. Theor. Electrophor.* **1**, 29 (1988).

conventional horizontal agarose gel electrophoresis apparatus and a suitable pulse controller.[27,28]

5. Gels are stained with ethidium bromide (0.1 μg/ml) and photographed under UV illumination.

Conventional Agarose Gel Electrophoresis

Conventional agarose gel electrophoresis, which is used to detect internucleosomal DNA degradation in isolated nuclei, cells, or tissues, can be performed according to any of a variety of protocols. One commonly utilized protocol involves lysis of cells in SDS and EDTA, digestion with proteinase K, extraction with phenol to remove peptide fragments, application of the resulting DNA to a gel with suitable separation properties [e.g., 1–2% (w/v) agarose], and staining with ethidium bromide (Fig. 2[28a]). Variations on this method include (1) lysis of cells in a nondenaturing buffer [e.g., 20 mM Tris (pH 7–8) containing EDTA and a neutral detergent] to extract the chromatin fragments below 10–20 kb,[29,30] which can then be quantitated and run on an agarose gel, thereby enhancing the sensitivity of the method for detection of low amounts of internucleosomal fragments, (2) direct end labeling of the free 3' ends in the SDS/proteinase K-treated samples with a single α-^{32}P-labeled dNTP and DNA polymerase or terminal deoxynucleotidyl transferase (TdT) prior to electrophoresis, with autoradiographic detection after electrophoresis to enhance fragment detection,[31,32] (3) transfer of the unlabeled DNA on the agarose gel to a nylon support followed by hybridization with a ^{32}P-labeled genomic DNA probe and autoradiography to detect nucleosomal fragments,[33] and (4) use of DNA-binding dyes with enhanced quantum yield [e.g., SYBR Green from Molecular Probes (Eugene, OR)] to increase the ability to detect nucleosomal ladders. Because of difficulties in quantitating the DNA fragments,[34] all these techniques should be viewed as qualitative rather than quantitative.

Materials

Suitable molecular weight standards, e.g., a 100-bp ladder or *Hind*III digestion products of λ phage (Life Sciences, Gaithersburg, MD)

[27] G. Roy, J. C. Wallenburg, and P. Chartrand, *Nucleic Acids Res.* **16,** 768 (1988).
[28] N. Denko, A. Giaccia, B. Peters, and T. D. Stamato, *Anal. Biochem.* **178,** 172 (1989).
[28a] S. H. Kaufmann, *Cancer Res.* **49,** 5870 (1989).
[29] T. Igo-Kemenes, W. Greil, and H. G. Zachau, *Nucleic Acids Res.* **4,** 3387 (1977).
[30] N. Kyprianou and J. T. Isaacs, *Endocrinology* **122,** 552 (1988).
[31] F. Rösl, *Nucleic Acids Res.* **20,** 5243 (1992).
[32] J. Tilly and A. Hsueh, *J. Cell. Physiol.* **154,** 519 (1993).
[33] P. W. Mesner, T. R. Winters, and S. H. Green, *J. Cell Biol.* **119,** 1669 (1992).
[34] P. Mesner and S. H. Kaufmann, *Adv. Pharmacol.* **41,** 57 (1997).

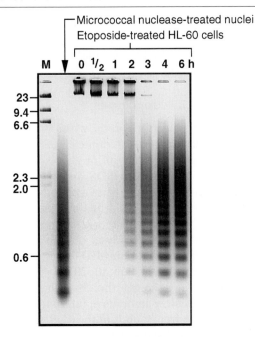

FIG. 2. Conventional agarose gel electrophoresis. HL-60 human leukemia cells were incubated with the topoisomerase-directed chemotherapeutic agent etoposide for the indicated length of time, then prepared for conventional agarose gel electrophoresis as described in text. Samples were separated on a 1.2% (w/v) agarose gel in TAE buffer and visualized by staining with ethidium bromide (0.1 μg/ml). Adjacent wells on the same gel contain λ HindIII molecular weight standards (M) and DNA from HL-60 nuclei that were isolated as previously described,[28a] digested with micrococcal nuclease *in vitro,* and prepared for electrophoresis in parallel with the cells.

Agarose (e.g., SeaKem LE agarose from FMC or ultrapure agarose from Life Sciences)

TE9S buffer [0.5 M EDTA, 10 mM NaCl, 10 mM Tris-HCl (pH 9), 1% (w/v) SDS]

A conventional horizontal gel electrophoresis apparatus and power supply

Procedure

1. Cells treated with a proapoptotic stimulus are gently sedimented and solubilized by vigorous vortexing in TE9S buffer containing proteinase K (1 mg/ml). If samples are lysed in a small volume (e.g., 20 μl/10^6 cells), it is possible to eliminate the ethanol precipitation step indicated below. Samples are incubated at 50° for 1–24 hr.

2. Protein contaminants are removed by two extractions of the aqueous phase with water-saturated 3:4 (v/v) phenol–chloroform. Peptides enter the organic phase, whereas DNA remains in the aqueous phase. Phenol is removed from the aqueous phase by one extraction with chloroform.

3. If necessary, the DNA can be precipitated by addition of ethanol to 70% (v/v) and NaCl (or sodium acetate) to 0.1 M final concentration. After centrifugation and washing with 70% ethanol, the DNA can be resuspended in a suitable volume of TE buffer.

4. Samples are digested with RNase A (1 mg/ml) for 1 hr at 20–22°.

5. After addition of tracking dye, DNA samples are applied to the wells of a 1.2–2% (w/v) agarose gel and subjected to conventional electrophoresis. Various buffer systems, including Tris–borate–EDTA and Tris–acetate–EDTA,[35] have been advocated for this electrophoresis.

6. Gels are stained with ethidium bromide (0.1 μg/ml) and photographed under UV light.

7. This separation method can be modified by use of alkaline electrophoresis conditions (e.g., 50 mM NaOH–1 mM EDTA) to detect alkali- and S1 nuclease-sensitive sites observed in some cells during the course of apoptosis.[8]

Terminal Transferase-Mediated dUTP Nick End-Labeling Method

To supplement the electrophoretic methods described above, DNA double-strand breaks can also be detected by several alternative approaches. The low molecular weight fragments that result from internucleosomal DNA degradation are readily extracted from ethanol-fixed cells by treatment with aqueous buffers,[36] providing the opportunity to distinguish apoptotic cells and nonapoptotic cells by flow microfluorimetry after staining with a DNA-binding dye such as propidium iodide.[37,38] Alternatively, the free DNA ends generated by internucleosomal cleavage can be labeled by enzymatic techniques.

Studies published to date have provided varying descriptions of the free ends that are generated during internucleosomal DNA degradation. Nucleosomal fragments generated during the course of glucocorticoid- or novobiocin-induced apoptosis in CEM human leukemia cells or rat thymo-

[35] L. G. Davis, W. M. Kuehl, and J. F. Battey, in "Basic Methods in Molecular Biology." Appleton & Lange, Norwalk, Connecticut, 1994.

[36] J. Gong, F. Traganos, and Z. Darzynkiewicz, *Anal. Biochem.* **218**, 314 (1994).

[37] V. N. Afanas'ev, B. A. Korol, Y. A. Mantsygin, P. A. Nelipovich, V. A. Pechantnikov, and S. R. Umansky, *FEBS Lett.* **194**, 347 (1986).

[38] Z. Darzynkiewicz, G. Juan, X. Li, W. Gorczyca, T. Murakami, and F. Traganos, *Cytometry* **27**, 1 (1997).

Termination buffer: 300 mM NaCl, 30 mM sodium citrate
Acetate buffer (pH 5.0), 50 mM
AEC reagent: Prepare fresh by dissolving 20 mg of 3-amino-9-ethylcar-
  bazole (Sigma, St. Louis, MO) in 2.5 ml of N,N-dimethylformamide,
  bringing to final volume of 50 ml with 50 mM acetate buffer (pH
  5.0), and adding 25 ml of 30% (w/v) H_2O_2 (Sigma) just before use

Method

1. All steps are performed at 20–22° unless otherwise specified.
2. Pretreatment: Cells mounted on slides (e.g., cytospins) or attached
to coverslips are fixed for 10 min in fresh 3.7% formaldehyde in PBS
followed by three 5-min washes in PBS. Paraffin-embedded tissue sections
should be baked at 65°; cooled; deparaffinized with two 5-min washes in
xylene; rehydrated by sequential washing with 100, 95, and 70% ethanol
(two 5-min washes in each) followed by two 10-min washes in PBS; treated
with proteinase K (20 mg/ml in PBS) for 15 min; and washed twice (5 min
each) with PBS.
3. If peroxidase will be used in the final staining step, endogenous
peroxidase activity must be quenched by reacting specimen with 3% (v/v)
H_2O_2 for 5 min followed by two 5-min washes in PBS.
4. If cells are mounted on a glass slide, the specimen can be circled
with a hydrophobic pen so that a small volume of solution placed atop the
specimen in future steps will not spread.
5. The specimen should be equilibrated in cacodylate buffer for 10 min.
Once equilibrated in cacodylate buffer, specimens should not be allowed to
dry.
6. After aspiration of excess cacodylate buffer, the specimen is immediately immersed in freshly prepared dUTP buffer and incubated at 37° for
1 hr in a humidified chamber. A mock reaction lacking TdT should be
included in each assay as a negative control.
7. The labeling reaction is terminated by immersing the specimen in
termination buffer for 15 min. Slides are then washed by immersion in PBS
for 10 min.
8. The specimen is overlaid with a small volume of peroxidase-labeled
streptavidin (1 : 100 in PBS). Anti-digoxigenin is substituted for streptavidin
if digoxigenin-labeled nucleotide is used. Samples are incubated at 37° for
30 min in a humidified chamber and then washed with several changes
of PBS.
9. If fluorochrome-tagged reagents are used, slides are mounted with
antifade reagent (e.g., Vectashield, Vector Laboratories, Burlingame, CA)
and visualized with a fluorescent microscope equipped with suitable filters.

10. If HRP-tagged reagents are used, specimens are immersed in a freshly prepared solution of AEC reagent for 5–10 min at 20–22°. They are then washed twice for 5 min in PBS and once for 5 min in H_2O. After samples are mounted in 50% glycerol, they are visualized using a suitable bright-field microscope.

Use of Caspase-Activated Deoxyribonuclease Inhibitor ICAD to Distinguish between Various Nucleases *In Vitro*

As indicated above, earlier experiments implicated DNase I, DNase II, and DNase γ in apoptotic cell death. More recently, the caspase-activated nuclease CAD/CPAN and its inhibitor ICAD/DFF45 have been reported to participate in this process.[16–19] The activities of these various nucleases can be distinguished *in vitro*. DNase I, DNase γ, and CAD/CPAN require divalent cations.[12,18,48] DNase II has an acid pH optimum.[49] Finally, CAD/CPAN is inhibited by the addition of exogenous ICAD/DFF45, whereas DNase I and DNase II are not.[17] This latter property can now be utilized as illustrated in Fig. 4 to probe cellular extracts for the presence of CAD activity. To prevent ICAD cleavage and CAD reactivation during experiments of this type, it is possible to add caspase inhibitors or use ICAD that has been mutated to remove the caspase cleavage site at Asp-117.[19]

Reagents

ICAD [conveniently prepared by affinity chromatography after expression as a fusion protein with glutathione transferase[19] or as a $(His)_6$-tagged protein (K. Samejima and W. C. Earnshaw, unpublished observations, 1998)]. This should be stored in CPAN buffer [10 mM HEPES (pH 7.4), 50 mM NaCl, 5 mM EGTA, 2 mM $MgCl_2$, 1 mM dithiothreitol (DTT)].

A suitable inhibitor of caspases, e.g., the broad-spectrum inhibitor carbonylbenzoxyvalinylalanylaspartylfluoromethyl ketone (zVAD-fmk)[50] or the inhibitor of caspase 3-like proteases acetylaspartylglutamylvalinylaspartylfluoromethyl ketone (DEVD-fmk).[51] In each case, it is important to use the free acid rather than the frequently supplied cell-permeant methyl ester, which inhibits caspase activity

[48] M. Kunitz, *J. Gen. Physiol.* **33,** 349 and 363 (1950).
[49] G. Bernardi and C. Sadron, *Biochemistry* **3,** 1411, (1964).
[50] N. Margolin, S. A. Raybuck, K. P. Wilson, W. Chen, T. Fox, Y. Gu, and D. J. Livingston, *J. Biol. Chem.* **272,** 7223 (1997).
[51] D. W. Nicholson, A. Ali, N. A. Thornberry, J. P. Vaillancourt, C. K. Ding, M. Gallant, Y. Gareau, P. R. Griffin, M. Labelle, and Y. A. Lazebnik, *Nature* (*London*) **376,** 37 (1995).

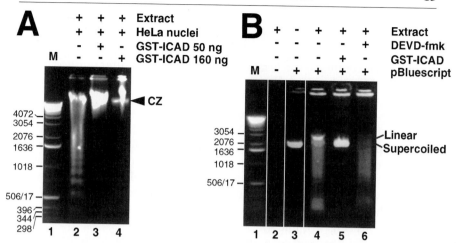

FIG. 4. Effect of ICAD on activity of a nuclease detected in extracts capable of inducing apoptotic changes in exogenous nuclei. (A) S/M extracts[53] were supplemented with increasing concentrations of GST–ICAD prior to incubation with exogenous HeLa nuclei as indicated in text. DNA was subjected to conventional agarose gel electrophoresis. M, Molecular weight markers; CZ, compression zone described in text. (B) Supercoiled pBluescript plasmid (lane 3) was incubated with untreated S/M extract (lane 4) or with S/M extract that had been pretreated with 160 ng of GST–ICAD (lane 5) or 100 μM DEVD-fmk (lane 6). The nuclease activity present in the S/M extracts linearizes the supercoiled substrate and produces shorter fragments (lane 4). Interestingly, this nuclease in extracts from preapoptotic chicken hepatoma cells is completely inhibited by mouse ICAD (lane 5), indicating conservation of the CAD/ICAD system across the phylogenetic tree.

poorly *in vitro* (L. M. Martins and W. C. Earnshaw, unpublished observation, 1998).

A suitable substrate, e.g., intact nuclei or purified plasmid.[17,52]

Method

1. Cellular extracts are treated with caspase inhibitor (e.g., 100 μM zVAD-fmk or DEVD-fmk) or diluent for 15 min at 37°. A high concentration of inhibitor is used to ensure that all detectable caspases are inhibited.

2. Purified ICAD [e.g., as the glutathione *S*-transferase (GST) fusion protein[19,52]] or corresponding buffer is added and the incubation is continued for an additional 10 min at 37°. Once again, a high concentration of

[52] K. Samejima, S. Toné, T. Kottke, M. Enari, H. Sakahira, C. A. Cooke, F. Durrieu, L. M. Martins, S. Nagata, S. H. Kaufmann, and W. C. Earnshaw, *J. Cell Biol.* **143**, 225 (1998).

the inhibitor (e.g., 30- to 100-μg/ml final concentration) should be used to ensure that all the CAD is complexed.

3. After substrate (e.g., 12 μg of the plasmid pBluescript) is added, incubation is continued for 30 min at 37°. Alternatively, after purified nuclei (e.g., 5 × 10^5 HeLa nuclei[53]) are added, incubation is continued for 1–2 hr at 37°.

4. If purified plasmid is used as a substrate, the reaction can be terminated with lysis buffer [50 mM Tris-HCl (pH 8.0), 10 mM EDTA, 0.5% sarkosyl, proteinase K (0.5 mg/ml)]. After incubation for 1 hr at 50°, plasmid can be extracted with phenol–chloroform and chloroform, ethanol precipitated, and applied to a 1% (w/v) agarose gel. Low levels of nuclease activity will be evident as cleavage of supercoiled substrate to linear plasmid, whereas higher levels of activity result in a smear of fragments below the linear species (Fig. 4B, line 4).

5. If nuclei are used as a substrate, they should be sedimented. All further incubations and extractions can then be performed as indicated in the description of conventional agarose gel electrophoresis (Fig. 2). Alternatively, sedimented nuclei can be resuspended in lysis buffer (see above) and incubated for 1 hr at 50°. DNA is then treated with RNase A (0.4 mg/ml) for 1 hr at 50°, extracted with phenol–chloroform and chloroform, ethanol precipitated, resuspended in TE buffer, and subjected to conventional agarose gel electrophoresis.

Concluding Remarks

DNA degradation has been widely observed in apoptotic cells. The electrophoretic techiques described in this chapter are useful for detecting this DNA degradation when it exists. It is important to realize, however, that internucleosomal DNA degradation appears to be a dispensable part of the apoptotic process. Several groups of investigators have described dying cells that display apoptotic morphological changes without any detectable internucleosomal DNA fragmentation.[3,19,54,55] It is not known at present whether the activity of the domain endonuclease is likewise dispensable. Accordingly, the DNA degradation detected by the techniques described in this chapter must be viewed as a marker of the apoptotic process rather than a critical component of the death process itself.

[53] Y. A. Lazebnik, S. Cole, C. A. Cooke, W. G. Nelson, and W. C. Earnshaw, *J. Cell Biol.* **123,** 7 (1993).
[54] D. S. Ucker, P. S. Obermiller, W. Eckhart, J. R. Apgar, N. A. Berger, and J. Meyers, *Mol. Cell. Biol.* **12,** 3060 (1992).
[55] D. G. Brown, X. M. Sun, and G. M. Cohen, *J. Biol. Chem.* **268,** 3037 (1993).

The identification of CAD as a caspase-activated deoxyribonuclease provides an explanation for the previous observation that caspase inhibitors prevent internucleosomal DNA degradation as well as protein cleavage in nuclei undergoing apoptosis.[15] In view of previous claims implicating other endonucleases in the apoptotic process (see Introduction), it is presently unclear whether CAD is responsible for internucleosomal DNA degradation in all cell types. Likewise, it is unclear whether CAD is responsible for generation of domain-sized fragments. The application of the ICAD-based inhibitor assay described above to suitable model systems should help answer these questions.

Acknowledgments

The authors gratefully acknowledge the kind gift of ICAD cDNA from Drs. M. Enari and S. Nagata and the secretarial assistance of Deb Strauss. Research in our laboratories has been supported by a grant from the NIH (CA 69008) as well as by funds from the Leukemia Society of America (S.H.K.) and the Wellcome Trust (W.C.E.).

[2] Detection of Apoptosis by Annexin V Labeling

By Ella Bossy-Wetzel *and* Douglas R. Green

Introduction

Apoptosis is associated with characteristic morphological and biochemical changes, including cell shrinkage, membrane blebbing, chromatin condensation, DNA fragmentation, and cell surface changes. Several techniques have been developed to measure DNA fragmentation, a late event in apoptosis. However, early detection of apoptosis is critical for the study of the central pathway leading to apoptosis. In this chapter we describe a sensitive and rapid method to detect early apoptosis. The assay is based on the observation that phosphatidylserine (PS), a phospholipid normally confined to the cytoplasmic face of the plasma membrane, translocates to the cell surface during apoptosis in most cell types and by many apoptotic stimuli. Externalization of PS to the cell surface marks the apoptotic cells to be recognized by neighboring cells or macrophages, facilitating the noninflammatory removal of dying cells by phagocytosis.

Once on the cell surface, PS can be detected by binding of fluorescein isothiocyanate (FITC)-labeled annexin V.[1-3] Annexin V protein belongs

[1] G. Koopman, C. P. M. Reutlingsperger, G. A. M. Kuijten, R. M. J. Keehnen, S. T. Pals, and M. H. J. van Oers, *Blood* **84**, 1415 (1994).

to a family of phospholipid-binding proteins that bind specifically, in the presence of calcium, to negatively charged phospholipids such as PS.[4] FITC-labeled annexin V-positive cells can be detected by flow cytometry or fluorescence microscopy. It is also possible to use annexin V labeling in combination with other dyes, such as propidium iodide and Hoechst 33342, which allows characterization of the progressive stages of apoptosis. In most cases, PS translocation to the cell surface occurs before DNA condensation, plasma membrane permeabilization, and membrane blebbing, thus serving as a rapid and convenient measure of early apoptosis.

Reagents and Buffers

Fluorescein isothiocyanate (FITC)-coupled annexin V (Clontech, Palo Alto, CA or Roche Molecular Biomedicals, Indianapolis, IN)

Annexin V-binding buffer: 10 mM HEPES (pH 7.4), 150 mM NaCl, 5 mM KCl, 1 mM MgCl$_2$, 1.8 mM CaCl$_2$

Propidium iodide stock solution: 100 μg/ml in 20 mM HEPES, pH 7.4

Hoechst 33342 stock solution: 100 μg/ml in 20 mM HEPES, pH 7.4

PBS–3% (w/v) bovine serum albumin (BSA): Sterilize by filtration and store at room temperature

Paraformaldehyde solution (4%): Prepare the fixative under a fume hood. Eight grams of paraformaldehyde (EM grade) is added to 100 ml of distilled H$_2$O and heated to 60° while mixing. To dissolve solids, add a few drops of 1 N NaOH until the solution becomes clear. The fixative is placed on ice to cool and 100 ml of 2× PBS is added. Last, the fixative is filtrated. The paraformaldehyde solution is prepared fresh each time or aliquots are stored at −20°

Antifade mounting reagent (Bio-Rad, Hercules, CA)

Equipment

Flow cytometer or fluorescence microscope

[2] C. H. E. Homburg, M. de Haas, A. E. G. K. von dem Borne, A. J. Verhoeven, C. P. M. Reutlingsperger, and D. Roos, *Blood* **85**, 532 (1995).

[3] S. J. Martin, C. P. M. Reutlingsperger, A. J. McGahon, J. A. Rader, R. C. van Schie, D. M. LaFace, and D. R. Green, *J. Exp. Med.* **182**, 1545 (1995).

[4] W. L. van Heerde, P. G. deGroot, and C. P. Reutlingsperger, *Thromb. Haemost.* **73**, 172 (1995).

Procedures

To determine apoptosis by annexin V binding, remove a 100-μl aliquot of cell suspension (approximately 1×10^5 cells) and transfer into a microtube. Pellet cells by centrifugation at 200g in a microcentrifuge for 5 min at room temperature. Aspirate the supernatant from the cell pellet and resuspend it in 200 μl of annexin V-binding buffer containing annexin V–FITC (1 μg/ml). For sensitive measure of apoptotic cell death by annexin V it is critical to use buffer containing 1.8 to 3 mM calcium. Incubate the sample for 10 to 15 min at 37°. The sample can be analyzed by flow cytometry without further wash steps. If no flow cytometer is available annexin V-labeled cells can be evaluated by fluorescence microscopy. The analysis should be done right after labeling, as longer incubation periods may increase the background.

If the annexin V-labeled cells cannot be analyzed immediately it is possible to preserve the staining by fixation with 4% paraformaldehyde in PBS, pH 7.4. To fix the annexin V–FITC-stained cells, cells are pelleted by centrifugation at 200g in a microcentrifuge and resuspended in 100 μl of PBS (pH 7.4)–3% (w/v) BSA. Cells are spun on glass slides, using a cytospin, at 300g for 5 min. The slides are air dried for 5 min and then fixed in 4% paraformaldehyde in PBS, pH 7.4, for 10 min, using a staining rack. After fixation wash the slides three times with PBS, pH 7.4, for 5 min. The slides are then mounted with glass coverslips and antifade mounting reagent. Apoptotic cells are recognized by FITC staining, using a fluorescence microscope. It is possible to store the slides for 4 weeks at 4°, if protected from light.

Note: Double staining with Hoechst 33342 dye (10 μg/mL) in HEPES, pH 7.4. Apoptotic cells can be recognized by their condensed, brightly stained chromatin. It is also possible to colabel cells with annexin V plus propidium iodide dye (10 μg/ml) in HEPES buffer. Propidium iodide stains cells in late apoptosis or undergoing secondary necrosis. Apoptotic cells are annexin V positive and propidium iodide negative. Necrotic cells have lost their plasma membrane integrity and are both annexin V and propidium iodide positive.

Note: Quantification of apoptosis by annexin V labeling by flow cytometry works best for cell suspension cells.[5,6] But it can also be used for adherent cell types. Label adherent cells prior to harvesting with annexin V–FITC.

[5] M. van Engeland, F. C. S. Ramaekers, S. Schutte, and C. P. M. Reutlingsperger, *Cytometry* **24,** 131 (1996).
[6] M. van Engeland, L. J. Nieland, F. C. Ramaekers, B. Schutte, and C. P. Reutlingsperger, *Cytometry* **31,** 1 (1998).

After annexin V binding do not harvest the cells by treatment with trypsin, as this will destroy the labeling. Instead, use a rubber policeman to remove the cells from the culture dish.

Acknowledgments

This work was supported by National Institutes of Health Grants A140646 and CA6938 to D.G. and by a fellowship by the Swiss National Science Foundation (823A-046638) to E.B.-W. We thank M. Pinkoski for critical reading of this chapter.

[3] Analysis of Apoptotic Cells by Flow and Laser Scanning Cytometry

By Zbigniew Darzynkiewicz and Elzbieta Bedner

A large number of flow cytometric methods to identify apoptotic cells and analyze morphological, biochemical, and molecular changes that occur during apoptosis have been developed. These methods are also applicable to the laser scanning cytometer (LSC), a microscope-based cytofluorometer that combines advantages of flow and image cytometry and that, by offering a possibility of assessment of cell morphology, is of particular utility in analysis of apoptosis. Apoptosis-related changes in cell morphology associated with cell shrinkage and condensation of cytoplasm and chromatin are detected by measurements of the intensity of light scatter of the laser beam in the forward and 90° angle directions. Changes in plasma membrane composition and function are analyzed by its altered permeability to certain dyes and by the appearance of phosphatidylserine, which reacts with annexin V–fluorochrome conjugates on the external surface of the membrane. Decrease in mitochondrial transmembrane potential is measured with several fluorochromes of the rhodamine or carbocyanine family. DNA fragmentation is detected either by measurement of cellular DNA content after elution of the degraded DNA from the cell before or during the staining procedure or by *in situ* labeling DNA strand breaks. Apoptotic cells are then recognized either on the basis of their reduced DNA-associated fluorescence as the cells with fractional DNA content ("sub-G_1 cells"), or as the cells with an extensive number of DNA breaks, respectively. Advantages and limitations of the preceding methods are discussed and their adaptation to LSC is presented.

Introduction

Flow and Laser Scanning Cytometry in Analysis of Apoptosis

Flow cytometry offers several advantages in analysis of apoptosis.[1–3] One of the virtues of this methodology is rapidity of cell measurement. Thousands of individual cells can be measured in seconds with high accuracy and reproducibility. Unlike assessment of apoptosis by microscopy, the choice of cells analyzed by flow cytometry is objective, unbiased by their visual perception. Cell subpopulations differing in the measured parameters can be identified and rare cells easily detected. The major advantage of flow cytometry, however, stems from the possibility of multiparameter analysis ("gating analysis") of the measurements that are recorded by the computer in a list mode. Such an analysis directly reveals a relationship between the measured cell features on a cell-by-cell basis. Given the above, it is not surprising that flow cytometry, more than any other instrumentation, is already widely used for quantitative estimates of the frequency of apoptosis in large cell populations and also in studies of molecular and metabolic changes that occur during apoptosis.

The laser scanning cytometer (LSC; CompuCyte, Cambridge, MA) is a microscope-based cytofluorometer that combines several advantages of flow and image cytometry.[4,5] The cells placed on slides are illuminated with one or two lasers and their fluorescence, together with forward light scatter signal, is measured by several photomultipliers with sensitivity and accuracy comparable to flow cytometry.[4–7] Multiparameter analysis is similar to that offered by flow cytometry. However, because cell staining and measurement on slides eliminate cell loss, which inevitably occurs during the repeated centrifugations necessary for sample preparation for flow cytometry, small cell number samples (e.g., fine needle biopsy specimens) can be analyzed. Furthermore, since the spatial x–y coordinates of each cell on the slide also are recorded, the cells selected after their initial LSC measurement can be relocated ("compusorted"), e.g., for visual microscopy after staining

[1] Z. Darzynkiewicz, S. Bruno, G. Del Bino, W. Gorczyca, M. A. Hotz, P. Lassota, and F. Traganos, *Cytometry* **13**, 795 (1992).
[2] W. G. Telford, L. E. King, and P. J. Fraker, *J. Immunol. Methods* **172**, 1 (1994).
[3] Z. Darzynkiewicz, G. Juan, X. Li, W. Gorczyca, T. Murakami, and F. Traganos, *Cytometry* **27**, 1 (1997).
[4] L. A. Kamentsky and L. D. Kamentsky, *Cytometry* **12**, 381 (1991).
[5] L. A. Kamentsky, D. E. Burger, R. J. Gershman, L. D. Kamentsky, and E. Luther, *Acta Cytol.* **41**, 123 (1997).
[6] R. J. Clatch and J. L. Walloch, *Acta Cytol.* **41**, 109 (1997).
[7] E. Bedner, P. Burfeind, W. Gorczyca, M. R. Melamed, and Z. Darzynkiewicz, *Cytometry* **29**, 191 (1997).

with another dye, or to carry out additional image analysis.[4–8] Because the geometry of the cells cytocentrifuged or smeared on the slide is more favorable for morphometric analysis than is the case for cells in suspension, more information on cell morphology can be obtained by laser scanning than by flow cytometry.[7,8]

Apoptosis was originally defined as a specific mode of cell death based on characteristic changes in cell morphology.[9] Cell morphology still remains the gold standard for the detection of apoptosis. Although individual features of apoptosis, which are discussed later in the chapter, may serve as markers for detection of apoptotic cells by flow cytometry, the mode of cell death must be confirmed by their inspection by light or electron microscopy. Because LSC offers the possibility of rapidly measuring large cell populations and also subjecting the selected cells to morphological inspection, it is uniquely suited for analysis of apoptosis.

Cellular Changes during Apoptosis Measured by Flow or Laser Scanning Cytometry

Morphological, biochemical, and molecular changes that occur during apoptosis[10–17] serve as specific markers to identify apoptotic cells by cytometry.[1–3] An early event of apoptosis is dehydration, which leads to cell shrinkage. This change is reflected by an alteration in the way the cells scatter the light of the laser beam in a flow cytometer. The intensity of light scattered by apoptotic cells in a forward direction along the laser beam, which correlates with cell size, is diminished. Late apoptotic cells or individual apoptotic bodies are characterized by a low intensity of the forward scatter signal. Chromatin condensation, which often is followed by nuclear fragmentation, is another characteristic feature of apoptosis. These changes increase the propensity of the cell to reflect and refract light, which may be manifested by a transient increase in the intensity of light scattered at a 90° angle in the direction of the laser beam (side scatter). However, at later stages of apoptosis, as the cells become small, the intensity of side scatter also decreases. Analysis of the forward and side light scatter signals

[8] E. Luther and L. A. Kamentsky, *Cytometry* **23**, 272 (1996).

[9] J. F. R. Kerr, A. H. Wyllie, and A. R. Curie, *Br. J. Cancer* **26**, 239 (1972).

[10] M. J. Arends, R. G. Morris, and A. H. Wyllie, *Am. J. Pathol.* **136**, 593 (1990).

[11] M. M. Compton, *Cancer Metast. Rev.* **11**, 105 (1992).

[12] T. G. Cotter, *Semin. Immunol.* **4**, 399 (1992).

[13] G. Majno and I. Joris, *Am. J. Pathol.* **146**, 3 (1995).

[14] S. Nagata, *Cell* **88**, 355 (1997).

[15] Z. N. Oltvai and S. J. Korsmeyer, *Cell* **79**, 189 (1994).

[16] A. H. Wyllie, M. J. Arends, R. G. Morris, S. W. Walker, and G. Evan, *Semin. Immunol.* **4**, 389 (1992).

[17] J. C. Reed, *J. Cell Biol.* **124**, 1 (1994).

of the cells thus provides the means to identify apoptotic cells by flow cytometry on the basis of their physical properties, without measurement of fluorescence.[18,19] It should be mentioned, however, that the light scatter changes alone are not a specific marker of apoptosis. Therefore, the analysis of light scatter should be combined with measurements that can provide a more definite identification of apoptotic cells.

At early stages of apoptosis the structural integrity of the plasma membrane and its transport function are, to a large degree, preserved. The permeability of the plasma membrane to such fluorochromes as 7-aminoactinomycin D (7-AAD) or Hoechst 33342 and 33258 dyes, however, is increased.[20–22] The early and most characteristic apoptosis-associated change in the plasma membrane is a breakdown of asymmetry of the phospholipids that leads to exposure of phosphatidylserine on the outer leaflet of the membrane.[23–26] This change can be detected by fluorochrome-tagged annexin V, the anticoagulant that avidly reacts with phosphatidylserine (Fig. 1; Ref. 23). A widely used method of identification of apoptotic cells relies on the detection of these changes of plasma membrane permeability or location of phosphatidylserine.[20–26]

Another early event of apoptosis is a decrease in mitochondrial transmembrane potential, which is reflected by a loss of the ability of the cell to accumulate the cationic fluorochrome rhodamine 123 (rh 123) or certain carbocyanine dyes in mitochondria.[27–31] This event is associated with a

[18] M. G. Ormerod, F. P. M. Cheetham, and X.-M. Sun, *Cytometry* **21**, 300 (1995).

[19] W. Swat, L. Ignatowicz, and P. Kisielow, *J. Immunol. Methods* **137**, 79 (1981).

[20] T. Frey, *Cytometry* **21**, 265 (1995).

[21] M. G. Ormerod, X.-M. Sun, R. T. Snowden, R. Davies, H. Fearhead, and G. M. Cohen, *Cytometry* **14**, 595 (1993).

[22] I. Schmid, W. J. Krall, Uttenbogaart, J. Braun, and J. V. Giorgi, *Cytometry* **13**, 204 (1992).

[23] V. A. Fadok, D. R. Voelker, P. A. Cammpbell, J. J. Cohen, D. L. Bratton, and P. M. Henson, *J. Immunol.* **148**, 2207 (1992).

[24] G. Koopman, C. P. M. Reutelingsperger, G. A. M. Kuijten, R. M. J. Keehnen, S. T. Pals, and M. H. J. van Oers, *Blood* **84**, 1415 (1994).

[25] M. van Engeland, F. C. S. Ramaekers, B. Schutte, and C. P. M. Reutelingsperger, *Cytometry* **24**, 131 (1996).

[26] M. van Engeland, L. J. W. Nieland, F. C. S. Ramaekers, B. Schutte, and C. P. M. Reutelingsperger, *Cytometry* **31**, 1 (1998).

[27] M. Castedo, T. Hirsch, S. A. Susin, N. Zamzani, P. Marchetti, A. Macho, and G. Kroemer, *J. Immunol.* **157**, 512 (1996).

[28] D. W. Hedley and E. A. McCulloch, *Leukemia* **10**, 1143 (1996).

[29] P. X. Petit, H. LeCoeur, E. Zorn, C. Dauguet, B. Mignotte, and M. L. Gougeon, *J. Cell Biol.* **130**, 157 (1995).

[30] S. Shimizu, Y. Eguchi, W. Kamiike, S. Waguri, Y. Uchiyama, H. Matsuda, and Y. Tsujimo, *Oncogene* **13**, 21 (1996).

[31] A. Cossariza, G. Kalashnikova, E. Grassilli, F. Chiapelli, S. Salvioli, M. Capri, D. Barbieri, F. Troiano, D. Monti, and C. Francheschi, *Exp. Cell Res.* **214**, 323 (1994).

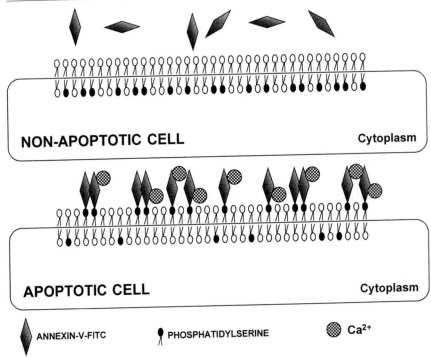

FIG. 1. The principle of detection of apoptotic cells by the assays utilizing annexin V. In live, nonapoptotic cells the plasma membrane phospholipid phosphatidylserine is located within the inner layer of the membrane, facing the cytoplasm. Early during apoptosis this asymmetry is broken and phosphatidylserine appears on the outer surface of the plasma membrane. Annexin V–FITC conjugate binds with high affinity to the exposed phosphatidylserine, thereby labeling apoptotic cells.

leakage of cytochrome c from mitochondria to cytoplasm, increased production of superoxide anions (reactive oxygen intermediates), and increased content of the reduced form of cellular glutathione.[28] The family of Bcl-2 and Bax proteins appears to play a critical role in preventing the loss of cytochrome c and mitochondrial transmembrane potential during apoptosis.[32]

DNA fragmentation, which is observed at somewhat later stages after the changes in mitochondria or plasma membrane, is so characteristic an event of apoptosis that it is considered to be a hallmark of this mode of cell death.[10,11,16] Initially, DNA is cleaved at the sites of attachment of chromatin loops to the nuclear matrix, which results in discrete 50- to

[32] S. L. Schendel, M. Montal, and J. C. Reed, *Cell Death Differ.* **5,** 327 (1998).

300-kb size fragments.[33] Subsequently, although not in every cell type, DNA is cleaved at the internucleosomal sections. As a result, the products of DNA cleavage are discontinuous, nucleosomal and oligonucleosomal DNA fragments of approximately 180 bp and multiples of this size. These products can be detected by a characteristic "laddering" pattern on agarose gels after electrophoresis.[10,11,16]

The fragmented DNA is easily extracted from apoptotic cells after their permeabilization with detergents or fixation with precipitating fixatives such as ethanol.[34] In late apoptotic cells, DNA is degraded to such an extent that it becomes extracted during the routine procedure of cell staining for flow or laser scanning cytometry. However, in early apoptotic cells, when DNA fragmentation is less extensive, its extraction is minimal but can be enhanced by washing cells in buffers of high molarity.[24] The analysis of cellular DNA content of apoptotic cells from which degraded DNA was extracted reveals them as cells with fractional DNA content, represented by the sub-G_1 peak on DNA content frequency histograms. This approach[35,36] is currently the most frequently used in flow cytometry to identify and quantify apoptotic cells. As mentioned, however, in some cell types (primarily of epithelial lineage) DNA cleavage during apoptosis does not proceed to internucleosomal sections but stops at 50- to 300-kb size fragments.[37–41] Because such large DNA fragments are not easily extractable from fixed cells, apoptosis of these cells may not be detected by the methods based on DNA extraction.

Fragmentation of DNA also can be detected by *in situ* labeling of DNA strand breaks in permeabilized cells with fluorochrome-tagged deoxynucleotides, using exogenous terminal deoxynucleotidyl transferase (TdT).[42–44] The presence of a large number of DNA strand breaks is a

[33] F. Oberhammer, J. M. Wilson, C. Dive, I. D. Morris, J. A. Hickman, A. E. Wakeling, P. R. Walker, and M. Sikorska, *EMBO J.* **12,** 3679 (1993).

[34] J. Gong, F. Traganos, and Z. Darzynkiewicz, *Anal. Biochem.* **218,** 314 (1994).

[35] I. Nicoletti, G. Migliorati, M. C. Pagliacci, F. Grignani, and C. Riccardi, *J. Immunol. Methods* **139,** 271 (1991).

[36] S. R. Umansky, B. R. Korol', and P. A. Nelipovich, *Biochim. Biophys. Acta* **655,** 281 (1981).

[37] G. M. Cohen, X.-M. Su, R. T. Snowden, D. Dinsdale, and D. N. Skilleter, *Biochem. J.* **286,** 331 (1992).

[38] R. J. Collins, B. V. Harmon, G. C. Gobe, and J. F. R. Kerr, *Int. J. Radiat. Biol.* **61,** 451 (1992).

[39] Z. F. Zakeri, D. Quaglino, T. Latham, and R. A. Lockshin, *FASEB J.* **7,** 470 (1993).

[40] L. Zamai, E. Falcieri, G. Marhefka, and M. Vitale, *Cytometry* **23,** 303 (1996).

[41] S. Hara, D. Halicka, S. Bruno, J. Gong, F. Traganos, and Z. Darzynkiewicz, *Exp. Cell Res.* **232,** 372 (1996).

[42] R. Gold, M. Schmied, G. Giegerich, H. Breitschopf, H. P. Hartung, K. V. Toyka, and H. Lassman, *Lab. Invest.* **71,** 219 (1994).

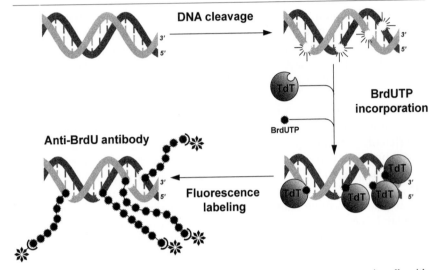

FIG. 2. Scheme illustrating the labeling of DNA strand breaks in apoptotic cells with BrdUTP, using exogenous terminal deoxynucleotidyltransferase (TdT) and anti-BrdU MAb.

characteristic feature of apoptosis, and the methods of strand break labeling [frequently denoted as TdT-mediated dUTP nick end labeling (TUNEL), or as tail or end labeling; Fig. 2] are widely used in basic research as well as in the clinic, e.g., to assess the effectiveness of chemotherapy in patients with leukemias.[45–48] DNA strand break labeling is generally combined with measurement of cellular DNA content, which allows the correlation of apoptosis with the phase of the cell cycle or DNA ploidy.[43,44]

The particular methods described in this chapter rely on identification of apoptotic cells by the specific features discussed above. Adaptation of these methods to LSC also is described. Additional techniques, which are less frequently used but may be suitable in certain cell systems, have been

[43] W. Gorczyca, S. Bruno, R. J. Darzynkiewicz, J. Gong, and Z. Darzynkiewicz, *Int. J. Oncol.* **1,** 639 (1992).

[44] W. Gorczyca, J. Gong, and Z. Darzynkiewicz, *Cancer Res.* **52,** 1945 (1993).

[45] W. Gorczyca, K. Bigman, A. Mittelman, T. Ahmed, J. Gong, M. R. Melamed, and Z. Darzynkiewicz, *Leukemia* **7,** 659 (1993).

[46] H. D. Halicka, K. Seiter, E. J. Feldman, F. Traganos, A. Mittelman, T. Ahmed, and Z. Darzynkiewicz, *Apoptosis* **2,** 25 (1997).

[47] X. Li, J. Gong, E. Feldman, K. Seiter, F. Traganos, and Z. Darzynkiewicz, *Leukemia Lymphoma* **13** (Suppl. 1), 65 (1994).

[48] K. Seiter, E. J. Feldman, F. Traganos, X. Li, H. D. Halicka, Z. Darzynkiewicz, C. A. Lederman, M.-B. Romer, and T. Ahmed, *Leukemia* **9,** 1961 (1995).

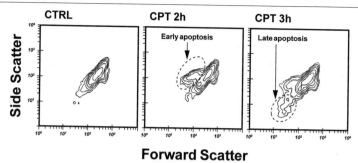

Forward Scatter

Fig. 3. Light-scattering properties of apoptotic cells. Apoptosis of approximately 40% of human promyelocytic HL-60 cells (S-phase cells) was induced by their incubation with 0.15 μM camptothecin (CPT), a DNA topoisomerase I inhibitor, for 3 and 4 hr, as described.[51] The light-scattering properties of the cells were measured using standard settings of the light scatter detectors of the flow cytometer (FACScan; Becton Dickinson, San Jose, CA). Note a decrease in the forward light scatter after 2 hr of treatment with CPT (early apoptosis), with no significant changes in side scatter, compared with control, untreated cells. Cells that have a decreased ability to scatter light in both the forward and 90° angle directions may be seen after 3 hr of incubation with CPT (late apoptosis).

described in many methodological articles.[49–52] The scope of this chapter does not allow the presentation of the applications of flow cytometry or LSC in the analysis of particular proteins involved in the apoptotic cascade, such as members of the Bcl-2 or caspase families, or in caspase enzymatic activation.

Light Scatter Analysis

Analysis of light scatter in the forward and right angle (90°; side scatter) directions is routinely done on every commercially available flow cytometer utilizing laser illumination. Because of great intercellular variability in light scatter, an exponential scale (logarithmic amplifiers) often must be used during analysis. Figure 3 illustrates typical changes in light scatter during

---

[49] Z. Darzynkiewicz and X. Li, Measurements of cell death by flow cytometry. *In* "Techniques in Apoptosis. A User's Guide" (T. G. Cotter and S. J. Martin, eds.), pp. 71–106. Portland Press, London, 1996.

[50] Z. Darzynkiewicz, X. Li, and J. Gong, *Methods Cell Biol.* **41,** 16 (1994).

[51] Z. Darzynkiewicz, X. Li, J. Gong, S. Hara, and F. Traganos, Analysis of cell death by flow cytometry. *In* "Cell Growth and Apoptosis. A Practical Approach" (G. P. Studzinski, ed.), pp. 143–168. University Press, Oxford, 1994.

[52] Z. Darzynkiewicz, X. Li, J. Gong, and F. Traganos, Methods for analysis of apoptosis by flow cytometry. *In* "Manual of Clinical Laboratory Immunology" (N. R. Rose, E. C. de Macario, J. D. Folds, H. C. Lane, and R. Nakamura, eds.), 5th Ed., pp. 334–344. ASM Press, Washington, DC, 1997.

apoptosis. Initially, a decrease in forward scatter is observed, with little change in side scatter. In some cell systems a transient increase in side scatter is observed. With time, the intensity of both the forward and side scatter signals is decreased. The necrotic mode of cell death is characterized by an initial cell swelling, which is manifested by an initial increase in forward scatter, followed shortly after by a rupture of the plasma membrane, which in turn is reflected by a rapid decrease in the ability of the cell to scatter light simultaneously in the forward and right angle directions.

The advantage of light scatter analysis is its simplicity and the possibility of combining it with cell analysis by flow cytometry, in particular with surface immunofluorescence, e.g., to identify the phenotype of the dying cell. However, the light scatter changes are specific neither to apoptosis nor necrosis. Mechanically broken cells, isolated nuclei, cell debris, and individual apoptotic bodies also have low light scatter properties. Furthermore, the distinction between necrotic and apoptotic cells, especially at later stages of apoptosis, is not always apparent. The light scatter analysis, therefore, requires several controls, and should be accompanied by another, more specific cytofluorometric assay, or by microscopy.

While both forward and side light scatter are generally measured by flow cytometry, only forward scatter can be measured by LSC.

Detection of Apoptotic Cells by Labeling with Annexin V–Fluorescein Isothiocyanate Conjugate

In live cells the plasma membrane phospholipids, phosphatidylcholine, and sphingomyelin are exposed on the external leaflet of the lipid bilayer while phosphatidylserine is almost exclusively on the inner surface.[23–26] As mentioned, the loss of phospholipid asymmetry leading to exposure of phosphatidylserine on the outside of the plasma membrane occurs early during apoptosis. This change is detected by annexin V–fluorescein isothiocyanate (FITC) conjugate, which preferentially binds to negatively charged phospholipids such as phosphatidylserine.[23–26] The staining is done in combination with propidium iodide (PI), which is excluded from live and early apoptotic cells but stains DNA and RNA in necrotic and late apoptotic cells, the plasma membranes of which are disrupted. Therefore, by staining cells with FITC–annexin V and PI, it is possible to detect live, nonapoptotic cells (annexin V negative/PI negative), early apoptotic cells (annexin V positive/PI negative), and late apoptotic or necrotic cells (both annexin V and PI positive) by cytometry.[23–26]

Reagents

Annexin V–FITC conjugate (1:1 stoichiometric complex; available, e.g., from Brand Applications, Maastricht, The Netherlands): Dissolve in binding buffer [10 mM HEPES (N-2-hydroxyethylpipera-zine-N'-2-ethanesulfonic acid)–NaOH, pH. 7, 140 mM NaCl, 2.5 mM CaCl$_2$] at a concentration of 1.0 μg/ml. This solution should be prepared fresh each time

PI stock solution: Prepared by dissoving 1 mg of PI (e.g., from Molecular Probes, Eugene, OR) in distilled water. This solution is stable for months when stored in the dark at 0–4°

Procedure for Flow Cytometry

1. Suspend 10^5 to 10^6 cells in 1 ml of FITC–annexin V in binding buffer for 5 min.
2. Add 10 μl of stock PI solution to the cell suspension prior to analysis.
3. Analyze cells by flow cytometry.
   a. Use excitation in blue light (e.g., 488-nm line of the argon ion laser).
   b. Forward light scatter may be used to trigger cell measurements.
   c. Measure green FITC–annexin V fluorescence at 530 ± 20 nm.
   d. Measure red (PI) fluorescence at >600 nm.

Analysis by Laser Scanning Cytometry

The cells can be stained in suspension as described above. A small volume (50–100 μl) of this suspension is then transferred into a shallow well prepared on the microscope slide, covered with a coverslip, and measured. The slides with wells are made by preparing a strip of Parafilm M (American National Can, Greenwich, CT) the size of the microscope slide, cutting a hole (2.0 × 0.5 cm) in the middle of this strip, placing the strip on the microscope slide, and heating the slide on a warm plate until the Parafilm starts to melt. Alternatively, the well can be made using a colorless nail polish, and allowing it to dry for at least 10 min.

The cells also can be stained directly on the slide. During the first step the cells must be electrostatically attached to microscope slides. This is done by centrifuging the cells, suspending the cell pellet in serum-free phosphate-buffered saline (PBS) (~2–4 × 10^5 cells/ml), placing a drop of this suspension within the well made of Parafilm on the microscope slide, and leaving the slide horizontally, at room temperature and at 100% humidity, for 15 min. PBS is then removed by vacuum suction, annexin V–FITC solution is added to the well for 5 min, followed by PI solution. The slide is then covered with a coverslip and transferred onto the LSC stage.

The fluorescence excitation and detection conditions are the same as described above for flow cytometry.

Note: To succeed in attaching cells to microscope slides it is essential to have the microscope slides clean, rinsed in ethanol, and dried. The cell suspension must be free of serum or any other protein (e.g., bovine serum albumin).

Results

Live nonapoptotic cells have minimal green (FITC) and essentially no red (PI) fluorescence (Fig. 4). Early apoptotic cells fluoresce green but have no red fluorescence. Late apoptotic or necrotic cells fluoresce strongly in both green and red. Broken cells, isolated nuclei, or cells with damaged membranes stain rapidly and brightly red. Cells with damaged membranes (e.g., cells scraped from flasks with rubber policeman or mechanically disaggregated tumor cells) may also be stained with annexin V–FITC. This assay can be combined with surface immunophenotyping, using fluorochromes emitting colors different than those emitted by FITC and PI.

Decrease in Mitochondrial Transmembrane Potential

Several membrane-permeable lipophilic cationic fluorochromes accumulate in mitochondria of live cells as a result of the transmembrane potential of this organelle. They are used, thus, as markers of the mitochondrial transmembrane potential. Among many available probes, rh123, carbocyanine $DiOC_6(3)$, and 5,5′,6,6′-tetrachloro-1,1′,3,3′-tetraethylbenzimid-

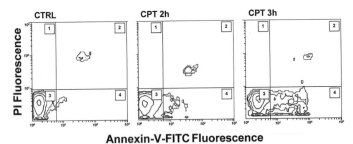

Annexin-V-FITC Fluorescence

Fig. 4. Detection of apoptotic cells by annexin V-binding assay. HL-60 cells were treated with CPT, as described in the caption to Fig. 1, for 2 and 3 hr. The cells were then stained with annexin V–FITC conjugate and PI, as described in the procedure. Their green (FITC) and red (PI) fluorescence was measured with an Elite ESP flow cytometer (Coulter, Miami, FL). Live, nonapoptotic cells stain neither with annexin V–FITC nor with PI, and this population is located in quadrant 3 of the bivariate display. Apoptotic cells have increased FITC fluorescence but show no significant increase in red fluorescence (quadrant 4). A few late apoptotic and/or necrotic cells that were present in untreated (control) and CPT-treated cultures fluoresce in both green and red wavelengths (quadrant 2). No cells are apparent in quadrant 1.

azolcarbocyanine (JC-1) are the most commonly used.[27–31] Cell incubation in the presence of rh123 or $DiO_6(3)$ leads to accumulation of these dyes in mitochondria, and the degree of their accumulation, measured by the intensity of cellular fluorescence, is a reflection of the transmembrane potential. JC-1, on the other hand, changes its color from green to orange, as a result of its aggregation on the polarized membrane, when bound in mitochondria.[31] The color shift of JC-1, thus, is a marker of the mitochondrial potential change. Because at early stages of apoptosis the cell mitochondrial transmembrane potential decreases,[27–31] the mitochondrial probes can monitor this change and therefore be used to identify cells undergoing apoptosis. The method presented below is based on the analysis of the green fluorescence of rh123. PI is added to identify necrotic and/or late apoptotic cells that have impaired integrity of plasma membrane.

Reagents

rh123 solution: Dissolve 1 mg of rh123 (e.g., from Molecular Probes) in 1 ml of distilled water
PI solution: Dissolve 1 mg of PI in 1 ml of distilled water
Both solutions can be stored at 0–4° in the dark for months.

Procedure for Flow Cytometry

1. Add 1 μl of rh123 stock solution to approximately 10^6 cells suspended in 1 ml of tissue culture medium (or PBS) and incubate for 5 min at 37°.
2. Add 20 μl of the PI stock solution and incubate for 5 min at room temperature.
3. Analyze cells by flow cytometry:
   a. Use excitation in blue light (e.g., 488-nm line of the argon ion laser).
   b. Use light scatter to trigger cell measurement.
   c. Measure green (rh123) fluorescence at 530 ± 20 nm.
   d. Measure red (PI) fluorescence at >600 nm.

Analysis by Laser Scanning Cytometry

As in the analysis of annexin V–FITC binding (see above), cells may be stained with rh123, $DiO_6(3)$ or JC-1, either in suspension or when electrostatically attached to slides. After cell staining with rh123 (1 μg/ml in PBS) for 10 min, the staining solution should be replaced by PBS containing PI at only 5 μg/ml, with no rh123 added.

Results

Live and early apoptotic cells stain green with rh123 and have minimal PI fluorescence. The intensity of rh123 fluorescence is diminished in apoptotic cells (Fig. 5). Isolated nuclei, necrotic or mechanically broken cells, and late apoptotic cells having a damaged mitochondrial and/or plasma membrane have no green fluorescence but stain intensely red with PI (not shown).

This is a simple method that combines two functional assays: the ability of the plasma membrane to exclude PI and of mitochondria to maintain their transmembrane potential. It should be stressed that exclusion of PI during apoptosis is not as efficient as in the case of live nonapoptotic cells; hence the method allows the detection of a transient phase of apoptosis when the cell stains with both PI and rh123, i.e., when exclusion of PI

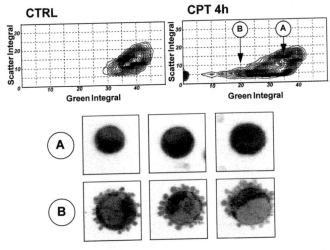

FIG. 5. Decrease in mitochondrial transmembrane potential during apoptosis, measured by rh123 uptake. HL-60 cells, untreated (control) or treated with 0.15 μM CPT for 4 hr, were incubated in the presence of rh123 and PI as described in text. Their green (rh123) and red (PI) fluorescence as well as forward light scatter were measured by LSC (CompuCyte, Cambridge, MA). A population of apoptotic cells shows decreased green fluorescence (integrated value) and also decreased light scatter (region A on the bivariate green integral versus scatter integral contour maps; region B represents cells with unchanged parameters). The possibility of a visual observation of the cells selected from regions A and B, as offered by LSC (Compusort feature) makes it possible to correlate cell morphology with the measured parameters. Note that the cells selected from region A show blebbing of the plasma membrane, a feature typical of apoptosis,[4–6] while the cells from region B show normal morphology. The red (PI) fluorescence of apoptotic cells is not markedly changed compared with nonapoptotic cells (not shown).

is already impaired but mitochondria still remain charged, although the transmembrane potential is decreased. The method can be combined with light scatter analysis (Fig. 5). This method is the best suited to discriminate between the apoptotic and necrotic mode of cell death and to analyze the changes in mitochondrial transmembrane potential rather than quantify apoptotic cells.

Detection of Apoptotic Cells on Basis of Their Fractional DNA Content (Sub-G_1 Cells)

Activation of an endonuclease results in cleavage of DNA, and the fragmented low molecular weight DNA can be extracted from the cells after their fixation and permeabilization. Having fractional DNA content, apoptotic cells stain with lesser intensity with any DNA fluorochrome.[34–36] Because the degree of DNA degradation varies depending on the stage of apoptosis, cell type, and often the nature of the apoptosis-inducing agent, the extent of extracted DNA during the staining procedure also varies. It has been noted that a high molarity phosphate–citrate buffer enhances extraction of the fragmented DNA.[34] This approach can be used to control the degree of DNA extraction from apoptotic cells to the desired level and to obtain their optimal separation by flow or laser scanning cytometry, as described in the procedure presented below.

Reagents

DNA extraction buffer: Mix 192 ml of 0.2 M Na$_2$HPO$_4$ with 8 ml of 0.1 M citric acid; the pH of this buffer is 7.8

DNA-staining solution: (1) dissolve 200 μg of PI in 10 ml of PBS; (2) add 2 mg of DNase-free RNase A (boil RNase for 5 min if it is not DNase free)

Note: Prepare fresh staining solution before each use.

Procedure for Flow Cytometry

1. Fix cells in suspension in 70% ethanol by adding 1 ml of cells suspended in PBS (1–5 × 10^6 cells) to 9 ml of 70% ethanol in a tube on ice. Note: Cells can be stored in fixative at 0–4° for several weeks.
2. Centrifuge the cells (200g, 3 min), decant the ethanol, suspend the cells in 10 ml of PBS, and centrifuge (300g, 5 min).
3. Suspend the cells in 0.5 ml of PBS, to which may be added 0.2–1.0 ml of the DNA extraction buffer.
4. Incubate at room temperature for 5 min; centrifuge (300g, 5 min).
5. Suspend the cell pellet in 1 ml of DNA-staining solution.

6. Incubate the cells for 30 min at room temperature.
7. Analyze the cells by flow cytometry.
   a. Use the 488-nm laser line (or a mercury arc lamp with a BG12 filter) for excitation.
   b. Measure red fluorescence (>600 nm) and forward light scatter.

Alternative Methods. Cellular DNA may be stained with fluorochromes other than PI, and other cell constituents may be counterstained in addition to DNA. The following is the procedure used to stain DNA with 4′,6-diamidino-2-phenylindole (DAPI).

1. After step 4, suspend the cell pellet in 1 ml of a staining solution that contains DAPI (Molecular Probes) at a final concentration 1 μg/ml in PBS. Keep on ice for 20 min.
2. Analyze cells by flow cytometry.
   a. Use excitation with UV light (e.g., 351-nm line from an argon ion laser, or mercury lamp with a UG1 filter).
   b. Measure the blue fluorescence of DAPI in a band from 460 to 500 nm.

Analysis by Laser Scanning Cytometry

Because the measurements are made on fixed cells, the cells may be deposited onto microscope slides either by cytocentrifugation (1000 rpm for 6 min; Shandon, Pittsburgh, PA) or electrostatically, as described earlier in this chapter. The slides are then fixed in 70% ethanol, in closed Coplin jars, for at least 2 hr. They may be stored in 70% ethanol at 0–4° for weeks. After fixation, the cells are rinsed with PBS and stained with a solution of PI containing RNase, prepared as described above for flow cytometry. If necessary, to enhance the separation between apoptotic and nonapoptotic cells, the slides may be rinsed with the DNA extraction buffer prior to staining with PI. The conditions for fluorescence excitation and measurement are the same as described above for flow cytometry.

Results

Apoptotic cells have a decrased PI (or DAPI) fluorescence compared with the cells in the main peak (G_1) (Fig. 6). It should be emphasized that the degree of extraction of low molecular weight DNA from apoptotic cells, and consequently the content of DNA remaining in the cell for flow cytometric analysis, may vary dramatically depending on the degree of DNA degradation (duration of apoptosis), the number of cell washings, and the molarity of the washing and staining buffers. Therefore, in step 3,

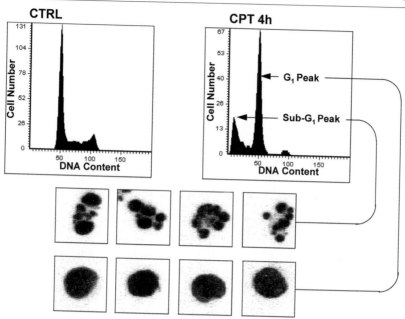

FIG. 6. Detection of apoptotic cells (sub-G_1 peak) based on DNA content measurement. HL-60 cells were either untreated (control) or exposed to 0.15 μM CPT, as described in the caption to Fig. 1, for 4 hr. These cells were fixed in ethanol, rinsed with phosphate–citrate buffer, stained with PI as described in the protocol, and their red fluorescence measured by LSC. Note the appearance of the cells with a fractional DNA content (sub-G_1 peak) in the culture treated with CPT. Assessment of morphology of the cells sorted from the G_1 and sub-G_1 peaks reveals that the former have normal-appearing nuclei while the latter show chromatin condensation and nuclear fragmentation, the typical features of apoptosis.

less or no extraction buffer need be added (e.g., 0–0.2 ml) if DNA degradation in apoptotic cells is extensive (late apoptosis) and more should be added (up to 1.0 ml) if DNA is not markedly degraded (early apoptosis), and there are problems with separating apoptotic cells from G_1 cells because of their overlap on DNA content frequency histograms.

Identification of Apoptotic Cells on Basis of *in Situ* Presence of DNA Strand Breaks

Endonucleolytic DNA fragmentation during apoptosis results in a large number of DNA strand breaks. The 3′-OH termini in the strand breaks can be detected directly or indirectly (e.g., via biotin or digoxigenin) with fluorochrome-labeled deoxynucleotides in a reaction catalyzed by exogenous TdT, as has been originally described.[43] Of all the markers of DNA strand breaks, BrdUTP appears to be the most advantageous with respect to sensitivity, low cost, and simplicity of the reaction (Fig. 2; Ref. 53). BrdU incorporated into DNA strand breaks is detected by an FITC-conjugated anti-BrdU antibody. It should be emphasized that for detection of DNA strand breaks cells must be prefixed with a cross-linking agent such as formaldehyde. DNA–protein cross-linking by this fixative prevents the extraction of the fragmented low molecular weight DNA from the cell, and thereby ensures that a large number of 3–OH ends, which serve as primers for the TdT reaction, remain preserved in apoptotic cells. Thus, despite subsequent cell permeabilization (with ethanol) and the cell washings during the procedure, the DNA content of early apoptotic cells (and with it the number of DNA strand breaks) is not markedly diminished compared with unfixed cells.

Reagents

First fixative: 1% formaldehyde (methanol-free, available from Polysciences, Warrington, PA) in PBS, pH 7.4. Prepare fresh before use

Second fixative: 70% ethanol

TdT reaction buffer (5× concentrated): Contains: 1 M potassium (or sodium) cacodylate, 125 mM Tris-HCl (pH 6.6), and bovine serum albumin (BSA, 1.25 mg/ml) (final concentrations)

Cobalt chloride (CoCl$_2$), 10 mM

TdT in storage buffer, 25 units in 1 μl

The buffer, TdT, and CoCl$_2$ are available from Boehringer Mannheim, (Indianapolis, IN).

BrdUTP stock solution: BrdUTP (2 mM, 100 nmol in 50 μl; Sigma, St. Louis, MO) in 50 mM Tris-HCl, pH 7.5

FITC-conjugated anti-BrdU monoclonal antibody (MAb) solution (per 100 μl of PBS): Combine 0.3 μg of anti-BrdU FITC-conjugated MAb (available from Becton Dickinson, San Jose, CA), 0.3% (v/v) Triton X-100, and 1% (w/v) BSA

Rinsing buffer: Dissolve 0.1% (v/v) Triton X-100 and BSA (5 mg/ml) in PBS

PI staining buffer: Dissolve PI (5 μg/ml) and DNase-free RNase A (200 μg/ml) in PBS

Procedure for Flow Cytometry

1. Fix cells in suspension in 1% formaldehyde for 15 min on ice.
2. Centrifuge (300g, 5 min), resuspend the cell pellet in 5 ml of PBS, centrifuge (300g, 5 min), and resuspend the cells (approximately 10^6 cells) in 0.5 ml of PBS.

---

[53] X. Li and Z. Darzynkiewicz, *Cell Prolif.* **28,** 571 (1995).

3. Add the above 0.5-ml aliquot of cell suspension into 5 ml of ice-cold 70% ethanol. *Note:* The cells can be stored in ethanol, at $-20°$, for several weeks.

4. Centrifuge (200g, 3 min), remove the ethanol, resuspend the cells in 5 ml of PBS, and centrifuge (300g, 5 min).

5. Resuspend the pellet (not more than 10^6 cells) in 50 μl of a solution which contains the following:

Reaction buffer 10 μl
BrdUTP stock solution 2.0 μl
TdT in storage buffer 0.5 μl (12.5 units)
CoCl$_2$ solution 5 μl
Distilled H$_2$O 33.5 μl

6. Incubate cells in this solution for 40 min at 37° (alternatively, incubation can be carried at 22–24° overnight).

7. Add 1.5 ml of the rinsing buffer and centrifuge (300g, 5 min).

8. Resuspend the cells in 100 μl of FITC-conjugated anti-BrdU MAb solution.

9. Incubate at room temperature for 1 hr or at 4° overnight. Add 2 ml of rinsing buffer and centrifuge (300g, 5 min).

10. Resuspend the cell pellet in 1 ml of PI staining solution containing RNase.

11. Incubate for 30 min at room temperature in the dark.

12. Analyze the cells by flow cytometry.

   a. Illuminate with blue light (488-nm laser line or BG12 excitation filter).

   b. Measure green fluorescence of FITC-anti BrdU MAb at 530 $\pm$ 20 nm.

   c. Measure red fluorescence of PI at >600 nm.

Commercial Kits. Phoenix Flow Systems (San Diego, CA) and PharMingen (San Diego, CA) provide kits (ApoDirect and APO-BRDU) to identify apoptotic cells on the basis of a single-step procedure utilizing TdT and either FITC-conjugated dUTP or BrdUTP, respectively. A description of the methods, which are nearly identical to the preceding method, is included with the kits. Another kit (ApopTag), based on two-step DNA strand break labeling with digoxigenin–16-dUTP by TdT, is provided by Oncor (Gaithersburg, MD).

Analysis of DNA Strand Breaks by Laser Scanning Cytometry

1. Add 300 μl of cell suspension in tissue culture medium (with serum) containing approximately 20,000 cells into a cytospin (e.g., Shandon Scientific, Pittsburgh, PA) chamber. Cytocentrifuge at 1000 rpm for 6 min. Alter-

natively, the cells may be electrostatically attached to the slides as described earlier in this chapter (see section describing annexin V–FITC labeling).

2. Without allowing the cytospins to dry completely, prefix the cells in 1% formaldehyde in PBS for 15 min on ice.

3. Transfer the slides to 70% ethanol and fix for at least 1 hr; the cells can be stored in ethanol for several days.

4–9. Follow the steps from 4 to 9 as described for flow cytometry. Small volumes (50–100 μl) of the respective buffers, rinses, or staining solutions are carefully layered on the cytospin area of the slides held horizontally. At appropriate times these solutions are removed with a Pasteur pipette (or vacuum suction pipette). Small pieces (2.5 × 1.0 cm) of thin polyethylene foil may be layered on slides atop the drops to prevent drying. The incubations should be carried out in a moist atmosphere to prevent drying at any step of the reaction.

10. Rinse the slide in PBS and mount the cells under a coverslip in a drop of the PI staining solution containing RNase A. If the preparations are to be stored for longer periods of time (hours to days, at 4°), the coverslips are mounted in a drop of a mixture of glycerol and PI staining solution (9 : 1, v/v).

11. Measure cell fluorescence with an LSC. Use the same fluorescence excitation and measurement wavelength as described above for flow cytometry.

Results

Identification of apoptotic cells is based on their intense labeling with FITC–anti BrdU MAb, which frequently requires use of an exponential scale (logarithmic photomultipliers) for data acquisition and display (Fig. 7). Simultaneous measurement of DNA content makes it possible to identify the cell cycle position of cells in both apoptotic and nonapoptotic populations. As cells enter the later stages of apoptosis they have lost some DNA stainability (despite formaldehyde fixation) and have also lost sites for BrdUTP incorporation.

Concluding Remarks

Each of the methods presented above has its advantages and suffers limitations. The advantages of light scatter measurement are its simplicity and the fact that it is not a fluorescent marker. The fluorescence detectors, thus, can be used to measure additional parameters that can be probed by fluorochrome markers. This provides the possibility of combining light

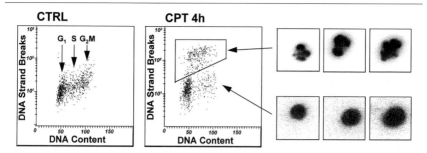

FIG. 7. Detection of apoptosis based on labeling DNA strand breaks. Untreated (control) or CPT-treated (15 μM, 4 hr) HL-60 cells were fixed in formaldehyde followed by ethanol and their DNA strand breaks labeled with BrdUTP (see Fig. 2) according to the procedure described. The red (PI) and green (FITC) fluorescence of the cells was measured by LSC. On the basis of differences in PI fluorescence (DNA content) one can discriminate cells in G_1, S, and G_2/M. Most cells with strand breaks have DNA content equivalent of that of S-phase cells. Their morphology shows chromatin condensation and nuclear fragmentation, features characteristic of apoptosis.

scatter analysis with other measurements, in particular with surface immunofluorescence (e.g., to identify the phenotype of dying cells) or with PI to identify cells with a ruptured plasma membrane (e.g., to identify necrotic, late apoptotic, or mechanically broken cells). The disadvantage of light scatter analysis is that it is not a specific marker of apoptosis. Mechanically broken cells, isolated nuclei, cell debris, and individual apoptotic bodies also have low light scatter properties. Light scatter also changes during cell fixation, sometimes in unpredictable ways. Furthermore, on the basis of differences in light-scattering properties late apoptotic cells cannot be distinguished from necrotic cells. The light scatter analysis, therefore, requires additional controls and should be accompanied by another, more specific cytofluorometric assay or by microscopy.

Binding of annexin V–FITC (or other fluorochrome) conjugate is considered to be a specific marker of apoptosis.[23–26] Another advantage of this assay is that by using antibodies labeled with different color fluorochromes, it can be combined with analysis of the cell surface immunophenotype. The disadvantage of the annexin V assay is that it requires live, unfixed cells. This requirement eliminates the possibility of transport or prolonged storage of cells prior to measurement. Furthermore, since annexin V binding is done on intact, nonpermeabilized cells, simultaneous staining of DNA and therefore analysis of a correlation between the cell cycle position and apoptosis (annexin V binding) requires use of dyes that penetrate plasma membrane. For the same reason the analysis of a correlation between annexin V binding and immunocytochemically detected intracellular proteins is not possible.

Measurement of the mitochondrial transmembrane potential is simple and does not require expensive reagents. Furthermore, the decrease in the mitochondrial potential is an early, and perhaps still to some extent, reversible event of apoptosis. Its analysis, therefore, in contrast to the changes detected by other methods presented in this chapter, may provide information about the state of the cell prior to the "point of no return" along the apoptotic pathway. By virtue of the spectral shift rather than the fluorescence intensity change, JC-1 appears to be the preferred mitochondrial probe.[31] A decrease in mitochondrial transmembrane potential, however, is not always a marker of apoptosis. This change, therefore, cannot be used as a sole criterion for identification of apoptotic cells.

Identification of apoptotic cells by their fractional DNA content has many virtues. The method is simple and inexpensive and simultaneously reveals DNA ploidy and/or the cell cycle distribution of the nonapoptotic cell population. Another advantage is its applicability to different DNA fluorochromes and instruments. Fragmented DNA, extracted from the ethanol-prefixed apoptotic cells, can be directly analyzed by gel electrophoresis for "laddering."[34] Being simple and low in cost, the method is most applicable for screening large numbers of samples to quantify apoptotic cells. It should be stressed, however, that the sub-G_1 peak on the DNA content frequency histograms, in addition to apoptotic cells, can also represent mechanically damaged cells, cells with lower DNA content (e.g., in samples containing cell populations with different DNA indices), or cells with different chromatin structure (e.g., cells undergoing erythroid differentiation), in which the accessibility of DNA to the fluorochrome (DNA stainability) is diminished.[54] Additional tests, therefore, should be done (e.g., by analysis of DNA laddering, the presence of DNA strand breaks, or cell morphology) to ensure that the presence of cells with diminished DNA stainability coincides with apoptotic cells detected by other means.

The presence of DNA strand breaks is a specific feature of apoptotic cells. Necrotic cells, or cells with primary DNA breaks caused by X-ray irradiation (up to the dose of 25 Gy) or DNA-damaging drugs, have, by an order of magnitude, fewer DNA strand breaks compared with apoptotic cells.[21] Because the cellular DNA content of both apoptotic and nonapoptotic cell populations is measured in this assay, the method offers the unique possibility of analyzing the cell cycle position, and/or DNA ploidy, of apoptotic cells. The method has been found to be useful for clinical material in leukemias, lymphomas, and solid tumors.[45–48] It also may be combined with cell surface immunophenotyping. Toward this end the cells are initially

[54] Z. Darzynkiewicz, F. Traganos, J. Kapuscinski, L. Staiano-Coico, and M. R. Melamed, *Cytometry* 5, 355 (1984).

immunophenotyped, then fixed with formaldehyde (to covalently attach the antibody to the cell surface) and subsequently subjected to DNA strand break assay using fluorochromes of another color than that used for immunophenotyping. The limitation of the DNA strand break assay is its complexity and high cost of the reagents. Furthermore, its utility is doubtful in situations when DNA fragmentation stops at 50- to 300-kb fragments and does not involve the internucleosomal sections.[37–41]

The selection of a particular method depends on the cell system, nature of the inducer of cell death, mode of cell death, information that is desired (e.g., specificity of apoptosis with respect to the cell cycle phase or DNA ploidy), and the technical restrictions (e.g., the need for sample transportation, type of flow cytometer available). Apoptosis can be recognized with greater certainty when more than a single viability assay is used. Furthermore, regardless of the assay used, the mode of cell death should be positively identified by inspection of cells by light or electron microscopy. Morphological changes during apoptosis are specific[9–13] and should be the deciding factor when ambiguity arises regarding the mechanism of cell death. The possibility of morphological examination of the cell after the measurement, as offered by LSC, makes this instrument uniquely suited for the detection of apoptosis.

The duration of apoptosis is relatively short (1–6 hr) and variable depending on cell type, inducer of apoptosis, whether it occurs *in vitro* or *in vivo*, etc. Prolongation of the apoptotic process (e.g., by drugs that inhibit the apoptotic effectors such as proteases or endonuclease) as well as an increased frequency of apoptosis are reflected by an increased apoptotic index (AI). An increased AI alone, therefore, should not be considered a measure of the increased frequency of apoptosis. Likewise, the "time window" through which a particular flow cytometric method "sees" this kinetic event, and thus can identify the apoptotic cell, also is variable. Hence, the AI values of the same cell population detected by different methods may not be identical. Simultaneous measurement of the absolute number of live cells in addition to quantitation of apoptosis or necrosis often are necessary to obtain an accurate estimate of cell death.

Acknowledgments

Supported by NCI Grant CA RO1 28704, the Chemotherapy Foundation, and "This Close" Foundation for Cancer Research. E.B., who is the recipient of an Alfred Jurzykowski Foundation fellowship, is on leave from the Department of Pathology, Pomeranian School of Medicine, Szczecin, Poland.

[4] Quantitative Measurement of Apoptosis Induced by Cytotoxic T Lymphocytes

By Michele Barry, Jeffrey Heibein, Michael Pinkoski, and R. Chris Bleackley

Cytotoxic T lymphocytes destroy virus-infected and malignant cells through the induction of apoptosis. This form of cell death is characterized by a number of cellular changes including cell shrinkage and membrane blebbing, chromatin condensation and DNA fragmentation, externalization of phosphatidylserine to the outer leaflet of the plasma membrane, and disruption of the inner mitochondrial transmembrane potential ($\Delta\Psi_m$). Cell death induced by cytotoxic T lymphocytes is associated with similar morphological and biochemical features. Here we demonstrate how methods typically employed to detect apoptotic cells can be adapted to monitor cell death mediated by cytotoxic T lymphocytes. We have specifically selected techniques that allow quantitative evaluation of death including membrance changes, DNA fragmentation, and mitochondrial depolarization.

Introduction

Cytotoxic T lymphocytes (CTLs) induce the death of target cells via two distinct and independent mechanisms: the exocytosis of granule components including perforin and granzymes and triggering of the Fas surface receptor.[1,2] Both methods initiate target cell destruction through the activation of members of the caspase family, which is accompanied by DNA fragmentation. As such, many of the routine methods that are currently employed for the detection of DNA fragmentation during programmed cell death or apoptosis can be applied to the detection of nuclear degradation mediated by CTLs with a few minor alterations. In our laboratory we routinely monitor the degradation of DNA in target cells via terminal deoxynucleotidyltransferase-mediated dUTP–fluorescein isothiocyanate (FITC) nick end labeling (TUNEL)[3] and via the release of metabolically labeled [³H]thymi-

[1] E. A. Atkinson and R. C. Bleackley, *Crit. Rev. Immunol.* **15**, 359 (1995).
[2] P. A. Henkart, *Immunity* **1**, 343 (1994).
[3] Y. Gavrieli, Y. Sherman, and S. A. Ben-Sasson, *J. Cell Biol.* **119**, 493 (1992).

dine DNA.[4] In addition, evidence has demonstrated that loss of inner mitochondrial potential is an early event in apoptosis,[5,6] and changes in the mitochondrial transmembrane potential can be monitored during CTL killing by using the potential sensitive dye $DiOC_6(3)$.[7,8] Phosphatidylserine externalization from the inner to the outer leaflet of the plasma membrane is a characteristic hallmark of apoptosis,[9,10] which also occurs during CTL-mediated death, allowing for the detection of apoptosis by CTLs. Target cell lysis can be readily quantitated by measuring the release of ^{51}Cr from prelabeled target cells after the addition of proapoptotic stimuli such as CTLs or anti-Fas.[11,12]

Quantitation of DNA Degradation

One of the most characteristic features displayed by cells undergoing apoptosis is the degradation of DNA into oligonucleosomal fragments, which can be observed as DNA laddering when genomic DNA is subjected to agarose gel electrophoresis.[13] DNA fragmentation in target cells can also be monitored by the terminal transferase-mediated incorporation of FITC-labeled dUTP onto the ends of fragmented DNA or the release of radioactively labeled DNA after the addition of CTLs. Both of these methods are used routinely in our laboratory with reproducible results.

^3H Release

To detect DNA fragmentation, target cells are grown overnight in growth medium containing [^3H]thymidine (1.5 μCi/ml).[4] The next day, cells are washed once in growth medium and further incubated in 10 ml of growth medium for 1 hr at 37°, 5% CO_2. The target cells are then washed

[4] R. S. Garner, C. D. Helgason, E. A. Atkinson, M. J. Pinkoski, H. L. Ostergaard, O. Sorenson, A. Fu, P. H. Lapchak, A. Rabinovitch, J. E. McElhaney, G. Berke, and R. C. Bleackley, *J. Immunol.* **153**, 5413 (1994).

[5] G. Kroemer, N. Zamzami, and S. A. Susin, *Immunol. Today* **18**, 44 (1997).

[6] G. Kroemer, *Cell Death Differ.* **4**, 443 (1997).

[7] M. Zoratti, and I. Szabò, *Biochim. Biophys. Acta* **1241**, 139 (1995).

[8] P. X. Petit, J. E. O'Connor, D. Grunwald, and S. C. Brown, *Eur. J. Biochem.* **220**, 389 (1990).

[9] V. A. Fadok, D. R. Voelker, P. A. Campbell, J. J. Cohen, D. L. Bratton, and P. M. Henson, *J. Immunol.* **148**, 2207 (1992).

[10] G. Koopman, C. P. M. Reutelingsperger, G. A. M. Kuijten, R. M. J. Keehnen, S. T. Pals, and M. H. J. van Oers, *Blood* **84**, 1415 (1994).

[11] K. T. Brunner, J. Mauel, J. C. Cerottini, and B. N. Chapuis, *Immunology.* **14**, 181 (1968).

[12] R. Duke and J. J. Cohen, *in* "Current Protocols in Immunology" (J. E. Coligan, A. M. Kruisbeek, D. H. Margulies, E. M. Shevach, and W. Strober, eds.), pp. 3.17.5–3.17.11. Green Publishing Associates and Wiley-Interscience, New York, 1992.

[13] A. H. Wyllie, *Nature* (*London*) **284**, 555 (1980).

three times in phosphate-buffered saline (PBS), counted, and resuspended in growth medium at 5×10^5 cells/ml. Target cells are aliquoted into 1.5-ml microcentrifuge tubes and an equal volume of CTLs is added at the designated effector cell-to-taget cell ratio. We routinely use effector-to-target ratios ranging from 5:1 to 1:1 to duplicate more precisely the *in vivo* situation, although it is possible to employ a vast excess of effector cells. The tubes are briefly vortexed to ensure efficient mixing and then incubated for at least 2 hr at 37°, 5% CO_2. After incubation, an equal volume of sample lysis buffer containing 0.5% Triton X-100 in PBS is added to all tubes except those designated for quantitating total ^3H incorporation, to which is added an equal volume of lysis buffer containing 2% (w/v) sodium dodecyl sulfate (SDS) and 0.2 M NaOH. All the tubes are vortexed vigorously and then only the tubes containing samples are centrifuged at 10,000g for 10 min at 4°. Half the supernatant from each tube is removed and ^3H quantitated by scintillation. The spontaneous release of ^3H-labeled DNA is monitored by incubating labeled target cells in the absence of CTLs. Percent fragmentation is calculated as follows: % DNA fragmentation = $100 \times [(\text{sample} - \text{spontaneous})/(\text{total} - \text{spontaneous})]$.

Flow Cytometric Analysis of DNA Fragmentation

Flow cytometry provides a rapid and quantitative means of assesssing the number of apoptotic cells in a given population. Although these techniques are best used for measuring apoptosis induced by purified components of cytolytic granules (e.g., granzyme B and perforin), they can be easily modified to measure cell death induced in targets by CTLs.

DNA fragmentation may be measured *in vitro* using terminal deoxynucleotidyltransferase-mediated dUTP–FITC nick end labeling (TUNEL).[3] For most applications, commercially available kits (Boehringer Mannheim, Indianapolis, IN) in provide all the necessary materials to perform TUNEL reactions.

The basic protocol involves incubating targets with granule components or whole CTLs for an appropriate time (typically 1–4 hr). After this incubation period, cells (0.5–2×10^5 cells) are fixed in 2% (w/v) paraformaldehyde for 15 min at room temperature. After fixing, cells may be stored at 4° for several days before further processing. Fixed cells are then permeabilized in PBS containing a detergent such as 0.1% (v/v) saponin or 0.1% (v/v) Triton X-100. After permeabilization, cells are washed in PBS and resuspended in a suitable volume of TUNEL mix [typically 25–50 μl; 200 mM potassium cacodylate, 25 mM Tris-HCl (pH 6.6), bovine serum albumin (0.25 mg/ml), 1.5 mM $CoCl_2$, 100 nM dATP, 300 pM dUTP–FITC, and terminal transferase enzyme]. Cells are then incubated for 1 hr in the dark at 37°, followed by a wash in PBS or PBS-containing detergent. Cells are resuspended in PBS and analyzed by flow cytometry. In this laboratory, TUNEL is quantitated by examining 10,000 events on a Becton Dickinson (San Jose, CA) FACScan with an excitation wavelength of 488 nm. The emission wavelength for dUTP–FITC (520 nm) is detected through the FL1 channel equipped with a 530-nm (20-nm bandpass) filter. Data are analyzed with CELLQuest software (Becton Dickinson).

In some cases, when using whole CTLs to induce apoptosis, there may be a sufficient difference in size between effector and target cell populations to permit exclusive gating on targets. However, this is not always an option. When cell populations are not easily discernible on the basis of forward and side scatter, it is necessary to label the cell mixture with an antibody to allow differentiation of cell types by the flow cytometer. Because target cell type may vary, and hence the makeup of corresponding surface molecules, it is generally more consistent to label CTLs with one of the numerous commercially available, directly conjugated anti-CD8 antibodies. Because FITC emits in the FL1 channel on most flow cytometers, the anti-CD8 antibody should contain a fluorophore that emits in another channel, such as FL2, which typically has a 580-nm (20-nm bandpass) filter; (R)-phycoerythrin (PE)-conjugated antibodies are generally suitable (excitation, 488 nm; emission maximum at 575 nm). CD8 labeling should be done before fixing by incubating cells on ice for 30 min with an appropriate dilution of anti-CD8–PE in PBS (the dilution must be determined empirically for each antibody and cell type). After incubation, cells are washed in ice-cold PBS and then fixed and TUNEL labeled as outlined above. CD8-positive cells may then be gated out during analysis by flow cytometry, allowing for TUNEL quantitation of the target cell population alone.

The TUNEL protocol outlined above also works well for cells mounted on slides for microscopic analysis and may be modified with a number of colorimetric, fluorimetric, and biotinylated dUTP conjugates, allowing more flexibility for analysis. To utilize other dUTP conjugates, it is necessary to perform TUNEL reactions with terminal transferase kits (Boehringer Mannheim), which include terminal transferase enzyme, reaction buffer, as well as $CoCl_2$. These kits need only be supplemented with dATP and the dUTP–conjugate of interest.

Quantitation of Membrane Changes

Chromium Release

The cytolytic activity induced by CTLs can be readily monitored by assessing chromium-51 (^{51}Cr) release from labeled target cells.[11,12] Before the addition of activated CTLs, 1×10^6 target cells are labeled by the direct addition of 100 μCi of ^{51}Cr to the cell pellet. The target cells are gently resuspended in the label and incubated at 37°, 5% CO_2 for 60 min in order to ensure efficient labeling of the cells. After labeling the target cells are washed once in 5.0 ml of growth medium [RMPI 1640 supplemented with 100 μM 2-mercaptoethanol, 10% (v/v) fetal calf serum, streptomycin (100 μg/ml), and penicillin (100 μg/ml)] to remove residual label and then resuspended in 10 ml of growth medium and further incubated at 37°, 5% CO_2 for 30 min. The target cells are then washed three times with PBS, counted, and resuspended in growth medium at 1×10^5 cells/ml. Labeled target cells are aliquoted into 96-well microtiter plates and an equal volume of activated CTLs is then added at various effector cell-to-target cell ratios. After addition of activated CTLs, the microtiter plate is gently centrifuged for 2 min at 200g to promote efficient cell-to-cell contact and the microtiter plate is then incubated at 37°, 5% CO_2 for 4 hr. After incubation the microtiter plate is centrifuged for 10 min at 400g in order to pellet the cells and ^{51}Cr release is directly quantitated with a γ counter to assess the amount of ^{51}Cr released into the supernatant. To monitor spontaneous release of ^{51}Cr, target cells are mixed with an equal volume of growth medium in the absence of CTLs. In addition, each assay is performed in triplicate to avoid error. Total cell counts are obtained by directly quantitating ^{51}Cr incorporation into target cells in the absence of CTLs, and cytolytic activity is calculated as follows: % cytolysis = 100 × [(sample − spontaneous)/ (total − spontaneous)].

Phosphatidylserine Externalization

Another feature of apoptosis that is readily observable is phosphatidyl-serine externalization.[9,10] Phosphatidylserine is normally localized exclusively to the inner leaflet of the plasma membrane. Early after an apoptotic stimulus phosphatidylserine relocalizes to the outer leaflet of the plasma membrane and may in part serve as a signal for neighboring cells to phagocytose an apoptotic cell.[14] Externalized phosphatidylserine on the surface of cells may be detected with annexin V, a protein with a high natural affinity for the lipid, which is conjugated to a fluorophore, typically FITC.[10,15,16] FITC–annexin V is commercially available from a number of suppliers.

Target cells (0.1–1 × 10^6 cells) are incubated with cytolytic granule components or whole CTLs as outlined above for TUNEL and then washed

[14] J. Savill, N. Hogg, Y. Ren, and C. Haslett, *J. Clin. Invest.* **90**, 1513 (1992).
[15] M. A. Swairjo and B. A. Seaton, *Annu. Rev. Biophys. Biomol. Struct.* **23**, 193 (1994).
[16] I. Vermes, C. Haanen, H. Steffens-Nakken, and C. Reutelingsperger, *J. Immunol. Methods* **184**, 39 (1995).

in 1× binding buffer [10 mM HEPES/NaOH (pH 7.4), 140 mM NaCl, and 2.5 mM CaCl$_2$]. Cells are resuspended in 195 μl of 1× binding buffer and 5 μl of commercial annexin V–FITC stock is added. Cells are incubated at room temperature for 5–15 min in the dark and subsequently washed in PBS. Cells may be analyzed by flow cytometry immediately or fixed with 2% (v/v) paraformaldehyde in PBS and stored at 4° for processing at later times. Annexin V–FITC is analyzed through the FL1 channel as outlined for TUNEL above. Phycoerythrin-conjugated annexin V is also available (PharMingen, san Diego, CA), allowing some flexibility for analysis. The annexin V–PE must be analyzed through the FL2 channel.

When annexin V analysis is to be used for CTL-treated targets, the target-specific annexin V signal must be separated from signal arising from effectors. As outlined for TUNEL above, this is accomplished by gating out CD8-positive cells with anti-CD8 antibodies. The anti-CD8 antibodies may be added to the annexin V mixture in 1× binding buffer and the reaction may be carried out on ice with no apparent detriment to annexin V binding (our unpublished observation, 1999). After a 15-min incubation, the cells may be analyzed immediately or alternatively may be fixed and analyzed at a later time. It is important to complement fluorophores. If annexin V–FITC is to be used, an anti-CD8 antibody emitting in the FL2 channel must be used to gate out CTLs whereas an anti-CD8–FITC must be employed for mixed populations treated with annexin V–PE.

Mitochondrial Membrane Depolarization

Evidence has emerged from a number of laboratories implicating mitochondrial changes as early events in apoptosis induced by a variety of stimuli.[5,6] One of these mitochondrial changes involves disruption of the inner membrane transmembrane potential ($\Delta\Psi_m$) through the opening of permeability transition pores. After pore opening, the normally impermeable inner membrane becomes permeable to ions and solutes and the negatively charged environment of the matrix is lost.[7] The loss in $\Delta\Psi_m$ can be readily measured by flow cytometry as a loss in fluorescence intensity of mitochondria-specific cationic lipophilic dyes.[8] These dyes include 3,3′-dihexyloxacarbocyanine iodide [DiOC$_6$(3)], Mito Tracker Green FM, and JC-1 (Molecular Probes, Eugene, OR). DiOC$_6$(3), for example, targets to the negative environment of the matrix in metabolically active mitochondria, where it emits intensely in the FL1 channel. After $\Delta\Psi_m$ loss, DiOC$_6$(3) leaks out of the matrix and this is seen as a shift in the dye-loaded population from the right-hand portion of a dot histogram [DiOC$_6$(3)^{high}] to the left [DiOC$_6$(3)^{low}]. As a positive control for $\Delta\Psi_m$ loss, carbonyl cyanide m-

chlorophenyl hydrazone (mClCCP) may be added to cells in conjunction with $DiOC_6(3)$ for 15 min at 37°.

Cells treated with purified cytolytic granule components may be processed as follows. Apoptosis is induced in targets. At a suitable interval after apoptosis induction (typically 1–2 hr) targets ($1-2 \times 10^5$ cells) are removed and centrifuged at 600g. The supernatant is removed and replaced with PBS containing 40 nM $DiOC_6(3)$ for 15 min at 37°. After this interval, cells are immediately analyzed by flow cytometry (FL1 channel). Alternatively, cells may be prelabeled with dye before treatment with the apoptotic stimuli. $DiOC_6(3)$ is stable in cells for at least 4 hr (our unpublished observation, 1999) and offers the advantage of allowing for real-time examination of $\Delta\Psi_m$ loss. Preloading of target cells is an essential step for the examination of killing induced by whole CTLs. Targets are treated with dye, as indicated above, followed by a wash step (PBS or serum-free medium) and resuspension in fresh medium (serum-free RPMI 1640 in this laboratory). Targets are then mixed with CTLs in an appropriate effector-to-target ratio for the cell types involved (e.g., 1 : 1 or 2 : 1 using Jurkat targets and human CTLs) and aliquots are withdrawn at specific intervals. Segregation of cell types for analysis by flow cytometry may be made through differences in size of the target and effector populations, determined using forward and side scatter provided the differences between populations are sufficient. Alternatively, mixed cell populations may be treated with an anti-CD8–PE antibody to specifically label CTLs. After a 15- to 20-min incubation on ice, cells are washed and kept on ice until analysis. It is important that cells not be fixed when using $DiOC_6(3)$, as fixing may affect retention of the dye and yield false-positive results.

Acknowledgments

Work in the authors' laboratory is funded by the Medical Research Council and National Cancer Institute of Canada. R.C.B. is a Distinguished Scientist of the Medical Research Council of Canada and a Howard Hughes International Research Scholar. M.B. is the recipient of a postdoctoral fellowship from the Alberta Heritage Foundation for Medical Research, and J.H. is the recipient of a studentship from the Medical Research Council of Canada.

[5] Apoptotic Nuclease Assays

By Francis M. Hughes, Jr. and John A. Cidlowski

One of the defining biochemical characteristics of apoptosis is the degradation of chromatin into regularly sized (oligonucleosomal and ~30- to 50-kb) fragments. Because destruction of the genome represents a clear commitment to death, considerable interest has focused on this component of apoptosis and numerous assays have been developed to assess the relevant nucleases involved. These assays fall into two major categories: (1) those independent of chromatin structure and (2) those dependent on chromatin structure. The chromatin-independent assays (plasmid degradation assay and radioactive gel assay) examine the ability to degrade naked DNA and are advantageous because of their simplicity and speed and ability to analyze single nucleases or mixtures of nucleases. However, these assays do not mimic the conditions present in normal cells and consequently do not assess the ability of an enzyme to function in apoptosis. In contrast, chromatin structure-dependent assays (nuclear autodigestion and HeLa nuclei assay) present intact chromatin to either endogenous or exogenous enzymes and assess the ability to degrade chromatin in a manner that recapitulates the genomic destruction seen *in vivo*. Detailed protocols are discussed for both classes of assays. These assays have been instrumental in the identification of several apoptotic nucleases.

Introduction

Apoptosis is associated with the activation of numerous enzymatic systems that serve both to propagate death signals and to degrade cellular macromolecules. One of the most studied of these systems is the nucleases, enzymes that degrade the dying cell genome into discrete fragments (Fig. 1). Specifically, during apoptosis chromatin is degraded into two distinct size classes: (1) nucleosomal or oligonucleosomal length fragments (multiples of 180–200 bp) and (2) large (~30- to 50-kb) fragments. At its most basic level, chromatin is packaged into nucleosomes, or "beads on a string," in which DNA is wrapped twice around a core group of histones. Between these structures lies an internucleosomal, or linker, DNA region that is specifically cleaved during apoptosis to release fragments of nucleosomal (180–200 bp) or oligonucleosomal (multiples of 180–200 bp) lengths. On an agarose gel these fragments form a characteristic banding pattern widely known as an "apoptotic ladder." The production of larger (~30- to 50-kb)

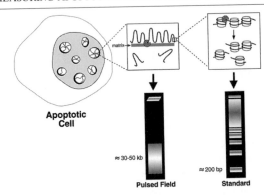

FIG. 1. The two known forms of DNA degradation during apoptosis. The internucleosomal degradation of chromatin (far right) is detected by standard agarose gel electrophoresis while the production of large (~30- to 50-kb) fragments, thought to arise from cleavage of chromatin "loops," is detected by pulsed-field electrophoresis.

fragments during apoptosis is hypothesized to arise from the cleavage of higher order chromatin structures. The nucleosomal substructure is first supercoiled to form a "solenoid," which is attached to the nuclear matrix at ill-defined sequences known as matrix attachment regions (MARs). The solenoid then "loops" out into the nucleoplasm and is reattached ~30–50 kb later at the next MAR. Nucleolytic release of ~30- to 50-kb fragments during apoptosis has been hypothesized to arise through cleavage of the chromatin at or near these MARs.

Chromatin degradation during apoptosis has attracted considerable attention since its discovery in 1980,[1] primarily because it represents a final commitment step beyond which effective repair is impossible and cell death becomes inevitable. Numerous assays have been developed to aid in identifying and characterizing this process. On the basis of the results of these assays, multiple enzymes have been put forth as candidates, including NUC18/cyclophilin,[2,3] nuc-40,[4] nuc-58,[4] DNase I,[5] DNase II,[6] DNase γ,[7]

[1] A. H. Wyllie, Nature (London) 284, 555 (1980).
[2] J. W. Montague, M. L. Gaido, C. Frye, and J. A. Cidlowski, J. Biol. Chem. 269, 18877 (1994).
[3] J. W. Montague, F. M. Hughes, Jr., and J. A. Cidlowski, J. Biol. Chem. 272, 6677 (1997).
[4] G. Deng and E. R. Podack, FASEB J. 9, 665 (1995).
[5] M. C. Peitsch, B. Polzar, J. Tschopp, and H. G. Mannherz, Cell Death Differ. 1, 1 (1994).
[6] A. Eastman, Cell Death Differ. 1, 7 (1994).
[7] D. Shiokawa, H. Ohyama, T. Yamada, K. Takahashi, and S.-I. Tanuma, Eur. J. Biochem. 226, 23 (1994).

ILCME,[8] a 25-kDa activity,[9,10] a 97-kDa molecule,[11] and a caspase-activated DNase (CAD).[12] The assays used to study apoptotic nucleases fall into two categories: (1) those independent of chromatin structure and (2) those dependent on chromatin structure. In this chapter we present protocols detailing two separate assays for each of these categories and discuss their advantages and shortcomings.

Chromatin Structure-Independent Assays

This class of nuclease assay can be extremely useful because of the speed and small volume (plasmid assay) or ability to assess the activity of a single nuclease present in a complex mixture (radioactive gel assay). Such attributes make these assays particularly suitable for analyzing and characterizing multiple parameters such as ion requirements and inhibitor specificity.

Plasmid Degradation Assay

One of the simplest and fastest assays of nuclease activity examines the digestion of a naked, linearized plasmid using a standard ethidium bromide-stained agarose gel. This technique is illustrated in Fig. 2 and details of the protocol are given below.

Protocol

1. Grow plasmid, isolate DNA, and linearize by standard molecular biological techniques. Dilute DNA to 1 μg/μl in H_2O.
2. Combine in a 1.5-ml microcentrifuge tube:
   A. 1 μl of linearized plasmid DNA (1 μg)
   B. 1 μl of 0.5 M Tris (pH 7.5) (50 mM final)
   C. 1 μl of 10 mM MgCl$_2$ (1 mM final)
Note: Stock concentration can be varied to examine the effects of Mg^{2+} concentration.
   D. 1 μl of 10 mM CaCl$_2$ (1 mM final)
Note: Stock concentration can be varied to examine the effects of Ca^{2+} concentration.

[8] N. N. Khodarev and J. D. Ashwell, *J. Immunol.* **156,** 922 (1996).
[9] R. A. Schwartzman and J. A. Cidlowski, *Endocrinology* **128,** 1190 (1991).
[10] R. A. Schwartzman and J. A. Cidlowski, *Endocrinology* **133,** 591 (1993).
[11] S. Pandey, P. R. Walker, and M. Sikorska, *Biochemistry* **36,** 711 (1997).
[12] M. Enari, H. Sakahira, H. Yokoyama, K. Okawa, A. Iwamatsu, and S. Nagata, *Nature* (*London*) **391,** 43 (1998).

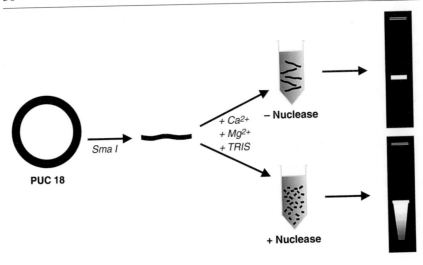

Fig. 2. The plasmid degradation assay. Linearized plasmid is incubated in the absence or presence of nucleases and appropriate ions, etc. Nuclease activity is detected as the appearance on an agarose gel of smaller (faster mobility) DNA fragments.

E. Crude or purified nuclease preparation

Note: Typically 10 μg of a nuclear or cytoplasmic extract or as little as 1 ng of a purified enzyme.

F. Experimental treatment (ions, inhibitors, etc.)

Note: Solvents [such as dimethyl sulfoxide (DMSO)] used to dissolve different inhibitors often interfere with the degradation of DNA in this assay because they are not sufficiently diluted. Consequently a sample containing only nuclease and solvent must be run in parallel to ensure the vehicle does not interfere with the reaction.

G. H$_2$O to a final volume of 10 μl.

3. Incubate for 1–5 hr at 37°.

4. Add 1 μl of proteinase K (20-mg/ml stock) and incubate at 55° for 1 hr.

Note: When using complex mixtures such as nuclear extracts this step is required to remove DNA-binding proteins that slow the migration of DNA in an agarose gel (similar to a gel shift assay). With purified enzymes this step may not be necessary.

5. Add 2 μl of a standard 6× DNA loading buffer,[13] electrophorese on

---

[13] T. Maniatis, E. F. Fritsch, and J. Sambrook, "Molecular Cloning: A Laboratory Manual." Cold Spring Harbor Laboratory Press, Cold Spring Harbor, New York, 1982.

a 1% agarose minigel, and visualize by ethidium bromide staining (0.5 μg/ml).

As shown in Fig. 3, in this assay linearized plasmid runs as a single band on the gel whereas active nuclease degrades this plasmid into smaller fragments that migrate faster. The major advantage of this technique is its speed. If the nuclease concentration can be adjusted to minimize the incubation, and the ethidium bromide is included in the agarose gel, the entire assay can be performed in less than 4 hr. Another major advantage is that the volume is small (10 μl), which reduces the amount of nuclease, inhibitor, etc., required. This is especially important when working with scarce and/or expensive reagents. The small volume can be detrimental, however, because solvents used as vehicles often interfere with nuclease activity because they are not sufficiently diluted. Finally, this assay measures the total activity present in a preparation. When prepared from whole cells, these extracts may contain various amounts of nonapoptotic nucleases that may normally be compartmentalized within organelles such as the lysosomes. To examine the activity of individual nucleases in a complex extract, the radioactive gel nuclease assay is employed.

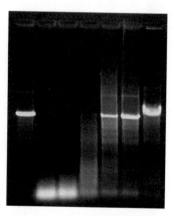

FIG. 3. An example of the plasmid degradation assay. Shown is an agarose gel depicting the integrity of linearized pUC18 plasmid after incubation with thymocyte nuclear extract (10 μg) and increasing concentrations of KCl for 5 hr. The "No Add" lane depicts samples treated exactly as the others without any added thymocyte extract. [Reprinted with permission from F. M. Hughes et al., J. Biol. Chem. **272,** 30567 (1997).]

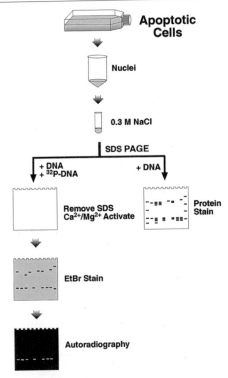

Fɪɢ. 4. The radioactive gel nuclease assay. Nuclear extracts are prepared from apoptotic cells and fractionated in an SDS–polyacrylamide gel with incorporated [^{32}P]-DNA. The gel is soaked to remove SDS and stimulate renaturation of the nucleases. The enzyme is then activated with Ca^{2+} and Mg^{2+}. After incubation the gel is dried and exposed to autoradiography. Active nucleases are detected as a cleared area, or "hole," in the background of radioactivity. [Reprinted with permission from J. A. Cidlowski *et al., Recent Prog. Horm. Res.* **51,** 457 (1996).]

Radioactive Gel Nuclease Assay

A chromatin structure-independent assay useful for distinguishing the activity of individual nucleases in a complex mixture is the radioactive gel nuclease assay. This assay is a modern modification of the Rosenthal–Lacks assay[14] and is illustrated in Fig. 4. In this assay, an SDS–polyacrylamide gel is polymerized with radiolabeled DNA and a nuclease preparation is then separated on the gel. Subsequent to renaturation and activation, the enzymes degrade the DNA in their immediate area. Degraded DNA fragments are then washed out of the gel and autoradiography reveals a "hole"

---

[14] A. L. Rosenthal and S. A. Lacks, *Anal. Biochem.* **80,** 76 (1977).

from which radioactivity has been lost. The original version of this assay used unlabeled DNA, stained the gel with ethidium bromide, and examined "holes" in fluorescence under transillumination. Although this technique does detect true nucleases, artifacts are a problem because DNA-binding proteins (such as histones) prevent intercalation of ethidium bromide into the double helix.[15] The use of radiolabeled DNA and autoradiography has circumvented this source of artifact.[16]

Protocol

1. Prepare ^{32}P-labeled DNA.

Note: Typically plasmid DNA is nick translated in the presence of [γ-^{32}P]dATP. However, virtually any source of DNA and any labeling technique that provides a uniform labeling (such as random priming) will suffice. Unlabeled double-stranded DNA can be added to adjust the sensitivity of the assay and prevent excessive degradation. The amount of unlabeled DNA required should be determined empirically.

2. Polymerize an 8 × 7 cm SDS–polyacrylamide minigel (1 mm thick) according to established techniques with the inclusion of ~750,000 cpm of ^{32}P-labeled DNA.

Note: This level of radioactivity was chosen to provide a uniform dark background after an overnight exposure.

3. Denature the samples and electrophorese through the radioactive SDS–polyacrylamide gel.

4. Remove the gel and place it in renaturation buffer [40 mM Tris (pH 7.6), 2 mM MgCl$_2$, 1 mM EDTA].

5. Wash the gel in copious amounts of renaturation buffer (typically 100 ml/gel) for
    A. 2 × 20 min
    B. 1 × overnight
    C. 6 × 30 min

Note 1: All washes are performed at room temperature on a shaker.

Note 2: This step removes SDS and facilitates renaturation of proteins, so extensive washing is recommended.

6. Place the gel in nuclease activation buffer [40 mM Tris (pH 7.6), 2 mM MgCl$_2$, 2 mM CaCl$_2$, 1 mM EDTA ± experimental treatment] and incubate at 37° for 1–24 hr (depending on the amount of nuclease activity present) with shaking.

Note 1: Ca^{2+} and Mg^{2+} levels can be varied to explore the effects of these ions. Similarly, inhibitors, activators, and other experimental modifiers of

[15] E. S. Alnemri and G. Litwack, *J. Biol. Chem.* **264,** 4104 (1989).
[16] M. L. Gaido and J. A. Cidlowski, *J. Biol. Chem.* **266,** 18580 (1991).

enzyme activity are included in this activation step. The reaction volume can be minimized to reduce the necessary amount of such modifiers.

Note 2: The progress of this reaction can be followed by including ethidium bromide (0.5 μg/ml) in the activation buffer and periodically examining the gel under UV transillumination. The gel will fluorescence brightly except in areas of nuclease activity (or DNA-binding activity), giving the appearance of a "hole."

7. Fix the gel in 40% (v/v) methanol–7% (v/v) acetic acid for 20 min, dry, and perform autoradiography.

Note: With the suggested level of radioactivity, the autoradiograph of the gel after an overnight exposure will appear dark (due to the ^{32}P) and areas of nuclease activity will be clear. This pattern can be compared with that seen with ethidium bromide staining (step 6) to differentiate between true nucleases (those producing "holes" with both ethidium bromide and autoradiography) and DNA-binding proteins, which produce "holes" only with ethidium bromide.

An example of a radioactive gel nuclease assay is shown in Fig. 5. One benefit of this assay is that it works equally well with purified enzymes (such as micrococcal nuclease) or more complex mixtures such as nuclear extracts from dexamethasone-treated (apoptotic) thymocytes. Moreover, it allows for an assessment of the number and relative molecular weights of each of the active nucleases.

A definitive limitation of this assay is that the methodology requires denaturation and renaturation of the enzyme in order to detect activity. Consequently, only relatively resilient enzymes are detected. However, given the importance of DNA degradation to the commitment of apoptosis, this would clearly be a desirable trait for an apoptotic nuclease. In addition, this assay separates proteins under denaturing conditions and, consequently, detects only single-chain nucleases, and not those requiring intact multimeric complexes. Finally, when working with complex mixtures such as nuclear extracts, the activities, or "holes," created by different nucleases may overlap, degrading all the DNA in the entire lane. It this case the sensitivity of the assay should be adjusted by adding unlabeled DNA during the gel polymerization step. In each system the amount of unlabeled DNA required should be determined empirically for optimal resolution.

The chromatin structure-independent assays have a number of attributes that have led to their widespread use in the study of apoptosis. However, it is important to remember that during apoptosis, nucleases cleave chromatin only at specific structures, so these assays are necessarily limited when attempting to assess the potential of an enzyme as an apoptotic nuclease. To aid in this assessment a second class of assay has been developed.

NUCLEASE
ACTIVITY GEL

PROTEIN
STAINING

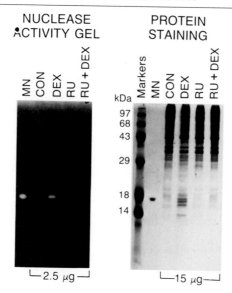

FIG. 5. An example of the radioactive gel nuclease assay, depicting the effect of dexamethasone (an apoptotic agent in thymocytes) on nuclease activity in nuclear extracts prepared from S49.1 cells (a thymocyte cell line). Cells were treated with vehicle (CON), dexamethasone (DEX; 0.1 μM), RU486 (RU; 1.0 μM), or RU plus DEX for 18 hr, extracts prepared, and nuclease activity assessed as described in text. Micrococcal nuclease (0.5 μg) was used as a positive control. The gel on the left is an autoradiograph displaying "holes" where nucleases are active. The gel on the right is a Comassie blue-stained gel showing the total protein content of these extracts. [Reprinted with permission from Caron-Leslie and J. A. Cidlowski, *Mol. Endocrinol.* **5**, 1169 (1991).]

Chromatin Structure-Dependent Assays

To assess nuclease activity in a manner that better recapitulates the *in vivo* situation, a series of assays that rely on the degradation of a chromatin substrate rather than naked DNA have been developed. In these assays, the chromatin maintains its *in vivo* three-dimensional structure and packaging. Consequently, the structural landmarks that are cleaved during apoptosis are presented intact to the nucleases under study, making these assays a direct assessment of the ability of an enzyme to function as an apoptotic nuclease.

There are two distinct types of these assays: (1) those that rely on activation of an endogenous nuclease to cleave an endogenous substrate (nuclear autodigestion) and (2) those that assess the ability of an active nuclease to cleave an exogenous substrate (typified here in the HeLa nuclei assay).

Nuclear Autodigestion Assay

In the autodigestion assay (illustrated in Fig. 6), nuclei are isolated and incubated in an environment known to activate the apoptotic nuclease(s) (i.e., elevated Ca^{2+} and Mg^{2+} levels). A latent endogenous nuclease(s), present within these nuclei, is activated and subsequently degrades the chromatin in a manner that reproduces what is found in an apoptotic cell (i.e., into ~30- to 50-kb and oligonucleosomal fragments). Many different cells are capable of this autodigestion reaction, although our experience has primarily been with thymocytes. Thus the assay described below is optimized for these cells. The major modification we have found necessary to adapt this protocol to different cell types is to alter the method for isolation of nuclei. It should be noted that not all cells are capable of this autodigestion reaction and caution is warranted when interpreting negative results.

Protocol

Isolation of Nuclei

1. Isolate thymocytes; wash, count, and pellet the cells (typically 10^9–10^{10} thymocytes are obtained per rat but other cell types may require as little as 5×10^6 cells/sample).[9,10]

2. Resuspend the pellet in 50 ml of 10 mM MgCl$_2$, 0.25% (w/v) Nonidet P-40 (NP-40).

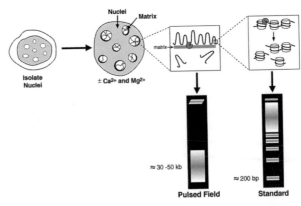

Fig. 6. The nuclear autodigestion assay. Nuclei are isolated and incubated with Ca^{2+} and Mg^{2+} to activate an endogenous nuclease. The activated nuclease then degrades the chromatin into large (~30- to 50-kb) and oligonucleosomal fragments, detected by pulsed field and standard electrophoresis, respectively.

Note: Suspension will take on a white hue as the plasma membranes lyse.

3. Pellet the nuclei and resuspend in 50 mM Tris (pH 7.4), 2 mM MgCl$_2$ at 4×10^8 nuclei/ml.

Note: When using thymocytes, the MgCl$_2$ concentration is critical to prevent lysis of nuclei. If nuclear lysis occurs in other cell types, our initial troubleshooting response is to increase the MgCl$_2$ concentration.

Assay

1. Combine in a 1.5-ml microcentrifuge tube:
    A. Experimental treatments (inhibitors, etc.)
    B. 20 μl of 1 M Tris (pH 7.4) (50 mM final)
    C. 8 μl of 0.1 M MgCl$_2$ (2 mM final)
    D. 4 μl of 0.1 M CaCl$_2$ (1 mM final)
    E. H$_2$O to a final volume of 300 μl

Note: The stock concentration of CaCl$_2$ can be varied to examine the dependency of Ca^{2+}. However, with thymocytes, if the MgCl$_2$ concentration is brought below ~1 mM, lysis ensues.

2. Add 100 μl of nuclei (4×10^7 nuclei; isolated in the preceding section) and rotate the samples at room temperature for 1.5 h.

Analysis

PULSED FIELD DNA ANALYSIS

1. For analysis of ~30- to 50-kb fragment formation, remove 200 μl and mix with 200 μl of Incert agarose (FMC Bioproducts, Rockland, ME) in PBS.

Note: Melt the agarose and cool to 37° in a water bath before use. Transfer the suspension to a 0.5-cm^2 plug mold and solidify at 4° for 3 min.

2. Extrude the plugs into 10 ml of pulsed field (PF) buffer [100 mM EDTA, 1% (v/v) N-lauroylsarcosine].

3. Incubate the plugs at 37° overnight.

4. Transfer the plugs into 1 ml of PF buffer containing 100 μg of proteinase K and incubate at 50° overnight.

5. Dialyze a 2-mm slice of each plug in 40 ml of 0.5× TBE (89 mM Tris, 89 mM boric acid, 2.5 mM EDTA) for 3 hr.

6. Run a pulsed field gel and visualize with ethidium bromide.

Note: We typically use a CHEF (contour-clamped homogeneous electric field) pulsed field system (Bio-Rad, Hercules, CA) run at 6.0 V/cm for 19 hr with a linear switch interval ramp from 0.5 to 45.0 sec. We have previously shown that these parameters linearly separate DNA fragments in the target range.[17]

[17] F. M. Hughes, Jr., and J. A. Cidlowski, *Cell Death Differ.* **4**, 200 (1997).

CONVENTIONAL DNA ANALYSIS

1. For analysis of oligonucleosomal fragment formation, stop the reaction in the remaining sample by adding
    A. 15 μl of 0.5 M EDTA
    B. 30 μl of 5.0 M NaCl
    C. 15 μl of 10% SDS
    D. 240 μl of TE buffer [10 mM Tris (pH 7.4), 1 mM EDTA]
2. Add 10 μl of proteinase K (20 mg/ml) and incubate for 1 hr at 55°.
3. Extract twice with phenol–chloroform–isoamyl alcohol (25:24:1, by volume) (500 μl/extraction).
4. Extract once with chloroform–isoamyl alcohol (24:1, v/v) (500 μl/extraction).
5. Precipitate the DNA by adding 10 μl of 5 M NaCl and 1.0 ml of ice-cold 100% (v/v) ethanol. Mix by inverting and incubate at −70° for ≥30 min.
6. Centrifuge for 20 min at 4°.
7. Pour off supernatant; dry the pellet for 20 min in a Speed-Vac.
8. Resuspend the DNA by adding 30 μl of TE buffer (do not resuspend) and 1 μl of DNase-free RNase A (10 mg/ml).
9. Incubate at 37° overnight (or for 2 hr, followed by gentle resuspension by pipetting).
10. Measure the DNA concentration by absorbance (260 nm) and analyze 15 μg by conventional agarose electrophoresis [1.8% (w/v) agarose gel]. Visualize by ethidium bromide staining.

An example of the results obtained with this assay is shown in Fig. 7. The major advantage of this assay is that it duplicates the DNA fragmentation patterns seen *in vivo* and thus measures the capacity of an enzyme to function as an apoptotic nuclease. The primary limitation of an autodigestion assay is that the enzyme is isolated with the nuclei and cannot be further purified and characterized. In addition, modification of the reaction conditions is somewhat limited because isolated nuclei are fragile and subject to lysis (particularly in thymocytes if the MgCl$_2$ levels are too low). Finally, there are two required steps in this assay: activation of a latent nuclease and the subsequent degradation of the chromatin. With this assay it is impossible to conclude if an inhibitory substance is preventing activation of the latent enzyme or suppressing the cleavage of substrate to product. To better explore the characteristics of an active enzyme in an environment conducive to purification of individual nucleases, we modified this assay system to allow for an active nuclease to be applied to an exogenous chromatin substrate. We have designated this assay the HeLa nuclear assay after the donor of the chromatin substrate.[9]

A

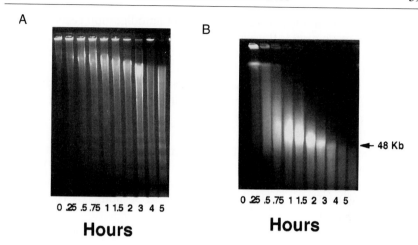

B

◄— 48 Kb

0 .25 .5 .75 1 1.5 2 3 4 5
Hours

0 .25 .5 .75 1 1.5 2 3 4 5
Hours

FIG. 7. An example of the nuclear autodigestion assay. The gels demonstrate the time course of chromatin degradation into oligonucleosomal (A) or large (~30- to 50-kb) fragments (B) when rat thymocytes nuclei are incubated with 1 mM Ca^{2+} and 2 mM Mg^{2+}.

HeLa Nuclear Assay

In this assay, illustrated in Fig. 8, nuclei are isolated from exponentially growing HeLa cells and incubated in the presence of an exogenous nuclease.

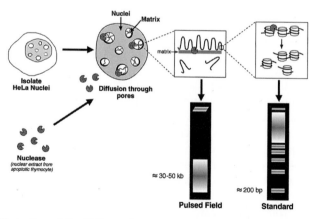

FIG. 8. Illustration of the HeLa nuclear assay. Nuclease preparations (such as extracts from apoptotic cells) are combined with isolated HeLa nuclei. The nuclease then degrades the chromatin in a manner specific to that enzyme. The activities of apoptotic enzymes are illustrated in which the DNA is degraded into large (~30- to 50-kb) fragments and oligonucleosomal fragments. [Reprinted with permission from J. A. Cidlowski *et al., Recent Prog. Horm. Res.* **51,** 457 (1996).]

Importantly, HeLa nuclei do not undergo an autodigestion reaction, even after 5 hr of incubation.[17] If we apply an exogenous nuclease to these donor nuclei, the enzyme then degrades the HeLa chromatin in a manner specific to the spatial constraints of that enzyme. In theory this protocol should be adaptable for any chromatin donor that does not undergo an autodigestion reaction, although our experience has beeen exclusively with HeLa cells.

Isolation of Nuclei

1. Harvest HeLa cells ($\sim$2 $\times$ 10^6 cells/sample); count and pellet the cells.
2. Resuspend the pellet in 50 ml of 10 mM MgCl$_2$, 0.25% (w/v) NP-40.
Note: The suspension will take on a white hue as the plasma membranes lyse.
3. Pellet the HeLa nuclei and resuspend in 50 mM Tris (pH 7.4) at 2 $\times$ 10^7 nuclei/ml.
Note: In contrast to the thymocyte nuclei, HeLa nuclei do not lyse in the absence of MgCl$_2$.

Assay

1. Combine in a 1.5-ml microcentrifuge tube:
   A. Nuclease preparation (typically 50 μg of a nuclear extract or as little as 5 ng of a purified enzyme)
   B. Experimental treatments (inhibitors, activators, etc.)
   C. 20 μl of 1 M Tris (pH 7.4) (50 mM final)
   D. 8 μl of 0.1 M MgCl$_2$ (2 mM final)
   E. 4 μl of 0.1 M CaCl$_2$ (1 mM final)
   F. H$_2$O to a final volume of 300 μl
Note: Stock concentration of CaCl$_2$ and MgCl$_2$ can be varied to examine the requirement of these ions
2. Add 100 μl of nuclei from the preceding section (2 $\times$ 10^6 nuclei) and rotate samples at room temperature for 1.5 hr.

DNA Analysis. Samples are analyzed for $\sim$30- to 50-kb and internucleosomal fragments exactly as described for the autodigestion assay.

An example of this assay is shown in Fig. 9. The HeLa assay has a number of advantages beyond recapitulation of apoptotic DNA fragmentation. For example, this assay works equally well with purified nucleases (such as micrococcal nuclease, DNase I, and cyclophilin[17]), crude cellular extracts, or fractions from a purification column.[17,18] Because this assay involves adding active enzyme to the HeLa chromatin, inhibitory influences can be reliably attributed to suppression of the active enzyme. Moreover, the

---

[18] F. M. Hughes, Jr., R. B. Evans-Storms, and J. A. Cidlowski, *Cell Death Differ.*, in press (1998).

A

B

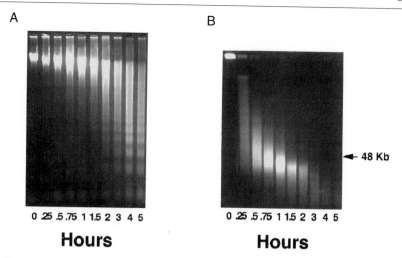

0 .25 .5 .75 1 1.5 2 3 4 5

0 .25 .5 .75 1 1.5 2 3 4 5

← 48 Kb

Hours　　　　　　**Hours**

FIG. 9. An example of the HeLa nuclear assay. The gels demonstrate the time course of chromatin degradation into large oligonucleosomal (A) or large (~30- to 50-kb) fragments (B) when HeLa nuclei are incubated with 100 μg of nuclear extract from apoptotic thymocytes. [Reprinted with permission from F. M. Hughes and J. A. Cidlowski, *Cell Death Differ.* **4,** 200 (1997).]

spatial constraints of an enzyme are clearly delineated by this assay. For example, our laboratory used the radioactive gel assay to implicate cyclophilin as a nuclease involved in apoptosis.[2] Although our initial hypothesis was that cyclophilin was responsible for internucleosomal DNA cleavage, the HeLa nuclear assay revealed that this enzyme only produces large (~30- to 50-kb) fragments and not internucleosomal DNA fragments.[3,17] The limitations of this assay are primarily those associated with the thymocyte autodigestion assay, such as its relatively large size (400-μl reaction volume) and processing time (typically 2–3 days). However, this assay clearly has the most utility in identifying, purifying, and characterizing apoptotic nucleases.

Conclusion

The study of DNA degradation during apoptosis has attracted considerable attention because it clearly represents a commitment to cell death. Despite the voluminous literature on this phenomenon, a consensus on a single enzyme responsible for DNA degradation in all dying cells has not yet been reached. This indecision suggests that multiple enzymes may be involved in this process and that they likely function in a cell type-specific and/or signal-specific manner. The use of the assays described here will continue to aid in the delineation of what appear to be several different but interrelated pathways of chromatin degradation.

Section II

Measuring Apoptosis in Lower Organisms

[6] Analysis of Programmed Cell Death and Apoptosis in *Drosophila*

By Po Chen *and* John M. Abrams

Programmed cell death plays an important role in the development and homeostasis of both vertebrates and invertebrates. In the fruit fly, *Drosophila melanogaster,* most, if not all, programmed cell deaths exhibit apoptotic features (e.g., Abrams *et al.*[1]). Numerous lines of evidence argue that core molecular components required for this process are highly conserved, and discoveries in *Drosophila* have identified important regulators and effectors of apoptotic cell death (for reviews see Refs. 2 and 3). This model system affords a powerful combination of genetic and molecular tools to study apoptosis within the context of normal and abnormal development. The patterning of tissues in the fly exhibits remarkable plasticity, including the ability to eliminate cells (by apoptosis) that do not successfully complete their developmental program and, in apparent contrast to the nematode *Caenorhabditis elegans,* many cell deaths in the fly are not strictly predetermined by lineage.[1,4] *Drosophila* is thus uniquely suited for studying how cell interactions or environmental stresses can specify or trigger the cell death fate.

In this chapter, we describe the methods that have been successfully used to identify genetic components required for apoptosis during normal and abnormal development. First, we describe techniques to detect programmed cell death in the embryo. These methods can also be used on imaginal disks,[5-7] ovaries,[8,9] and larval midgut or salivary gland preparations.[10] The second section describes the assays to study cell death proteins using cultured *Drosophila* cells. Finally, we describe methods to visualize macrophages in live and fixed preparations.

[1] J. M. Abrams, K. White, L. Fessler, and H. Steller, *Development* **117,** 29 (1993).
[2] A. Rodriguez, P. Chen, and J. M. Abrams, *Am. J. Hum. Genet.* **62,** 514 (1998).
[3] K. McCall and H. Steller, *Trends Genet.* **13,** 222 (1997).
[4] T. Wolff and D. F. Ready, *Dev. Biol.* **113,** 825 (1991).
[5] T. E. Spreij, *Netherlands J. Zool.* **21,** 221 (1971).
[6] N. M. Bonini, W. M. Leiserson, and S. Benzer, *Cell* **72,** 379 (1993).
[7] M. Milan, S. Campuzano, and A. Garciabellido, *Proc. Natl. Acad. Sci. U.S.A.* **94,** 5691 (1997).
[8] K. McCall and H. Steller, *Science* **279,** 230 (1998).
[9] K. Foley and L. Cooley, *Development* **125,** 1075 (1998).
[10] C. Jiang, E. H. Baehrecke, and C. S. Thummel, *Development* **124,** 4673 (1997).

General References

There are several excellent references that provide detailed information on this experimental system. For general information, Ashburner's *Drosophila: A Laboratory Handbook*[11] and the accompanying *Drosophila: A Laboratory Manual*[12] are outstanding resources. For a concise description of general methods, *Drosophila: A Practical Approach,*[13] and *Drosophila melanogaster: Practical Uses in Cell and Molecular Biology*[14] are also useful. The most recent and comprehensive description of fly development is *Development of Drosophila melanogaster,* volumes I and II from Bate and Martinez Arias.[15] Also Peter Lawrence's *The Making of a Fly*[16] is an outstanding and easy-to-read book that summarizes the molecular genetics of development in this organism.

Visualizing Apoptosis in Whole-Mount Embryos

General Buffers

Phosphate buffer, pH 7.2: Mix 72 ml of 0.1 M Na_2HPO_4 with 28 ml of 0.1 M NaH_2PO_4

Phosphate-buffered saline (PBS): See p. 380 in Ref. 12

Bleach (50%, v/v): This simple 1:1 mix of bleach and water is used to remove the chorion membrane that surrounds the embryo. Most procedures require this as a first step

Collection and Dechorionation of Embryos. To collect embryos, we place healthy populations of flies into plastic bottles to which egg collection plates have been attached. Recipes for egg collection plates can be found in the references mentioned above. Just before use, a small amount of yeast paste is placed onto each plate. To obtain large batches of staged embryos, frequent replacement of these plates is advised. For a broad representation of most stages in which cell death occurs, we typically collect embryos for 7 hr (at 25°) and then age these collections for another 14 hr (at 18°). This schedule allows us to examine embryos from about ~7 hr of developmental

[11] M. Ashburner, "*Drosophila:* A Laboratory Handbook." Cold Spring Harbor Laboratory Press, Cold Spring Harbor, New York, 1989.

[12] M. Ashburner, "*Drosophila:* A Laboratory Manual." Cold Spring Harbor Laboratory Press, Cold Spring Harbor, New York, 1989.

[13] D. B. Roberts, "*Drosophila:* A Practical Approach." IRL Press, Oxford, 1986.

[14] L. S. B. G. A. E. A. Fyrberg, *Methods Cell Biol.* **44** (1994).

[15] M. Bate and A. M. Arias, "Development of *Drosophila melanogaster,*" Vols. I and II. Cold Spring Harbor Laboratory Press, Cold Spring Harbor, New York, 1993.

[16] P. A. Lawrence, "The Making of a Fly: The Genetics of Animal Design." Blackwell Scientific Publications, London, 1992.

age (stage 12), when prominent cell deaths are first observed, to ~14 hr of developmental age (stage 16).

To prepare embryos for most staining techniques, embryos are dechorionated in 50% (v/v) sodium hypochlorite (bleach) and washed extensively. This procedure is detailed in the section on handling embryos in Ref. 13.

Staining of Embryos with Vital Dye Acridine Orange or Nile Blue

An easy method to visualize dying cells exploits vital dyes. These dyes selectively stain apoptotic forms of cell death and they are convenient because they can be used on whole-mount preparations. However, these preparations are practical only for limited periods of time (about 0.5 hr) because the tissue is only transiently stained. Acridine orange and Nile blue are two vital dyes that have proved to be useful for the detection of cell death.[1] Acridine orange is preferred in situations calling for broad visual depth whereas Nile blue offers better resolution within limited focal planes. For examples of typical images, see Fig. 1 and Abrams *et al.*[1] It is important to note that both staining procedures detect apoptosis even after

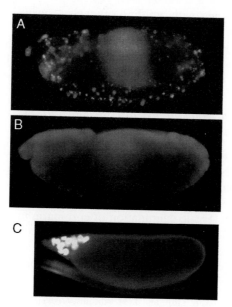

Fig. 1. Visualization of programmed cell death in *Drosophila*. Acridine orange staining detects apoptosis (see text) in a wild-type embryo (A) and a cell death-defective H99 mutant at the same stage (B). H99 embryos are deleted for essential death activators and defective for programmed cell death. TUNEL (C) detects the programmed death of nurse cells in developing ovaries. [Parts (A) and (B) are from P. Chen, *et al.*, *Genes Dev.* **10,** 1773 (1996).]

engulfment has occurred, permitting visualization of cell corpses within large phagocytic macrophages.[1] Interestingly, transcripts encoding important death activators [e.g., *reaper (rpr)*, *grim*, and *head involution defective (hid)*] also persist during the later stages of apoptosis and are frequently associated with phagocytosed corpses residing inside macrophages.[17–19]

Reagents

Acridine orange (A-6014; Sigma, St. Louis, MO), 5 μg/ml in 0.1 *M* phosphate buffer

Nile blue A (N0766; Sigma), 100 μg/ml in 0.1 *M* phosphate buffer

Halocarbon oil, series 700 (Halocarbon Products, River Edge, NJ)

1. Dechorionate the embryos in bleach and wash thoroughly.

2. Place the embryos in a tube with equal volumes of heptane and either acridine orange or Nile blue staining solution. Shake for 5 min.

3. Carefully remove the embryos (they will accumulate at the interface) and place them on a glass slide. Allow the heptane to evaporate briefly (avoid overdrying) and cover the embryos with a small amount of halocarbon oil. Gently place a coverslip over the sample.

4. Samples stained with acridine orange are viewed on a fluorescent microscope, using either the rhodamine or fluorescein filter. Generally, these images are similar. However, it is our impression that dying cells are first visible in the red channel. As time passes, dying cells become visible under both filters and, finally, staining in the green channel predominates.

Terminal Deoxynucleotidyltransferase-Mediated dUTP Nick End Labeling of Embryos and Ovaries

The terminal deoxynucleotidyltransferase (TdT)-mediated dUTP nick end labeling (TUNEL) technique was developed for the detection of apoptotic cells in histological sections[20] and was adapted for use on whole-mount *Drosophila* tissues by Robinow *et al.*[21] and White *et al.*[19] The procedure exploits the characteristic degradation of chromatin in apoptotic nuclei to label free DNA ends *in situ* with terminal deoxynucleotidyltransferase. One can examine the expression pattern of a gene and apoptosis in the same embryo by double labeling by *in situ* hybridization and TUNEL.[19] An

[17] P. Chen, W. Nordstrom, B. Gish, and J. M. Abrams, *Genes Dev.* **10**, 1773 (1996).
[18] M. E. Grether, J. M. Abrams, J. Agapite, K. White, and H. Steller, *Genes Dev.* **9**, 1694 (1995).
[19] K. White, M. Grether, J. M. Abrams, L. Young, K. Farrell, and H. Steller, *Science* **264**, 677 (1994).
[20] Y. Gavrieli, Y. Sherman, and S. A. Ben-Sasson, *J. Cell Biol.* **119**, 493 (1992).
[21] S. Robinow, T. A. Draizen, and J. W. Truman, *Dev. Biol.* **190**, 206 (1997).

example of this labeling technique can be seen in Fig. 1C, which images the apoptotic death of ovarian nurse cells.

Reagents

Reaction buffer for embryos: 1× Terminal transferase buffer (Boehringer Mannheim Biochimica, Mannheim, Germany), 2.5 mM CoCl$_2$, 0.3% (v/v) Triton X-100, 10 μM dUTP mix (1:2 ratio of 10 μM biotin–16-dUTP and 10 μM dUTP from Boehringer Mannheim Biochimica), and terminal deoxynucleotidyltransferase (0.5–15 U/ml)
Reaction buffer for ovaries: The same as above, except that a final concentration of 60 μM biotin–16-dUTP is used and Triton is not included
PBS-T: Phosphate-buffered saline with 0.3% (v/v) Triton X-100
Fluorescein isothiocyanate (FITC)–avidin DN or Vectastain kit (Vector Laboratories, Burlingame, CA)
3,3'-Diaminobenzidine (DAB) at 5 mg/ml
H$_2$O$_2$ (3%, v/v)

1. Fix the embryos with 4% (w/v) paraformaldehyde and remove from vitelline membrane either by hand dissection or by "methanol cracking" as described in Ref. 12, pp. 190–191. For ovaries, hand dissect tissue in PBS. Fix in 2% (w/v) paraformaldehyde in PBS, supplemented with 0.1% (v/v) Tween 20 and agitate gently together with 3 vol of heptane for 30 min.
2. Rinse the embryos in PBS-T and then in reaction buffer without dUTP or enzyme.
3. Incubate the embryos in reaction buffer with dUTP and enzyme at 37° for 3 hr.
4. Wash extensively in PBS-T. Labeled nucleotides can be detected with avidin conjugates (e.g., peroxidase or fluorescein) available from Vector Laboratories. [To visualize with FITC–avidin conjugate, use a 1:200 dilution. To visualize for peroxidase, use components A and B from the Vectasatin kit and develop according to the supplier instructions; this will require a solution of DAB (0.5 mg/m) and 3% (v/v) H$_2$O$_2$.]

Note: We sometimes experience difficulty in the labeling, apparently because of poor penetrance. This problem can be solved by light digestion of embryos with protease K as described for *in situ* hybridizations.[14] After this, continue with step 2.

Studying Apoptotic Functions of Genes in *Drosophila* Cell Culture

Drosophila cell culture provides a convenient complement to the powerful genetic and molecular tools available in this model organism. Consistent with genetic data, central death activators (*rpr, hid,* and *grim*) induce apoptotic cell death when expressed in SL2[17,22,23] and Mbn-2 cells (J. Varkey, P. Chen, and J. M. Abrams, unpublished observations, 1999). The apoptotic cells manifest characteristic morphological changes that are easily identified and scored (see Fig. 2[23a]). Apoptotic death induced in these cultured cells can be blocked by coexpression of viral antiapoptotic protein p35, or by the presence of peptide caspase inhibitors.[17,22,23]

We typically maintain SL2 cells in D-SFM, which is a defined medium to which we have adapted our cells. SL2 cells can alternatively be cultured in Schneider's medium supplemented with 10% fetal calf serum.[14] Both media are available from GIBCO-BRL (Gaithersburg, MD). While Mbn-2 cells are quite adherent, SL2 cells are semiadherent. However, when used in the following protocols at the recommended cell density, more than 95% SL2 cells stay attached to the plate as long as the cells are gently handled.

Propidium Iodide Staining and Flow Cytometry

A direct method to quantitate cell death is simply to count the incidence of apoptotic and nonapoptotic cells in a given field. Alternatively, one can use flow cytometric assays to quantitate "hypodiploid" nuclei[24] as a measure of apoptotic cell death (Fig. 2).

Reagents

PI solution: Propidium iodide (50 μg/ml), 0.1% (w/v) sodium citrate, 0.1% (v/v) Triton X-100

1. Plate exponentially growing cells at 10^6/ml in six-well plates or 35-mm plates, 2 ml/well, and incubate for several hours to overnight.
2. Treat as desired.
3. Gently wash the cells off the plates, save 1 ml for protein gel if needed, and put 1 ml into an Eppendorf tube.
4. Spin down cells at 5000 rpm for 5 min at room temperature.
5. Aspirate the supernatant and resuspend the cells with 1 ml of 2% (v/v) Formaldehyde in PBS; fix for 15 min.
6. Spin at 5000 rpm for 5 min; resuspend the cell pellet in 0.5 ml of PI solution.
7. Keep the cells at 4° overnight in the dark, and analyze within 24 hr.

[22] G. J. Pronk, K. Ramer, P. Amiri, and L. T. Williams, *Science* **271,** 808 (1996).
[23] W. Nordstrom, P. Chen, H. Steller, and J. M. Abrams, *Dev. Biol.* **180,** 213 (1996).
[23a] P. Chen, P. Lee, L. Otto, and J. M. Abrams, *J. Biol. Chem.* **271,** 25735 (1996).
[24] O. Tounekti, J. Belehradek, and L. M. Mir, *Exp. Cell Res.* **217,** 506 (1995).

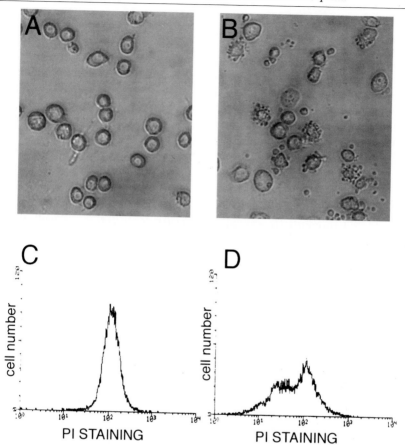

FIG. 2. Detection of apoptosis in SL2 cells. (A and B) Photomicrographs of SL2 cells stably transfected by pMT-rpr. In these cells, conditional *reaper* expression is directed by the metallothionein promoter. (C and D) Flow cytometric analyses of these cells. (A) and (C) are uninduced control cells (no copper). (B) shows cells induced for *reaper* expression (treated with copper for 13 hr) and (D) shows results of flow cytometry from samples induced for *reaper* after 16 hr. Note overt membrane blebbing and apoptotic morphology of cells in (B) that are represented as a distinct population of "hypodiploid" nuclei in (D). [(A) and (B) are from P. Chen, *et al., J. Biol. Chem.* **271,** 25735 (1996).]

[*Note:* We typically analyze samples on a Becton Dickinson (San Jose, CA) FACScan flow cytometer using LYSIS II software. The settings we use for the FACScan are as follows: FSC, linear E.00; Amplifier, 9.99; SSC, log 260; FL1, OFF; FL2, OFF; FL3, log 334. All resolutions are set at 1024.]

Assays for Promotion or Inhibition of Apoptosis by Transfection

Studies of gene products that encode pro- or antiapoptotic functions can be pursued through transient expression assays. Constitutive expression from the actin promoter or conditional expression from the metallothionein promoter are typically used. Cell killing can be directly visualized by microscopy, and/or quantitated by flow cytometry. Alternatively, we frequently evaluate cell killing measured as the loss of cotransfected reporter gene activity (*lacZ*). Described below is the protocol we use for conditional expression from a metallothionein promoter. Typical transfection efficiencies range between 50 and 60%.

Day 1

1. Plate exponentially growing cells at 0.6×10^6/ml in six-well plates or 35-mm plates in D-SFM containing gentamicin (or Schneider's medium for Mbn-2 cells), 2 ml/well; allow the cells to attach for at least 1 hr.

2. Prepare the following solutions in Eppendorf tubes:

   Solution A: For each transfection, dilute the DNA into 100 μl of D-SFM without antibiotics (we generally use 0.2 μg of pact-lacZ, 2 μg of apoptotic gene plasmid, and/or 3 μg of antiapoptotic gene plasmid)

   Solution B: For each transfection, dilute 5.5 μl of CellFECTIN (GIBCO-BRL) into 100 μl of D-SFM without antibiotics

3. Combine the two solutions, mix gently, and incubate at room temperature for 25 min.

4. Wash the cells once with 2 ml of D-SFM without antibiotics.

5. For each transfection, add 0.8 ml of D-SFM without antibiotics to each tube containing lipid–DNA complexes. Mix gently. Aspirate the wash medium, and overlay the diluted lipid–DNA complexes onto the washed cells.

6. Incubate the cells for 5 hr at 25°.

7. Remove the transfection mixtures and add 2 ml of D-SFM containing gentamicin (or Schneider's medium for Mbn-2 cells). Incubate the cells at 25° for 36–48 hr.

Day 3

8. If testing conditional expression directed by the metallothionein promoter, split the cells into two groups (we use 12-well plates), add 0.7 mM CuSO₄ to one set, and incubate for the desired period, generally 2–16 hr.

9. Process the cells for FACScan, or β-galactosidase (β-Gal) staining, etc.

For a potent death activator (e.g., *grim*) we typically observe ~25–30% apoptotic cells within 5 hr, and a 10-fold decrease in LacZ$^+$ cells compared with "empty vector" controls.

Note: The same transfection procedure can also be used for the production of stable cell lines by including an appropriate selection vector. For this purpose, cells can be cotransfected with pCohygro,[25] and transferred to medium supplemented with hygromycin B (300 μg/ml; Sigma) 48 hr after transfection. Approximately 3–5 weeks is required to establish stable lines.

Assay for Caspase Activity

Preparation of S-100 Cytosol

This procedure is adapted from methods used for mammalian cells.[26]

Reagents

Buffer A: 20 mM HEPES-KOH, (pH 7.5), 10 mM KCl, 1.5 mM MgCl$_2$, 1 mM sodium EDTA, 1 mM sodium EGTA, 1 mM dithiothreitol, supplemented with pepstatin (5 μg/ml), leupeptin (10 μg/ml), aprotinin (2 μg/ml), 0.1 mM phenylmethylsulfonyl fluoride (PMSF)

1. Pellet 1×10^8 cells at 3000g for 5 min at room temperature.
2. Wash the pellets once with ice-cold phosphate-buffered saline (PBS) and resuspend with 1 ml of buffer A containing 250 mM sucrose for 15 min on ice.
3. Homogenize the cells in a Teflon homogenizer with 10 strokes.
4. Spin the homogenate twice at 750g for 10 min at 4°, and then centrifuge the supernatant at 10,000g for 15 min at 4°.
5. Centrifuge the supernatant from step 4 at 100,000g for 1 hr at 4°; the resulting supernatant is S-100 cytosol. The S-100 cytosol can be aliquoted and stored at −80°.

(*Note:* The pellet from step 4 is the heavy membrane fraction that is rich in mitochondria.)

Enzyme Assays

METHOD I: CLEAVAGE OF PEPTIDE SUBSTRATES. Incubate S-100 cytosol containing 50 μg of protein with 100 μM Ac-DEVD-pNA in a total volume of 500 μl at room temperature, then determine the OD$_{410}$ value.

[25] H. Johansen, A. van der Straten, R. Sweet, E. Otto, G. Maroni, and M. Rosenberg, *Genes Dev.* **3,** 882 (1989).
[26] J. Yang, X. S. Liu, K. Bhalla, C. N. Kim, A. M. Ibrado, J. Y. Cai, T. I. Peng, D. P. Jones, and X. D. Wang, *Science* **275,** 1129 (1997).

METHOD II: CLEAVAGE OF POLY(ADP-RIBOSE) POLYMERASE

1. Mix S-100 cytosol containing 16 μg of protein with 100 ng of bovine poly(ADP-ribose) polymerase (PARP) (Biomol Research Laboratories, Plymouth Meeting, PA) in a total volume of 20 μl.

2. Incubate at 25° for 1 hr.

3. Stop the reaction by adding 25 μl of 2× sodium dodecyl sulfate (SDS) sample buffer, and detect PARP protein by Western analysis using anti-PARP (Biomol Research Laboratories).

Figure 3 illustrates "signature" PARP cleavage derived from fly cells against this mammalian substrate.

Methods to Visualize Macrophages

Macrophages are found at zones where cell death occurs and it is sometimes useful to visualize them independently. These phagocytic cells are usually easy to identify because they are large cells that contain multivesicular inclusions. Furthermore, they are often engorged with cell corpses and it is therefore simple to distinguish phagocytes in tissue sections. Below are several methods that can be used to detect these cells that are not dependent on their association with apoptotic cells.

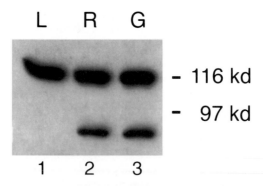

FIG. 3. Apoptotic signaling in *Drosophila* cells activates caspase activity. L2 cells (L) transfected and induced for expression of death activators RPR (R) or GRIM (G) were harvested before apoptosis, and fractionated into cytosolic (S-100) and mitochondrial fractions as described in text. These samples were assayed for caspase-like proteolytic activity, using bovine PARP, and analyzed by immunoblotting with anti-PARP antibody. The cytosol of RPR- and GRIM-expressing cells exhibits caspase activity, detected as a "signature" 85-kDa cleavage fragment (lanes 2 and 3). This proteolytic activity is not found in the cytosol of parental L2 cells (lane 1).

Detection of Macrophage Scavenger Receptors in Live Embryos

This method exploits the activity of scavenger receptors on the surface of *Drosophila* macrophages.[27] These receptors are remarkably similar to a class of endocytic receptors found on mammalian macrophages and, as such, they are readily detected by using a fluoresceinated ligand, DiI-AcLDL. For examples of typical images, see Abrams *et al.*[27]

Materials and Reagents

1,1-Dioctadecyl-3,3,3′3′-tetramethylindocarbocyanine perchlorate-labeled acetylated LDL (DiI-AcLDL): This reagent can be prepared according to Ref. 28 or may be purchased from Molecular Probes (Eugene, OR). Prepare a 25-mg/ml solution in injection buffer (0.1 mM phosphate buffer, 5 mM KCl, pH 6.8)
Double-stick tape
Charcoal agarose blocks: These are made from 2% (w/v) agarose, mixed with enough charcoal to give a deep black color

1. Dechorionate the embryos, wash them thoroughly, and prepare them for microinjection as described in Ref. 12, protocol 100. Briefly, ~100 embryos are oriented in a line on black agar along the anterior–posterior axis. Apply a strip of double-stick tape to a glass slide and lightly touch this to the line of embryos. Desiccate the embryos in a container of Drierite for ~10–14 min (precise time should be empirically determined).
2. Microinject DiI-AcLDL into the embryos.
3. Allow the embryos to age and then view.

Notes: We have found that the best accumulation of ligand occurs when embryos are injected at stage 10 or 11 (age 4.5–7.5 hr) and then viewed at stages 15/16 (age 12.5–15.5 hr). For viewing, particularly under a confocal microscope, we often refrigerate the samples to retard movement of the embryo inside the egg case. Although the ligand appears to disperse throughout the embryo, preferential uptake near the injection site is often observed.

Detection of Macrophages in Fixed Preparations

There are several markers that can be used in standard procedures[11–13] for this purpose. Antibodies directed against Peroxidasin[29] (termed anti-X

---

[27] J. M. Abrams, A. Lux, H. Steller, and M. Krieger, *Proc. Natl. Acad. Sci. U.S.A.* **89,** 10375 (1992).

[28] M. Krieger and D. M. Kingsley, *Proc. Natl. Acad. Sci. U.S.A.* **81,** 5454 (1984).

[29] R. E. Nelson, L. I. Fessler, Y. Takgi, B. Blumberg, D. R. Keene, P. F. Olson, C. G. Parker, and J. H. Fessler, *EMBO J.* **13,** 3438 (1994).

in Abrams *et al.*[1]) appear to selectively label phagocyte membranes in the embryo. Alternatively, macrophage nuclei can be labeled with an enhancer trap strain referred to as p197.[27] This line shows specific *lacZ* expression that can be detected histochemically or by using an anti-LacZ antibody.

Acknowledgments

We thank Johnson Varkey for use of unpublished data and are grateful to Joe Chapo and Su-Inn Ho for excellent technical assistance. This work was supported by Grant AG12466 from the National Institutes of Health.

[7] Analysis of Programmed Cell Death in the Nematode *Caenorhabditis elegans*

By Duncan Ledwich, Yi-Chun Wu, Monica Driscoll, and Ding Xue

The nematode *Caenorhabditis elegans* has been shown to be an excellent model organism with which to study the mechanisms of programmed cell death because of its powerful genetics and the ability to study cell death with single-cell resolution. In this chapter, we describe methods that are commonly used to examine various aspects of programmed cell death in *C. elegans*. These methods, in combination with genetic analyses, have helped identify and characterize many components of the *C. elegans* cell death pathway, illuminating the mechanisms by which these components affect programmed cell death.

Introduction

Programmed cell death is a widespread phenomenon in the animal kingdom and a prominent feature in the development of the nematode *C. elegans*.[1] During the development of an adult hermorphradite, 131 of 1090 somatic cells generated undergo programmed cell death.[2-4] The ability to observe and monitor cell division or cell death of individual cells in transparent living animals, using Nomarski differential interference contrast (DIC) optics, facilitated the determination of the complete *C. elegans* cell lin-

[1] R. E. Ellis, J. Yuan, and H. R. Horvitz, *Annu. Rev. Cell. Biol.* **7,** 663 (1991).
[2] J. E. Sulston and H. R. Horvitz, *Dev. Biol.* **56,** 110 (1977).
[3] J. Kimble and D. Hirsh, *Dev. Biol.* **70,** 396 (1979).
[4] J. E. Sulston, E. Schierenberg, J. G. White, and J. N. Thomson, *Dev. Biol.* **100,** 64 (1983).

eage[2-4] and the determination of the exact time and position of the 131 cells that die, which are invariant from animal to animal.[2,4] The advantage of studying cell death at the single-cell level in combination with the powerful genetics available in *C. elegans* has led to the identification of many mutations that affect different aspects of programmed cell death and has helped define a genetic pathway that regulates and executes programmed cell death in *C. elegans*.[1] Included in this pathway are genes that specify which cells should die (e.g., *ces-1* and *ces-2*),[5] genes that execute the cell death program (e.g., *ced-3, ced-4,* and *ced-9*),[6,7] genes that are involved in cell corpse engulfment (e.g., *ced-1, ced-2, ced-5, ced-6, ced-7,* and *ced-10*),[8,9] and genes that clean up cellular debris (e.g., *nuc-1*).[10]

A second advantage of studying cell death in *C. elegans* is the availability of a detailed genetic map, the physical map of cosmid and yeast artificial chromosome (YAC) clones that cover the entire genome,[11,12] and the almost completely sequenced genome.[12] Such tools greatly facilitate the cloning and characterization of cell death genes of interest. Furthermore, the emergence of the powerful double-stranded RNA (dsRNA) interference technique to produce potential loss-of-function phenotypes in specific genes[13] and the ease of isolation of deletion mutations from libraries of mutagenized worms, using polymerase chain reaction (PCR)-based screening,[14] make reverse genetics a practical approach to determine the functions of *C. elegans* homologs of mammalian cell death proteins. In this chapter we review and describe the methods that can be used to study various features of programmed cell death in *C. elegans*. General techniques for culturing and handling of *C. elegans* have been described in detail in other references.[15-17]

[5] R. E. Ellis and H. R. Horvitz, *Development* **112**, 591 (1991).
[6] H. Ellis and H. R. Horvitz, *Cell* **44**, 817 (1986).
[7] M. O. Hengartner, R. E. Ellis, and H. R. Horvitz, *Nature* (*London*) **356**, 494 (1992).
[8] E. M. Hedgecock, J. E. Sulston, and J. N. Thomson, *Science* **220**, 1277 (1983).
[9] R. E. Ellis, D. M. Jacobson, and H. R. Horvitz, *Genetics* **129**, 591 (1991).
[10] J. E. Sulston, *Philos. Trans. R. Soc. London Ser. B.* **275**, 287 (1976).
[11] A. Coulson, J. Sulston, S. Brenner, and J. Karn, *Proc. Natl. Acad. Sci. U.S.A.* **83**, 7821 (1986).
[12] Consortium, T.C.e.S., *Science* **282**, 2012 (1998).
[13] A. Fire, S. Q. Xu, M. K. Montgomery, S. A. Kostas, S. E. Driver, and C. C. Mello, *Nature* (*London*) **391**, 806 (1998).
[14] G. Jansen, E. Hazendonk, K. L. Thijssen, and R. H. Plasterk, *Nature Genet.* **17**, 119 (1997).
[15] S. Brenner, *Genetics* **77**, 71 (1974).
[16] M. Driscoll, *Methods Cell Biol.* **46**, 323 (1995).
[17] W. B. Wood, "The Nematode *Caenorhabditis elegans*," pp. 587–606. Cold Spring Harbor Laboratory Press, Cold Spring Harbor, New York, 1988.

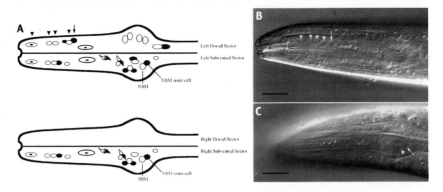

FIG. 1. Assay for scoring extra nuclei in the anterior pharynx of *C. elegans*. (A) Schematic representation of the nuclei located within the anterior pharynx examined in the assay. Nuclei (open) of living cells and nuclei (shaded) of 16 cells that normally undergo programmed cell death in wild-type animals are shown. The two NSM neurons and their sister cells are indicated. (B) Nomarski photomicrograph of the anterior pharynx of a *ced-3(n717)* mutant animal (left side). Several nuclei of the cells that normally live in wild-type animals are indicated by arrowheads. One extra nucleus is indicated by an arrow. (C) Nomarski photomicrograph of the anterior pharynx of a *ced-3(n717)* mutant animal (right side). An extra nucleus from the NSM sister cell located posterior and dorsal to the nucleus of the NSM neuron (arrowhead) is indicated by an arrow. The positions of the nuclei indicated in (B) and (C) are also pointed out in (A), using similar labels. The scale bars represent 10 μm.

Description of Methods

Assays for Scoring Cell Survival, Using Nomarski Optics

The transparent nature of *C. elegans* cuticle allows cells to be observed by Nomarski optics. Because the *C. elegans* cell lineage has been defined and is invariant from animal to animal, it is possible to identify and characterize mutations that block programmed cell death by scoring extra surviving cells.[5–7] The anterior pharynx, a well-defined structure,[18] is especially suited for counting extra surviving cells (Fig. 1).[9] Nuclei of the cells in the anterior pharynx are easily recognized as rippled indents in the smooth wall of the pharynx (Fig. 1B and C). Sixteen cells in the anterior pharynx of wild-type animals undergo programmed cell death during midembryogenesis; their corpses are rapidly engulfed and disappear.[4] In cell death-defective mutants, such as *ced-3* and *ced-4* mutants, some of the 16 cells survive and the resulting extra nuclei are readily visible and can be counted under Nomarski optics (Fig. 1B and C). The number of extra nuclei in the anterior pharynx reflects the severity of the cell death defect in mutant animals. For example,

[18] D. G. Albertson and J. N. Thomson, *Phil. Trans. R. Soc. Lond.* **275,** 299 (1976).

in a strong ced-3(n2433) mutant, an average of 12.4 extra nuclei (extra cells) is seen in the anterior pharynx, while in a weak ced-3(n2427) mutant, an average of only 1.2 extra cells is seen.[19]

To score pharyngeal nuclei, the third (L3) or fourth (L4) stage larvae are anesthetized by mounting on an agar pad with a drop of 10 mM sodium azide (NaN$_3$) solution and the number of extra nuclei in the anterior pharynx is counted by Nomarski microscopy. Figure 1A schematically shows the positions of the 16 cells (nuclei) that undergo programmed cell death in the anterior pharynx of wild-type animals.[9,18]

Two serotonergic pharyngeal neurosecretory motoneuron (NSM) neurons are located at the posterior part of the anterior pharynx. Their sister cells normally undergo programmed cell death (Fig. 1).[4,5] Gain-of-function mutations in the ces-1 gene or a loss-of-function mutation in the ces-2 gene prevent two NSM sister cells from undergoing programmed cell death.[5] In these mutants or other cell death-defective mutants (e.g., ced-3 and ced-4 mutants), an extra nucleus of the NSM sister cell can be observed just posterior and dorsal to each of two NSM neurons (Fig. 1C). This survival versus death assay for NSM sister cells presents an example of how programmed cell death can be studied in a single cell in *C. elegans*.

Cell Corpse Assay for Defects in Cell Death

Most somatic cell deaths (113/131) in *C. elegans* hermaphrodites occur during embryogenesis, mainly between 250 and 450 min after fertilization.[2,4] The remainder of the cell deaths occur during the first (L1) and second (L2) larval stages outside the head region of the animals.[2] Like somatic cells, some cells in the hermaphrodite germline also undergo programmed cell death. The germline is syncytial; however, during programmed cell death, germline nuclei that are destined to die become cellularized.[20] Cell deaths in the germline appear to occur in the region where nuclei are arrested in the pachytene stage of meiosis, near the turn of the gonadal arm (Fig. 2B), but not in the mitotic region.[20] Occasionally, mature oocytes have been observed to undergo programmed cell death.

When cells in *C. elegans* undergo programmed cell death, they adopt refractile and raised button-like morphology that persists for 10 to 30 min as observed under Nomarski optics (Fig. 2). Since the same cells always die at characteristic times in every wild-type animal, mutations that affect

---

[19] D. Xue, S. Shaham, and H. R. Horvitz, *Genes Dev.* **10,** 1073 (1996).
[20] M. O. Hengartner, *C. elegans II, in* "Cell Death" (D. L. Riddle, T. Blumenthal, B. J. Meyer, and J. R. Priess, eds.), Vol. II, pp. 383–417. Cold Spring Harbor Laboratory Press, Cold Spring Harbor, New York, 1997.

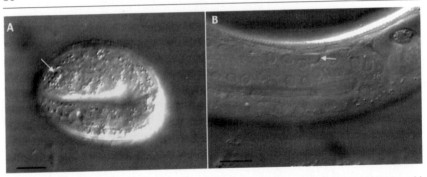

Fig. 2. Cell corpses in wild-type animals show a characteristic morphology. (A) Nomarski micrograph of an embryo at twofold stage. One raised button-like somatic cell corpse is indicated by an arrow. (B) Nomarski micrograph of a posterior gonadal arm. One germline cell corpse is indicated by an arrow. The scale bars represent 10 μm.

cell death can be easily identified and characterized on the basis of altered patterns or morphology of cell corpses.

Many mutations have been found to generate abnormal cell corpse patterns. For example, mutations in the *ced-1, ced-2, ced-5, ced-6, ced-7*, or *ced-10* gene result in persisting cell corpses, suggesting that these genes are involved in either the recognition or the phagocytosis of cell corpses during the engulfment process.[8,9] The persistent cell corpse phenotype is easily distinguished in the head region of early larvae as no larval cell deaths normally occur in this region; corpses observed are those that have been generated during embryogenesis and persist through early larval stages (Fig. 3). In contrast, in animals that are defective in cell death execution, such as *ced-3* and *ced-4* mutants, the number of cell corpses at all developmental stages is reduced because cells that normally die now survive. Therefore, the cell corpse assay is useful for assessing defects in the killing or the engulfment process of programmed cell death.

Cell corpses in a specific region and at a particular developmental stage can be counted to quantitate the severity of cell death defects. For example, the expressivity and the penetrance of the engulfment-defective mutants have been quantitated by scoring the number of cell corpses in the pharynx of L1 larvae.[9]

Because cell corpses in *C. elegans* all have a characteristic raised button-like morphology (Fig. 2), mutations that affect the morphology of cell corpses can be identified. Furthermore, mutations that affect the timing or kinetics of cell death may change the time course of corpse appearance.[20]

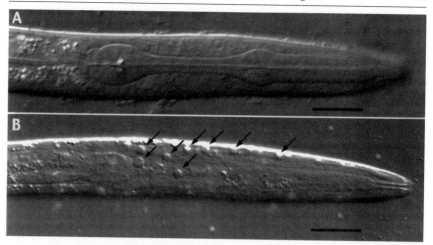

FIG. 3. Engulfment-defective mutants exhibit persistent cell corpses. (A) Nomarski micrograph of the head of a wild-type L1 larva that lacks cell corpses. (B) Nomarski micrograph of the head of a *ced-1(e1735)* mutant at L1 larval stage with many cell corpses (indicated by arrows). The scale bars represent 10 μm.

Methods for Viewing Caenorhabditis elegans by Nomarski Differential Interference Contrast Optics and Recovery of Animals from Slides

The preceding methods for scoring extra surviving cells and cell corpses involve viewing *C. elegans* by Nomarski microscopy. Animals are placed on top of an agar pad containing a drop of 10 mM NaN$_3$ solution on a microscope slide. Sodium azide anesthetizes the animals so that individual cells or cell corpses can be observed with high magnification objectives without being disturbed by worm movement. This procedure also allows individual animals or embryos to be rescued alive off the agar pad. Using this method, populations of nematodes can be screened for subtle cell death-defective phenotypes that normally would not be detected under a dissecting microscope.

To make an agar pad, melt 5% (w/v) agar in M9 solution (40 mM Na$_2$HPO$_4$, 22 mM KH$_2$PO$_4$, 8.5 mM NaCl, 18.7 mM NH$_4$Cl) and keep it at 65° in a heating block. A tape of approximately 0.1 mm in thickness is applied to two slides, which are used as spacer slides. The specimen slide is placed in parallel between the two spacer slides, flat on a desk, and a small drop of melted agar is then spotted in the middle of the specimen slide. A fourth slide is immediately pressed down perpendicular to all three slides, onto the tape of the spacer slides. The drop of agar is squashed flat, and the specimen slide is then carefully slid out from underneath the fourth slide (Fig. 4). One to 2 μl of 10 mM NaN$_3$ in M9 solution is then spotted on the agar pad, and embryos or worms are placed in the sodium azide solution. Sodium azide anesthetizes the worms without killing them, so that they can be rescued later if desired. To view embryos, M9 buffer is usually used instead of the sodium azide solution, as eggs are immobile. A coverslip is gently laid on top of the agar pad. To view living cells or cell corpses, Nomarski DIC optics is required, usually at ×1000 magnification.

To rescue live worms or embryos from the agar pad, the positions of the worms or embryos of interest on the microscope slide are recorded, using the positions of other worms or embryos as references, and the slide is taken to the dissecting scope. The coverslip is carefully removed until the desired worm or embryo is exposed. A mouth pipette filled with M9 solution is used to transfer the worm or embryo to a clean plate seeded with bacteria. A small amount of M9 is drawn into the fine end of the mouth pipette via capillary action. The mouth piece that comes with the capillary tubes is then used to expel a small drop of M9 directly onto the worm. The M9 solution is then drawn back into the mouth pipette along with the worm or embryo and expelled again onto a clean plate. To make a mouth pipette, a 50-μl capillary tube is pulled into a needle shape by heating the middle of the tube and then quickly pulling the two ends of the capillary tube apart from each other. The capillary tube is then broken at the middle and the fine end of the tube is used to transfer worms.

Assays for Defects in DNA Degradation

One conserved feature of programmed cell death is the degradation of genomic DNA of dying cells.[1] The degradation of DNA in dying cells of *C. elegans* can be monitored by staining animals *in situ* with a DNA-labeling dye, such as diamidinophenolindole (DAPI) or Feulgen.[8,10]

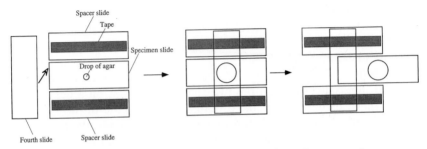

FIG. 4. Schematic demonstration of the making of an agar pad.

Cell deaths in *C. elegans* ventral cord occur only during the first (L1) and second (L2) larval stages of development.[2] When wild-type larvae at developmental stage L3 or L4 are stained with Feulgen, living cells but not the cells that have died earlier in the ventral cord show Feulgen staining.[8] In contrast, in mutants that are defective in cell corpse engulfment, some cell corpses in the ventral cord persist through late larval stages and show extra Feulgen staining.[8] This observation indicates that engulfment of cell corpses is a prerequisite for complete degradation of genomic DNA of dead cells. Like engulfment-defective mutants, late larvae of *nuc-1* mutant animals also exhibit extra Feulgen staining at positions where cell deaths have occurred. In *nuc-1* mutants, both cell killing and engulfment processes proceed normally but the pycnotic DNA of dead cells is not completely degraded and persists as a compact mass of DAPI- or Feulgen-reactive material.[8,10] The *nuc-1* gene is also required for the degradation of genomic DNA from germline cell corpses: large DAPI-positive DNA masses accumulate in the sheath cells (the engulfing cells for germline cell corpses) of *nuc-1* mutant animals but not in those of wild-type animals.[20] In addition, *nuc-1* mutant animals show persistent DAPI staining in the intestinal lumen because the genomic DNA from the bacteria on which worms are fed is not completely degraded in the intestine of the *nuc-1* mutants.[8,10]

To perform a DAPI DNA-staining assay, animals are harvested from plates with M9 buffer and washed twice with the same buffer to remove bacteria. They are then transferred to a microcentrifuge tube containing 1 ml of fixative solution [60% (v/v) ethanol, 30% (v/v) acetic acid, and 10% (v/v) chloroform] and fixed for at least 1.5 hr with mild agitation. The fixed worms are then washed twice with M9 buffer and stained in DAPI staining solution (DAPI, 1.0 μg/ml in M9 buffer) for at least 30 min. The stained animals are mounted on an agar pad for observation with a fluorescence microscope with a DAPI filter. The Feulgen staining assay is more laborious and thus less frequently used. The procedures for Feulgen staining have been described in detail by Sulston and Horvitz.[2]

The DNA-staining assays are useful for identifying and characterizing phenotypes involving abnormal DNA degradation during programmed cell death. These assays also provide the means by which to probe the relationship of DNA degradation to other aspects of programmed cell death, such as cell corpse engulfment.

Assays for Gene Functions in Cell Death, Using Transgenic Techniques

Transgenic animals are straightforward to generate in *C. elegans.* Transgene constructs are injected directly into the gonad of young adult hermaphrodites, usually with a marker plasmid such as pRF4, which creates a dominant Roller phenotype in transgenic animals.[21] The germline in *C. elegans* gonad develops as a syncytium. As oocytes form, the injected DNA is packaged into the oocytes. The injected DNA does not usually integrate into chromosomes, but rather is maintained as an extrachromosomal array containing hundreds of copies of the gene injected.[21] To generate animals with the transgene array integrated into one of the chromosomes, X-ray or UV light can be used to initiate breaks in chromosomes where the array can integrate.[22] F$_2$ progeny of irradiated animals are then screened for 100% inheritance of the transgene, which should segregate in crosses in a Mendelian fashion. Detailed protocols for generating transgenic nematodes can be found in other references.[16,21]

Transgenic techniques have been used to test whether expression of a gene (either endogenous or exogenous) under the control of specific promoters can interfere with the cell death process in *C. elegans*. Genes whose activities are expected to cause cell death can be expressed in cells that are not essential for the viability of the animals. One commonly used promoter is the *mec-7* promoter, which drives gene expression in six mechanosensory neurons (AVM, ALML/R, PVM, and PLML/R) that mediate touch response in *C. elegans*.[23,24] The killing effect of a transgene is assessed by the death of the mechanosensory neurons in which the transgene is ectopically expressed. For example, expression of the cell-killing gene *ced-3* or *ced-4* in touch cells, using the *mec-7* promoter, causes the death of mechanosensory neurons.[25] The death of these neurons can be scored by Nomarski optics (e.g., deaths of the left and right ALM cells can be scored at the early L1 stage for the absence of ALM nuclei) and can be further confirmed by lineage analysis.[25] An alternative way to detect the death of the mechanosensory neurons is to coexpress green fluorescent protein (GFP)[26] with the test gene in mechanosensory neurons, using the *mec-7* promoter. The presence of GFP can then be used as an easy cell survival marker (D. Ledwich and D. Xue, unpublished results, 2000). Other promoters that have been used for similar cell death assays include the *unc-30* promoter,[25] which drives gene expression in VD and DD neurons at the ventral cord.[27]

For genes that are implicated in inhibiting cell death, ectopic expression studies can be performed with broadly expressed promoters, such as the

[21] C. M. Mello, J. M. Kramer, D. Stinchcomb, and V. Ambros, *EMBO J.* **10**, 3959 (1991).
[22] J. C. Way, L. L. Wang, J. Q. Run, and A. Wang, *Genes Dev.* **5**, 2199 (1991).
[23] M. Chalfie and M. Au, *Science* **243**, 1027 (1989).
[24] M. Hamelin, I. M. Scott, J. C. Way, and J. G. Culotti, *EMBO J.* **11**, 2885 (1992).
[25] S. Shaham and H. R. Horvitz, *Genes Dev.* **10**, 578 (1996).
[26] M. Chalfie, Y. Tu, G. Euskirchen, W. W. Ward, and D. C. Prasher, *Science* **263**, 802 (1994).
[27] Y. S. Jin, R. Hoskins, and H. R. Horvitz, *Nature (London)* **372**, 780 (1995).

heat-shock promoters and the *dpy-30* promoter,[28,29] as lack of cell death has no apparent deleterious effects on *C. elegans*.[1,6,7] *Caenorhabditis elegans* heat-shock promoters are ideal for such a purpose because they are heat inducible and are expressed at high levels.[28] The combination of two different heat-shock promoters (*hsp16-2* and *hsp16-41*) can drive gene expression in most *C. elegans* somatic cells.[28] After the heat-shock treatment, transgenic animals are examined for extra surviving cells as described above to assess the death inhibitory activity of the transgene. This method has been used in *C. elegans* to test several endogenous or exogenous cell death inhibitors, including *C. elegans* cell death inhibitor CED-9[30] and its human homolog Bcl-2,[31] the long transcript of the *C. elegans ced-4* gene,[29] and the baculovirus cell death inhibitor p35.[32,33] These proteins have been shown to prevent cell death in *C. elegans* to different extents, with CED-9 having the strongest death-protective activity. Various *C. elegans* ectopic expression vectors, including the expression vectors containing the *mec-7* promoter and the heat-shock promoters, can be requested from the laboratory of A. Fire at the Carnegie Institute of Washington (ftp://ciw2.ciwemb.edu/pub/FireLab Vectors).

To carry out a cell survival assay on transgenic animals carrying heat-shock constructs, gravid adults are transferred to freshly seeded plates to lay eggs for 1 hr at 20°. These plates are then moved to a 33° incubator for the heat-shock treatment. After a 45-min heat shock, the adult animals are returned to 20° for recovery and to continue egg laying for 75 min before they are removed from the plates. L3 or L4 larvae derived from the eggs remaining on the plates are scored for extra cells in the anterior pharynx.

In Vitro Assays for Death Proteases

One key member of the *C. elegans* cell death pathway is the *ced-3* gene. Mutations in *ced-3* prevent most if not all programmed cell deaths in *C. elegans. ced-3* encodes a protein that is similar to members of the mammalian caspase family, which are highly specific cysteine proteases that cleave their substrates after aspartate.[34,35] CED-3 and other caspases are first synthesized as inactive protease precursors and later activated through

[28] E. G. Stringham, D. K. Dixon, D. Jones, and E. P. Candido, *Mol. Biol. Cell* **3**, 221 (1992).
[29] S. Shaham and H. R. Horvitz, *Cell* **86**, 201 (1996).
[30] M. O. Hengartner, and H. R. Horvitz, *Cell* **76**, 665 (1994).
[31] D. L. Vaux, I. L. Weissman, and S. K. Kim, *Science* **258**, 1955 (1992).
[32] A. Sugimoto, P. D. Friesen, and J. H. Rothman, *EMBO J.* **13**, 2023 (1994).
[33] D. Xue and H. R. Horvitz, *Nature* (*London*) **377**, 248 (1995).
[34] J. Yuan, S. Shaham, S. Ledoux, H. M. Ellis, and H. R. Horvitz, *Cell* **75**, 641 (1993).
[35] E. S. Alnemri, D. J. Livingston, D. W. Nicholson, G. Salvesen, N. A. Thornberry, W. W. Wong, and J. Yuan, *Cell* **87**, 171 (1996).

proteolysis to generate active proteases, which are heterodimers containing a large subunit of about 20 kDa and a small subunit of about 10 kDa.[35] One of the central questions in cell death research concerns how the activities of the CED-3/caspases are regulated, and identifying the targets of these death proteases that mediate the cell-killing process. To address these questions, it is important to have an *in vitro* assay that can be used to quantitate the activities of the death proteases and to test potential substrates of these proteases.

Active death proteases, including CED-3, can be easily produced by overexpressing protease precursors in bacteria, using robust bacterial protein expression vectors such as pET vectors (Novagen, Madison, WI).[19] To facilitate the purification of active proteases from bacteria, an epitope tag such as the His$_6$ tag or the FLAG tag (DYKDDDDK) is added to the carboxyl terminus of the protease precursor, and the active protease complex can be purified by affinity chromatography through appropriate affinity columns.[19]

Two methods have been developed to measure the activities of purified proteases. One involves using peptide substrates that are covalently linked to a fluorogenic group such as AMC (7-amido-4-methylcoumarin). When fluorogenic substrates are incubated with the death proteases, cleavage of the substrates by the proteases releases the fluorogenic group, which emits fluorescence when excited at an appropriate wavelength. The resulting fluorescence can then be quantitated with a fluorescence spectrophotometer. The fluorescence intensity is proportional to the amount of peptide substrate cleaved and the activity of the caspase can be determined accordingly. For example, Ac-DEVD-AMC is a good fluorogenic substrate for the CED-3 protease (D. Xue, unpublished results, 2000). This assay is best used for studying the activities of death proteases (purified or unpurified) and the regulation of protease activities by potential regulators. A typical fluorometric reaction for the CED-3 protease contains 10 ng of purified CED-3 protease and 5 μM Ac-DEVD-AMC in 1 ml of CED-3 buffer [50 mM Tris-HCl (pH 8.0), 0.5 mM EDTA, 0.5 mM sucrose, and 5% (v/v) glycerol]. The assay is conducted at 30° for 1 hr. The reaction is then excited at 360 nm, and the intensity of emission is measured at 460 nm. The readout is compared with that of 5 μM AMC control solution to determine the percentage of the fluorogenic peptide that is cleaved by CED-3 as a measurement of protease activity. Potential protease regulators (either purified or unpurified) can be added directly into the reaction to test whether they can alter the protease activity.

The second method involves incubating [^{35}S]methionine-labeled proteins with death proteases to measure the protease activities or to determine whether the labeled protein is a protease substrate.[19] The procedures include synthesizing [^{35}S]methionine-labeled proteins in an *in vitro* transcription and translation-coupled system (Promega, Madison, WI), incubating labeled proteins with death proteases, resolving protease reactions on sodium dodecyl sulfate (SDS)–polyacrylamide gels, and detecting labeled protein products by autoradiography. This assay is best used for examining whether a protein is a substrate of death proteases. It can also be applied to measure activities of death proteases, albeit it is less quantitative than the fluorometric assay. A typical reaction of such an assay contains 1.0 μl (10 ng) of purified CED-3 protease, 1.0 μl of [^{35}S]methionine-labeled protein, and 2.0 μl of CED-3 buffer. The assay is conducted at 30° for 1 hr.[19] The protease reaction is terminated by adding 4.0 μl of SDS gel loading buffer and then resolved on a 10 or 15% (w/v) SDS–polyacrylamide gel. The gel is dried and then subjected to autoradiography to determine the cleavage of the labeled substrate by the protease.

Concluding Remarks

The methods described above for studying programmed cell death in *C. elegans* have been instrumental for elucidating the functions of the cell death genes and the mechanisms by which they affect programmed cell death. The opportunity to observe and monitor the death of a single cell, such as the death of NSM sister cells, in live animals allows cell death to be studied *in vivo* at a level and resolution that no other systems can currently match. The assay for scoring extra cells in the anterior pharynx of *C. elegans* provides a quantitative measurement of defects in cell death. The cell corpse assay helps to identify and characterize engulfment-defective mutants and the studies of such mutants will shed light on how dying cells interact with engulfing cells to coordinate the tightly regulated corpse-engulfment process. The *in vitro* protease assays enable genetic pathway of cell death to be analyzed at the biochemical and mechanistic levels.[19,33,36] The completed *C. elegans* genome sequencing project and the availability of reverse genetic approaches, such as *C. elegans* deletion library screens for deletion mutations and the simple but powerful dsRNA interference technique, open new avenues to study candidate genes that may be involved in *C. elegans* cell death (by their sequence homology to other cell death proteins) but have escaped previous mutant hunts.

It should be noted, however, that some assays commonly used to study apoptosis in other organisms are not yet available or remain to be developed in *C. elegans*. For example, it is not yet possible to visualize the formation of DNA ladders (the hallmark of apoptosis) during *C. elegans* programmed

[36] D. Xue and H. R. Horvitz, *Nature (London)* **390**, 305 (1997).

cell death. Perhaps the percentage of cells that undergo programmed cell death at any given developmental stage is too low for the internucleosomal DNA fragmentation to be clearly detected. Other methods to detect chromosomal fragmentation in dying cells, such as the terminal deoxynucleotidyltransferase-mediated dUTP nick end-labeling (TUNEL) assay,[37] are plausible and are currently under development (Y.-C. Wu, G. M. Stanfield, and H. R. Horvitz, personal communication, 2000). Assays that assess the integrity of plasma membrane, such as the trypan blue exclusion assay,[38] or assays that assess the polarity of the lipid bilayer of plasma membrane, such as the annexin V labeling assay,[39] remain to be tested in *C. elegans*. Development of these cell death assays and other new assays can further facilitate the study of the mechanisms of programmed cell death in *C. elegans* and will certainly contribute to the understanding of apoptosis in other organisms.

Acknowledgments

We thank J. Lee for comments on this manuscript. D. Xue is a recipient of a Burroughs Wellcome Fund Career Award in the Biomedical Sciences.

[37] Y. Gavrieli, Y. Sherman, and S. A. Ben-Sasson, *J. Cell Biol.* **119**, 493 (1992).
[38] A. Glucksmann, *Biol. Rev. Cambridge Philos. Soc.* **26**, 59 (1950).
[39] G. Koopman, C. P. Reutelingsperger, G. A. Kuijten, R. M. Keehnen, S. T. Pals, and M. H. van Oers, *Blood* **84**, 1415 (1994).

Section III

Proteases Involved in Apoptosis and Their Inhibitors

[8] Caspase Assays

By HENNING R. STENNICKE and GUY S. SALVESEN

Introduction

Proteases, i.e., peptide bond hydrolases, constitute a large heterogeneous group of enzymes. The active site of proteases is divided into a catalytic site and a substrate-binding site. The catalytic site consists of a small number of amino acid residues directly involved in the breakage of the peptide bond. On the basis of the critical features of the catalytic apparatus, proteases fall into one of four categories: the aspartic proteases, the metalloproteases, the serine proteases, and the cysteine proteases.[1] The substrate-binding site is composed of a fairly large number of amino acid residues that form a cleft on the enzyme surface to secure proper alignment of the substrate and promote catalysis. In all protease families, the binding site may be subdivided into a number of pockets, of which one often determines the primary substrate specificity. The caspases are cysteine proteases that exhibit primary specificity for aspartate; they use a polarized cysteine side chain to attack the peptide bond following an aspartate residue in their substrates, thus the origin of their name (cysteine-dependent aspartate-specific proteases).[2–4] In this chapter we describe methods for determining caspase activity, using synthetic substrates and natural substrates, and determination of active caspase concentrations.

Substrate and Inhibitor Binding

Although the function of proteases is to hydrolyze peptide bonds in natural proteins, most assays designed to characterize or discover proteases rely on synthetic peptide substrates that are engineered to adapt to specific sites in the enzyme active site (Fig. 1). Usually a reporter group is attached to the peptide on the C-terminal side of the scissile peptide bond. Cleavage of the synthetic substrate releases a chromogenic or fluorogenic group that is readily detected by spectroscopic instruments. Thus, these synthetic substrates are conventionally used to define the activities of proteases. Similar strategies are used to synthesize specific protease inhibitors, where

[1] N. D. Rawlings and A. J. Barrett, *Biochem. J.* **290,** 205 (1993).
[2] G. M. Cohen, *Biochem. J.* **326,** 1 (1997).
[3] G. S. Salvesen and V. M. Dixit, *Cell* **91,** 443 (1997).
[4] D. W. Nicholson and N. A. Thornberry, *Trends Biochem. Sci.* **22,** 299 (1997).

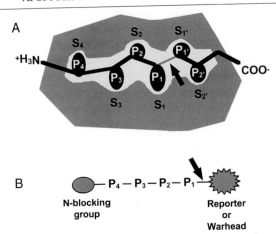

FIG. 1. The general mode of substrate and inhibitor binding. (A) The peptide chain of a substrate (or inhibitor) lies across the substrate-binding cleft of the protease. The side chains of each residue of the substrate are numbered sequentially, with the "P" designation, from the scissile bond (arrowed). Those toward the N terminal of the substrate are called the "unprimed" side, and those toward the C terminal of the substrate are called the "primed" side. Complementary pockets on the surface of the protease are given the "S" designation. Although the P sites constitute a single side chain, the S sites are each usually composed of several different portions of the enzyme, since they define a binding surface. (B) The reporter of a substrate is adjacent to the scissile bond and may be one of a number of moieties that are detectable by fluorescence or colorimetry. The warhead of an inhibitor occupies the same position, but cannot be released and binds the catalytic machinery. The peptide side chains (P_1–P_4) deliver a certain degree of specificity, and the N-blocking group is frequently added to prevent digestion by aminopeptidases and increase binding since many endopeptidases do not catalyze as readily when a free α-amine is present.

the leaving group of the substrate is replaced by a group that reacts with the catalytic machinery of the protease.

Measuring Caspase Activity

Activity of the caspases, either purified or in a crude mixture, can be readily measured with synthetic substrates. A number of different synthetic tetrapeptide substrates are commercially available from various suppliers [Enzyme Systems Products (Livermore, CA), Alexis (San Diego, CA), Bachem (Bubendorf, Switzerland), PharMingen (San Diego, CA), Calbiochem (San Diego, CA), and Boehringer Mannheim (Indianapolis, IN)].

TABLE I
SYNTHETIC CASPASE SUBSTRATES

Substrate[a]	Caspase[b]
VAD	1
DEVD	3, 6, 7, 8, CED-3
VEID	6, 8
IETD	8, 9, 10
WEHD	1, 4, 5
YVAD	1, 4, 5
VDVAD	2

[a] Compilation of the core peptide sequences of some commercially available substrates and the caspases with which they can be successfully used, based on published findings.[5-7] Blocking groups are either acetyl- (Ac-) or benzoxycarbonyl- (Z-), and reporter groups can in principle be pNa, AFC, or AMC.
[b] We emphasize that the indications of the caspase describes only preferred substrates for use *in vitro*. They do not provide a way to distinguish the activity of the different caspases *in vivo*.

Table I[5-7] shows some of these substrates and the caspases for which they are suitable. In general, the substrates are N-blocked tetrapeptides containing a reporter group at the C terminus, which is adjacent to the specificity-determining aspartate and is released by the active protease. The most commonly used reporter groups are *p*-nitroanilide (pNA; colorimetric detection by absorbance at 405–410 nm), 7-amino-4-methylcoumarin (AMC; flourometric detection by excitation at 380 nm and emission at 460 nm), and 7-amino-4-trifluoromethylcoumarin (AFC; fluorometric detection by excitation at 405 nm and emission at 500 nm, or colorimetric detection by absorbance at 380 nm). The colorimetric assays are less sensitive than the fluorimetric assays, but more practical if the laboratory does not possess a fluorimeter.

Samples for assay can be purified recombinant caspases, or crude cell lysates. For the purified caspases the assay presents no problem in interpre-

[5] R. V. Talanian, C. Quinlan, S. Trautz, M. C. Hackett, J. A. Mankovich, D. Banach, T. Ghayur, K. D. Brady, and W. W. Wong, *J. Biol. Chem.* **272**, 9677 (1997).
[6] N. A. Thornberry, T. A. Rano, E. P. Peterson, D. M. Rasper, T. Timkey, M. Garcia-Calvo, V. M. Houtzager, P. A. Nordstrom, S. Roy, J. P. Vaillancourt, K. T. Chapman, and D. W. Nicholson, *J. Biol. Chem.* **272**, 17907 (1997).
[7] H. R. Stennicke and G. S. Salvesen, *J. Biol. Chem.* **272**, 25719 (1997).

tation, but for crude cell lysates there will be hydrolysis of some of the substrates by noncaspase proteases. For example, Z-IETD–AFC, but not Z-DEVD–AFC, is cleaved by the proteasome in detergent extracts of human 293 cells (Q. Deveraux, personal communication, 1999). Thus, care must be taken to consider other activities in crude extracts. Although DEVD-based substrates are sometimes called "caspase 3 specific" they are actually cleaved by all known caspases, but most efficiently by caspase 3.

Synthetic Substrates

To conserve enzyme and substrate we recommend that assays be performed in small volumes (100–200 μl) in 96-well microtiter plates.

Reagents

Substrate as appropriate for the caspase of interest (see Table I[5-7]); however, we routinely use substrates based on the tetrapeptide YVAD to assay caspase 1[8] and DEVD to assay caspases 3, 6, 7, and 8[7]

Stock solution: Prepare stock solution in dry dimethyl sulfoxide (DMSO) at 10 mM for AMC and AFC substrates and at 20 mM for pNA substrate. Stock solutions are stable for at least 1 year at $-20°$

Caspase buffer: 20 mM piperazine-N,N'-bis(2-ethanesulfonic acid) (PIPES), 100 mM NaCl, 10 mM dithiothreitol (DTT), 1 mM EDTA, 0.1% (w/v) 3-[(3-cholamidopropyl)-dimethyl-ammonio]-1-propane-sulfonate (CHAPS), 10% (w/v) sucrose, pH 7.2. The 10 mM DTT may be substituted with 20 mM 2-mercaptoethanol

Colorimetric Assay Procedure

1. Preactivate the recombinant caspase of interest by incubation in caspase buffer for 15 min at 37°. Caspases from freshly prepared cell lysates rarely need preactivation.

2. Add preactivated caspase to the caspase buffer supplemented with a suitable substrate from the stock described under Reagents (Table I). Typically, substrate concentration is 0.2 mM to allow for detection in the linear range of substrate utilization, before substrate depletion can play a role in the rate of hydrolysis. However, the substrate concentration range may be varied as long as (1) the concentration of DMSO in the assay is <2%; (2) the substrate-to-enzyme ratio is at least 50; and (3) substrate depletion is not significant. All assays are performed at 37°.

---

[8] N. A. Thornberry, *Methods Enzymol.* **244,** 615 (1993).

3. The initial rate of hydrolysis is determined from the observed progress curves. It is important to ensure that the rates are calculated from the linear portion of the progress curve, and so single time point assay should be avoided unless it has previously been determined that the assay response is linear in the measuring range. If substrate depletion occurs too rapidly to allow measurement of a rate, dilute the caspase sample and repeat the assay.

4. Usually just an accurate calculation of the rate of hydrolysis is all that is required, because precise calculation of caspase activity is best determined by active site titration (see below). However, kinetic parameters may be determined with any of the well-described derivations of the Michaelis–Menten equation, except double-reciprocal plots, which tend to be insensitive to imprecise data. When determining K_m values it is important to ensure that the substrate concentration ranges between 0.2 and five times the K_m. Kinetic parameters can be determined only for pure caspases, not complex mixtures.

5. For systems where only a determination of the catalytic activity is required, use a standard assay in which the caspase buffer contains 0.2 mM substrate.

Fluorometric Assay Procedure

The fluorometric assay is similar to the colorimetric assay described above. The fluorometric assay provides an approximately 100-fold increase in sensitivity and is therefore particularly useful in assaying low caspase concentrations, such as when measuring caspase activity in cytosolic extracts and when investigating natural caspase inhibitors. The AFC and AMC reporter groups are about equally sensitive to fluorometric detection; however, we prefer AFC because it can also be quantitated colorimetrically at 380 nm if desired, and has less interference from cell extracts. Individual preferences will depend on the detection equipment available for excitation and emission wavelengths. Filter instruments in general are useful if they are within 10 nm of the excitation and emission maxima. Because of the higher sensitivity of the fluorogenic substrates the substrate concentration in the standard assay may be decreased to 10 mM in order to conserve substrate. However, assays with less substrate will be depleted faster, and so it is essential to ensure that activity is measured in the linear range of substrate hydrolysis.

Protein or Natural Substrates

Homogeneous caspase preparations may also be used to examine cleavage of natural substrates. This technique is used to identify the most likely caspase involved in specific protein cleavages,[9] or to determine the kinetics of natural substrate cleavage by particular caspases.[10,11] To perform these procedures one must have a readily detectable protein cleavage assay, and purified, characterized caspases.

There are two ways to perform such assays and the choice of assay depends to a large degree on the desired level of sophistication. One is to monitor the time dependence of hydrolysis, which requires several time points to obtain a rate, at a variety of substrate concentrations, in order to obtain a set of catalytic parameters.[10] A simpler alternative, which is more crude but can be effectively used to obtain an estimate for k_{cat}/K_m for a particular reaction, is described below.

Reagents. The reagents used in the protein cleavage assay are the same as those used for the synthetic substrates, with the exception of the protein substrate solutions. Substrates can be either purified proteins, or radiolabeled *in vitro*-transcribed/translated proteins.

Purified Proteins

The concentration of the protein stock solution depends on the availability and solubility of the protein of interest as well as the sensitivity of the detection method of choice. Generally, Western blots provide good sensitivity, but antibodies may react differentially with the uncleaved protein and its reaction products; it is therefore preferable to quantitate the decrease in the full-length protein, rather than an increase in the cleavage products. It is essential to keep the substrate protein concentration below the K_m of the reaction, so that first-order approximations for substrate cleavage are valid. The K_m values for good substrates are frequently as low as 100 nM, and so we recommend keeping substrate below this level.

Translated Proteins

Translated proteins, with a radiolabeled amino acid incorporated for ease of detection, may also be used as substrates. In this case calculating k_{cat}/K_m values can be misleading because of the presence of other potential competing substrates in the transcription/translation extracts, which are invisible to the detection system. One can, however, determine the rank order of caspases in the cleavage of the test substrate. A note of caution with translation systems: rabbit reticulocyte lysates contain caspase zymo-

[9] C. Caulin, G. S. Salvesen, and R. G. Oshima, *J. Cell Biol.* **138,** 1379 (1997).
[10] D. D'Amours, M. Germain, K. Orth, V. M. Dixit, and G. G. Poirier, *Radiat. Res.* **150,** 3 (1998).
[11] L. Casciola-Rosen, D. W. Nicholson, T. Chong, K. R. Rowan, N. A. Thornberry, D. K. Miller, and A. Rosen, *J. Exp. Med.* **183,** 1957 (1996).

gens that can be activated by the test caspase to give erroneous cleavages. Therefore we prefer *Escherichia coli* transcription/translation systems (Promega, Madison, WI) to circumvent this problem, although wheat germ systems should also be appropriate.

Procedure

1. Preactivate the caspase of interest by incubation in caspase buffer for 15 min at 37°.

2. Add varying concentrations of preactivated caspase to caspase buffer supplemented with a standard concentration of the protein substrate of interest. The desirable caspase concentrations depend on how good the substrate is; however, we have found that a range of caspase concentrations from 1:10 to 1:6500 of the protein substrate concentration in steps of 3-fold dilutions is an excellent starting point, providing good coverage in a small number of samples.

3. Incubate at 37° for 15 min to 1 hr, as desired. The time will depend on the pattern of cleavage under the chosen experimental conditions, but should not be longer than 1 hr because of the instability of caspases over long incubations.[7]

4. Stop the reaction by boiling in sodium dodecyl sulfate (SDS) and separate the reaction products by SDS–polyacrylamide gel electrophoresis (PAGE) (see Fig. 2 for an example).

5. When using a pure protein substrate, a crude measure of the second-order rate constant (k) may be obtained by determining the caspase concentration at which 50% conversion occurs within the time frame used according to Eq. (1):

$$k = \ln 2/E \cdot t \tag{1}$$

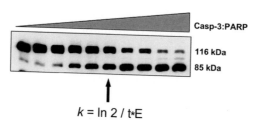

FIG. 2. Cleavage kinetics of poly(ADP-ribose) polymerase by caspase 3. Shown is a comparison of the influence of the caspase-to-substrate ratio on the amount of PARP converted in 30 min at 37° in caspase buffer. The arrow indicates the caspase 3 concentration at which 50% substrate conversion is achieved.

Where E is the active caspase concentration required to deplete half of the substrate in the incubation period t (Fig. 2). On the assumption that the concentration of protein substrate is below its K_m, and with the proviso that proteolysis follows the normal Michaelis relationship, the value of k is equal to k_{cat}/K_m.

Importantly, this relation is valid only in systems with a single cleavage site, or where a single precursor/product relationship can be clearly seen. Thus, it cannot be used for more complex fragmentation because k_{cat}/K_m describes a single chemical reaction, in this case a single peptide bond cleavage.

Determination of Active Site Concentration of Caspases

Although enzyme activities are often cited as specific activities, our experience has been that these vary tremendously between laboratories. This probably reflects different assay buffers, substrate concentrations, instruments, etc. Therefore we recommend determining the active site concentration of the caspases as the most objective measurement. This is conventional now for proteolytic enzymes. In the absence of direct active site titrants, which are not currently available for the caspases, covalently binding inhibitors may be used.[12] In particular the inhibitor benzoxycarbonyl-Val-Ala-Asp-fluoromethyl ketone (Z-VAD-FMK) is a suitable inhibitor for this purpose because it can be used for all the presently known caspases. Caution must be exercised in the choice of inhibitors for determining the active concentration of the caspases, because several of the currently commercially available inhibitors are methylated in order to facilitate cellular uptake—these inhibitors cannot be used for active site titration of the caspases.

The actual procedure for the titration of the caspases is straightforward. The irreversibility of proteinase inactivation by Z-VAD-FMK and stability of the covalent complex allow the functional concentration to be determined from the equivalent concentration of inhibitor. Because the FMK inhibitor forms a covalent adduct, the caspase may be incubated with varying concentrations of the inhibitor in a small volume, resulting in high concentrations of both caspase and inhibitor, which facilitates the reaction. Residual caspase activity relates directly to the active enzyme because Z-VAD-FMK is depleted only by the active form for caspases 3, 6, 7, and 8. Generally, the

---

[12] Q. Zhou, J. F. Krebs, S. J. Snipas, A. Price, E. S. Alnemri, K. J. Tomaselli, and G. S. Salvesen, *Biochemistry* **37**, 10757 (1998).

caspases are incubated with inhibitor ranging in concentration from 0 to 2–4 times the provisional caspase concentration. The enzyme and inhibitor are then incubated in a small volume of caspase buffer for 30 min at 37°, and then diluted into caspase buffer containing peptide substrate to a final concentration of 0.1–0.2 mM. The linear rate of hydrolysis is plotted against the inhibitor concentration and the total concentration of the enzyme is determined from the intercept with the x axis as outlined in Fig. 3.

Reagents. The most important part of the procedure is to accurately weigh out the Z-VAD-FMK, since all calculations are based on the absolute concentration of the compound. The compound is assumed to be 95%, but this should be checked by consulting the data sheet provided by the supplier. Prepare stock solution at 10 mM of the inhibitor titrant, Z-VAD-FMK (it is pivotal to use the unmethylated derivative), and store as 10-μl aliquots in dry DMSO. The inhibitor is stable in DMSO for at least 3 months at −20°, at least 3 days at room temperature, and survives at least three freeze–thaw cycles without losing activity. Therefore it is sufficiently stable for the purpose of active site titration. Substrate stock and caspase buffer are prepared as described above.

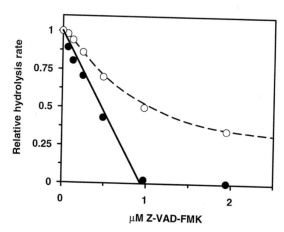

Fig. 3. Active site titration of the caspases. Residual rates of substrate hydrolysis are plotted against the concentration of Z-VAD-FMK. The intercept on the x axis gives the active caspase concentration of the sample. In the solid circles the reaction is sufficiently complete—the curve is linear—to determine the caspase concentration. In the open circles the inhibition has not gone to completion and the caspase concentration cannot be reliably calculated. This can be overcome by increasing the inhibition time, or increasing the caspase concentration to drive the velocity of inhibition.

Procedure

The assays are carried out in a total volume of 100 μl for detection in a 96-well plate reader, preferably operating in the kinetic mode. For analysis requiring larger volumes, increase proportionally.

1. Prepare working solutions of inhibitor titrant at 1–10 μM in caspase buffer, as required, from the 10 mM stock. The inhibitor is not fully soluble in aqueous solvents above 1 mM, and should be diluted in DMSO when working above 100 μM, before final dilution into caspase buffer.

2. Activate the enzyme to be measured in the usual way in assay buffer; 15 min at 37° is usually sufficient.

3. Add 20 μl of each working solution of inhibitor titrant to 80 μl of caspase solution (about 1.0 μM provisional concentration based on protein content) and incubate for 30 min at 37°.

4. Assay samples from each reaction mixture, using an appropriate substrate in assay buffer to determine the residual activity. Plot residual activity as a function of inhibitor concentration. As with all enzyme assays, the rate of substrate hydrolysis must be in the linear range. If the reaction is too fast, and substrate is exhausted too rapidly, dilute the caspase/Z-VAD-FMK reaction mixture to obtain a linear rate.

5. The plot should be linear with an intercept on the *x* axis equal to the concentration of active enzyme (Fig. 3). The curvature of the plot as it approaches this axis indicates that the reaction has not gone to completion. This can be overcome by longer incubation or by increasing the concentrations of enzyme and inhibitor. The concentration of the enzyme can only rarely be accurately determined by extrapolating a straight line drawn from partial inhibition.

[9] Determination of Caspase Specificities Using a Peptide Combinatorial Library

By Nancy A. Thornberry, Kevin T. Chapman, and Donald W. Nicholson

Introduction

The caspase family of cysteine proteases is composed of at least 14 mammalian family members. Phylogenetically they can be divided into two distinct subgroups having markedly different functions *in vivo*. Members of the interleukin 1β-converting enzyme (ICE) subfamily [ICE/caspase 1 was originally identified as the protease responsible for the proteolytic maturation of proinflammatory interleukin 1β (proIL-1β)] cleave and activate multiple cytokines (e.g., proIL-1β, proIL-18) and thus participate in mounting an inflammatory response. Members of the CED-3 subfamily, on the other hand, mediate a cascade of proteolytic cleavage events that culminate in cell death and the apoptotic phenotype (CED-3 was originally identified as a protease in *Caenorhabditis elegans* that is necessary for developmental cell deaths). The central role that these proteases play in both of these important physiological processes has been resolved in part by inhibitor studies, knockout animals, substrate identification, and extensive *in vitro* studies with purified or recombinant caspase family members.[1,2] In this chapter we describe a strategy for defining the precise substrate specificity of individual caspases and discuss how this information can contribute to our understanding of the function that each enzyme serves in its particular biochemical pathway.

Two of the most biologically distinct caspases, caspase 1 and caspase 3, have been purified from natural sources, cloned, and their structures determined by X-ray crystallography.[3-7] The results of these studies, together with a comparison of the sequences of all caspases, indicate that these enzymes are synthesized as zymogens that contain three distinct domains: an N-terminal polypeptide (3 to 25 kDa), a large subunit ($\sim$20 kDa), and a small subunit ($\sim$10 kDa). Activation involves proteolytic pro-

[1] D. W. Nicholson and N. A. Thornberry, *Trends Biochem. Sci.* **22**, 299 (1997).

[2] N. A. Thornberry and Y. Lazebnik, *Science* **281**, 1312 (1998).

[3] J. Rotonda, D. W. Nicholson, K. M. Fazil, M. Gallant, Y. Gareau, M. Labelle, E. P. Peterson, D. M. Rasper, R. Ruel, J. P. Vaillancourt, N. A. Thornberry, and J. W. Becker, *Nature Struct. Biol.* **3**, 619 (1996).

[4] D. W. Nicholson, A. Ali, N. A. Thornberry, J. P. Vaillancourt, C. K. Ding, M. Gallant, Y. Gareau, P. R. Griffin, M. Labelle, Y. A. Lazebnik, N. A. Munday, S. M. Raju, M. E. Smulson, T.-T. Yamin, Y. L. Yu, and D. K. Miller, *Nature (London)* **376**, 37 (1995).

[5] N. A. Thornberry, H. G. Bull, J. R. Calaycay, K. T. Chapman, A. D. Howard, M. J. Kostura, D. K. Miller, S. M. Molineaux, J. R. Weidner, J. Aunins, K. O. Elliston, J. M. Ayala, F. J. Casano, J. Chin, G. J.-F. Ding, L. A. Egger, E. P. Gaffney, G. Limjuco, O. C. Palyha, S. M. Raju, A. M. Rolando, J. P. Salley, T.-T. Yamin, T. D. Lee, J. E. Shively, M. MacCross, R. A. Mumford, J. A. Schmidt, and M. J. Tocci, *Nature (London)* **356**, 768 (1992).

[6] N. P. Walker, R. V. Talanian, K. D. Brady, L. C. Dang, N. J. Bump, C. R. Ferenz, S. Franklin, T. Ghayur, M. C. Hackett, L. D. Hammill, L. Herzog, M. Hugunin, W. Houy, J. A. Mankovich, L. McGuiness, E. Orlewicz, M. Paskind, C. A. Pratt, P. Reis, A. Summani, M. Terranova, J. P. Welch, L. Xiong, A. Moller, D. E. Tracey, R. Kamen, and W. W. Wong, *Cell* **78**, 343 (1994).

[7] K. P. Wilson, J. A. Black, J. A. Thomson, E. E. Kim, J. P. Griffith, M. A. Navia, M. A. Murcko, S. P. Chambers, R. A. Aldape, S. A. Raybuck, and D. J. Livingston, *Nature (London)* **370**, 270 (1994).

cessing between domains, removal of the N-terminal polypeptide, and association of the large and small subunits to form a heterodimer. The active site, which is formed by amino acids from both subunits, contains a catalytic diad composed of cysteine and histidine, suggesting that these enzymes employ a typical cysteine protease mechanism.

Studies of caspase specificity by traditional approaches indicated that two highly conserved properties of these enzymes dictate hydrolysis of substrates. First, they have a near-absolute requirement for aspartic acid in the P_1 position, and second, they require at least four amino acids N terminal to the cleavage site. These observations, together with the finding that substrates with the general structure Ac-XXXD-aminomethylcoumarin (AMC) are efficiently hydrolyzed by caspases, led to the design of the approach described below for the systematic determination of caspase specificity (see Fig. 1).

Positional Scanning Synthetic Combinatorial Library

Positional scanning peptide libraries (PS-SCLs) are used for the identification of peptide sequences that have high affinity for a target protein. These libraries are generally composed of several sublibraries. In each sublibrary, one position is defined with an amino acid, while the remaining positions contain a mixture of amino acids present in equimolar concentrations. Analysis of the library results in an understanding of the specificity of the target protein for amino acids in each position. Positional scanning libraries have been used for the identification of receptor ligands, enzyme inhibitors, specific antigens, and protease substrates.[8–10]

Design of Caspase Positional Scanning Synthetic Combinatorial Library

The PS-SCL used in this study, with the general structure Ac-$[P_4]$-$[P_3]$-$[P_2]$-Asp-AMC (Fig. 2), permits the determination of caspase amino acid preferences in P_2, P_3, and P_4 positions. The decision to keep aspartic acid invariant at P_1 was based on the stringent P_1-Asp specificity of all caspases. The fluorescent group AMC is incorporated in the P_1' position; hydrolysis of the compounds in the library is monitored simply by monitoring the increase in fluorescence that results from release of the AMC moiety. This

[8] N. A. Thornberry, T. A. Rano, E. P. Peterson, D. M. Rasper, T. Timkey, M. Garcia-Calvo, V. M. Houtzager, P. A. Nordstrom, S. Roy, J. P. Vaillancourt, K. T. Chapman, and D. W. Nicholson, *J. Biol. Chem.* **272**, 17907 (1997).

[9] T. A. Rano, T. Timkey, E. P. Peterson, J. Rotonda, D. W. Nicholson, J. W. Becker, K. T. Chapman, and N. A. Thornberry, *Chem. Biol.* **4**, 149 (1997).

[10] C. Pinilla, J. R. Appel, P. Blanc, and R. A. Houghten, *BioTechniques* **13**, 901 (1992).

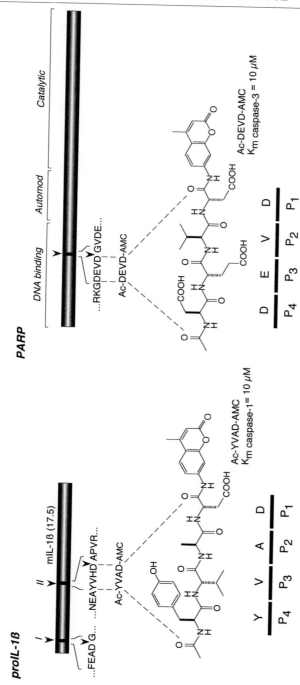

FIG. 1. Key distinguishing features of caspase recognition motifs. Two examples are shown: pro-IL1β, which is cleaved by caspase 1 (ICE), and poly(ADP-ribose) polymerase (PARP), one of the substrates for caspase 3 and related proteases. The core recognition motif contains a P_1 Asp, which is a near-absolute requirement, and four residues to the left of the scissile bond (P_4–P_1). Potent inhibitors and efficient fluorogenic substrates for caspases (e.g., the aminomethylcoumarins shown) can be developed on the basis of these tetrapeptides. These features can be taken advantage of in the design of positional scanning combinatorial substrate libraries: (1) a tetrapeptide is sufficient; (2) the P_1 residue can be fixed as Asp; and (3) AMC is tolerated as a fluorogenic leaving group in P_1.

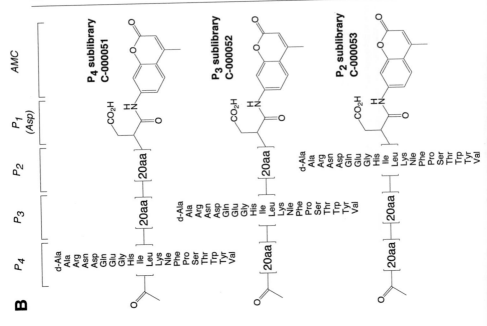

PS-SCL is composed of three separate sublibraries, each containing 20 mixtures (Fig. 2B). In each of the mixtures, one position (P_2, P_3, or P_4) contains 1 of 20 amino acids, whereas the other two contain a mixture of amino acids present at approximately equimolar concentrations. The entire library thus contains 60 total mixtures of 400 compounds (20 × 20), thus yielding 8000 distinct peptides.

Synthesis

As a representative example, the P_3 spatially addressed library is prepared as follows[9]: N-allyloxycarbonyl-L-aspartic acid-α-AMC is loaded onto a Rapp (Tuebingen, Germany) Polymere TentaGel S NH_2 resin containing the 4-(4-hydroxymethyl-3-methoxyphenoxy)butyric acid (HMPB) handle via the Mitsunobu reaction [diisopropyl azodicarboxylate (DIAD)/triphenylphosphine (TPP)]. The allyloxycarbonyl (Alloc) group is removed with $Pd(Ph_3)_4$ and 1,3-dimethylbarbituric acid (DMBA) in dichloromethane (DCM). The isokinetic mixture of protected amino acids is then prepared by dissolving the requisite amounts of each monomer in N,N-dimethylacetamide (DMA) along with 1-hydroxybenzotriazole hydrate (HOBT) followed by addition of 1-(3-dimethylaminopropyl)-3-ethylcarbodiimide hydrochloride (EDC). The isokinetic mixture is added to the resin, followed by agitation for 2 hr. The resin is washed with DMA and the procedure is repeated. The resin is washed with DMA, DCM, and N,N-dimethylformamide (DMF). Elimination of Fmoc (25% piperidine in DMF for 15 min) is followed by washing with DMA, tetrahydrofuran (THF), isopropyl alcohol (IPA), and DCM. The resin is transferred into 20 individual reaction vessels by the isopycnic slurry method and washed with DMA. Position P_3 is "spatially addressed" by preactivating the 20 individual amino acids with EDC/HOBT as described above, followed by addition to the 20 reaction vessels. After 2 hr the resin is washed with DMA and the procedure re-

---

FIG. 2. Strategy for synthesizing a positional-scanning combinatorial substrate library for caspases. Asp-aminomethylcoumarin (AMC), tethered to a solid support, is combined with an isokinetic mixture of proteinogenic amino acids ([20aa]) to establish all 20 amino acids in P_2 coupled to the P_1 Asp-AMC in a single reaction mixture (A). The mixture is then recombined with a fresh isokinetic mixture of the same amino acids to establish all 20 amino acids in P_3, resulting in a single solid-phase reaction mixture containing all 400 permutations (20 × 20) of P_3-P_2-Asp-AMC tripeptides. This mixture is then separated into 20 individual aliquots and coupled with a known single amino acid in P_4 (the positionally defined amino acid). Thus, after release from the solid support, each well of the P_4 sublibrary contains a known amino acid in P_4 coupled to all 400 possible permutations of adjacent amino acids (in P_3-P_2) linked to the P_1 Asp-AMC (a total of 8000 fluorogenic tetrapeptides). Similar strategies are used to generate the equivalent P_3 and P_2 sublibraries (B).

peated. After Fmoc removal and washing, P_4 is installed by adding the isokinetic mixture of amino acids to each vessel. After Fmoc removal, the N terminus is acetylated with Ac_2O/pyridine/DMF (1:2:3, by volume) for 1 hr. The acetylation is repeated, followed by washing with DMA, H_2O, THF, IPA, and DCM. The resin-bound mixtures are then twice cleaved for 30 min with using trifluoroacetic acid (TFA)/H_2O/phenol (PhOH)/triisopropylsilane (TIS) (88:5:5:2, by volume). The cleavage solution is aged for 1 hr and 20 min before solvent removal *in vacuo*. The tetrapeptide-AMC derivates are twice precipitated from cold Et_2O before being lyophilized from CH_3CN/H_2O (2:1, v/v). The yields for the individual wells range from 30 to 49%. Each of the 60 samples is prepared as approximately 10 mM stocks in dimethyl sulfoxide (DMSO) in a 96-well plate format.

Preparation of Caspases

The positional scanning approach to defining caspase substrate specificity is dependent on a preparation of enzyme that is (1) homogeneous with respect to other caspase family members (e.g., analysis of a crude cell extract containing multiple active caspases would be of limited value), (2) is free of degradative proteases that would nonspecifically degrade peptide–AMC substrates, and (3) is sufficiently concentrated to provide a robust fluorogenic signal. The source of the enzyme can be either recombinant[11] or purified.[4,5] Methods for preparing caspases suitable for positional substrate scanning have been extensively detailed elsewhere.[4,5,11]

Specificity Determinations

To determine protease specificity, enzyme is added to reaction mixtures containing 100 μM substrate mix, 100 mM HEPES, 10 mM dithiothreitol (DTT), pH 7.5, in a total volume of 100 μl. Under these conditions the final concentration of each individual compound is approximately 0.25 μM. The amount of enzyme required depends on the catalytic efficiency of the caspase of interest for cleavage of tetrapeptide fluorogenic substrates. In general, the final concentration of enzyme required is comparable to that used for an assay employing the best tetrapeptide substrate. Production of AMC is monitored continuously at ambient temperature in a Tecan (Hombrechtikon, Switzerland) Fluostar 96-well plate reader, using an exci-

[11] M. Garcia-Calvo, E. P. Peterson, D. M. Rasper, J. P. Vaillancourt, R. Zamboni, D. W. Nicholson, and N. A. Thornberry, *Cell Death Differ.* **6**, 362 (1999).

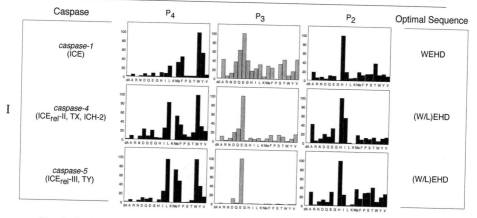

FIG. 3. Optimal sequences of human group I caspases as determined by the PS-SCL.

tation wavelength of 380 nm and an emission wavelength of 460 nm. Representative data from the analysis of human caspases[8] are shown in Figs. 3–5.

Interpreting Results and Functional Significance

A comparison of the specificities of the human caspase family reveals several key features. The results of this PS-SCL divide these enzymes

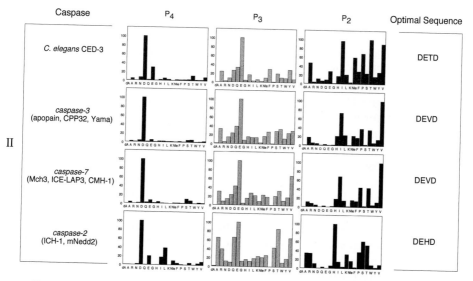

FIG. 4. Optimal sequences of human group II caspases as determined by the PS-SCL.

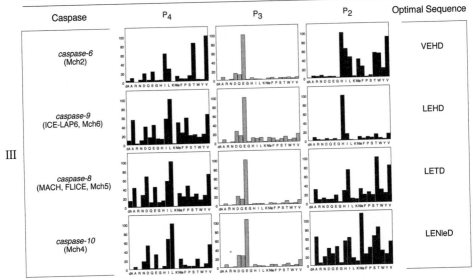

Fig. 5. Optimal sequences of human group III caspases as determined by the PS-SCL.

into three major groups, and indicate that S_4 is the single most important determinant of specificity among caspases. Group I caspases (1, 4, and 5), all favor hydrophobic amino acids in S_4, with an optimal sequence of WEHD. Group II enzymes (2, 3, 7, *C. elegans* CED-3) have a strict requirement for Asp in S_4, preferring the sequence DEXD. Caspases in group III tolerate many amino acids in S_4, but have a marked preference for those with branched, aliphatic side chains, and an optimal sequence of (I, V, L)EXD. Within each group the specific amino acid preferences are exceedingly similar, in some cases identical, implying that at least some of these enzymes (1) have redundant functions or (2) are cell type- or tissue-specific isoforms.

It is clear from several lines of evidence that tetrapeptide specificity extends to macromolecules. It is most compelling that, when endogenous substrates for particular caspases are known, the tetrapeptide sequence at the cleavage sites is similar or identical to the optimal sequence determined by the PS-SCL. In addition, for caspases 1 and 3, the k_{cat}/K_m for cleavage of tetrapeptide substrates ($>10^6\ M^{-1}\ \text{sec}^{-1}$) is greater than or equal to the corresponding rate of cleavage of their authentic macromolecular substrates. Consequently, the tetrapeptide specificities provide important clues regarding the biological functions and relationships between these enzymes. These results, together with those from numerous biochemical and genetic studies, suggest that the caspases may be functionally grouped as follows:

(1) Group I caspases function primarily as mediators of inflammation, where they are involved in the proteolytic activation of proinflammatory cytokines; (2) group II caspases, and probably caspase 6, are mediators in the effector phase of apoptosis, where they are responsible for cleavage of key structural and homeostatic proteins; (3) group III caspases, with the possible exception of caspase 6, are involved in signaling pathways, where they function as upstream activators of the effector caspases.

Virtues and Limitations of Positional Scanning Synthetic Combinatorial Library

Several methods have been described for determinations of protease specificity. Traditional approaches typically involve synthesis of large numbers of peptides and/or peptide-based inhibitors, which are individually evaluated for hydrolysis. More recently, strategies for the systematic study of protease specificity have emerged. The most notable examples are substrate phage display, and the positional scanning approach described here.

The virtues of positional scanning as a method to determine protease specificity are severalfold. First and most important, it produces an accurate description of specificity, as demonstrated by the findings that the optimal sequences obtained for the caspases are similar or identical to the cleavage site sequences found in known endogenous substrates for these enzymes (described above). Second, it is extremely rapid; using 96-well plate technology, a complete analysis can be obtained in the length of time required to run an enzyme assay. Third, it requires only catalytic quantities of enzymes.

The most significant limitation of this approach is the relatively small number of positions that can be investigated simultaneously. This number is determined by a number of factors, including the selectivity of the enzyme, the solubility of the compounds, and the sensitivity of the detection method employed. For example, in the PS-SCL described here, where 3 positions are explored, each mixture contains 400 compounds, each present at a final concentration of 0.25 μM. Thus, in the event that only one compound in the mixture is efficiently hydrolyzed by the enzyme of interest, the method of detection must be sensitive enough to detect <0.25 μM product (AMC in this case). The utility of the method can also be limited by the design of the library. For example, this PS-SCL, in which a fluorescent leaving group is incorporated in P_1', restricts its utility to proteases that can tolerate this substitution, and only permits analysis of specificity N terminal to the cleavage site. This method will also fail to provide insights into substituent effects that may be significant in substrate and inhibitor binding. Finally, chemical synthesis of such libraries are labor intensive, a practical obstacle to widespread use of this technology.

Despite these caveats, the results described here with the caspases clearly demonstrate the power of this approach for obtaining an intimate understanding of specificity that can be exploited to identify substrates and inhibitors, and to provide clues to the biological functions of these enzymes. Systematic methods such as these are expected to facilitate efforts to elucidate functions for the many orphan proteins that are identified in genome sequencing projects.

[10] Criteria for Identifying Authentic Caspase Substrates during Apoptosis

By SOPHIE ROY and DONALD W. NICHOLSON

During apoptotic cell death, members of a discrete and highly limited subset of cellular polypeptides are cleaved by caspases.[1,2] The collective contributions of these proteolytic events manifest the apoptotic phenotype. Despite the enormous biochemical and morphological changes that accompany this form of cell death, most cellular polypeptides escape at least the preengulfment stages of apoptosis unscathed; a survey of protein-banding patterns by sodium dodecyl sulfate (SDS)–polyacrylamide or two-dimensional (2-D) gels, for example, reveals that only a small percentage of the cellular proteome is cleaved during apoptosis. This chapter outlines a series of experimental approaches that can be used to identify and authenticate legitimate caspase substrates and to delineate which caspase family member or subtype is likely to account for proteolysis *in vivo*.

An initial identification of cellular protein constituents that are cleaved by caspases during apoptosis can be undertaken by several approaches. Validation, by the criteria described below, can then aid in substantiating the protein as a legitimate caspase substrate and define the biochemical characteristics of its proteolysis during cell death. Caspases recognize within their polypeptide substrates a core tetrapeptide motif that, in every case, contains an essential aspartic acid in the P_1 position (see Fig. 1). The caspase superfamily contains three specificity subgroups, which differentiate their substrates largely on the basis of the residue in P_4 (Fig. 1A).[3] Tetrapeptides

[1] N. A. Thornberry and Y. Lazebnik, *Science* **281,** 1312 (1998).
[2] D. W. Nicholson, *Cell Death Differ.* **6,** 1028 (1999).
[3] N. A. Thornberry, T. A. Rano, E. P. Peterson, D. M. Rasper, T. Timkey, M. Garcia-Calvo, V. M. Houtzager, P. A. Nordstrom, S. Roy, J. P. Vaillancourt, K. T. Chapman, and D. W. Nicholson, *J. Biol. Chem.* **272,** 17907 (1997).

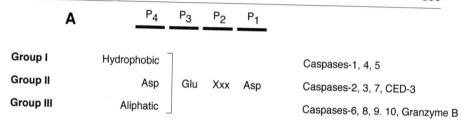

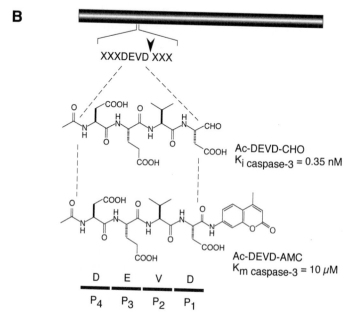

FIG. 1. Caspase proteolytic specificities. Caspases recognize a core tetrapeptide motif within proteolytic substrates. The specificities of the caspases fall into three subgroups, with the major specificity determinant being P_4 (A). All the caspases prefer Glu in P_3, are promiscuous in P_2, and absolutely require Asp in P_1. These motifs have been valuable for the development of potent inhibitors and fluorogenic substrates for the caspases (B). The example shown is based on the caspase 3 cleavage site within PARP and DNA-PK$_{cs}$.

based on these motifs have been useful for generating caspase inhibitors and fluorogenic substrates (Fig. 1B) and are important tools for the experiments described below. A database search for these motifs indicates that every protein contains at least one reasonable consensus site (three-dimensional context is clearly an important factor), and therefore a database survey for proteins containing these sites is of little value.

Candidate substrate approach. New caspase substrates are often identified in laboratories that do not specifically work on apoptosis. This occurs because proteins of interest to that laboratory are sometimes seen to undergo proteolysis in unhealthy cells. Although this approach may not appear to be systematic, it has uncovered the majority of those proteins that we now know to be cleaved by caspases during cell death.[4]

Lessons from autoimmunity. Human patients with autoimmune disorders, such as lupus erythematosus, typically and surprisingly tend to have serum autoantibodies that recognize only one endogenous (self) protein. These proteins are mostly polypeptides that are cleaved by caspases during apoptosis. (This relationship suggests that unresolved apoptosis and the neo-epitopes generated by caspase cleavage contribute to the molecular basis of autoimmunity.) The sera from these patients can be used to track down caspase substrates by expression cloning and other immunoidentification procedures. Examples of caspase substrates that have been identified by this approach include U1-70K, DNA-PK$_{cs}$, and NuMA among several others.[5]

Protein purification. Classic protein purification techniques can be used to identify the components that account for specific biochemical processes that occur during apoptosis. In some cases, a caspase-mediated proteolytic event is involved in the process. Two good examples are the identification of ICAD, the inhibitor of CAD (caspase-activated DNase), which must be cleaved to relinquish the active CAD endonuclease,[6,7] and BID, which is cleaved by caspases and converted to a molecule that participates in mitochondrial cytochrome *c* release during apoptosis.[8–10]

Genomic panning. Genomic panning examines the total expressed cellular mRNAs and whether any of their corresponding proteins are cleaved by caspases *in vitro*. cDNAs can be derived from total cellular poly(A$^+$) RNA that are first size fractionated into manageable pools. The pools are translated into radiolabeled proteins by coupled *in vitro* transcription/translation and then combined with recombinant caspases (at physiologically relevant concentrations) or extracts from apoptotic cells. Any bands that are shifted in molecular mass are potential caspase substrates. The pool can be further subfractionated to obtained a single clone representing the cleaved protein, which can be identified by DNA sequencing and further

[4] C. Stroh and K. Schulze-Osthoff, *Cell Death Differ.* **5,** 997 (1998).
[5] A. Rosen and L. Casciola-Rosen, *Cell Death Differ.* **6,** 6 (1999).
[6] M. Enari, H. Sakahira, H. Yokoyama, K. Okawa, A. Iwamatsu, and S. Nagata, *Nature (London)* **391,** 43 (1998). [Published erratum appears in *Nature (London)* 1998; 393:396]
[7] H. Sakahira, M. Enari, and S. Nagata, *Nature (London)* **391,** 96 (1998).
[8] A. Gross, X. M. Yin, K. Wang, M. C. Wei, J. Jockel, C. Milliman, H. Erdjument-Bromage, P. Tempst, and S. J. Korsmeyer, *J. Biol. Chem.* **274,** 1156 (1999).
[9] H. Li, H. Zhu, C. J. Xu, and J. Yuan, *Cell* **94,** 491 (1998).
[10] X. Luo, I. Budihardjo, H. Zou, C. Slaughter, and X. Wang, *Cell* **94,** 481 (1998).

examined as a potential caspase substrate *in vivo*. This strategy has been successful in identifying gelsolin and PRK2 as caspase substrates, for example.[11–13] An alternative method, which also picked up gelsolin, is a modified yeast two-hybrid system.[14]

Proteomics. The cellular proteome represents the total protein constituents of a given cell type (usually cultured) and it can be examined for proteolytic changes in the absence or presence of an apoptotic stimulus by high-resolution 2-D gel electrophoresis. Positive identification of protein dots that are shifted during apoptosis, and are therefore potential caspase substrates, can be accomplished by peptide-mass fingerprinting and sequencing by mass spectroscopy paired with computational proteomics. Several new caspase substrates have been tentatively identified by this strategy as well.[15,16]

Having made the initial identification, a new potential caspase substrate can be validated and characterized by the following experiments, which are further detailed in the remainder of this chapter.

1. The candidate polypeptide should be cleaved at a discrete site after the induction of apoptosis in intact cells, and the extent and rate of proteolysis should roughly correspond to that of known caspase substrates. This serves as an initial indication that the polypeptide may be a caspase "victim" during apoptosis and establishes a basis for proceeding to the following *in vitro* experiments.

2. Extracts made from apoptotic cells (but not healthy controls) should reproduce the cleavage event *in vitro*.

3. Using these apoptotic cell extracts, a protease inhibitor profile should indicate that an E-64-(N-(N-(L-3-trans-carboxirane-2-carbonyl)-L-leucyl)-agmatine)-insensitive cysteine protease is responsible for the cleavage activity.

[11] K. D. Lustig, P. T. Stukenberg, T. J. McGarry, R. W. King, V. L. Cryns, P. E. Mead, L. I. Zon, J. Yuan, and M. W. Kirschner, *Methods Enzymol.* **283**, 83 (1997).
[12] V. L. Cryns, Y. Byun, A. Rana, H. Mellor, K. D. Lustig, L. Ghanem, P. J. Parker, M. W. Kirschner, and J. Yuan, *J. Biol. Chem.* **272**, 29449 (1997).
[13] S. Kothakota, T. Azuma, C. Reinhard, A. Klippel, J. Tang, K. Chu, T. J. McGarry, M. W. Kirschner, K. Koths, D. J. Kwiatkowski, and L. T. Williams, *Science* **278**, 294 (1997).
[14] S. Kamada, H. Kusano, H. Fujita, M. Ohtsu, R. C. Koya, N. Kuzumaki, and Y. Tsujimoto, *Proc. Natl. Acad. Sci. U.S.A.* **95**, 8532 (1998).
[15] E. C. Muller, M. Schumann, A. Rickers, K. Bommert, B. Wittmann-Liebold, and A. Otto, *Electrophoresis* **20**, 320 (1999).
[16] E. Brockstedt, A. Rickers, S. Kostka, A. Laubersheimer, B. Dorken, B. Wittmann-Liebold, K. Bommert, and A. Otto, *J. Biol. Chem.* **273**, 28057 (1998). [Published erratum appears in *J. Biol. Chem.* 1998; 273:33884]

4. Potent inhibition of proteolysis should occur with caspase-selective inhibitors and an analysis of three of these inhibitors will indicate which caspase subtype (I, II, or III) is involved.

5. Cleavage of the candidate polypeptide with the appropriate recombinant caspase should faithfully reproduce the events seen in apoptotic cells and in extracts derived from these cells. A kinetic analysis (k_{cat}/K_m) should confirm that cleavage occurs with relevant kinetic efficiency.

6. Mutation of the essential P_1 Asp in the candidate polypeptide should entirely eliminate cleavage as well as confirm the site of proteolysis itself.

7. The consequences of cleavage in altering the function of the candidate caspase substrate might be obvious, but this can be confirmed by a variety of approaches that are largely dependent on each substrate.

Core Protocols

Materials. The methods described in this section assume that both an antibody and a cDNA clone for the candidate caspase substrate are available. Reversible caspase inhibitors (tetrapeptide aldehydes) can be purchased from several commercial sources. Recombinant caspases can be generated from appropriately engineered cDNA clones,[17] but can generally be obtained from several of the laboratories active in this field of research.

Protocol 1: Induction of Apoptosis

Choosing the appropriate apoptotic stimulus is an empirical process and depends on the cell type used. Jurkat cells, for example, readily undergo apoptotic cell death after a brief incubation (2–4 hr) with a number of apoptotic stimuli, including anti-CD95 (Fas/Apo-1) antibodies (MBL, Nagoya, Japan) at a concentration of 1 μg/ml, the topoisomerase inhibitor camptothecin at a concentration of 1 μg/ml, and the nonspecific kinase inhibitor staurosporine at a concentration of 1 μg/ml. Other cell types are relatively resistant to apoptotic cell death. K562 cells, for example, will die only after an 18-hr incubation with camptothecin at a concentration of 5 μg/ml. HeLa and 3T3 L1 cells are rendered apoptotic by treatment with 1 μM staurosporine for 2 hr. It is generally our practice to initiate the apoptotic process by replacing the culture medium with fresh medium supplemented with the appropriate concentration of the apoptotic stimulus. Also, we have observed that the apoptotic response of some cell types (e.g., Jurkat) is more vigorous if cells are incubated in a serum-free defined

[17] M. Garcia-Calvo, E. P. Peterson, D. M. Rasper, J. P. Vaillancourt, R. Zamboni, D. W. Nicholson, and N. A. Thornberry, *Cell Death Differ.* **6,** 362 (1999).

medium for 48 hr before the addition of the apoptotic stimulus. A number of assays can be performed to ensure that the treatment is not compromising the integrity of the plasma membrane, as this occurs in necrotic but not apoptotic cell death: (1) The culture medium can be assayed for release of the housekeeping enzyme lactate dehydrogenase; (2) the cells can be analyzed by fluorescence-activated cell sorting (FACS) after dual staining with annexin V (to monitor the plasma membrane remodeling characteristic of apoptotic cells) and propidium iodide (a plasma membrane-impermeable DNA dye); (3) DNA laddering can be visualized directly by agarose gel electrophoresis (see the next section); and (4) cell extracts can be analyzed by immunoblotting for cleavage of the known proteolytic victims of apoptosis, such as poly(ADP-ribose) polymerase (PARP; see Protocol 3: Immunoblotting, below).

Protocol 2: DNA Laddering

Cells (use scraping to remove adherent cells from the dish) and apoptotic bodies present in the culture medium are collected by centrifugation at 10,000g for 10 min. Cell pellets (2×10^5 cells/pellet) are first resuspended in 0.5 ml of 0.6% (w/v) SDS and 10 mM EDTA (pH 7.5), after which 125 μl of 5 M NaCl is added (final [NaCl], 1 M). After overnight incubation at 4°, the supernatants are clarified by centrifugation at 14,000g for 20 min at 4° and treated with DNase-free RNase (50 μg/ml) for 45 min at 37°. The DNA is extracted once with phenol–chloroform–isoamyl alcohol (25:24:1, v/v/v), isopropanol precipitated, resuspended in loading buffer [30% (v/v) glycerol, 10 mM EDTA (pH 7.5), 0.05% (w/v) bromophenol blue], and analyzed by electrophoresis on a 1.2% (w/v) agarose gel in 1× TAE (40 mM Tris–acetate, 1 mM EDTA, pH 8.3). Bands are visualized after ethidium bromide staining. The key tricks to obtaining good apoptotic ladders are as follows: (1) the velocity of the initial centrifugation must be high enough to ensure recovery of the apoptotic bodies from the culture medium; and (2) it is critical to limit the number of cells from which the DNA is extracted (2×10^5 cells/pellet, maximum), as the excess DNA will hinder the migration and resolution of the DNA fragments in the agarose gel.

Protocol 3: Immunoblotting

There are a number of effective protocols for preparing cell extracts for immunoblot analysis. We find the following protocol most useful, as it allows the extraction of virtually all proteins, including those that have poor solubility (e.g., PARP). Moreover, the cell extracts can be stored at −20° indefinitely and freeze–thawed numerous times without significant protein degradation. Cells (use scraping to remove adherent cells from the dish) and apoptotic bodies present in the culture medium are collected by centrifugation at 4000g for 10 min and lysed in a buffer containing 4 M urea, 10% (v/v) glycerol, 2% (w/v) SDS, 0.003% (w/v) bromophenol blue, and 5% (v/v) 2-mercaptoethanol (added to the buffer immediately before use). We typically resuspend 2×10^6 cells/ml of buffer. Extracts are subjected to brief sonication (45 sec) on ice to shear the genomic DNA (which would otherwise interfere with SDS–PAGE), boiled, analyzed by SDS–PAGE, and transferred to nitrocellulose by electroblotting. The blots are first incubated in blocking buffer [Tris-buffered saline, pH 7.4 (TBS); 5% (w/v) milk (blotting-grade blocker nonfat dry milk; Bio-Rad, Hercules, CA); 0.05% (v/v) Tween 20] for 1 hr at room temperature and then incubated for an additional 1 hr in primary antibody diluted in blocking buffer (for immunoblotting of PARP, we use the commercially available monoclonal anti-PARP ascites fluid from clone C-2-10 at a dilution of 1 : 5000). After washing three times in 1× TBS with 0.1% (v/v) Tween 20 for 5 min, blots are incubated for 1 hr at room temperature in the appropriate secondary antibody coupled to horseradish peroxidase diluted 1 : 3000 in blocking buffer. Blots are washed three times in 1× TBS–0.3% (v/v) Tween 20 for 5 min and three times in 1× TBS–0.1% (v/v) Tween 20 for 5 min. Detection is performed by enhanced chemiluminescence (ECL system; Amersham, Arlington Heights, IL).

Protocol 4: Preparation of Catalytically Competent Extracts from Apoptotic Cells

Cells (use scraping to remove adherent cells from the dish) and apoptotic bodies in the culture medium are collected by centrifugation at 4000g for 10 min. Approximately 1×10^8 cells are resuspended in 1 ml of lysis buffer [50 mM HEPES-KOH, pH 7.0; 2 mM EDTA; 0.1% (w/v) 3-[(3-cholamidopropyl)-dimethy-lammonio]-1-propanesulfonate (CHAPS); 10% (w/v) sucrose; 1 mM dithiothreitol (DTT)] and incubated on ice for 30 min. The cell extracts are then clarified by centrifugation at 10,000g for 10 min at 4° and the supernatants are aliquoted and stored at −80°. Note that the cell extracts should be prepared in the absence of protease inhibitors until the protease inhibitor sensitivity profile of the cleavage event under study is determined. Once this is established, cell extracts can be prepared in an inhibitor cocktail containing aprotinin (10 μg/ml), 1 mM phenylmethylsulfonyl fluoride, pepstatin A (10 μg/ml), and leupeptin (20 μg/ml).

Protocol 5: In Vitro Cleavage Assays

The cDNA for the candidate polypeptide is subcloned downstream of an RNA polymerase promoter (T3, T7, or SP6) in the sense orientation.

The radioloabeled polypeptide is prepared by coupled *in vitro* transcription/ translation in commercially available reticulocyte or wheat germ lysate in the presence of either [^{35}S]methionine or [^{35}S]cysteine. In some instances, the efficiency of cleavage is influenced by whether the candidate substrate is transcribed/translated in reticulocyte or wheat germ lysate. Purification of the radiolabeled candidate substrate by fast protein liquid chromatography (FPLC) on a Superdex-75 column (Pharmacia, Piscataway, NJ) is sometimes warranted, as reticulocyte lysates contain small amounts of rabbit caspase 3 that, if activated, could participate in substrate cleavage. Cleavage of the *in vitro*-transcribed/translated radiolabeled candidate substrate is performed in lysis buffer (as described in protocol 4, except the DTT concentration is increased to 5 m*M*) in the presence of catalytically competent cell extract (see protocol 4) or purified recombinant caspases. The final volume of the reaction is 25 μl. After a 1-hr incubation at 37°, the cleavage reaction is terminated by the addition of SDS Laemmli loading buffer and analyzed by SDS–PAGE and fluorography. Recombinant caspases are purified as described[17] and used at a concentration no greater than 30 n*M*. To characterize the protease inhibitor sensitivity profile of the cleavage reaction, the protease inhibitors are preincubated for 20 min at room temperature with an amount of catalytically competent extract that results in approximately 25% substrate cleavage. The radiolabeled substrate is then added and the cleavage reaction is performed as described above.

Protocol 6: Kinetic Evaluation of Cleavage

The *in vitro* cleavage assay is performed as described above in the presence of serial dilutions of recombinant caspase. The cleavage products are separated by SDS–PAGE, visualized by fluorography, and quantitated by laser densitometry or phosphorimaging. All reactions are carried out using levels of substrate well below K_m, where the appearance of product is a first-order process. Values for K_{cat}/K_m are calculated from the relationship $S_t/S_0 = e^{-k_{obs} \cdot t}$ where S_t is the concentration of substrate remaining at time t, S_0 is the initial substrate concentration, and $k_{obs} = k_{cat} \cdot [\text{enzyme}]/K_m$.

Experimental Steps for Authenticating Caspase Substrates

Step 1: Proteolytic Cleavage Should Occur during Apoptosis

Cultured cells are made apoptotic by any one of a number of stimuli (see protocol 1 for examples). Apoptosis can be confirmed by several criteria such as DNA laddering (protocol 2), FACS analysis, or proteolysis of known caspase substrates (protocol 3). A useful control is to include the nonselective pancaspase inhibitor ZVAD [(Z)-VAD(OMe)CH$_2$F; 10–100 μM], which should prevent both the apoptotic phenotype and biochemical events that accompany it. Caution should be used with this inhibitor, however, because it has several limitations: (1) The inhibitor is available as an aspartyl methyl ester, which aids cell permeability, but once the ester is removed the stability of the inhibitor is poor (<50 min); (2) the esterified version of the inhibitor is unsuitable for *in vitro* studies because the ester blocks the Asp carboxylate, which is essential for caspase recognition; (3) not all caspases are effectively inhibited by this compound (e.g., caspase 2)[18]; and (4) the nature of the electrophile allows it to attack other noncaspase proteases.[19] Despite these caveats, this inhibitor is a useful tool for the initial implication of a caspase in cell death processes. An example of expected results is shown in Fig. 2, where camptothecin treatment results in the classic DNA ladder with concomitant maturation of procaspase 3 and cleavage of PARP, a well-substantiated caspase 3 substrate. The candidate caspase substrate should be cleaved at a discrete site (or very few) and caspase inhibition with ZVAD should block all these processes. The degree and timing of proteolysis vary between different caspase substrates, requiring some judgment as to whether cleavage falls within reasonable limits. If the new potential caspase substrate meets these criteria, it is then suitable for *in vitro* analysis. At this junction, direct testing with recombinant caspases could be performed, particularly if these pure enzymes recapitulated the cleavage events seen in intact cells; however, we favor studies with cell extracts first (steps 2–4 below) since (1) it provides more stringent criteria for the involvement of a caspase; and (2) the type of caspase involved (group I, II, or III; see Fig. 1A) can be delineated.

Step 2: In Vitro Cleavage Should Be Reproducible with Apoptotic Cell Extracts

Cytosolic extracts can be prepared from apoptotic versus nonapoptotic cells (protocol 4). These extracts should contain the proteolytic activity that accounts for cleavage of the candidate substrate in intact cells, particularly if it is a caspase. The extracts are then combined with radiolabeled candidate protein derived by coupled transcription/translation in either reticulocyte or wheat germ lysates. The choice of translation system depends on which yields the cleanest band (which varies surprisingly between the two systems

---

[18] M. Garcia-Calvo, E. P. Peterson, B. Leiting, R. Ruel, D. W. Nicholson, and N. A. Thornberry, *J. Biol. Chem.* **273**, 32608 (1998).

[19] P. Schotte, W. Declercq, S. Van Huffel, P. Vandenabeele, and R. Beyaert, *FEBS Lett.* **442**, 117 (1999).

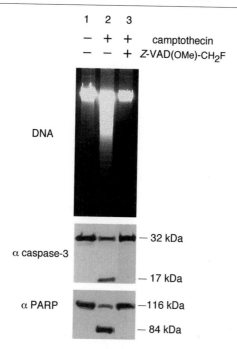

FIG. 2. Caspase activation and proteolysis during apoptosis. Cells can be stimulated to undergo apoptosis with a variety of agents, resulting in caspase activation (seen by conversion of the 32-kDa caspase 3 proenzyme to its active 17-kDa/12-kDa heterodimeric form), proteolysis of caspase substrates such as poly(ADP-ribose) polymerase, and morphological changes such as the genomic breakdown that can be seen as the classic oligonucleosomal ladder. All these events are prevented by the presence of a nonselective caspase inhibitor.

and differs for individual cDNA clones) and it is useful to perform a careful plasmid titration before proceeding further. It is also useful to purify the reticulocyte lysate-derived radiolabeled protein by gel filtration because endogenous rabbit procaspase 3 is present and if activated during the experiment can complicate the interpretation of results. The candidate caspase substrate should be cleaved when combined with extracts derived from apoptotic cells but not by extracts from healthy (nonapoptotic) cells. Furthermore, the cleavage pattern should resemble the bands seen by immunoblotting of cells as described in the preceding section. (On rare occasions, extra bands may appear, particularly in transmembrane proteins, indicating that membrane topology may affect the accessibility of some sites to the proteases in intact cells but not *in vitro*.) If *in vitro* cleavage faithfully reproduces the pattern seen in intact cells undergoing apoptosis, then the extracts should be titrated for the studies described in the following two sections, such that cleavage is limited to <25% (preferably 10%) of the input radiolabeled candidate protein.

Step 3: Cleavage Should Be Dependent on an E-64-Insensitive Cysteine Protease

Caspases are cysteinyl proteases (e.g., they use a cysteine residue as the active site nucleophile) and are thus sensitive to cysteine alkylators such as iodoacetamide and N-ethylmaleimide. Unlike other known cysteine proteases, however, they are not inhibited by E-64. Therefore, if a spectrum of protease inhibitors indicates that the cleavage of the candidate protein by an activity in apoptotic cell extracts is an E-64-insensitive cysteine protease, then it is likely due to a caspase. Apoptotic cell extracts should be prepared as described in protocol 4 except that the protease inhibitor cocktail must be omitted. *In vitro* incubations should be performed as described in protocol 5 except that the DTT concentration should be limited to 1 mM, otherwise it will interfere with the cysteine protease inhibitors. A reasonably complete spectrum of inhibitors is described in Table I, although a more limited subset representing each protease inhibitor subfamily, such as those described in Fig. 3, is often sufficient. If an experiment implicates a caspase (e.g., it has the E-64-insensitive cysteine protease fingerprint), then the inhibitor studies with caspase-directed inhibitors as described in the following section will both confirm it as well as indicate which type of caspase is involved.

Step 4: Caspase Inhibitors Should Prevent in Vitro Cleavage

Each of the three caspase subgroups has a preferred substrate specificity (see Fig. 1A) and this has been used to develop potent peptidyl inhibitors. Owing to the promiscuity of the enzymes, however, these inhibitors are not exceedingly selective. It is therefore important to titrate three representative inhibitors to gain a composite pattern of inhibition that will indicate which caspase subgroup is involved in the cleavage of the candidate caspase substrate. Apoptotic cell extracts (protocol 4) are used to cleave the radiolabeled caspase substrate *in vitro* (protocol 5) in the presence of varying concentrations of Ac-YVAD-CHO or Ac-WEHD-CHO (group I), Ac-DEVD-CHO (group II), or Ac-IETD-CHO (group III). The concentration range covered should be between 0.1 nM and 10 μM (e.g., 1:10 serial dilutions across this range). One of the inhibitors should have an IC_{50} value of about 0.1–1 nM and the rank order of potency between the three inhibitors will indicate which caspase subgroup is involved in the cleavage of the candidate caspase substrate. It is extremely important to use reversible inhibitors for these studies (e.g., aldehydes, but not fluoro- chloro-, or

TABLE I
GENERAL PROTEASE INHIBITORS USEFUL FOR IMPLICATING CASPASES[a]

Spectrum	Inhibitor	Final concentration	Solvent
Broad	α_2-Macroglobulin	1 mg/ml	Buffer
Serine	pAPMSF	100 μM	H_2O
	Aprotinin	2 μg/ml	Buffer
	Elastinal	100 μM	Buffer
	PMSF	1 mM	Ethanol
	TLCK	100 μM	Methanol
	TPCK	100 μM	Methanol
	Soybean trypsin inhibitor	1 mg/ml	Buffer
Ser/Cys	Antipain	100 μM	Buffer
	Chymostatin	100 μM	DMSO
	Leupeptin	100 μM	Buffer
Cysteine	E-64	10 μM	Buffer
	Iodoacetamide	5 mM	Buffer
	N-Ethylmaleimide	5 mM	Buffer
Metallo	Amastatin	10 μM	Ethanol
	Bestatin	10 μM	Methanol
	Diprotin A	50 μM	Buffer
	EDTA	5 mM	Buffer
	Phosphoramidon	8.5 μM	Buffer
Aspartyl	Pepstatin	1 μM	Methanol
Caspase	Ac-XXXD-CHO	0.1 nM–10 μM	DMSO

Abbreviations: pAPMSF, (p-amidino-phenyl)-methane-sulfonyl fluoride; PMSF, phenyl-methylsulfonyl fluoride; TLCK, L-1-chloro-3-(4-tosyl-amido)-7-amino-2-heptanone; TPCK, tolylsulfonyl phenylalanyl chloromethyl ketone.

[a] Each protease inhibitor should be made as a 50× stock in the indicated solvent ("buffer" being the aqueous assay buffer used for the cleavage reactions) and preincubated with the protease source (20 min) prior to the addition of substrate. The DTT concentration in the assay buffer should be reduced to 1 mM so as not to interfere with IAA and NEM alkylation. The final concentration indicated for each inhibitor is the high end of the known effective concentration (the low end of the range is generally 1/10 the final concentrations indicated here). For the caspase inhibitors, X represents any amino acid since there are several of these tetrapeptide aldehydes available.

acyloxymethyl-ketones), otherwise a distinction cannot be made between the subgroups.

Step 5: Recombinant Caspases Should Reproduce in Vitro and Cellular Cleavage Events with Realistic Kinetics

At this point it will be reasonably clear whether a caspase is involved in the cleavage of the new candidate substrate. The next step is to reproduce the cleavage events seen in apoptotic cells and in apoptotic cell extracts using recombinant caspases. Several caspases can be tried, but the protease

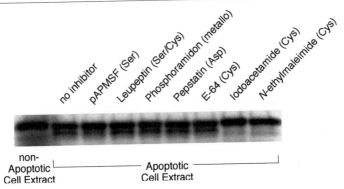

FIG. 3. Protease inhibitor profile. Caspases are E-64-insensitive cysteine proteases. An inhibitor profile can implicate caspases or other protease families. The example shown is Gas2. A more extensive list of appropriate protease inhibitors is described in Table I.

most likely involved in cleaving the protein under study will be narrowed down by the inhibitor studies described in the preceding section. It is critical to do this in a quantitative manner since recombinant caspases can produce many illegitimate cleavage events under high, nonphysiological conditions. The best way to establish the kinetic parameters for caspase-mediated proteolysis is to perform a titration with varying concentrations of the recombinant caspase as described in Fig. 4. Our practice is to make serial dilutions such that the concentration of the recombinant caspase in the *in vitro* incubation mixtures ranges between 0.01 and 10 nM (this also approximates the range of estimated cellular concentrations). Under defined incubation conditions, a kinetic value (the k_{cat}/K_m) can be established using the calculations described in protocol 6. This value can be used to compare the efficiency of cleavage of the candidate caspase substrate with that of authenticated caspase substrates. For example, most caspase 3 substrates have k_{cat}/K_m values in the range of 10^{-5} to 10^{-6} M^{-1} sec^{-1}. Values in the 10^{-4} M^{-1} sec^{-1} range should be viewed with caution whereas values in the 10^{-7} M^{-1} sec^{-1} range reflect extremely efficient proteolysis (PARP, e.g., is in the 10^{-6} M^{-1} sec^{-1} range).

Step 6: Mutation of the P_1 Asp Should Abolish Proteolysis

Caspases are strictly dependent on the presence of Asp in the P_1 position of substrates (see Fig. 1). Mutation of this residue, preferably to Ala, should abolish cleavage in all of the above-mentioned assays. This helps firmly establish the involvement of a caspase in the cleavage event itself, plus it confirms the actual site of proteolysis. Definition of the cleavage site, however, can sometimes be complicated by cleavage site nesting and multiple

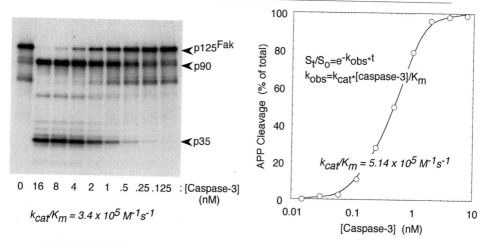

FIG. 4. Establishing cleavage kinetics. Determining whether the *in vitro* cleavage of a candidate substrate with recombinant caspases occurs under physiologically relevant conditions is an important step in defining whether the cleavage kinetics are realistic. This is established with an enzyme titration curve that is numerically transformed into a value for k_{cat}/K_m. The example on the left is the raw data for focal adhesion kinase (FAK) and the example on the right is the plotted and calculated data for the cleavage of the amyloid-β precursor protein.

sites, which presumably help guarantee that proteolysis will occur (see Fig. 5). In most cases, however, cleavage occurs at a single site that is highly conserved among mammalian species and often across evolution. Uncleavable Asp-to-Ala mutants also have some utility in defining the functional consequences of proteolysis, or lack thereof, during apoptosis as discussed in the following section.

Step 7: Cleavage Should Have a Functional Consequence

The apoptotic phenotype of the dying cell is not a consequence of widespread proteolysis[20] but rather the result of cleavage of a subset of proteins at discrete sites that are, with few exceptions, conserved across species. Caspase cleavage sites are generally located between functional domains, possibly because amino acids within these linkers are surface exposed and thus accessible to caspases. In most cases, the consequence of caspase cleavage is the separation of functional domains, which can result either in loss of function [e.g., caspase cleavage of ICAD, an inhibitor of the caspase-activated DNase (CAD)[6]], gain of function (e.g., caspase

[20] S. Wilhelm, and G. Hacker, *Eur. J. Cell Biol.* **78,** 127 (1999).

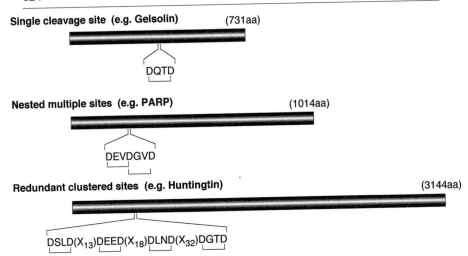

FIG. 5. Caspase cleavage sites. Three examples of caspase cleavage sites with different levels of complexity are shown. Mutagenesis of the preferred site (e.g., the DEVD in PARP) often results in alternative sites being used by caspases to ensure proteolysis (e.g., the nested DGVD site in the PARP example). The cluster of group II caspase sites in Huntingtin is contained within a short segment (<3%) of the polypeptide length.

cleavage of PKC δ separates the regulatory subunit from the catalytic subunit and activates the protein kinase activity[21]), or altered metabolism (e.g., caspase cleavage of the amyloid-β precursor protein separates the cytosol-exposed carboxy terminus from the extracellular amino terminus and shunts the protein to an amyloidogenic degradative pathway[22]). The functional consequence of caspase cleavage can be determined in a number of ways: (1) by expressing a substrate rendered uncleavable by mutating the caspase cleavage site, (2) by expressing a deletion mutant corresponding to a fragment generated by caspase cleavage, and (3) by analyzing cell lines that lack the specific substrate of interest. These approaches are further illustrated by the following three examples: (1) Caspase cleavage of lamins was shown to be required for chromatin condensation and nuclear shrinkage, as the expression of uncleavable mutants of lamin A or B abolished these nuclear events[23]; (2) the expression of deletion mutants of focal

[21] M. F. Denning, Y. Wang, B. J. Nickoloff, and T. Wrone-Smith, *J. Biol. Chem.* **273,** 29995 (1998).
[22] F. G. D. X, Gervais, G. S. Robertson, J. P. Vaillancourt, Y. Zhu, J. Huang, A. LeBlanc, D. Smith, M. S. Shearman, E. E. Clarke, H. Zheng, L. H. T. Van der Ploeg, S. C. Ruffolo, N. A. Thornberry, S. Xanthoudakis, R. J. Zamboni, S. Roy, and D. W. Nicholson, *Cell* **97,** 1 (1999).
[23] L. Rao, D. Perez, and E. White, *J. Cell Biol.* **135,** 1441 (1996).

adhesion kinase (FAK) corresponding to the carboxy-terminal fragments generated by caspase cleavage exhibited a dominant-negative phenotype and suppressed the phosphorylation of endogenous FAK, much like the natural variant FRNK[24]; and (3) caspase cleavage of gelsolin was reported to play a critical role in mediating the characteristic morphological changes of apoptosis, as cell blebbing and DNA fragmentation were delayed after induction of apoptosis in gelsolin-deficient cells.[13] Noteworthy are the considerable caveats associated with all three approaches. Transfection studies are fraught with liabilities related to the fact that the transfected protein is often overexpressed and not necessarily in its native location or under the control of the normally operating regulatory mechanisms. Moreover, the endogenous protein is generally still present, which may affect the outcome of the experiment. Finally, the importance of certain substrates in mediating the systematic disassembly of the apoptotic cell may be cell type dependent. Neutrophils, for example, do not express the caspase substrates PARP, NuMA, U1-70K, and DNA-PK$_{cs}$, but yet exhibit the biochemical and morphological changes characteristic of apoptosis.[25] The value of these types of studies must be assessed on a case per case basis, keeping in mind the many potential complications that are introduced from these perturbations.

[24] F. G. Gervais, N. A. Thornberry, S. C. Ruffolo, D. W. Nicholson, and S. Roy, *J. Biol. Chem.* **273**, 17102 (1998).

[25] D. M. Sanghavi, M. Thelen, N. A. Thornberry, L. Casciola-Rosen, and A. Rosen, *FEBS Lett.* **422**, 179 (1998).

[11] Purification and Use of Granzyme B

By Lianfa Shi, Xiaohe Yang, Christopher J. Froelich, and Arnold H. Greenberg

Granzyme B (GrB) is the primary molecular mediator of apoptosis by cytotoxic T lymphocytes (CTLs) and natural killer (NK) cells. It is a unique mammalian aspartic acid-cleaving serine protease. On T cell receptor activation, GrB is released from the CTL cytoplasmic granules by exocytosis, enters the target cells and, in the presence of the granule pore-forming protein perforin, it initiates the processing of caspases and apoptosis. GrB apoptosis is also activated by adenovirus, which can effectively replace perforin. Methods for the purification and quantitation of GrB and perforin, and the preparation and titration of adenovirus, are described. In addition, methods for application of these reagents to the initiation of apoptosis in tumor target cells, with several assays for detecting GrB apoptotic activity, are detailed.

I. Introduction

CTL- and NK cell-mediated killing represents an immunologic defense against viral pathogens, parasites, and tumor cell proliferation. CTLs and NK cells are important mediators of graft rejection and autoimmune disease. They use two primary mechanisms to induce target cell death: granule-based[1] and Fas/Apo-1/CD95-based[2] killing. The granule exocytosis pathway involves two types of molecules: the pore-forming protein perforin and lymphocyte-specific serine esterases termed *granzymes*.

Perforin is a 65- to 70-kDa glycoprotein that is synthesized only in CTLs and NK cells.[3] Perforin is similar to the C9 component of complement; on target membranes it produces, in the presence of calcium, pores (up to 20 nm in diameter) that can act as ion channels. Both membrane damage and apoptosis inflicted by CTLs and NK cells from perforin-deficient mice are profoundly suppressed.[4] Purified perforin at high doses induces plasma membrane lysis of virtually all target cells.

Four granzymes have been purified including granzyme A (GrA) and GrB, GrK/human tryptase 2, and GrM (metase 1).[5] Purified GrA, GrB, and GrK induce apoptosis in target cells in the presence of sublytic amounts of perforin[6] or adenovirus.[7] GrB is the most powerful apoptosis inducer, while the tryptase granzymes are slow acting. Neither the granzymes nor perforin are able to induce apoptosis on their own.[8] CTLs and NK cells from GrB homozygous null mutants exhibit a defect in their ability to induce rapid DNA fragmentation and apoptosis in target cells.[9,10] The

[1] P. A. Henkart, M. S. Williams, and H. Nakajima, *Curr. Top. Microbiol. Immunol.* **198,** 75 (1995).

[2] S. Nagata, *Cell* **88,** 355 (1997).

[3] C. C. Liu, C. M. Walsh, and J. D. E. Young, *Immunol. Today* **16,** 194 (1995).

[4] D. Kagi, B. Ledermann, K. Burki, R. M. Zinkernagel, and H. Hengartner, *Annu. Rev. Immunol.* **14,** 207 (1996).

[5] M. J. Smyth and J. A. Trapani, *Immunol. Today* **16,** 202 (1995).

[6] L. Shi, R. P. Kraut, R. Aebersold, and A. H. Greenberg, *J. Exp. Med.* **175,** 553 (1992).

[7] C. J. Froelich, K. Orth, J. Turbov, P. Seth, R. Gottlieb, B. Babior, G. M. Shah, R. C. Bleackley, V. M. Dixit, and W. Hanna, *J. Biol. Chem.* **271,** 29073 (1996).

[8] L. Shi, C. M. Kam, J. C. Powers, R. Aebersold, and A. H. Greenberg, *J. Exp. Med.* **176,** 1521 (1992).

[9] J. W. Heusel, R. L. Wesselschmidt, S. Shresta, J. H. Russell, and T. J. Ley, *Cell* **76,** 977 (1994).

[10] S. Shresta, D. M. MacIvor, J. W. Heusel, J. H. Russell, and T. J. Ley, *Proc. Natl. Acad. Sci. U.S.A.* **92,** 5679 (1995).

residual apoptosis observed in longer assays in GrB-deficient mice is further abrogated in GrB/GrA double-deficient mice,[11] arguing that both granzymes participate in CTL/NK cell apoptosis. However, CTL-induced membrane damage appears to be partly mediated by perforin.[11]

GrB from humans (27–32 kDa), mice (27/29 kDa), and rats (32/34 kDa) have been purified and characterized. GrB has a unique substrate site specificity for aspartic acid.[12] Processing of an amino-terminal prodipeptide is required for activity of native and recombinant murine GrB[13] and this cleavage is mediated by the granule thiol protease, dipeptidyl-peptidase I.[14]

The proteolytic specificity of GrB is shared with members of the cysteine protease caspases. The caspases are strongly implicated as a common mediator of apoptotic cell death including that induced by GrB.[15] The caspases are expressed as zymogens and are activated by proteolytic cleavage at specific Asp residues. Several caspases have the consensus GrB cleavage site with an isoleucine or valine in the P_4 position[16] and are processed by GrB *in vitro*. Current models propose that GrB enters the target cell by crossing the plasma membrane and then, after an apoptosis-activating event mediated by perforin, directly cleaves and activates one or more caspases to initiate apoptosis and promote translocation of GrB to the nucleus.[7,17,18]

A. Methods for Purification and Use of Granzyme B and Perforin

Several methods are now available for purification of GrB and perforin, each with particular advantages and disadvantages. We detail two methods that offer practical alternatives, considering the specific application and the limitation of the laboratory setting.

1. *Rat GrB:* This method is based on the use of the rat LGL leukemia cell line RNK. The cells are expanded in rat peritoneal cavity and then

[11] M. M. Simon, M. Hausmann, T. Tran, K. Ebnet, J. Tschopp, R. ThaHla, and A. Muellbacher, *J. Exp. Med.* **186**, 1781 (1997).

[12] M. Poe, J. T. Blake, D. A. Boulton, M. Gammon, N. H. Sigal, J. K. Wu, and H. J. Zweerink, *J. Biol. Chem.* **266**, 98 (1991).

[13] A. Caputo, R. S. Garner, U. Winkler, D. Hudig, and R. C. Bleackley, *J. Biol. Chem.* **268**, 17672 (1993).

[14] M. J. McGuire, P. E. Lipsky, and D. L. Thiele, *J. Biol. Chem.* **268**, 2458 (1993).

[15] A. J. Darmon, D. W. Nicholson, and R. C. Bleackley, *Nature (London)* **377**, 446 (1995).

[16] N. A. Thornberry, T. A. Ranon, E. P. Pieterson, D. M. Rasper, T. Timkey, M. Garcia-Calvo, V. M. Houtzager, P. A. Nordstrom, S. Roy, J. P. Vaillancourt, K. T. Chapman, and D. W. Nicholson, *J. Biol. Chem.* **272**, 17907 (1997).

[17] D. A. Jans, P. Jans, L. J. Briggs, V. Sutton, and J. A. Trapani, *J. Biol. Chem.* **271**, 30781 (1996).

[18] L. F. Shi, S. Mai, S. Israels, K. Browne, J. A. Trapani, and A. H. Greenberg, *J. Exp. Med.* **185**, 855 (1997).

purified from isolated granules. The advantage of this method is the ability to simultaneously purify GrB, perforin, and the tryptase granzymes A and K. Methods for GrA and GrK purification are described elsewhere.[8] All granzymes and perforin are active across species. The disadvantage of this method is the need to use rats and the relatively lower yields of GrB compared with the YT cell line method.

2. *Human GrB:* In this procedure, GrB is isolated from YT cells, an autonomously proliferating NK line devoid of granzyme A and K activity.[19] The process is rapid and has high GrB yields, and thus is preferred for GrB purification alone.

Until recently it has not been possible to express granzymes in typical high-yield systems. Two publications have outlined methods for expression and purification of active GrB in yeast[20] and by baculovirus expression.[21] GrB purified by these methods may ultimately replace the purification methods described below. However, it has not yet been possible to express active perforin. At this time the RNK perforin purification system described below is best suited for high-quality and high-yield protein, although other methods are reported.[22] The use of adenovirus to deliver GrB allows, in some circumstances, the replacement of perforin and is also described here.

II. Purification of Rat Granzyme B and Perforin from RNK LGL Leukemia Cells

A. Growth of RNK-16 Cells in Rats

Cytotoxic granules are isolated from rat RNK-16 NK-like leukemia cells.[23] The cells are serially passaged *in vivo* as ascites in Fischer (F344) rats.

Protocol. F344 rats (200–225 g) are injected intraperitoneally with 1.5 ml of Pristane (2,6,10,14-tetramethylpentadecane; Sigma, St. Louis, MO) 4–6 days before cells are to be injected. About $5–10 \times 10^7$ RNK cells are washed twice in phosphate-buffered saline (PBS) or Hanks' balanced salt solution (HBSS), resuspended in 1 ml of PBS, and injected intraperitoneally into each rat. After 12–15 days, cells are harvested. Routinely, each rat yields $1–6 \times 10^9$ RNK cells. Aliquots of cells that appear healthy and give

[19] W. L. Hanna, X. Zhang, J. Turbov, J. Winkler, D. Hudig, and C. Froelich, *Protein Purif. Exp.* **4**, 398 (1993).

[20] C. T. N. Pham, D. A. Thomas, J. D. Mercer, and T. J. Ley, *J. Biol. Chem.* **273**, 1629 (1998).

[21] Z. Xia, C. M. Kam, C. Huang, J. C. Powers, R. J. Mandle, R. L. Stevens, and J. Lieberman, *Biochem. Biophys. Res. Commun.* **243**, 384 (1998).

[22] U. Winkler, T. M. Pickett, and D. Hudig, *J. Immunol. Methods* **191**, 11 (1996).

[23] C. W. Reynolds, E. W. Bere, and J. M. Ward, *J. Immunol.* **132**, 534 (1984).

a high yield without blood contamination are used for reinjection or frozen for future use.

B. Granule Isolation from RNK Cells

Solutions

HHH buffer: HBSS (GIBCO-BRL, Gaithersburg, MD) containing 10 mM HEPES buffer solution (HEPES; GIBCO-BRL) and heparin (100 U/ml; Sigma)

Sucrose (2M): 136.92 g dissolved in 100 ml of distilled water, top up to 200 ml, filter to sterilize.

Disruption buffer (DB) stock solution: 2 M sucrose (125 ml), 1 M HEPES (10 ml), 0.5 M EGTA (8 ml), distilled H_2O (857 ml). Add heparin to a final concentration of 150 U/ml before use. The 0.5 M EGTA stock solution is made by increasing the pH with solid NaOH (Sigma) until it is completely dissolved in water

Adjusted Percoll: Make stock solution [2 M sucrose (125 ml), 10 ml of 1 M HEPES, 4 ml of 0.5 M EDTA, 434 ml of H_2O, adjust to pH 7.4 with 1 N HCl], then add 427 ml of Percoll after filtration. It is convenient to filter a measured volume into bottles, leaving space to add Percoll (Sigma) later

Protocol

1. Harvest cells from a rat, using HHH buffer. To do this, kill the rat by CO_2 inhalation. Pin the rat to a pad and slit the skin over the abdomen; then, with a 30-cm^3 syringe and 18-gauge needle, inject 30 ml of HHH buffer at ambient temperature, mix in the peritoneal cavity, and withdraw slowly. Keep the harvested cells on ice. Repeat the procedure, using a total of about 200 ml/rat. Count the cells, and remove an aliquot for reinjection or storage at $-80°C$, then centrifuge at 190g for 5 min. When removing the buffer take the top Pristane layer first. Resuspend the cells in disruption buffer (DB) at a concentration of 10^8 cells/ml and add heparin to a final concentration of 1000 U/ml.

2. Put a nitrogen cavitation bomb (Parr Instrument, Moline, IL) in a cold room overnight before use. Transfer the harvested RNK cells to a 125-ml Erlenmeyer flask with a stir bar, and place in the chilled nitrogen bomb. Pump in N_2 to a pressure of 450 psi and then equilibrate at this pressure while stirring on ice for 20 min. Slowly release the pressure by opening the valve, allowing the contents to ooze thickly into a clean 125-ml Erlenmeyer flask. Open the bomb and recover any traces of sample in the original Erlenmeyer flask.

3. Add MgCl$_2$ to 5 mM and Dnase I (800 U/ml; Sigma). Slowly stir the homogenate for 30 min at room temperature. The homogenate will be thick at first but will gradually thin. Keep on ice for this and all subsequent steps. Nuclei are removed by filtration sequentially through a 5-μm then a 3-μm filter (Nuclepore, Pleasanton, CA). Push the sample through the filter with a 10-ml syringe. If the filter is blocked, just change the filter and continue.

4. Put 19 ml of adjusted Percoll in the bottom of a 26-ml rigid polycarbonate centrifuge tube (Beckman, Fullerton, CA). Carefully layer 5 ml of sample on top, then centrifuge in a 70 Ti rotor at 20,000 rpm for 10 min (L8-70 M; Beckman). Harvest the granules by withdrawing the bottom 5 ml from each tube, using a 20-gauge spinal needle and a 5-ml syringe. The highest granule titer is in the first 4 ml, with the peak usually between 2 and 3 ml. Pool the granule fractions from all the tubes. Centrifuge at 33,000 rpm in the 70 Ti rotor for 3 hr at 4°. Granules are recovered as a loose pellet just above the Percoll. Store the isolated granule at −80° or use immediately for further purification.

The granules prepared this way are fully active and by solubilization in 2 M NaCl with 1 mM EGTA, pH 7.2, granzymes and perforin will be released from the granule matrix. After returning the solubilized granules to 0.145 M NaCl with buffer HE (see Section VI,A), the material can be used directly in the apoptosis assays described below.

C. Purification of Granzyme B

All columns for purification of GrB are attached to a fast protein liquid chromatography (FPLC) system (Pharmacia LKB Biotechnology, Uppsala, Sweden) and all the columns are purchased from the same company. The isolated granules (see Section II,B) are solubilized in 2 M NaCl by adding dry NaCl with three freeze–thaw cycles and then centrifuged at 100,000g for 1 hr at 4°. The supernatant is harvested and filtered through a sterile 0.22-μm pore size filter (Millipore, Bedford, MA) to remove membrane fragments.

1. Phenyl Superose Chromatography, Flow Rate 0.5 ml/min

Phenyl Superose HR 5/5 (1-ml) column
Buffer A: 20 mM Tris-HCl (pH 7.2), 1 mM EGTA, 0.02% (w/v) NaN$_3$, 2 M NaCl
Buffer B: Buffer A without 2 M NaCl
All buffers and fractions are kept on ice. The solubilized, filtered granule material (8 × 10^9 RNK cell equivalents in a 4- to 5-ml volume) is loaded on the column. The running program is as follows: sample loading for 10 min (5 ml), then washing for 24 min with buffer A, then eluting with a

2–0 M NaCl (0–100% buffer B) linear gradient for 20 min. Fractions (1 ml) are collected and the eluate monitored by measurement of OD_{280}. GrB is in the flowthrough fractions, while perforin activity is found in the fractions eluted from the column. The active perforin fractions are pooled and stored at −80°. The flowthrough fractions are pooled and concentrated with a Centriprep-10 concentrator (Amicon, Danvers, MA) for application to the heparin column.

2. Heparin Chromatography, Flow Rate 0.5 ml/min

Column: HiTrap heparin, 1 ml (Pharmacia)
Buffer A: 20 mM Tris-HCl (pH 7.2), 0.1 mM EGTA, 0.02% (w/v) NaN$_3$
Buffer B: Buffer A plus 2 M NaCl

Dilute the concentrated flowthrough fractions from Phenyl Superose chromatography in buffer A to give a final NaCl concentration of 0.1 M or less, and then load the sample onto the heparin HiTrap column. The running program is as follows: sample loading for 30 min, washing for 20 min with buffer A, and eluting for 20 min with a linear gradient of 0.1–1 M NaCl (0–50% buffer B) and then with a 10-min linear gradient of 1–2 M NaCl (50–100% buffer B). One-milliliter fractions are collected and screened for activity. The active fractions are pooled and concentrated with a Centricon-10 microconcentrator (Amicon) to a final volume of 1 ml.

3. Mono S Chromatography, Flow Rate 1 ml/min

Mono S column: Mono S HR 5/5 column, 1 ml (Pharmacia)
Buffer A: 10 mM bis-Tris (pH 6.0), 50 mM NaCl
Buffer B: Buffer A plus 1 M NaCl

The active concentrated fraction from the heparin column is diluted to 10 ml (at least a 10-fold dilution to give a suitable buffer change and low salt concentration) with 12 mM bis-Tris, pH 5.5, and then applied to the Mono S cation column. The running program is as follows: sample loading for 10 min, column washing for 10 min, and eluting for 5 min with 0–0.5 M NaCl (0–50% buffer B) and then for 20 min with 0.5–1.0 M NaCl (50–100% buffer B). GrB elutes at about 0.7 M NaCl. The purified GrB is pooled, aliquoted, and stored at −80°. The protein concentration is measured with a micro-BCA protein assay kit (Pierce Life Sciences, Rockford, IL), which is sensitive in a range of 0.5 to 20 μg/ml in a microplate format with a microplate reader. We usually recover about 150 μg of GrB from 10^{10} RNK cells.

D. Purification of Perforin

The initially purified perforin fractions from the phenyl Superose column, often stored at −80°, are thawed and concentrated with a Centricon-

30 microconcentrator (Amicon) to a small volume (~1 ml). This material is diluted at least 10-fold in buffer A, to give a final NaCl concentration of 0.1 M. The sample is loaded onto a 1-ml heparin HiTrap column.

Heparin Chromatography, Flow Rate 0.5 ml/min

Buffer A: 20 mM Tris-HCl, (pH 7.2), 1 mM EGTA, 0.02% (w/v) NaN$_3$
Buffer B: Buffer A plus 1 M NaCl

The running conditions are as follows: sample loading for 20 min, column washing for 10 min, and eluting for 20 min with 0–1 M NaCl (0–100% buffer B). Perforin is eluted at about 0.4 M NaCl. The purified active perforin is pooled, aliquoted, and stored at −80°. Activity is quantitated as described below (Section IV). Protein is measured as described for GrB (Section II,C). Recovery of perforin is usually 20–30 μg for 10^{10} cells.

III. Purification of Human Granzyme B from YT Natural Killer Cells

Reagents

YT medium: 5% (v/v) fetal bovine serum (FBS) and 5% (v/v) cosmic calf serum (CCS) in IDMEM (Iscove's modification of Dulbecco's modified Eagle's medium; GIBCO-BRL) with penicillin–streptomycin (Pen/Strep) and L-glutamine

HBSS–BSA: 500 ml of 1% (w/v) bovine serum albumin (BSA) in HBSS without Ca^{2+}/Mg^{2+} (GIBCO-BRL); sterile filter

Relaxation buffer: 10 mM piperazine-N,N'-bis(2-ethanesulfonic acid (PIPES; Sigma), 1 mM ATP, 1.5 mM EGTA, 0.1 M KCl, 1.25 mM MgCl$_2$ adjusted to 285 mOsm with 4 M NaCl in Milli-Q (MQ) H$_2$O (Millipore)

Extraction buffer: 50 mM morphinoethanesulfonic acid (MES; Sigma), 25 mM NaCl, 0.5 mM EDTA, 0.5% (v/v) Triton-X, pH 6

GrB solubilization buffer: 381 mM NaCl in 20 mM sodium acetate buffer (pH 4.5), 2 mM EDTA

Buffer A: 50 mM MES, 25 mM NaCl, 2 mM CaCl$_2$, pH 6.0

Buffer B: 50 mM MES, 2.0 M NaCl, 2 mM CaCl$_2$, pH 6.0

Neutralization buffer: 0.2 M sodium acetate, pH 3.5

Dialysis buffer: 150 mM NaCl in water

A. Nitrogen Bomb Cavitation Procedure for Isolation of YT Cell Granules

Approximately 1–3 × 10^9 cells, cultured in flasks, bags, or bioreactors, are collected, washed twice in HBSS plus 1% (w/v) BSA (Sigma), and resuspended at 1×10^9 cells/ml in ice-cold relaxation buffer. The bomb chamber (Parr Instrument) is chilled by filling with 2 in. of ice and placed on a stir plate. The cells, resuspended in relaxation buffer, are dispensed to a 100- to 150-ml beaker containing a medium-sized stir bar and the beaker is placed on the ice in the cavitation chamber. The stir plate is adjusted to a speed that minimizes foaming. The bomb is assembled and connected to a nitrogen tank. The pressure is gradually increased to 250 lb/in^2 and the valve on the nitrogen tank is closed. After 9 min, the valve on the left side of the bomb is slowly opened and adjusted such that a slow, steady stream of cavitated material is directed into a 250-ml plastic beaker. This is the whole cavitate (WC).

The WC is distributed in two 50-ml tubes and centrifuged at 400g for 7 min. Postnuclear supernatant (PNS) is saved and the nuclear pellet is resuspended with relaxation buffer to 50% of the original volume, followed by a second centrifugation. Saving an aliquot for GrB enzymatic assays, the PNS is added to 50-ml high-speed tubes and centrifuged at 10,000g for 18 min. The supernatant is aspirated without removing the sedimented material, which is subjected to extraction.

B. Extraction Procedure

The design here takes advantage of the fact that GrB will aggregate with its cognate binding protein, serglycin, under hypotonic conditions while the detergent solubilizes lipid-associated proteins. The granule-enriched pellet is extracted three times with the detergent containing hypotonic buffer. The granule-enriched pellet is then solubilized with a nondetergent, hypertonic buffer and centrifuged again. The final supernatant contains the GrB.

For extract A, the granule-enriched pellet may be solubilized with 2 ml of cold extraction buffer for each 1×10^9 cells. The procedure can be stopped here, and aliquots frozen at $-80°$ for subsequent extractions. Extract A is vortexed intermittently for 15–20 min. After saving a small aliquot to measure enzymatic activity, the remainder of extract A is centrifuged in microcentrifuge tubes for 7 min at 10,000g. For extract B, add the original volume of extraction buffer to the centrifuged pellet, vortex intermittently for 10–15 min, and centrifuge as described above. Repeat this process for extract C, solubilizing the pellet for 10 min. The granzyme B is extracted at this step by adding cold GrB solubilization buffer for 30 min. After 30 min, transfer into two 50-ml high-speed tubes and centrifuge at 10,000g for 15 min (no brake), then retrieve the supernatant for GrB purification.

The yield of GrB is markedly influenced by the cavitation process. Excessive cavitation will lead to rupture of the cytotoxic granules and loss of GrB, which is purified from the granule-enriched preparation by differential extraction. The quality of the preparation is established by measuring the GrB activity in extract A. Detection of GrB activity in the WC will yield spurious results because of the presence of other cytosolic proteinases that react with the chromogenic substrate. A satisfactory cavitation will yield approximately 3500 units per 10^9 cells. See Section V,A for a GrB enzymatic assay method using the Boc-Ala-Ala-Asp-SBzl substrate.

C. Preparation for Fast Protein Liquid Chromatography Fractionation

Before connecting the Mono S cartridge (Mono S HR10; Pharmacia) to the FPLC [Rainin (Walnut Creek, CA) Dynamax model SD-200 and Mono S HR10 column], purge the system with buffers A and B together for 3 min, followed by buffer A alone for 3 min with the injection loop open for the last minute. The Mono S column is then connected and the salt gradient and UV detectors are activated. Buffer B is then started and when the gradient monitor reaches maximum value (~2.0 mS), buffer B is run for 5 min. Buffer A is run through the column until the gradient monitor returns to baseline (~100 mS). A collection rack is filled with 13 × 100 mm test tubes and 400 μl of sodium acetate buffer (pH 3.50) is added to test tubes 55–75.

The extract (see Section III,C) is filtered through a 0.45-μm pore size membrane filter and diluted with buffer A up to 50 ml. The dilute extract is poured into a 50-ml Superloop (Amersham-Pharmacia, Inc.). To ensure that no air bubbles remain, turn on the pump at 0.6 ml/min with the top unattached. When no bubbles remain, stop the pump and attach the Superloop to the injection port.

D. Purification of Granzyme B

The fraction collector is set to 99.99 min/tube to collect the flowthrough. The run is started by opening the injection port and setting the flow rate at 2 ml/min (0% buffer B). The gradient detector and data display on the computer should read a maximum of 650 mS to ensure binding of GrB. After the Superloop has emptied, the injection port is closed and the run is continued by passing buffer A through the column until the protein peak has been at baseline for approximately 5 min. After placing the test tube rack in a tray of ice water on the fraction collector, the fraction collector is changed to collect sample at 1 min/tube. The gradient is started and approximately 80 fractions are collected [GrB usually peaks at fraction 65 (590 mM NaCl)].

E. Concentration, Dialysis, and Determination of Granzyme B Protein Concentration

The samples containing the highest GrB activity are usually tubes 55–75. The enzymatic assay should be performed on the flowthrough (50 μl), the GrB-containing fractions (2 μl), and starting material (2 μl). The fractions containing GrB are pooled and concentrated with two Centricon-10 units (Amicon) that have been previously soaked overnight in 0.5% Tween [0.02% (w/v) sodium azide]. The concentrators are thoroughly rinsed with distilled water by pipetting until no bubbles are apparent. The concentrators are centrifuged at 5500 rpm until the volume is less than 1 ml. The eluate is saved as a precaution should the membrane be defective. Samples are concentrated to approximately 500 μl and dialyzed against 0.15 M NaCl. To determine the final protein concentration, the enzymatic activity is measured on doubling dilutions of 1 μl of concentrated sample. With the assumption that specific activity will be approximately 15 units/μg, estimate microliters of sample per microliter of solvent to load onto the gel. The unknown and increasing amounts of previously isolated granzyme B (-1, 2, and 4 μg) are subjected to sodium dodecyl sulfate–polyacrylamide gel electrophoresis (SDS–PAGE) and stained with Coomassie. The gel is photographed and the latter is scanned for densitometric measurement of protein concentration, using the NIH Image program.

IV. Perforin Activity Detection

Perforin induces cytoplasmic membrane damage by polymerization and insertion into sensitive target cell membrane in the presence of calcium. Commonly used methods for detecting perforin activity are ^{51}Cr release, trypan blue staining, and hemoglobin release. These methods are used to detect perforin during purification or for determining lytic levels for a given cell in establishing the conditions for support of GrB-induced apoptosis. We briefly describe two of these methods as follows.

A. ^{51}Cr Release Assay

Target cells are labeled by incubating 100 μCi of labeled sodium chromate (^{51}CrNa$_2$CrO$_4$; Amersham, Arlington Heights, IL) with 2 × 10^6 cells in 0.5 ml of RPMI 1640–10% (v/v) FCS at 37° for 1 hr. Lymphoid target cells such as YAC-1, Jurkat, etc., are good targets for this assay. The labeled cells are washed three times with cold HBSS and resuspended in HH–BSA (see Section VI,A below) at 1 × 10^5 cells/ml. Perforin isolated by FPLC usually contains a high concentration of NaCl and should be diluted to

0.145 M NaCl with HE before use, then further diluted with HE-Na for a perforin titration. The assay is easily performed in a 96-well V-bottom microplate. An aliquot of 100 μl containing 1×10^4 target cells is added to 100 μl of diluted perforin. Several disposable test tubes, each with 100 μl of cells, are set aside to measure the total ^{51}Cr label added to each well. After 1 hr of incubation in a CO_2 incubator at 37°, the plate is centrifuged at 400g for 5 min. A 100-μl aliquot of supernatant from each well is carefully removed and placed in tubes to be counted in a γ scintillation counter. The percentage (%) of specific ^{51}Cr release is calculated by doubling the counts in the supernatant to adjust for volume and then dividing by the total label added to each well.

B. Hemoglobin Release Assay

Rabbit or sheep red blood cells are washed three times with cold PBS and suspended at 1% (v/v) in 10 mM HEPES, 0.145 M NaCl, BSA (10 μg/ml), and 4 mM $CaCl_2$, pH 7.4. An aliquot of 100 μl of cells is added to 100 μl of diluted perforin in a 96-well plate as described above. The plate is incubated at room temperature for 20 min and the reaction is stopped by acidification with 25 μl of 0.15 M MES buffer (Sigma), pH 6.0, containing 0.145 M NaCl. The plate is centrifuged for 5 min at 400g and 150 μl of cell-free supernatant is transferred from each well to a 96-well flat-bottomed microplate. The hemoglobin released into the supernatant is detected with an enzyme-linked immunosorbent assay (ELISA) reader at a wavelength of 412 nm. The percent release is calculated against the maximal hemolysis determined by adding 0.01% (v/v) saponin (Sigma) to the red blood cells.

V. Use of Granzyme B in Apoptosis and Enzymatic Assays

The GrB activity is measured by methods that assess its proteolytic and apoptosis-inducing ability. The measurement of peptide substrate hydrolysis is particularly useful during the purification of GrB. GrB-induced apoptosis can be detected by all the assays traditionally used to measure apoptosis. Most commonly, GrB apoptosis is measured by Hoechst or 4′,6-diamidino-2-phenylindole (DAPI) dye staining for visualization of chromatin condensation, or DNA fragmentation assays. Since GrB-induced apoptosis *in vitro* depends on the dose of perforin or adenovirus, the assays for perforin activity (Section IV) and adenovirus preparation and titration (Section VII) are also described.

A. Protease Activity Assay

To measure granzyme B activity, a stock solution of Boc-Ala-Ala-Asp-SBzl (Enzyme Systems Products, Livermore, CA) in DMSO is used at a final concentration of 0.1 mM with buffer [0.2 M HEPES, 0.3 M NaCl, 1 mM EDTA, 0.5% (v/v) Triton X-100, pH 7.0].[12] Substrate hydrolysis is measured by colorimetric absorbance changes, using a UVmax microplate reader (Molecular Devices, Menlo Park, CA). Enzymatic activity is reported either as the rate of hydrolysis (milliabsorbance change per minute [mA/min]) or as units, where 1 unit of hydrolytic activity is defined as the amount of enzyme that could hydrolyze 1 nmol of substrate per minute. We use an extinction coefficient of 13,100 M^{-1} cm^{-1} at 405 nm for the 3-carboxy-4-nitrophenoxide ion, produced from the reaction between Ellman's reagent [5,5'-dithiobis(2-nitrobenzoic acid)] and the freed benzyl mercaptan.[24]

VI. Granzyme B Apoptosis Assays

The choice of an apoptosis detection method depends partly on the target cell. DNA fragmentation assays using labeled DNA or ethidium-stained DNA ladders are suitable for some cells such as HeLa, Jurkat, Rat-1, YAC-1, and U937, but Hoechst/DAPI staining can be used on a much wider variety of targets, as some do not easily form low molecular weight soluble DNA fragments. Other assays of apoptosis including terminal deoxynucleotidyl transferase (TdT)-mediated dUTP nick end labeling (TUNEL) and annexin V binding can also be used and are described in [1, 2] in this volume.

A. Granzyme B- and Perforin-Induced Apoptosis

The principles of the method for activating GrB apoptosis are similar for each of the apoptosis detection assays, while labeling procedures and harvesting cells for quantitation differ and are outlined separately. GrB is stable and will not deteriorate with simple handling or even a couple of freeze–thaw cycles. Perforin, on the other hand, is extremely labile and will self-aggregate and inactivate on exposure to calcium. It loses activity on prolonged incubation at room temperature or refreezing. For this reason we store perforin in small aliquots (~20–50 μl) suitable for a single assay. Any perforin remaining after an assay is saved, and then later pooled and retitred. Perforin must be thawed and handled on ice, and always be diluted

[24] G. L. Ellman, *Arch. Biochem. Biophys.* **74,** 443 (1958).

in an EGTA-containing medium until it is activated in the presence of target cells by free calcium added in excess of the EGTA. The optimum concentration of perforin that supports GrB apoptosis varies between target cells and thus must be titred. Perforin activates GrB apoptosis best at lower levels of lytic activity (0–20%). At high concentrations in some cells it induces a necrotic cell death before the activation of apoptosis. For apoptosis assays, titrate the perforin starting from sublytic amounts, using a constant amount of GrB (e.g., 1 μg/ml) for HeLa, Rat-1, or MCF-7 cells and most other cells. This level of perforin can then be used routinely in subsequent assays when starting from the same stock of perforin.

Buffers and Materials

HBSS: Hanks' balanced salt solution

HE: 10 mM HEPES and 1 mM EGTA, pH 7.2. HE is used to dilute GrB and perforin to 0.145 M NaCl

HE-Na 10 mM HEPES, 1 mM EGTA, 0.145 M NaCl, pH 7.2; for further GrB and perforin dilution

HH–BSA: HBSS, 10 mM HEPES, 2 mM CaCl$_2$, 0.4% (w/v) BSA for target cell suspension. The calcium is 1 mM in excess of EGTA and is sufficient to activate perforin

GrB and perforin-containing materials are diluted in a 96-well V-bottomed microtiter plate. GrB and perforin, as they come from the FPLC purification, usually contain a high concentration of NaCl and must be adjusted by dilution before use. This is done by dilution with buffer HE to a final NaCl concentration of 0.145 M. Further dilution of perforin must be carried with buffer HE-Na.

In each microwell 80 μl of GrB and perforin is aliquoted in the dilutions of interest. To this 80 μl of target cells (1 × 10^5/ml to 1 × 10^4/ml, depending on assay) is added and then incubated at 37° for 1 to 4 hr. A buffer control (80 μl of target cells plus 80 μl of HE-Na), GrB control (without perforin), and perforin control (without GrB) should be included in each experiment.

B. Hoechst/4',6-Diamidino-2-phenylindole Dye Staining

Hoechst 33342 (Sigma) and 4',6-diamidino-2-phenylindole (DAPI; Sigma) are used to stain DNA, which are visualized as a bright blue color by a fluorescence microscope with UV light and filters. Both reagents are suitable but we find DAPI staining more stable.

Method. Prepare a stock Hoechst or DAPI solution (1 mg/ml in PBS; store at 4°). Target cells are prepared at 1 × 10^5 cells/well. After a 1- to 4-hr incubation, the cells are fixed by adding 18 μl of formaldehyde (37% solution), with shaking at room temperature for 10 min. Add Hoechst or

DAPI to each well at a final concentration of 1 μg/ml for another 10 min (20 μl of 100× diluted stock Hoechst or DAPI works well). The plate is centrifuged at 400g for 5 min to pellet the cells. The supernatant is carefully removed without touching the cell pellet. Add 10 μl of FluoroGuard (Bio-Rad, Hercules, CA) to each well, mix the cells with a pipette, and mount the cell mixture on glass slides for flourescence microscopy. The nuclear chromatin in apoptotic targets is highly condensed or broken into apoptotic bodies, which are easily distinguished from normal nuclei. The normal and apoptotic nuclei in 200–300 cells are counted, using at least 4 or 5 high-power fields per sample for statistical accuracy, and the percentage of apoptotic cells calculated.

C. [^{125}I]Iododeoxyuridine or [^{3}H]Thymidine Release Assay

Target cell DNA damage induced by GrB and perforin can be detected by release of soluble [^{125}I]iododeoxyuridine ([^{125}I]UdR) from cells ([^{125}I]UdR is a thymidine analog). A short-term pulse label with [^{125}I]UdR selectively labels those cells in S phase (~50% of the population). Longer labeling (overnight) with [^{3}H]thymidine ([^{3}H]TdR) is possible because it is nontoxic. Both labeling methods are equally effective and the release of [^{125}I]UdR or [^{3}H]thymidine from parallel cultures is indistinguishable. This method has the advantage of using an automated scintillation counter, compared with visual counting methods. It is best suited to targets of lymphoid origin, which readily solubilize DNA after GrB or other apoptotic stimuli in short-term assays. Longer assays are less reliable as DNA degradation can be secondary to necrotic cell death.

Materials

TET: 50 mM Tris-HCl, pH 7.2, containing 25 mM EGTA and 1% (v/v) Triton X-100 (Sigma); for [^{125}I]UdR solubilization

5-[^{125}I]Iodo-2'-deoxyuridine ([^{125}I]UdR) or [5'-^{3}H]thymidine ([^{3}H]TdR) (Amersham)

Target cell and culture medium: Commonly used target cells for [^{125}I]UdR or [^{3}H]TdR release are YAC-1, EL4, L5178Y, P815, and SL2 cells. They are grown in RPMI 1640 (GIBCO-BRL) containing 10% (v/v) fetal bovine serum

Method. Target cells in the log phase of growth are used for labeling. About 5–10 μCi of [^{125}I]UdR is added to 2–5 × 10^6 target cells in 0.5 ml of RPMI 1640–10% (v/v) FBS. The cells are incubated at 37° in 5% CO_2 for 90 min. For [^{3}H]TdR labeling, target cells (1 × 10^5/ml) are incubated for 16 hr in culture medium containing [^{3}H]TdR at 0.1 μCi/ml. After incubation, the cells are washed three times with HBSS, and resuspended at 1.25×10^5 cells/ml in HH–BSA. Each test well will receive an aliquot of 80 μl, which should be labeled with an equivalent of 1000 cpm for 10^4 cells. Higher labeling will increase the background. Three or four 80-μl aliquots are set aside in disposable tubes to serve as the measure of total counts per minute placed in each well.

After the incubation period with GrB and perforin, 40 μl of TET is added to the reaction well to stop the reaction. The contents are then carefully mixed (to prevent foaming) with a multichannel pipette and the plates are centrifuged at 400g for 5 min. Half (100 μl) of the supernatant is carefully harvested from each well into a small disposable tube. Soluble [^{125}I]UdR (or [^{3}H]TdR) in the supernatant is determined in a scintillation counter. Percentage (%) of [^{125}I]UdR release is calculated as follows:

Percent [^{125}I]UdR release

$$= \left(\frac{\text{cpm of test well} - \text{mean cpm of buffer control}}{\text{mean total cpm} - \text{mean cpm of buffer control}} \right) \times 100$$

Each test point should be carried out in triplicate and standard errors are usually less than 5% of the mean.

D. DNA Fragmentation

Oligonucleosomal DNA fragmentation (180- to 200-bp DNA ladder) is a common feature of apoptosis, including that induced by GrB and perforin.[6] This method uses much larger cell numbers. Samples of 1×10^6 cells in 1 ml of reaction buffer containing 500 μl of GrB/perforin (e.g., 2 μg of GrB and 0.2 μg of perforin) are added to an Eppendorf tube. After incubation, cells are pelleted at 3000 rpm for 1 min in a microcentrifuge.

The cell pellets are then resuspended in 35 μl of lysis buffer [10 mM EDTA, 50 mM Tris-HCl (pH 8.0) containing 0.5% (w/v) sodium-N-lauryl sarkosin (Sigma)] and proteinase K (0.5 mg/ml; Sigma) and incubated at 50° for 1 hr. DNase-free ribonuclease A (Rnase A; Worthington, Freehold, NJ) is added to a final concentration of 0.5 mg/ml. The sample is incubated for a further 1 hr, and then 6× gel-loading buffer, 0.25% (w/v) bromphenol blue (Sigma), 0.25% (w/v) xylene cyanol FF (Sigma), and 40% (w/v) sucrose in water are added to each sample to give a 1× gel-loading buffer. A 1% (w/v) agarose gel [1 g of agarose in 100 ml of 0.5× TBE (0.045 M Tris-borate, 1 mM EDTA) containing ethidium bromide (0.5 μg/ml)] in 0.5× TBE buffer is run, applying a voltage of 1–5 V/cm (measured as the distance between the electrodes). Continue until the bromphenol blue and xylene cyanol FF have migrated the appropriate distance through the gel. The DNA fragmentation bands are visualized under UV light.

VII. Induction of Apoptosis by Delivering Granzyme B with Type 2 Replication-Deficient Adenovirus

Although the mechanism of action by which adenovirus can replace perforin has not been firmly established, this method is capable of introducing exogenous proteins to the cytosol of any cell that possesses membrane receptors for adenovirus (AD).[7] The technique has been experimentally verified with the following human and murine cell lines: Jurkat, YAC-1, HeLa, MCF-7, and 293 cells. Untested lines may be examined for the presence of receptors for AD by using appropriate antibodies.

A. Generation of Adenovirus Stock for Large-Scale Production

The 293 cell line (1×10^6 in complete medium) is seeded in 100-mm petri dishes. Cells are ready for infection at 80–90% confluency. The subconfluent cells are prepared for infection by rinsing with 10 ml of PBS. A stock of AD is diluted to 1×10^9 PFU/ml of basal DMEM. To provide a multiplicity of infection (MOI) of approximately 10 PFU/cell, add 1 ml of the dilute AD to each monolayer, rocking the dishes every 15 min for 1 hr at 37°. Subsequently, 10 ml of complete medium is added and the plates incubated at 37°. After 48 hr cells are detached by pipetting and centrifuged at 500g for 10 min. The supernatant is saved as the source of virus for large-scale production of AD. A volume sufficient to infect 15 to 20 dishes (150 mm) of 293 cells should be available.

B. Infection and Harvest

The 293 cells are prepared for infection as described above, except that the cells are seeded in 150-mm dishes. The AD stock prepared above (2.5 ml) is added to each dish, followed by incubation at 37° for 1 hr with rocking every 15 min. Complete medium (20 ml) is then added to each dish followed by incubation at 37°. After 48 hr cells are removed by pipetting and centrifuged at 500g for 10 min in 50-ml polypropylene tubes. The cellular pellet in each tube is resuspended in 10 ml of PBS and the resuspended cells are combined, adjusting to 5×10^7/ml. The cells are then centrifuged at 500g for 25 min in the 50-ml polypropylene tubes. After removing the supernatant, each pellet is resuspended in 15 ml of 10 mM Tris buffer (20 mM MgCl$_2$, pH 7.8) and the suspension is freeze–thawed four times by alternate exposure to dry ice and a water bath maintained at 37°. The ruptured cells are treated with DNase (10 μg/ml) for 30 min at 37° and an equal volume of Freon (1,1,2-trichloro-1,2,2-trifluoroethane; Sigma) is added.

To extract the virus, the tubes are shaken by hand for 10 min and centrifuged at 850g for 10 min. The upper aqueous phase is removed and transferred to a new tube, taking care to avoid the white interface, which is cellular protein precipitate. The retrieved aqueous phase is layered onto a gradient that consists of 10 ml of CsCl at a density of 1.4 g/ml plus 10 ml at 1.2 g/ml. CsCl is dissolved in 10 mM $MgCl_2$, 50 mM Tris buffer (pH 7.8). The gradients are developed in an SW28 centrifuge (Beckman, Fullerton, CA) at 67,000g for 3 hr at 4°. The banded material containing the viral particles is removed (1.5–2 ml) and mixed with an equal volume of the Tris buffer used above. After layering the material on a second gradient, which consists of 4 ml of CsCl at a density of 1.4 g/ml plus 4 ml at 1.2 g/ml, the gradient is developed in an SW41 centrifuge (Beckman) at 62,500g overnight at 4°. The band of virus is removed (0.5–1.5 ml) and mixed with a fourfold excess of freezing solution [0.1% (w/v) BSA, 50% (v/v) glycerol, 10 mM Tris, 100 mM NaCl]. This is the purified virus stock. Before freezing at −80°, dilute the stock 1 : 50 with H_2O and measure the OD_{260} (1 OD_{260} = ~10^{12} particles/ml = 38 μg of DNA).

C. Plaque-Forming Assay to Determine Quantity of Adenovirus

The 293 cells (3 × 10^6) are seeded in 60-mm dishes and cultured overnight. Viral stock is serially diluted in basal DMEM and 0.2 ml of dilute virus is added to each dish. The cells are infected for 1 hr with rocking every 15 min. The medium is aspirated and cells are covered with 5 ml of agarose. The agarose mixture is prepared as follows: equal volumes of 2% (w/v) low melting point agarose and 2× DMEM plus 4% (v/v) IgG-free calf serum containing 2× Pen/Strep and 50 mM $MgCl_2$. Cultures are fed every 5 days with 3 ml of the agarose mixture and plaques are counted 2 weeks after infection. Plaque-forming units of the stock is calculated on the basis of the dilution of virus added to the dish. For example, if 0.2 ml of medium diluted 1 × 10^8 from the stock results in 20 plaques, the plaque-forming units of the viral stock is 1 × 10^{10}.

D. Procedure for Granzyme B/Adenovirus Apoptosis Assay

To induce apoptosis in 1 × 10^6 target cells (100 μl) with granzyme B at 1 μg/ml, proceed as follows. Add 1 × 10^6 cells in 94 μl of RPMI plus 0.5% (w/v) BSA to microcentrifuge tube. Take the desired volume of granzyme B stock and dilute 1 : 10 in medium containing 0.5% (w/v) BSA. For example, if the experiment contains 10 samples, each of which contains 1 × 10^6 cells, then dilute 1 μl of stock to 10 μl. Add 1 μl of newly diluted granzyme B stock (0.1 μg) to a microcentrifuge tube. Add 5 μl of AD stock (5 × 10^7 PFU) to provide an MOI of 100 : 1 Incubate at 37° for 4–6 hr and perform a readout for apoptosis. If cell lysates will be analyzed for other proteins, it is recommended that the specific granzyme B oligopeptide inhibitor (Ac-Ile-Glu-Thr-Asp-CHO; Alexis Biochemical, San Diego, CA) be added to minimize adventitious proteolysis.

[12] Viral Caspase Inhibitors CrmA and p35

By QIAO ZHOU and GUY S. SALVESEN

Introduction

To replicate and take over the synthetic machinery of host cells viruses must evade the normal apoptotic response to infection. Consequently, some viruses have developed mechanisms to avoid apoptosis of infected cells by expressing potent caspase inhibitors. Currently, two viral proteins are known to directly inhibit caspases: cytokine response modifier A (CrmA) from cowpox virus,[1] and p35 from *Autographa californica* nuclear polyhedrosis virus.[2] Closely related viruses (e.g., vaccinia virus, which is related to cowpox virus, and *Spodoptera litura* nuclear polyhedrosis, which is related to *A. californica* nuclear polyhedrosis virus) contain orthologs that may have similar activities. In this chapter we describe expression of CrmA and p35 in *Escherichia coli* and how to characterize the purified proteins.

CrmA

CrmA is a 38-kDa single-chain protein directed to the cytosol of virally infected mammalian cells. Originally identified as an inhibitor of interleukin 1β (IL-1β) activation mediated by caspase 1,[1] CrmA has been shown to be a strong inhibitor of apoptosis induced by some, but not all, death stimuli (reviewed in Ref. 3). These include overexpression of caspase 1, serum withdrawal,[4] nerve growth factor withdrawal,[5] ligation of some death receptors,[6] and detachment from the extracellular matrix.[7] Gene ablation experi-

[1] C. A. Ray, R. A. Black, S. R. Kronheim, T. A. Greenstreet, P. R. Sleath, G. S. Salvesen, and D. J. Pickup, *Cell* **69**, 597 (1992).
[2] R. J. Clem, M. Fechheimer, and L. K. Miller, *Science* **254**, 1388 (1991).
[3] G. S. Salvesen and V. M. Dixit, *Cell* **91**, 443 (1997).
[4] M. Miura, H. Zhu, R. Rotello, E. A. Hartwieg, and J. Yuan, *Cell* **75**, 653 (1993).
[5] V. Gagliardini, P.-A. Fernandez, R. K. K. Lee, H. C. A. Drexler, R. J. Rotello, M. C. Fischman, and J. Yuan, *Science* **263**, 826 (1994).
[6] M. Tewari and V. M. Dixit, *J. Biol. Chem.* **270**, 3255 (1995).
[7] N. Boudreau, C. J. Sympson, Z. Werb, and M. J. Bissell, *Science* **267**, 891 (1995).

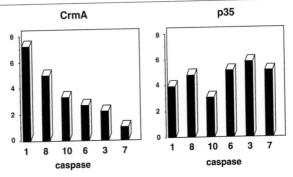

FIG. 1. Selectivity of CrmA and p35 for different caspases. Selectivity is described by comparing the respective rate constants for inhibition as log k_{ass}, with the longest columns being the fastest inhibitors.[8,12,13] Note that CrmA has a large range of inhibition rate constants, but that p35 tends not to discriminate between caspases. Inhibition of caspase 7 by CrmA is barely detectable and of the other caspases shown here, only caspases 1, 8, and possibly 10 are considered physiologic targets of CrmA.

ments suggest that caspase 1 is unlikely to be essential in apoptosis, and therefore not the target in these death pathways; thus the prime target of CrmA in apoptosis is most probably another caspase.[8] Studies using purified CrmA and purified caspases demonstrate the quantitative order of effective inhibition described in Fig. 1.[8,12,13] Although not quantitated in the same way, CrmA also inhibits caspases 4 and 5.[9] The selectivity of CrmA for different caspases can be useful in narrowing down the question of which caspase is involved in a particular scenario. CrmA can also potently inhibit granzyme B,[10] which is a serine protease that shares the primary Asp specificity of the caspases.[11]

p35

Similarly, the 35-kDa single-chain protein of baculovirus known as p35 inhibits apoptosis in virally infected host insect cells.[2,14,15] Moreover, ectopic

[8] Q. Zhou, S. Snipas, K. Orth, V. M. Dixit, and G. S. Salvesen, *J. Biol. Chem.* **273,** 7797 (1997).
[9] S. Kamada, Y. Funahashi, and Y. Tsujimoto, *Cell Death Differ.* **4,** 473 (1997).
[10] L. T. Quan, A. Caputo, R. C. Bleackley, D. J. Pickup, and G. S. Salvesen, *J. Biol. Chem.* **270,** 10377 (1995).
[11] C. J. Froelich, V. M. Dixit, and X. Yang, *Immunol. Today* **19,** 30 (1998).
[12] T. Komiyama, C. A. Ray, D. J. Pickup, A. D. Howard, N. A. Thornberry, E. P. Peterson, and G. Salvesen, *J. Biol. Chem.* **269,** 19331 (1994).
[13] Q. Zhou, J. F. Krebs, S. J. Snipas, A. Price, E. S. Alnemri, K. J. Tomaselli, and G. S. Salvesen, *Biochemistry* **37,** 10757 (1998).
[14] R. J. Clem and L. K. Miller, *Mol. Cell. Biol.* **14,** 5212 (1994).
[15] N. E. Crook, R. J. Clem, and L. K. Miller, *J. Virol.* **67,** 2168 (1993).

expression of p35 also prevents apoptosis in *Caenorhabditis elegans, Drosophila melanogaster,* as well as in mammalian cells.[2,14,16–20] The ability of p35 to prevent apoptosis has been attributed to direct inhibition of the caspases. So far, it has been shown that human caspases 1, 2, 3, 4, 6, 7, 8, and 10,[13,21] the *C. elegans* homolog Ced-3,[18] and the *Spodoptera frugiperda* caspase 1[22] can all be potently inhibited. Significantly, granzyme B is not inhibited.[13,21] Weak or no inhibition is observed with representative members of other protease families we have tested. Inhibition by p35 requires an unblocked active site on the caspase. Thus, p35 is an excellent caspase-specific inhibitor with high affinity and is useful in demonstrating caspase involvement in various experimental systems. However, the relative lack of discrimination among the caspases makes it difficult to dissect the presumed caspase cascade using p35.

Expression in Animal Cells

The outcome of ectopic expression of CrmA and p35 in various cell lines is generally consistent with their predicted specificity against caspases. Thus, CrmA rescues cells from apoptosis triggered primarily by signals that are promoted via death receptors, where caspase 8 (and possibly caspase 10) are the initiators.[3] CrmA cannot rescue from death triggered by stress induction and genotoxic damage. On the other hand, p35 rescues from apoptosis following almost all triggers, including stress induction and genotoxic damage, matching well with its broad specificity for caspases, and confirming the central role of caspases in apoptosis. Both CrmA[23,24] and p35[24,25] cDNAs have been expressed in transgenic animals using tissue-specific promoters.

[16] B. A. Hay, T. Wolff, and G. M. Rubin, *Development* **120,** 2121 (1994).
[17] S. Rabizadeh, D. J. LaCount, P. D. Friesen, and D. E. Bredesen, *J. Neurochem.* **61,** 2318 (1993).
[18] D. Xue and H. R. Horvitz, *Nature (London)* **377,** 248 (1995).
[19] D. R. Beidler, M. Tewari, P. D. Friesen, G. Poirier, and V. M. Dixit, *J. Biol. Chem.* **270,** 16526 (1995).
[20] F. F. Davidson and H. Steller, *Nature (London)* **391,** 587 (1998).
[21] N. J. Bump, M. Hackett, M. Hugunin, S. Seshagiri, K. Brady, P. Chen, C. Ferenz, S. Franklin, T. Ghayur, P. Li *et al., Science* **269,** 1885 (1995).
[22] M. Ahmad, S. Srinivasula, L. Wang, G. Litwack, T. Fernandes-Alnemri, and E. S. Alnemri, *J. Biol. Chem.* **272,** 1421 (1997).
[23] C. M. Walsh, B. G. Wen, A. M. Chinnaiyan, K. O'Rourke, V. M. Dixit, and S. M. Hedrick, *Immunity* **8,** 439 (1998).
[24] K. G. Smith, A. Strasser, and D. L. Vaux, *EMBO J.* **15,** 5167 (1996).
[25] W. Du, J. E. Xie, and N. Dyson, *EMBO J.* **15,** 3684 (1996).

This chapter does not deal with transfection of animal cells. The basic techniques are comparable to other transfection strategies and, in common with most investigations of antiapoptotic proteins, should be optimized for the particular cell line and death stimulus being contemplated.

Expression in *Escherichia coli*

Pure, fully active CrmA or p35 can be readily obtained from expression of His-tagged proteins in *E. coli*. The presence of the tag does not interfere in any known way with the activity of CrmA, and His-tagged p35 is fully active as a caspase inhibitor. However, p35 of native sequence has not been rigorously tested to see if it differs from His-tagged material. For convenience the tag is N terminal in CrmA, but it must be C terminal in p35 because it has not proved possible yet to produce active p35 with an N-terminal His tag.

Materials

Buffers

Buffer A: 50 mM Tris, 100 mM NaCl, pH 8.0
Buffer B: 50 mM Tris, 500 mM NaCl, pH 8.0
Buffer C: 50 mM Tris, 100 mM NaCl, 200 mM imidazole, pH 8.0
Buffer D: 50 mM Tris, 100 mM NaCl, 50 mM imidazole, pH 8.0

Reagents

Luria broth
Isopropyl-β-D-thiogalactopyranoside (IPTG, 0.5 M stock solution): Store at $-20°$
Ampicillin (Amp, 50-mg/ml stock solution): Store at $-20°$
Chloramphenicol (25 mg/ml in ethanol): Store at $-20°$
Chelating Sepharose Fast Flow (Pharmacia, Piscataway, NJ): Store at 4°
NiSO$_4$ (0.1 M)

Hardware

Baffled shaker flasks or Erlenmeyer flasks
Orbital shaker
Sonicator
Small chromatography columns
Peristaltic pump
Gradient maker
Stirrer
Fraction collector

Spectrophotometer
Gel apparatus for sodium dodecyl sulfate–polyacrylamide gel electrophoresis (SDS–PAGE)

Expression and Purification of Recombinant CrmA

The *crmA* gene of cowpox virus[1] is cloned as an *NcoI–XhoI* fragment
into a derivative of the pFLAG-1 expression vector (IBI), under the control
of a *tac* promoter.[10] The CrmA cDNA is preceded by a sequence encoding
a His_6 tag with an *NdeI* site in front of the His_6 sequence. The construct
can be obtained from the American Type Culture Collection (Rockville,
MD). In principle, any simple expression system could be used because
CrmA production seems insensitive to varied expression conditions, and
the protein exhibits minimal toxicity in *E. coli*. Expression is directed to
the cytosol, and results in high concentrations of soluble protein. Transcription of the His_6–CrmA-encoding gene starts from ATG at the *NdeI* site.
Expressed His_6-CrmA protein is purified by immobilized nickel affinity
chromatography, with an average yield of at least 5 mg/liter of culture.
CrmA has a molecular mass of 38 kDa, and the estimated extinction coefficient according to the Edelhoch relationship[26] is 32,940 M^{-1} cm^{-1}, from
which the protein concentration can be calculated after determining absorbance at A_{280}.

We use *E. coli*. TG1 as the host for expression. TG1 cells are transformed
with the His_6–CrmA construct by standard procedures. Transformed TG1
cells are maintained on LB–Amp plates or stored as glycerol stocks [10%
(v/v) glycerol concentration].

Expression

All growth is at 37° in a shaking incubator (300 rpm).
The following protocol is for a 500-ml culture, which can be scaled up
or down as necessary.

1. Pick a single colony from the plate and inoculate 3 ml of 2× TY
plus ampicillin (100 μg/ml); grow the culture overnight.
2. Dilute 0.5 ml of the overnight culture into 50 ml of 2× TY plus
ampicillin (100 μg/ml); grow to an A_{600} of 0.5 (about 1.5 hr).
3. Dilute 25 ml into 500 ml of fresh 2× TY plus ampicillin (100 μg/
ml); grow to an A_{600} of 0.5–0.6 (1.5 to 2 hr).
4. Induce with 1 mM IPTG for 3 hr.

[26] H. Edelhoch, *Biochemistry* **6**, 1948 (1967).

5. Put the culture on ice for 10 min. Pellet the cells by centrifugation at 5000g for 5 min at 4°. Proceed to purification or store the pellet temporarily at −20°.

Purification

1. Resuspend the pellet in 100 ml of buffer A. Sonicate the suspension in short bursts to disintegrate the cells. The time and power of sonication will depend on the instrument. Keep the cell suspension on ice while sonicating.

2. Centrifuge at 12,000g for 60 min at 4°. Keep the supernatant, which contains the soluble CrmA protein. While centrifuging, go to step 4 to prepare the affinity column.

3. Filter the supernatant through a 0.22-μm pore size bullet filter (Millipore, Bedford, MA). Keep the material on ice. Save a 50-μl portion of this starting material for later analysis by SDS–PAGE.

4. Set up a column with 2 ml of chelating Sepharose Fast Flow (Pharmacia). We use Econo-Columns from Pharmacia. Wash with 10 ml of distilled H_2O. Charge with 1 ml of 0.1 M NiSO$_4$. Wash with 20 ml of distilled H_2O. Equilibrate with 20 ml of buffer A.

Steps 5–7 should be carried out at 4°.

5. Load the filtered supernatant onto the Ni column. Save the flowthrough.

6. Wash the column with 100 ml of buffer B, then wash with 10 ml of buffer A.

7. Elute with buffer D in five 1-ml fractions. Alternatively, an imidazole gradient can be used if there is contamination by *E. coli* proteins adsorbed to the affinity matrix (see the next section, on p35 purification).

8. Perform SDS–PAGE, using 10 μl of each fraction as well as the starting material, flowthrough, and the wash, to determine the recovery and purity. The purity as judged by SDS–PAGE and should be >95%. If substantial material elutes in the flowthrough, this can be reapplied to a fresh Ni-charged chelating column, with the washing and imidazole elution repeated.

9. Determine the A_{280} of each fraction. Pool the fractions with high CrmA concentrations and little impurity, as judged by A_{280} and SDS–PAGE (Fig. 2).

10. Dilute the pooled fractions with buffer A, if necessary, so that the final A_{280} is about 1.0. Concentrated CrmA proteins with A_{280} values of more than 2.0 tend to polymerize on long-term storage. Store at −80° in small aliquots. Use a fresh aliquot for each experiment and discard unused material.

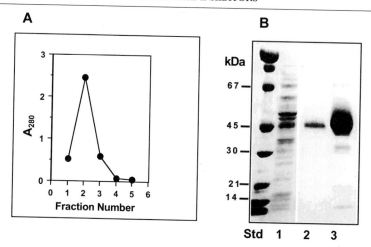

Fig. 2. Purification of CrmA. (A) Typical elution profile of CrmA, using the protocol described. Each fraction is 1 ml, eluted with a pulse of buffer D. Fractions 1–3 were pooled and analyzed by SDS–PAGE, as shown in (B). Lane 1 is starting material (*E. coli* filtered lysate); lanes 2 and 3 are different loadings of the pooled purified material.

Expression and Purification of Recombinant p35

The p35 cDNA with the stop codon removed is cloned into pET21b(+), using the *Nde*I/*Xho*I sites, in frame with a C-terminal His$_6$ tag followed by a stop codon, and expressed in *E. coli* BL21(DE3)pLysS. Expressed protein is purified by affinity chromatography, using chelating Sepharose charged with NiSO$_4$. The average yield of this method is 3–4 mg/liter of culture. The estimated extinction coefficient[26] according to the protein sequence of this 35-kDa protein is 40,800 M^{-1} cm^{-1}. This protein lacks stability, with a half-life of about 2 days at 4°. Activity also declines on freeze–thawing, so frozen storage is not recommended. Assays should be conducted with freshly prepared material.

Escherichia coli BL21(DE3)pLysS is used as the host for expression. Transformed cells are stored as glycerol stocks [8% (v/v) glycerol concentration] and plated on LB plates containing both ampicillin and chloramphenicol. This *E. coli* host cannot be stored on plates at 4°, and so fresh plates are used within 3 days of storage at room temperature.

Expression

Initial growth is at 37°, and induction is at 30°, both in a shaking incubator (300 rpm).

The following protocol is for a 500-ml culture, which can be scaled up or down as necessary.

1. Inoculate 3 ml of LB or 2× TY containing ampicillin (100 μg/ml) and chloramphenicol (34 μg/ml) with a single colony from the plate. Grow overnight at 37° with shaking (300 rpm).

2. Transfer 0.5 ml into 50 ml of fresh medium containing antibiotics. Grow to an A_{600} of 0.5 (about 2 hr).

3. Transfer 20 ml into 500 ml of fresh medium containing antibiotics. Grow to an A_{600} of 0.6–0.8 (about 3 hr).

4. Decrease the incubator temperature to 30°. Add IPTG to a final concentration of 1 mM. Induce for 3–4 hr.

5. Put the culture on ice for 10 min. Harvest the cells by centrifugation at 5000g for 5 min at 4°.

6. Discard the supernatant. Store the pellet at $-20°$ temporarily or continue purification.

Purification

1. Resuspend the pellet in buffer A. Freeze–thaw once to break the cells.

2. Sonicate the suspension in short bursts for 2 min. The time and power of sonication will depend on the instrument. Keep the sample on ice all the time. Repeat if the material is still viscous. Viscosity can be decreased by passing the lysate through a 27-gauge needle.

3. Centrifuge at 16,000g for 90 min at 4°. Collect the supernatant.

4. While centrifuging, prepare the Ni-Sepharose column as described in the section on CrmA purification.

5. Filter the supernatant through a 0.45-μm pore size filter, and then with a 0.22-μm pore size bullet filter. Material from *E. coli* BL21(DE3) pLysS is more viscous than cell lysates from TG1 and will need a few more filter units. Keep the material on ice. Save 50 μl for later analysis by SDS–PAGE.

Steps 6–8 should be carried out at 4°.

6. Load the filtered starting material onto a 2-ml Ni-Sepharose column equilibrated with 10 ml of buffer A.

7. Wash with 125 ml of buffer B, followed by 20 ml of buffer A.

8. Elute with a linear gradient of 0–200 mM imidazole at a flow rate of 1 ml/min. The gradient can be achieved with a specialized instrument (e.g., a dual pump protein purification instrument) or simply by using a gradient maker holding buffers A and C in two compartments, with flow directed through a peristaltic pump. Collect 40 fractions (1 ml each). A fraction collector would be useful.

9. Read the A_{280} value of each fraction. Perform SDS–PAGE with selected fractions. The p35 protein usually elutes the in the latter half of the gradient, and most impurities elute in the first half of the gradient (Fig. 3). The p35 peak should be >95% pure on SDS–PAGE.

10. Pool fractions with high p35 concentration and little impurity. The pooled protein can be further subjected to buffer exchange, dialysis, or concentration as required. Concentration in a stepwise manner together with buffer exchange may prevent precipitation that can happen with single-step concentration. Imidazole can be removed by extensive dialysis against buffer A.

Assays of Inhibitory Activity

After purification the proteins are tested for their ability to inhibit caspases. CrmA is tested with caspase 8, and p35 is tested with caspase 3 for convenience. The object is to determine the absolute inhibitory concentration of CrmA or p35 by titrating against a caspase of known activity. The concentration of the caspase(s) must first be determined (see [8] in

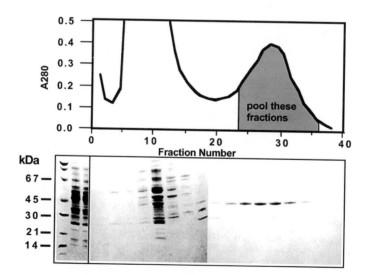

Fig. 3. Typical elution profile of p35, using a 0–200 mM continuous imidazole gradient. Forty fractions, each 1 ml in volume, were collected at a flow rate of 1 ml/min. Purification of His-tagged proteins from *E. coli* BL21 (DE3)pLysS tends to be more susceptible to contamination, as seen in fractions 5–20. Thus we prefer the gradient elution strategy, to allow resolution of p35 (fractions 24–38) from the earlier eluting contaminants.

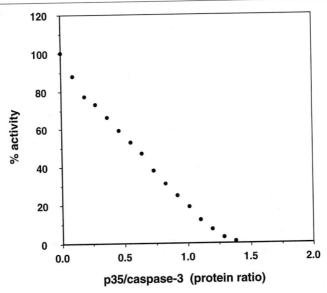

Fig. 4. Titration of caspase 3 by p35. The active site concentration of caspase 3 was determined by titration with Z-VAD-FMK, and then used to calculate the inhibitory concentration of freshly isolated p35. Reactants were incubated for 30 min at 37°, after which Ac-DEVD-pNA was added to determine residual caspase activity. The titration point is close to unity, indicating that the p35 preparation is almost fully active (for discussions about the slight deviation from unity, see Ref. 13). [Reprinted from permission from Q. Zhou, et al., Biochemistry 37, 10757 (1998). Copyright 1998 American Chemical Society.]

this volume[27]), and then this standardized caspase is used. Because both inhibitors form tight complexes with their targets, the caspase are incubated with varying concentrations of the inhibitor in a small volume, resulting in high concentrations of both caspase and inhibitor, which facilitates the reaction. Generally, the caspases are incubated with inhibitor ranging in concentration from 0 to 10 times the caspase concentration. The enzyme and inhibitor are incubated in a small volume of caspase buffer for 30 min at 37°, and residual activity is assayed in caspase buffer containing peptide substrate to a final concentration of 0.1–0.2 mM. The linear rate of hydrolysis is plotted against the inhibitor concentration and the active concentration of the inhibitor is determined from the intercept with the x axis as outlined in Fig. 4.

[27] H. R. Stennicke and G. S. Salvesen, Methods Enzymol. 322, [8], 2000 (this volume).

Reagents

Stock p35 or CrmA solution: Prepare at a provisional concentration of 10 μM (based on absorbance at 280 nm) in caspase buffer: 20 mM piperazine-N,N'-bis(2-ethanesulfonic acid (PIPES), 100 mM NaCl, 10 mM dithiothreitol (DTT), 1 mM EDTA, 0.1% (w/v) 3-[(3-chol-amidopropyl)-dimethyl-ammonio]-1-propanesulfonate (CHAPS), 10% (w/v) sucrose, pH 7.2. The 10 mM DTT may be substituted with 20 mM 2-mercaptoethanol if desired

Stock solution of the caspase reporter substrate Ac-DEVD-pNA: Prepare in dry DMSO at 20 mM

Stock solutions are stable for at least 1 year at $-20°$. Dilute the substrate to a working concentration of 0.2 mM in caspase buffer. The working substrate solution is stable for 8 hr at room temperature

Working solutions of caspase 3 or 8: Prepare at 1.0 μM in caspase buffer

Procedure

The assays are carried out in a total volume of 100 μl for detection in a 96-well plate reader, preferably operating in the kinetic mode. For analysis requiring larger volumes, increase proportionally.

1. Prepare working solutions of inhibitor at 0.1–10 μM in caspase buffer, as required, from the 10 μM stock. Incubate the enzyme and inhibitor separately for 15 min at 37° in caspase buffer to allow reduction of any disulfide bonds that may have formed during isolation or storage. Intermolecular disulfides occur during CrmA production, and can interfere with complex formation.

2. Add 10 μl of each working solution of inhibitor to 40 μl of caspase buffer per well or tube, followed by 10 μl of enzyme solution, and incubate for 30 min at 37° to allow complex formation.

3. Assay residual caspase activity in samples from each reaction mixture by adding 50 μl of working substrate solution. Plot residual activity as a function of inhibitor concentration. In the case of caspase 3, it is necessary to dilute the enzyme–inhibitor incubation mixture 10-fold before assay because this enzyme is more active on the substrate than caspase 8. Simply remove 10 μl from the 50-μl reaction mixture into 90 μl of working substrate solution to assay residual activity.

4. The plot should be linear, with an intercept on the x axis equal to the concentration of active inhibitor (Fig. 4). Curvature of the plot as it approaches this axis indicates that the reaction has not gone to completion. This can be overcome by longer incubation or by increasing the concentrations of enzyme and inhibitor. The concentration of the inhibitor can only

rarely be accurately determined by extrapolating a straight line drawn from partial inhibition.

Reaction Kinetics

Both CrmA and p35 have high affinity for their target caspases and low dissociation rates.[8,13] The stability of the tight complexes allows the functional inhibitor concentration to be determined from the equivalent concentration of caspase, as described above. More detailed reaction kinetics can be performed to determine the rate and strength of binding. The second-order rate constant of inhibition (k_{ass}) can be readily determined from progress curve analysis.[8,13,28] However, with tight-binding inhibitors such as CrmA and p35, or where peptide bond cleavage may be part of their inhibitory mechanism, determination of the inhibition constant, K_i, can be problematic. This is particularly so when reversibility is not readily demonstrable, owing to low dissociation rates. In such cases, the estimated K_i serves as a useful parameter in appreciating the overall inhibition potency, but a physical meaning cannot be easily assigned to it. Assuming a simple bimolecular process, this constant can be estimated from steady state rates of caspase activity in the absence and presence of various concentrations of inhibitor. Readers interested in determining reaction kinetics are referred to more detailed methods.[8,13,28]

[28] J. F. Morrison and C. T. Walsh, *Adv. Enz. Relat. Areas Mol. Biol.* **59,** 201 (1988).

[13] Purification and Use of Recombinant Inhibitor of Apoptosis Proteins as Caspase Inhibitors

By Quinn L. Deveraux, Kate Welsh, and John C. Reed

Introduction

Cell death proteases, known as "caspases," are highly conserved and integral components of both initiation and execution of apoptotic programs in diverse species.[1-5] Caspase inhibitors, including (1) the aldehyde and

[1] S. Kumar, *Trends Biochem. Sci.* **20,** 198 (1995).
[2] M. Whyte, *Cell Biol.* **6,** 245 (1996).
[3] N. A. Thornberry, A. Rosen, and D. W. Nicholson, Control of apoptosis by proteases, *in* "Apoptosis: Pharmacological Implications and Therapeutic Opportunities," Vol. 41, pp. 155–177. ■■■, ■■■, 1997.

fluoro- or chloromethyl ketone versions of synthetic peptide substrates and (2) the viral caspase inhibitors CrmA and p35, have been powerful tools for the investigation of cell death programs.[6–17] Several human proteins from the inhibitor of apoptosis family (IAPs) were shown to directly inhibit specific caspases.[18–22] Thus, the IAPs represent the only known endogenous caspase inhibitors known in mammals. To date, five human IAPs have been identified: NAIP, cIAP1/HIAP-2/hMIHB, cIAP2/HAIP-1/hMIHC, XIAP/hILP, and survivin.[23–25] IAPs were first discovered in insect viruses (baculo-viruses) by Miller and colleagues, where they were found to be necessary for suppression of the insect host cell death response (reviewed in Ref. 26). Interestingly, ectopic expression of some of the baculoviral IAPs was reported to block cell death induced by a variety of apoptotic stimuli in cultured cells from humans, implying that IAPs target an evolutionarily conserved step(s) in the cell death pathway.[23,24] Subsequent studies found that XIAP, cIAP1 and cIAP2 can bind and potently inhibit specific caspases reviewed in Ref. 26. XIAP, cIAP1, and cIAP2 were shown to exhibit high specificity toward caspases 3, 7, and 9 but not caspases 1, 6, 8, and 10 or the *Caenorhabditis elegans* caspase CED3.[18–20] Similar results have subse-

---

[4] G. S. Salvesen and V. M. Dixit, *Cell* **91**, 443 (1997).

[5] N. A. Thornberry and Y. Lazebnik, *Science* **281**, 1312 (1998).

[6] A. Sugimoto, P. D. Friesen, and J. H. Rothman, *EMBO J.* **13**, 2023 (1994).

[7] B. A. Hay, T. Wolff, and G. M. Rubin, *Development* **120**, 2121 (1994).

[8] T. Komiyama, *et al., J. Biol. Chem.* **269**, 19331 (1994).

[9] V. Gagliardini, *et al., Science* **263**, 826 (1994).

[10] M. Tewari, *et al., Cell* **81**, 801 (1995).

[11] N. J. Bump, *et al., Science* **269**, 1885 (1995).

[12] D. R. Beidler, M. Tewari, P. D. Friesen, G. Poirier, and V. M. Dixit, *J. Biol. Chem.* **270**, (1995).

[13] M. Tewari, W. G. Telford, R. A. Miller, and V. M. Dixit, *J. Biol. Chem.* **270**, 1 (1995).

[14] S. M. Srinivasula, M. Ahmad, T. Fernandes-Alnemri, G. Litwack, and E. S. Alnemri, *Proc. Natl. Acad. Sci. U.S.A.* **93**, 14486 (1996).

[15] Q. Zhou, *et al., J. Biol. Chem.* **272**, 7797 (1997).

[16] K. Orth and V. M. Dixit, *J. Biol. Chem.* **272**, 8841 (1997).

[17] Q. Zhou, *et al., Biochemistry* **37**, 10757 (1998).

[18] Q. L. Deveraux, R. Takahashi, G. S. Salvesen, and J. C. Reed, *Nature (London)* **388**, 300 (1997).

[19] N. Roy, Q. L. Deveraux, R. Takashashi, G. S. Salvesen, and J. C. Reed, *EMBO J.* **16**, 6914 (1997).

[20] Q. L. Deveraux, *et al., EMBO J.* **17**, 2215 (1998).

[21] R. Takahashi, *et al., J. Biol. Chem.* **273**, 7787 (1998).

[22] I. Tamm, *et al., Cancer Res.* **59**, 5315 (1998).

[23] C. S. Duckett, *et al., EMBO J.* **15**, 2685 (1996).

[24] P. Liston, *et al., Nature (London)* **379**, 349 (1996).

[25] G. Ambrosini, C. Adida, and D. Altieri, *Nature Med.* **3**, 917 (1997).

[26] Q. L. Deveraux and J. C. Reed, *Genes Dev.* **13**, 239 (1999).

quently been obtained for *Drosophila* and baculoviral IAPs, demonstrating their ability to bind and inhibit some caspases but not others.

This chapter provides methods for the expression, purification, and use of recombinant IAPs as caspase inhibitors in assays using purified caspases and cytosolic extracts, which have been exploited as cell-free systems for studying apoptosis. Recombinant XIAP is used as an example; however, other IAPs such as cIAP1 and cIAP2 have been expressed and purified under similar conditions, although these recombinant IAPs are less potent caspase inhibitors relative to XIAP.

Methods

Expression and Purification of GST–XIAP

The XIAP-encoding cDNA (GenBank accession number U32974) is subcloned into pGEX4T-1 (Pharmacia, Piscataway, NJ). Plasmid pGEX4T-1-XIAP (conferring ampicillin resistance) is then transformed into *Escherichia coli* BL21(DE3) harboring the plasmid pT-Trx (conferring chloramphenicol resistance).[27] For expression and purification, inoculate a single colony into 5 ml of LB medium containing ampicillin (100 μg/ml) and chloramphenicol (35 μg/ml) and incubate overnight ($\sim$16 hr) at 30°. Using 3 ml of the overnight culture, inoculate 250 ml of LB containing ampicillin (100 μg/ml) and chloramphenicol (35 μg/ml) in a 2-liter Erlenmeyer flask. Incubate the culture at 37° with vigorous shaking until the optical density reaches 1.0 at 600 nm. Add 250 ml of LB medium ($\sim$22°) and isopropyl-β-D-thiogalactopyranoside (IPTG) to a final concentration of 50 μM to the culture and continue incubation at room temperature ($\sim$22°) with vigorous shaking for 16 hr. Bacterial cells are then centrifuged at $\sim$6000g for 15 min at 4° and the packed pellet is placed in a freezer at $-20°$.

All steps in the purification of glutathione S-transferase (GST)–XIAP are conducted at 4°. Relative to the pellet volume, 20 vol of lysis buffer [phosphate-buffered saline (PBS) containing 20 mM 2-mercaptoethanol (2-ME), 1 mM phenylmethylsulfonyl fluoride (PMSF), and lysozyme (100 μg/ml)] is used to suspend the bacterial pellet. Lysis is achieved by sonication for eight repeats of a 30-sec pulse cycle followed by 30 sec of cooling on ice. Typically we employ a 7-mm sonicator probe tip with an intensity setting of 20 on a 0–100 scale. The resulting lysate is clarified by centrifugation at 48,000g for 30 min, using a fixed-angle rotor (SS34; Sorvall, Newtown, CT) and 35-ml polypropylene centrifuge tubes. Transfer the supernatant to 50-ml Corning polypropylene conical centrifuge tubes containing a 1/20 vol of packed glutathione resin (preequilibrated with PBS

---

[27] T. Yamakawa, *et al., J. Biol. Chem.* **270,** 25328 (1995).

and 20 mM 2-ME). Incubate the lysate with the resin for 1 hr with gentle rotation. Pellet the glutathione–Sepharose by a 10-min centrifugation at 1000g in a swinging-bucket rotor and wash the resin twice with at least 20 vol of ice-cold PBS containing 20 mM 2-ME. Elute the GST–XIAP from the resin by incubation with 2 vol of elution buffer [50 mM Tris (pH 8.8), 10 mM reduced glutathione, 20 mM 2-ME] for 30 min at 4° with mild rotation. Remove the supernatant from the resin by centrifugation at 1000g in a swinging-bucket rotor and repeat the elution step with fresh elution buffer. Aliquots are then immediately frozen at −80°. Eluted protein can be analyzed by sodium dodecyl sulfate–polyacrylamide gel electrophoresis (SDS–PAGE) with staining by Gelcode (Pierce, Rockford, IL). A typical purification of GST–XIAP yields protein concentrations ranging from 0.2 to 0.8 mg/ml (3–12 μM) before concentration procedures. The resultant protein is maximally 20% homogeneous because of partial degradation products. Thrombin cleavage of the GST from the recovered GST–XIAP generally yields low amounts of XIAP because of nonspecific cleavage.

The GST–XIAP can be further purified by chromatography on a fast protein liquid chromatography (FPLC) MonoQ column (Amersham-Pharmacia). The MonoQ column (10/10) is equilibrated with 50 mM Tris at pH 8.8 with 20 mM 2-ME. The eluant from the glutathione–Sepharose resin is loaded onto the column. Elution is achieved with a linear gradient of 50 mM Tris at pH 8.8 with 20 mM 2-ME and 50 mM Tris at pH 8.8 with 20 mM 2-ME and 1 M NaCl. The GST–XIAP elutes at approximately 250 mM NaCl.

GST–XIAP eluted from the MonoQ column can be further purified on an 8-ml hydroxyapatite column. This column is equilibrated with 50 mM Tris at pH 7.5 with 20 mM 2-ME. After adding the GST–XIAP, the column is washed with 20 ml of the equilibrating buffer. GST–XIAP is eluted with a linear gradient of 50 mM Tris at pH 7.5 with 20 mM 2-ME to 500 mM potassium phosphate at pH 7.5 with 20 mM 2-ME. The desired protein elutes at approximately 65% of the gradient. The GST–XIAP is quite homogeneous as it elutes from this column but it continues to degrade.

Preparation and Use of Extracts Representing a Cell-Free Apoptotic System

Several "cell-free" apoptotic systems have previously been described.[18,28–32] This section provides a simple procedure that can be used

[28] S. J. Martin, et al., EMBO J. **15**, 2407 (1996).
[29] X. Liu, C. N. Kim, J. Yang, R. Jemmerson, and X. Wang, Cell **86**, 147 (1996).
[30] H. M. Ellerby, et al., J. Neurosci. **17**, 6165 (1997).
[31] D. Newmeyer, D. M. Farschon, and J. C. Reed, Cell **79**, 353 (1994).

for a variety of cultured cells and may be adaptable to certain tissue samples. Generally, at least four (150 × 25 mm) culture plates containing 60–80% confluent cells, such as 293 epithelial cells (~1 × 10^8), are removed by gentle scraping with a rubber policeman and then pelleted by centrifugation at 400g for 5 min, washed with PBS, and centrifuged again to pellet the cells. This usually produces a packed cell pellet volume of 500 μl, which can be manipulated in a 1.6-ml polypropylene Eppendorf-type centrifuge tube, using a microcentrifuge. Alternatively, nonadherent cells such as Jurket T cells can be employed (2–3 × 10^8). The procedure can be scaled accordingly, depending on the amount of cell extract required. Gently suspend the cell pellet in at least 5 vol (relative to the cell pellet) of ice-cold hypotonic lysis buffer [20 mM HEPES (pH 7.2), 10 mM KCl, 1.5 mM MgCl$_2$, 1 mM EDTA, 1 mM DTT, and 0.1 mM PMSF] and wash once in the same buffer. Centrifuge at 400g for 5 min. Suspend the cells in 2–3 vol (relative to the packed cell pellet) of hypotonic lysis buffer and incubate on ice for 20 min. Disrupt the cells by passage through a 26-gauge needle (~15 times for larger 293 cells and ~30 times or more for smaller cells such as Jurkat T cells) attached to a syringe. Cell lysis can be monitored by light microscopy. A Dounce homogenizer B-type pestle or similar device may also be used; however, for small volumes (<500 μl) a 1-ml syringe may be more efficient with less loss of material. Clarify the cell extracts by centrifugation at 16,000g for 30 min. Extracts should be adjusted to ~10 mg of total protein per milliliter and can be used immediately or stored at −80°.[15]

For initiating caspase activation, either 10 μM horse heart cytochrome c (Sigma, St. Louis, MO) together with 1 mM dATP or 100 nM purified recombinant caspase 8 or granzyme B, can be added to cell extracts.[18,28,29,33,*] The reactions are then incubated at 30–37°. Samples of 1–2 μl (10–20 μg) may be removed and assayed for caspase activity by using chromogenic or fluorogenic peptide substrates such as DEVD–AFC. Under the conditions described above, extracts from many cell types (e.g., 293, HeLa, and Jurkat T) achieve maximal DEVD–AFC cleavage activity 10–15 min at 30°. Samples can be assayed in a 96-well format with a fluorometer [Molecular Devices (Menlo Park, CA) or similar model]. Usually a 10-μg

[32] S. J. Martin, et al., EMBO J. **14,** 5191 (1995).
[33] L. T. Quan, et al., Proc. Natl. Acad. Sci. U.S.A. **93,** 1972 (1996).
[34] M. H. Cardone, et al., Science **282,** 1318 (1998).
[35] D. H. Burgess, et al., Oncogene **6,** 256 (1999).
* Some tumor cell lines have defects in the regulation or components of their cytochrome c-mediated caspase activation machinery.[22,34] Also, some normal tissues such as skeletal muscle do not possess competent cytochrome c pathways.[35] Extracts produced from these sources will not support cytochrome c-mediated activation of caspase.

sample is sufficient for measurement of the hydrolysis of DEVD–AFC under the conditions described. Reactions are measured in a final volume of 100 μl of caspase buffer [50 mM HEPES (pH 7.4), 100 mM NaCl, 10% (w/v) sucrose, 1 mM EDTA, 0.1% (w/v) CHAPS, and 10 mM DTT] containing 100 μM fluorogenic peptide substrate such as DEVD–AFC (Enzyme Systems, Dublin, CA). Thus many reactions can be measured simultaneously. Aliquots of 2–3 μl ($\sim$25 μg) from extracts may also be used for Western blot analysis to correlate caspase processing with peptide cleavage activities. Purified active caspases or samples from intact cells can be measured in a manner similar to that described above. (See the following references for further details: Refs. 18–20, 28, 30, and 32.)

Use of GST–IAPs as Caspase Inhibitors

Purified recombinant IAPs can be used as inhibitors of caspases in extracts or against purified recombinant caspases.[18–21,26,36] For inhibition of recombinant caspases, typical reactions for measuring caspase 3 or 7 contain 100–300 pM purified recombinant caspase in 100 ml of caspase buffer containing 100 μM DEVD–AFC as a substrate and the reactions are carried out at 37°.[33] Direct addition of a 10-fold molar excess of GST–XIAP results in significant inhibition of caspase 3 and 7 activity as shown in Fig. 1, whereas addition of recombinant GST or denatured GST–XIAP (DN-XIAP) has little effect. Employing similar assays, inhibition constants (K_i) for XIAP were estimated at 0.2 and 0.7 nM against caspase 3 and caspase 7, respectively.[18] Up to a 50-fold molar excess of recombinant XIAP has no significant effect on caspase 1, 6, 8, or 10 activities, indicating XIAP is specific for its caspase targets. Similar results were obtained for recombinant cIAP1 and cIAP2, although they are less potent caspase inhibitors at least as recombinant molecules.[18–21]

Likewise, recombinant IAPs may be used as caspase inhibitors in cell-free apoptotic systems. Caspase activation in cell extracts may be achieved by several methods as described above and elsewhere.[18,28,29,33,37] Direct addition of 0.2 μM recombinant GST–XIAP prepared as described above to cell-free extracts severely suppresses the ability of these extracts to be activated with 10 μM cytochrome c and 1 mM dATP or 100 nM active caspase 8 with respect to the cleavage of DEVD–AFC. It is critical that the IAP be added in excess relative to the caspase target. The caspases and the order of caspase activation can vary depending on the stimulus used. However, we have found in extracts prepared as described above that on average the concentration of caspases such as caspases 3, 7, and 9

[36] J. M. Jurgensmeier, et al., Proc. Natl. Acad. Sci. U.S.A. **95,** 4997 (1998).
[37] M. Muzio, G. S. Salvesen, and V. M. Dixit, J. Biol. Chem. **272,** 2952 (1997).

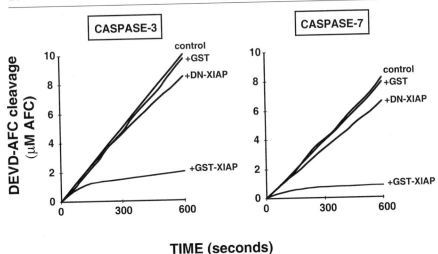

TIME (seconds)

Fig. 1. Inhibition of caspases 3 and 7 by recombinant GST–XIAP. Recombinant active caspases 3 (A) and 7 (B) were incubated with 100 μM DEVD-AFC in the presence or absence of recombinant GST–XIAP, GST, or denatured GST–XIAP (DN-XIAP). Reactions were incubated at 37° for the indicated times with continuous monitoring of AFC. Caspase 3 (A) and caspase 7 (B) were employed at 100 and 300 pM, respectively, using a 10-fold molar excess of the recombinant proteins. Denatured XIAP (DN-XIAP) was prepared by incubating the recombinant GST–XIAP preparation at 80° for 10 min. The sample was then centrifuged at 16,000g for 10 min and used as a control (DN–XIAP). Other recombinant GST fusion proteins were also used as controls (not shown).

is approximately 10–100 nM. Thus 200 nM XIAP is a reasonable starting concentration for inhibition assays in cell lysates. Recombinant GST or other proteins purified in a similar manner may be added as experimental controls. Similar experiments were used to dissect both the caspase targets of XIAP and to order caspases involved in both the Fas- and Bax-induced apoptotic programs.[20]

Recombinant GST fusion proteins can also be recovered from cell lysates to determine interacting proteins as previously described.[18,20] Briefly, 5 μg of recombinant GST–XIAP is added to 100 μl of cell lysate and can be recovered under a variety of experimental conditions by addition of 20 μl of packed glutathione beads, which were previously equilibrated in the same buffer in which the cells were lysed. After incubation for 30 min at 4° on ice, the glutathione beads are collected by centrifugation and washed three times in 1 ml of ice-cold lysis buffer. Proteins can be eluted from the beads by incubation at 80° for 10 min in 40 μl of sodium dodecyl sulfate (SDS) sample buffer before electrophoresis.

*Determination of Inhibition Constants (K_i) for Caspase Suppression
by IAPs*

Inhibition rates and equilibria are determined by measurement of progress curves, where increasing concentrations of recombinant inhibitor, such as GST–XIAP, are added to individual reactions in which the caspase enzyme is held at the same concentration.[15,17,18,38] It is important that the inhibitor be in a molar excess relative to the enzyme to render the reaction pseudo-first order for convenience of analysis. For example, if caspase 3 is used at 100 pM then the XIAP concentration might range from 5- to 20-fold molar excess (0.5–2 nM). Generally the enzyme (caspase) and substrate (100 μM DEVD–AFC) are added first and the reactions are initiated by the addition of inhibitor (XIAP). We generally determine inhibition constants (K_i values) in a final volume of 100 μl of caspase buffer as described above. Calculations are based on a single-step reaction scheme where K_i(app) is given by $[I]/(V_0/V_i - 1)$, where $[I]$ is the inhibitor concentration, V_i is the steady state of substrate hydrolysis in the presence of I, and V_0 is the uninhibited rate. The overall equilibrium constant can be obtained by taking into account the substrate concentration S and the K_m of enzyme for substrate: $K_i = K_i(\text{app})/(1 + S/K_m)$.

Concluding Remarks

By employing the methods described here, recombinant XIAP and other IAPs can be used as tools for blocking specific caspases and thereby dissecting apoptotic pathways. With estimated inhibitory constants in the picomolar range, recombinant XIAP is a useful reagent for inhibiting specific caspases 3, 7, and 9. Moreover, because XIAP is a natural caspase inhibitor, it is of biological interest. In this regard, these assays provide the basis for screening mutants of both the IAPs and of caspases for structure–function analysis of their interactions. In addition, these methods can be used for the identification of antagonists of the IAPs. For example, GST–XIAP can be incubated with small molecules or peptides before addition to caspase 3 or 7. GST–XIAP alone strongly inhibits caspase 3 or 7; however, if incubated with an XIAP antagonist, caspase activity is not inhibited.

Acknowledgments

Quinn L. Deveraux is a fellow of the Leukemia Society of America. We thank the NIH (GM-60554) for their generous support.

[38] H. R. Stennicke and G. S. Salvesen, *Biochim. Biophys. Acta* **1387,** 17 (1998).

[14] Monitoring Activity of Caspases and Their Regulators in Yeast Saccharomyces cerevisiae

By CHRISTINE J. HAWKINS, SUSAN L. WANG, and BRUCE A. HAY

Introduction

Caspases are a family of site-specific proteases that play important signaling and effector roles in most apoptotic death pathways.[1] Caspases are made as zymogens and become activated after cleavage. Caspase activity is regulated by both activators and inhibitors. Yeast provide a useful system in which to characterize the function of caspases and their regulators because yeast appear to lack endogenous caspases as well as other components of the apoptotic machinery. Also, yeast are genetically the most tractable eukaryote with which to work. Finally, transformation occurs at a high frequency, making them suitable for library screening or large-scale mutagenesis. In addition to caspases, yeast systems have also been used to analyze other cell death pathway components.[2-11] These approaches are discussed in [27] and [28].

Yeast, Media, and Plasmids

Saccharomyces cerevisiae W303α (*MATα can1-100 leu2-3,112 his3-11,-15 trp1-1 ura3-1 ade2-1*) is a convenient strain for most experiments.

Dropout medium: Autoclave yeast nitrogen base without amino acids (6.7 g/liter; Difco, Detroit, MI) (with agar at 20 g/liter for solid medium). After sterilization, add 0.1 vol of a 10× filter-sterilized

[1] N. A. Thornberry and Y. Lazebnik, *Science* **281**, 1312 (1998).
[2] T. Sato, M. Hanada, S. Bodrug, S. Irie, N. Iwama, L. H. Boise, C. B. Thompson, E. Golemis, L. Fong, H.-G. Wang, and J. C. Reed, *Proc. Natl. Acad. Sci. U.S.A.* **91**, 9238 (1994).
[3] M. Hanada, C. Aimé-Sempré, T. Sato, and J. C. Reed, *J. Biol. Chem.* **270**, 11962 (1995).
[4] W. Greenhalf, C. Stephan, and B. Chaudhuri, *FEBS Lett.* **380**, 169 (1996).
[5] B. Ink, M. Zornig, B. Baum, N. Hajibagheri, C. James, T. Chittenden, and G. Evan, *Mol. Cell. Biol.* **17**, 2468 (1997).
[6] J. M. Jurgensmeier, S. Krajewski, R. Armstrong, G. M. Wilson, T. Oltersdorf, L. C. Fritz, J. C. Reed, and S. Ottilie, *Mol. Biol. Cell* **8**, 325 (1997).
[7] H. Zha and J. C. Reed, *J. Biol. Chem.* **272**, 31482 (1997).
[8] W. Tao, C. Kurshner, and J. Morgan, *J. Biol. Chem.* **272**, 15547 (1997).
[9] Q. Xu and J. C. Reed, *Mol. Cell* **1**, 337 (1998).
[10] C. James, S. Gschmeissner, A. Fraser, and G. Evan, *Curr. Biol.* **7**, 246 (1997).
[11] W. Tao, D. W. Walke, and J. I. Morgan, *BBRC* **260**, 799 (1999).

stock of amino acid mix (L, W, H, U; Sigma, St. Louis, MO) with the appropriate additional amino acids or uracil [i.e., W, H, U (20-mg/liter final), L (30-mg/liter final)] and sugar source, either glucose or galactose, to 2% (w/v). For faster growth on inducing medium, raffinose can also be added (1%, w/v). For induction of copper promoters, add copper sulfate (usually 3 μM).

Complete medium: Autoclave a solution of 1% (w/v) yeast extract (Difco) and 2% (w/v) Bacto-Peptone (Difco) (with agar at 20 g/liter for solid medium). Add either glucose (2%, w/v) or galactose (2%, w/v) after sterilization to give YPglc or YPgal. For induction of copper promoters, add copper sulfate (usually 3 μM).

Low-copy yeast plasmids based on the pRS31X series[12] are convenient vectors for protein expression. Versions of these containing either the *GAL*1/10, *ADH*1, or *CUP*1 promoters to drive gene expression are available.[13]

Small-Scale Yeast Transformation

Yeast can be easily and efficiently transformed by incubation in lithium acetate followed by heat shock. A convenient protocol, based on Ito *et al.*[14] provides enough competent yeast for 10 small-scale transformations. The volumes of the starter culture and expanded culture can be increased for larger numbers of transformations. This method can be efficiently used for introducing up to three plasmids with different nutritional selections simultaneously. To generate transformants bearing four or more plasmids, two plasmids can be initially introduced using this protocol. A starter culture from one double transformant is grown in selective medium and then expanded into complete medium and transformed with the subsequent plasmids as described below.

1. Inoculate 5 ml of YPglc with a single colony of the parent yeast strain and grow to stationary phase: for 12–16 hr at 30° with shaking at 230 rpm in YPglc. *Note:* If starting with previously transformed yeast, inoculate a transformant colony into the appropriate dropout medium with glucose. These cultures require longer growth, about 18 hr.

2. Add the starter culture to 30 ml of YPglc, and grow for another 3 hr.

3. Centrifuge at 4000 rpm for 5 min, and discard the supernatant.

[12] R. S. Sikorski and P. Hieter, *Genetics* **122,** 19 (1989).
[13] C. J. Hawkins, S. L Wang, and B. A. Hay, *Proc. Natl. Acad. Sci. U.S.A.* **96,** 2885 (1999).
[14] H. Ito, Y. Fukada, K. Murata and A. J. Kimura, *Bacteriology* **153,** 163 (1983).

4. Resuspend in 35 ml of LiAc/TE [100 mM lithium acetate, 10 mM Tris-HCl (pH 7.5), 1 mM EDTA].

5. Centrifuge as described above; discard the supernatant.

6. Resuspend the pellet in 1 ml of LiAc/TE.

7. For each transformation, mix 10 μl of carrier DNA (boiled, chilled salmon sperm DNA, 10 mg/ml) with about 0.3 μg of each plasmid to be transformed. Add 100 μl of the yeast solution and 600 μl of PEG/LiAc/TE [40% (w/v) polyethylene glycol 3350, LiAc/TE] per tube. Mix by inversion.

8. Incubate at 30° for 30 min.

9. Add 70 μl of dimethyl sulfoxide (DMSO) per tube; mix by inversion.

10. Incubate at 42° for 6 min.

11. Pellet the yeast by centrifuging at 14,000 rpm for about 15 sec (start the microcentrifuge and stop it when it gets to full speed).

12. Discard the supernatant and resuspend the pellet in 70 μl of TE [10 mM Tris-HCl (pH 7.5), 1 mM EDTA] to pellet; resuspend and plate on selective plates.

13. Incubate the plates at 30° for 2–3 days, until colonies are visible.

Monitoring Caspase Activity in Yeast Using a Transcription-Based Reporter

A number of techniques for visualizing the activity of site-specific proteases have been developed based on assays for the presence of a specific protease cleavage event.[15–19] Two approaches have been developed to monitor caspase activity in yeast: one based on cleavage-dependent activation of a transcription-based reporter, and a second based on caspase-dependent yeast cell killing.[13,20] The essential feature of a transcription-based reporter is that caspase cleavage leads to transcriptional activation of a reporter such as the bacterial *lacZ* gene.[13] One version of such a reporter involves the use of a chimeric protein (known as CLBDG6) that consists of three parts: a type 1 transmembrane protein (human CD4) truncated to remove the cytoplasmic domain, then a linker containing various caspase target

[15] J. O. McCall, S. Kadam, and L. Katz, *Biotechnology* **12**, 1012 (1994).

[16] H. J. Sices and T. M. Kristie, *Proc. Natl. Acad. Sci. U.S.A.* **95**, 2828 (1998).

[17] I. Stagljar, C. Korostensky, N. Johnsson, and S. te Heesen, *Proc. Natl. Acad. Sci. U.S.A.* **95**, 5187 (1998).

[18] X. Xu, A. L. Gerard, B. C. Huang, D. C. Anderson, D. G. Payan, and Y. Luo, *Nucleic Acids Res.* **26**, 2034 (1998).

[19] S. Mizukami, K. Kikuchi, T. Higuchi, Y. Urano, T. Mashima, T. Tsuruo, and T. Nagano, *FEBS Lett.* **453**, 356 (1999).

[20] J. J. Kang, M. D. Schaber, S. M. Srinivasula, E. S. Alnemri, G. Litwack, D. J. Hall, and M.-A. Bjornsti, *J. Biol. Chem.* **274**, 3189 (1999).

sites, followed by a transcription factor (LexA-B42). When this molecule is expressed in yeast that also carry a plasmid in which a LexA-dependent promoter drives *lacZ* expression, the transcription of *lacZ,* and thus β-galactosidase activity, depends on cleavage events that liberate the transcription factor from the membrane.[13] This system has been developed and used for detection of caspase activity, but in theory should be applicable to any protease for which the cleavage/recognition site is known. Because the screens are function based, relying only on the ability of an expressed protein to cleave at the caspase target site linker, it enables cloning of proteases that recognize caspase cleavage sites but that are unrelated by sequence to caspases (such as the serine protease granzyme B).

Caspase-Dependent Toxicity

Enforced expression of some caspases does kill yeast.[13,20] This discovery formed the basis of a second system to analyze the activity of caspases and their inhibitors, and is discussed below. However, to overcome this complication in a reporter-based assay, replica plating is used. Yeast are grown initially on noninducing plates to allow growth of yeast carrying caspase expression plasmids, and the colonies are then filter-lifted onto inducing plates. After detection of the reporter (β-galactosidase) activity on the filter, the corresponding original yeast colony from the noninducing plate can be located and grown for analysis. β-Galactosidase is a useful reporter for these assays because it is a stable protein and is enzymatically active in the presence of active caspases in dying cells.[21]

Nonspecific Cleavage

For the caspase reporter system to be effective the chimeric transmembrane protein that constitutes the caspase substrate must be translocated efficiently to the membrane and remain intact in the absence of introduced caspases. Background cleavage or failure to translocate can be measured simply by monitoring the level of β-galactosidase produced (as described below) in cells that contain only the transmembrane reporter and the LexA-responsive reporter plasmid. If background is a problem it can be dealt with to some extent by lowering the level of reporter expression through the use of variable-level inducible promoters, such as the copper-inducible *CUP*1 promoter.

---

[21] M. Miura, H. Zhu, R. Rotello, E. A. Hartwieg, and J. Yuan, *Cell* **75,** 653 (1993).

CD4   cleavage sites   LexA
*...CVRCRHRR*DEVDG-WEHDG-IEHDG-IETDG-DEHDG-DQMD G*TMKALTARQQ...*

FIG. 1. The amino acid sequence of a target site linker region that incorporates consensus sites for all caspases for which specificity information is available, and the flanking CD4 and LexA sequences.

Target Site Linker

To maximize the probability of being able to detect an active caspase in yeast the reporter should contain caspase target sites that bracket the specificities of known caspases. Figure 1 shows the amino acid sequence of a target site linker region that incorporates consensus sites for all caspases for which specificity information is available[22] and the flanking CD4 and LexA sequences. This linker contains the four known caspase consensus target sites, a granzyme B site, as well as the pseudosubstrate cleavage site for baculovirus p35, a broad-specificity caspase inhibitor. This combination of target sites should be cleaved by all known caspases, and is therefore likely to be recognized by other, novel caspases. Yeast proteins can undergo ubiquitin-mediated degradation, with the half-life of the protein determined largely by the amino-terminal residue.[23] To ensure that the cleaved LexA-B42 transcription factor does not undergo ubiquitin-dependent degradation, a glycine residue, which acts as a stabilizing residue in the N-end rule pathway of protein degradation in yeast, was introduced after each cleavage site.

Weeding out False Positives

Two major classes of false positives are library proteins that activate transcription by binding the *lexA* operator sequences directly, and proteases that cleave the reporter at noncaspase target sites. Either one of these events will induce *lacZ* expression. These false positives can be eliminated through the use of a modified version of the membrane reporter protein, a "false-positive reporter," in which the essential P_1 aspartate residues of the caspase target sites are replaced by glycines. Such a change should prevent caspase-mediated proteolysis of the substrate. Library plasmids isolated during a caspase screen can be transformed into yeast bearing the *lacZ* reporter and this false-positive reporter and the resulting transformants assayed for β-galactosidase activity. Plasmids that activate *lacZ*

---

[22] N. A. Thornberry, T. A. Rano, E. P. Peterson, D. M. Rasper, T. Timkey, M. Garcia-Calvo, V. M. Houtzager, P. A. Nordstrom, S. Roy, J. P Vaillancourt, K. T. Chapman, and D. W. Nicholson, *J. Biol. Chem.* **272,** 17907 (1997).
[23] A. Varshavsky, *Proc. Natl. Acad. Sci. U.S.A.* **93,** 12142 (1996).

expression in this assay are unlikely to encode proteases with specificity for caspase target sites.

Library Screening for Caspase-Like Proteases

To screen for proteases that cleave caspase target sites, library plasmids in which insert expression is driven by the inducible *GAL* promoter are transformed into caspase reporter yeast that carry a plasmid with a *GAL*-driven version of the fusion protein reporter and the LexA-responsive transcriptional cassette, with the following modifications. This technique can also be used to assay reporter activation in single yeast clones streaked onto plates.

1. After the heat shock step, resuspend the yeast in 5 ml of TE and plate on selective plates with glucose at 200 μl/15-cm plate. Also plate 100 μl of 1:10 and 1:100 dilutions to determine the transformation frequency.

2. After about 36 hr, when small colonies are visible, lay nitrocellulose filters carefully on the plates, and punch asymmetric holes through the filter and the agar using needles, for subsequent orientation and colony identification.

3. Place the filters yeast side up on YPgal plates, and incubate at 30° for 18 hr.

4. Submerge each filter in liquid nitrogen until bubbling stops. Remove and thaw, yeast side up.

5. Place the filters yeast side up on filter paper that has been soaked in stain solution: 15 μl of 5-bromo-4-chloro-3-indolyl-β-D-galactopyranoside (X-Gal) [20 mg/ml in dimethyl formamide] and 2.5 μl of 2-mercaptoethanol (2-ME) per milliliter of Z buffer (60 mM Na$_2$HPO$_4$, 40 mM NaH$_2$PO$_4$, 10 mM KCl, 1 mM MgSO$_4$, pH 7.0).

6. Incubate the filters in a dark humidified chamber at 37° for 4 to 20 hr.

7. Line up blue colonies on the stained filters with the original plates, pick the corresponding original colonies and streak out on fresh plates, and repeat the induction and staining process to clone positive transformants.

Liquid β-Galactosidase Assays

Although less sensitive than X-Gal assays, liquid β-galactosidase assays using *o*-nitrophenyl-β-D-galactopyranoside (ONPG) provide a more quantitative estimate of *lacZ* reporter gene activity.

1. Inoculate colonies to be assayed (in triplicate) into 2 ml of noninducing, selective medium and grow for 14–18 hr at 30°, with shaking.

2. Centrifuge the cultures at 14,000 rpm for 15 sec, and resuspend in TE.

3. Centrifuge again, resuspend at 1:10 dilution (in 2 ml) in selective inducing medium (i.e., with galactose and any other inducers such as CuSO$_4$).

4. Incubate at 30°C, with shaking, for 10 hr.

5. Centrifuge the yeast, resuspend in Z buffer, and recentrifuge.

6. Resuspend in 400 μl of Z buffer, and measure the OD$_{600}$.

7. Place 100 μl of each sample into a fresh tube and snap freeze in liquid nitrogen to disrupt the cells. Thaw by moving the tube to a 37° water bath for 1 min.

8. Add 700 μl of Z buffer with 2-ME (0.27%, v/v) and 160 μl of ONPG (4 mg/ml in Z buffer, made fresh before the start of the experiment); start the timer.

9. Incubate the tubes at 30° until yellow color develops.

10. When the solution turns yellow, add 400 μl of 1 M Na$_2$CO$_3$ and record the time.

11. When all reactions have been stopped, centrifuge all the tubes (10 min, 14,000 rpm) to pellet the contents.

12. Carefully transfer the top 1 ml from each tube to a clean tube and measure the OD$_{420}$.

13. Correct for the different number of yeast cells in the different samples by calculating β-galactosidase units[24,25] as follows:

$$\beta\text{-Galactosidase units} = \frac{2500 \times \text{OD}_{420}}{t \times \text{OD}_{600}}$$

where t equals the time until the yellow color appears.

Variations on Transcription-Based Caspase Reporter System

Screening for Caspase Activators

The simplest version of the reporter system permits the detection of autoactivating caspases. Some caspases do not autoactivate at the levels expressed in yeast, and are thus unlikely to be detected unless an activating agent is also provided.[13,20] Caspase 9 is an example of such a caspase. When expressed alone, caspase 9 does not activate the reporter. However, coexpression of an activated form of Apaf-1, which binds to and promotes caspase 9 activation, results in caspase 9-dependent reporter activation.[13]

[24] J. H. Miller, "Experiments in Molecular Genetics." Cold Spring Harbor Laboratory, Cold Spring Harbor, New York, 1972.

[25] J. H. Miller, in "A Short Course in Bacterial Genetics," p. 4. Cold Spring Harbor Laboratory Press, Cold Spring Harbor, New York, 1972.

Thus the reporter system described above could be used to screen for activating proteins.

Cell death pathways often involve caspase cascades, in which upstream family members cleave and activate downstream caspases. Expression of downstream caspases in yeast does not lead to reporter activation or (as described below) cell killing.[13,20] This failure to activate is probably because these caspases lack the ability to autoactivate in yeast.[20] It may be possible to screen for caspases that do not autoactivate in yeast by providing a known caspase that can promote activation. Clearly for such a system to be effective, the known "activator" caspase must not activate the reporter by itself. It may be possible to achieve a low background of reporter activation by the activator caspase by expressing it at low levels, or by altering cleavage sites in the reporter to remove any recognized by the activator caspase.

The unprocessed forms of many proteases contain sequences that impair their ability to autoactivate. Some caspases, such as caspase 7, are translated with a prodomain that impairs their ability to autoactivate.[26,27] Granzyme B, an unrelated serine protease that shares target site similarities with some members of the caspase family,[22] is also expressed as an inactive precursor with an amino dipeptide whose removal leads to enzymatic activity. It may be possible to screen for active forms of proteases that contain such inactivating prosequences by screening libraries that are unlikely to contain full-length coding sequences. These can be generated by using random primers rather than oligo(dT) during the initial cDNA synthesis.

Substrate Libraries

The transcription-based caspase reporter system described above can also be modified to screen for substrates of known caspases by replacing the target site linker in CLBDG6 with cDNA fragments from random-primed libraries. It should also be possible to carry out screens for preferred peptide substrates by introducing a randomized sequence into the target site linker region. False positives arising from cleavage by yeast proteases can be identified by virtue of their ability to activate the reporter in the absence of an introduced protease. Other false positives can be identified by virtue of their ability to cleave and activate a false-positive reporter that consists of CD4 and LexA-B42 sequences without an intervening target site linker.

[26] H. J. Duan, A. M. Chinnaiyan, P. L. Hudson, J. P. Wing, W. W. He, and V. M. Dixit, *J. Biol. Chem.* **271,** 1621 (1996).
[27] J. A. Lippke, Y. Gu, C. Sarnecki, P. R. Caron, and M. S.-S. Su, *J. Biol. Chem.* **271,** 1825 (1996).

Identifying Caspase Inhibitors

Known inhibitors of caspase activity should reduce reporter activity in the transcription-based reporter system described above. Baculovirus p35, a pancaspase inhibitor, and *Drosophila* and mammalian members of the IAP family of caspase inhibitors are in fact able to inhibit reporter expression, but this inhibition is incomplete and does not create a strong enough differential signal to be useful for screening purposes.[13] Caspase inhibitors can be identified, however, by virtue of their ability to inhibit a caspase-dependent growth arrest or cell death phenotype seen when some caspases are expressed at high levels.

Expression of many caspases that autoactivate results in lethality in yeast. However, coexpression of a caspase inhibitor rescues yeast viability and growth, permitting colony formation.[13,20,28] This fact creates a powerful screen for caspase inhibitors, because in the presence of an active caspase only those cells that also express a caspase inhibitor will produce colonies. False positives can arise in this sort of screen for several reasons: (1) the yeast are mutant in genes required to induce expression from the *GAL* promoter, which drives expression of the caspase; and (2) the caspase expression plasmid has undergone a mutation that renders the caspase nonfunctional or that blocks caspase expression. These false positives can be eliminated by isolating the library plasmid, transforming it into fresh caspase-bearing yeast, and assessing growth of transformants on inducing plates.

Library Screening for Caspase Inhibitors

A variation of the small-scale transformation protocol (above) can be used for library screening for caspase inhibitors.

1. Grow yeast containing the caspase expression plasmid in 10 ml of selective medium with glucose, then expand into 70 ml of YPglc for 3 hr.

2. Transform the yeast, by the small-scale procedure, performing 10 transformations with 5 μg of each library plasmid.

3. After the heat shock and centrifugation, pool the yeast by resuspension in 250 ml of prewarmed YPglc.

4. Grow the transformed yeast for 4 hr at 30°, with shaking at 230 rpm.

5. To remove the glucose and amino acids, centrifuge the yeast and discard the supernatant. Resuspend the pellet in 250 ml of TE, recentrifuge, and resuspend in 10 ml of TE.

6. To determine the transformation frequency, plate 100 μl of a 1:10 and 1:100 dilution on dropout plates with glucose to select for cells that carry the library plasmid.

---

[28] P. G. Ekert, J. Silke, and D. L. Vaux, *EMBO J.* **18**, 330 (1999).

7. Plate the remainder of the suspension (200 μl/plate) on 15-cm drop-out plates with galactose (2%, w/v) and raffinose (1%, w/v).

8. Check the plates daily from the fourth day posttransformation for colony growth.

9. Streak colonies onto dropout medium plates with galactose/raffinose, then grow in 2 ml of liquid dropout selective medium with galactose/raffinose to obtain enough cells for plasmid DNA extraction and retesting.

Verification of Positive Clones

1. Streak colonies onto selective inducing plates. This step verifies that the colonies can survive and turn pink (indicating that they possess the *ade2* mutation that is present in the W303α strain). This step also selects for the library plasmid that encodes the caspase inhibitor, hopefully enriching for it over other library plasmids that may have been cotransformed.

2. Inoculate 2 ml of liquid selective inducing medium with a colony and grow for 14–18 hr at 30° with shaking.

3. Centrifuge the yeast for 15 sec at 14,000 rpm and resuspend in complete medium (YPglc).

4. Grow the yeast for 3 hr. This step increases the growth rate of the yeast and reduces the strength of the cell wall.

5. Pellet the yeast and discard the supernatant.

6. Add 200 μl of lysis buffer [2% (v/v) Triton X-100, 1% (w/v) sodium dodecyl sulfate (SDS), 100 mM NaCl, 10 mM Tris-HCl (pH 8.0), 1 mM EDTA, pH 8.0], 200 μl of phenol–chloroform–isoamyl alcohol (25:24:1, v/v/v) and 200 μl of glass beads.

7. Ensure that the tube is fully closed (and wear gloves). Vortex at top speed for 2 min.

8. Centrifuge for 5 min at 14,000 rpm.

9. Transfer the top layer to a fresh tube, add 20 μl of 3 M sodium acetate and 200 μl of isopropanol, and mix.

10. Precipitate by centrifugation (14,000 rpm, 10 min).

11. Wash the pellet in cold 70% (v/v) ethanol, centrifuge for 1 min at 14,000 rpm, and remove traces of ethanol.

12. Resuspend the pellet in 200 μl of TE.

13. Use 2 μl to transform *Escherichia coli*.

14. Isolate library plasmid DNA from the transformed bacteria. This can be done either by identifying colonies that contain library plasmids, using colony lifts and a probe for the plasmid auxotrophic marker, or by transforming bacteria that are auxotrophic for the plasmid-encoded nutritional marker. Once plasmid DNA is isolated from *E. coli* it is retransformed into yeast together with the caspase. Transformants are plated onto

selective noninducing medium. Colonies are then streaked onto selective inducing medium plates to determine if the presence of a library plasmid suppresses caspase-dependent death.

Obviously not all library plasmids that suppress caspase-dependent death need encode caspase inhibitors. For example, such plasmids might encode proteins that disrupt expression of the caspase or that suppress caspase-dependent death downstream of caspase cleavage. Other approaches, including *in vitro* assays for caspase inhibitor function using purified proteins, will be necessary to confirm that the library plasmid encodes a caspase inhibitor.

Screens for Molecules that Disrupt Productive Caspase–Caspase Inhibitor Interactions

Members of the IAP (inhibitor of apoptosis) family of proteins are the only known cellular caspase inhibitors.[29] Work in *Drosophila* shows that IAP function is essential for cell survival, and that an important mechanism for promoting cell death involves the expression of death-activating proteins such as Reaper, HID, and Grim, which disrupt IAP–caspase interactions.[30] Given the general conservation of cell death signaling mechanisms throughout evolution, it seems likely that proteins that induce cell death or sensitize cells to other death-inducing signals by disrupting productive IAP–caspase interactions also exist in mammals.

As a first step, the expression of the caspase inhibitor (e.g., the IAP) is first titrated such that a minimal level required for caspase inhibition and rescue from lethality is achieved. The copper-inducible CUP1 promoter is ideal for these purposes because different levels of activation can be achieved by varying the levels of copper in the medium. Cells that survive because they express a caspase inhibitor as well as a death-inducing caspase can then be used as a background against which to identify inhibitors of IAP function.

To carry out a screen for IAP inhibitors a replica-plating strategy must be used. Cells carrying caspase and IAP expression plasmids are grown on noninducing medium and transformed with library plasmids as described above. Colonies are then replica plated onto inducing medium and scored for growth over the next 3–4 days. Those colonies that fail to grow on inducing medium encode candidate inhibitors of IAP function. Retesting of the library plasmid can be carried out as described below, and illustrated in Fig. 2.

[29] L. K. Miller, *Trends Cell Biol.* **9,** 323 (1999).
[30] S. L. Wang, C. J. Hawkins, S. J. Yoo, H.-A. J. Muller, and B. A. Hay, *Cell* **98,** 453 (1999).

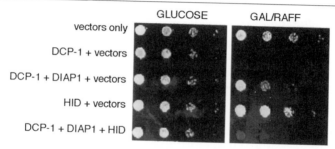

FIG. 2. *Drosophila* HID blocks the ability of DIAP1 to suppress caspase-dependent cell death. Library plasmids that are good candidates to encode specific inhibitors of IAP function have no growth phenotype when present in yeast in isolation (compare *HID + vectors* with *empty vectors*). DIAP1 protects cells from caspase-dependent cell death (compare *DCP-1 + vectors* with *DCP-1 + DIAP1 + vectors*). But HID suppresses colony formation when co-expressed in cells with DCP-1 and DIAP1 (compare *DCP-1 + DIAP1 + vectors* with *DCP-1 + DIAP1 + HID*).

1. Cells are transformed with different combinations of three expression plasmids with different auxotrophic markers that are either empty vectors, or that express a caspase (DCP-1), an IAP (DIAP1), or a putative IAP inhibitor, HID (see Fig. 2).

2. Single colonies carrying different combinations of plasmids are inoculated into selective medium containing 2% (w/v) glucose and grown at 30° overnight or to an OD_{600} of 1.0–2.0. It is likely that the densities of the cultures will be similar. If not, adjust with TE.

3. Spin down 300 μl of each culture (14,000 rpm, 1 min) and wash the cells free of glucose by resuspending the pellet in TE and spinning as described above. Resuspend the pellet in 200 μl of TE.

4. Serially dilute each sample 10-fold with TE four times. Dilutions may be performed directly in a 32-well frogger tray (Dan-Kar, Wilmington, MA; *http://www.dan-kar.com*). The volume of sample in each well is 100 μl.

5. Plate the dilutions onto selective inductive medium plates (i.e., selective plates containing 2% (w/v) galactose, if using a GAL-inducible promoter) and noninducing selective growth medium plates [2% (w/v) glucose] using a 32-spoke frogger (Dan-Kar). The volume of each spot is about 5–10 μl. Cell growth on the noninducing glucose medium provides an indication of cell number, while growth on inducing medium provides a reporter for cell survival.

6. Allow the spots to dry and incubate the plates at 30° until colonies are visible. Colonies will grow more slowly on galactose-containing medium.

An example of such an assay for an inhibitor of IAP function is shown in Fig. 2, using the *Drosophila* IAP inhibitor HID as an example.[30]

Acknowledgments

C.J.H. was supported by a fellowship from the Human Frontiers Science Program. B.A.H. is a Searle Scholar. This work was supported by grants to B.A.H. from the National Institutes of Health (GM057422-01), the Ellison Medical Foundation and the Burroughs Wellcome Fund (New Investigator Award in the Pharmacological Sciences).

Section IV

Cell-Free Systems for Monitoring Steps in Apoptosis Pathways

[15] *In Vitro* Assays for Caspase-3 Activation and DNA Fragmentation

By Xuesong Liu and Xiaodong Wang

Introduction

Apoptosis is a morphologically distinct form of cell death executed by an evolutionarily conserved biochemical pathway.[1,2] The best recognized biochemical hallmarks of apoptosis are the activation of caspases, condensation of chromatin, and fragmentation of genomic DNA into nucleosomal fragments.[3,4] Caspases are a family of cysteine proteases that cleave protein substrates specifically after aspartic acid residues.[5] Caspases involved in apoptosis exist as inactive zymogens that become activated proteolytically when cells are undergoing apoptosis. The activation of caspase cleaves many cellular proteins, leading to the final disassembly of the apoptotic cells.[6]

Because caspases are activated only when cells are undergoing apoptosis, *in vitro* assays that measure caspase activation have been used to detect apoptosis. Furthermore, similar assays can also be set up *in vitro* to study DNA fragmentation and chromatin condensation by adding normal nuclei to apoptotic extracts.

Methods that are described in this chapter include the preparation of HeLa cell 100,000 g supernatant (S-100),[7] preparation of cytochrome *c*-free cytosol from HeLa cells,[7] an *in vitro* assay for caspase-3 activation,[7] expression and purification of recombinant caspase-3 from bacteria,[8] preparation of hamster liver nuclei,[8] an assay for DNA fragmentation,[8] and chromatin condensation *in vitro*.[9]

---

[1] M. D. Jacobson, M. Weil, and M. C. Raff, *Cell* **88**, 347 (1997).

[2] D. L. Vaux, *Cell*, **90**, 389 (1997).

[3] W. D. Nicholson, A. Ali, N. A. Thornberry, J. P. Vaillancourt, C. K. Ding, M. Gallant, Y. Gareau, P. R. Griffin, M. Labelle, Y. A. Lazebnik, N. A. Munday, S. M. Raju, M. E. Smulsom, T.-T. Yamin, V. L. Yu, and D. K. Miller, *Nature* (*London*) **376**, 37 (1995).

[4] A. H. Wyllie, *Nature* (*London*) **284**, 555 (1980).

[5] E. S. Alnemri, D. J. Livingston, D. W. Nicholson, G. Salvesen, N. A. Thornberry, W. W. Wong, and J. Yuan, *Cell* **87**, 171 (1996).

[6] W. D. Nicholson and N. A. Thornberry, *Trends Biochem. Sci.* **257**, 299 (1997).

[7] X. Liu, C. N. Kim, J. Yang, R. Jemmerson, and X. Wang, *Cell* **86**, 147 (1996).

[8] X. Liu, H. Zou, C. Slaughter, and X. Wang, *Cell* **89**, 175 (1997).

[9] X. Liu, P. Li, P. Widlak, H. Zou, X. Luo, W. T. Garrard, and X. Wang, *Proc. Natl. Acad. Sci. U.S.A.* **95**, 8461 (1998).

Buffers

Buffer A: 20 mM HEPES-KOH (pH 7.4), 10 mM KCl, 1.5 mM MgCl$_2$, 0.5 mM EDTA, 0.5 mM EGTA, 1 mM dithiothreitol (DTT), 1 mM phenylmethylsulfonyl fluoride (PMSF)

Buffer B: 20 mM HEPES-KOH (pH 7.4), 1 M NaCl, 10 mM KCl, 1.5 mM MgCl$_2$, 0.5 mM EDTA, 0.5 mM EGTA, 1 mM DTT, 1 mM PMSF

Buffer C: 20 mM HEPES-KOH (pH 7.4), 10 mM KCl, 1.5 mM MgCl$_2$, 250 mM sucrose, 0.5 mM EDTA, 0.5 mM EGTA, 1 mM DTT, 1 mM PMSF

Buffer D: 10 mM HEPES-KOH (pH 7.6), 2.4 M sucrose, 15 mM KCl, 2 mM sodium EDTA, 0.15 mM spermine, 0.15 mM spermidine, 0.5 mM DTT, 0.5 mM PMSF

Buffer E: 10 mM piperazine-N,N'-bis(2-ethanesulfonic acid) (PIPES, (pH 7.4), 80 mM KCl, 20 mM NaCl, 5 mM sodium EGTA, 250 mM sucrose, 1 mM DTT

Buffer F: 100 mM Tris-HCl (pH 8.5), 5 mM EDTA, 0.2 M NaCl, 0.2% (w/v) sodium dodecyl sulfate (SDS), proteinase K (0.2 mg/ml)

Buffer G: 10 mM Tris-HCl (pH 7.5), 1 mM sodium EDTA, DNase-free RNase A (200 μg/ml; Worthington, Freehold, NJ)

TBE buffer: 90 mM Tris–borate, 2 mM EDTA

Preparation of Extracts

Preparation of S-100 from HeLa Cells

1. Grow human HeLa S3 cells as described.[10]
2. Collect the cells (5 × 10^5/ml) by centrifugation at 1800g for 10 min at 4° and wash once with ice-cold phosphate-buffered saline (PBS).
3. Resuspend the cell pellet in 5 vol of ice-cold buffer A supplemented with protease inhibitors [pepstatin A (5 μg/ml), leupeptin (10 μg/ml), aprotinin (2 μg/ml)].
4. Leave the resuspended cell pellet on ice for 15 min.
5. Disrupt the cells by homogenization (15 strokes with the B pestle in a 100-ml homogenizer: Kontes Glass, Vineland, NJ).
6. Centrifuge at 1000g for 10 min at 4°. Retain the supernatant.
7. Centrifuge the supernatant further at 10^5g for 1 hr at 4° in a Beckman (Fullerton, CA) SW 28 rotor.

---

[10] X. Wang, M. R. Briggs, X. Hua, C. Yokoyama, J. L. Goldstein, and M. S. Brown, *J. Biol. Chem.* **268**, 14497 (1993).

8. Store the resulting supernatant (S-100 fraction) (containing cytochrome *c*) at −80° in multiple aliquots and use it for the *in vitro* apoptosis assay.

Preparation of Cytochrome c-Free Cytosol from HeLa Cells

1. Grow human HeLa S3 cells as described.[10]
2. Collect the cells (5 × 10⁵/ml) by centrifugation at 1800*g* for 10 min at 4° and wash once with ice-cold PBS.
3. Resuspend the cell pellet in 5 vol of ice-cold buffer C supplemented with protease inhibitors [pepstatin A (5 μg/ml), leupeptin (10 μg/ml), aprotinin (2 μg/ml)].
4. Disrupt the cells by homogenization [three strokes in a 5-ml Wheaton (Millville, NJ) homogenizer, with the pestle polished with sand paper].
5. Centrifugate the lysate in a microcentrifuge for 5 min at 4°.
6. Centrifuge the supernatant at 10⁵*g* for 30 min in a tabletop ultracentrifuge (Beckman). The resulting supernatant, which contains little cytochrome *c*, is designated S-cytosol and used for the *in vitro* apoptosis assay.

Assay of Caspase-3 Activation *In Vitro*

In Vitro Translation of Caspase-3

1. Clone the polymerase chain reaction (PCR) fragment encoding amino acids 29–277 of hamster caspase-3[11] into *Nde*I and *Bam*HI sites of pET 15b vector (Novagen, Madison, WI).
2. Translate the resulting construct in a TNT T7 transcription/translation kit (Pharmacia, Piscataway, NJ) in the presence of [³⁵S]methionine according to the following formulation: 8 μl of TNT buffer, 4 μl of amino acid mixture (minus methionine), 4 μl of RNAsin, 100 μl of rabbit reticular lysate, 2 μl of T7 polymerase, 10 μg of plasmid DNA and 5 μl of [³⁵S]methionine.
3. Pass the translated protein through a 1-ml nickel affinity column (Qiagen, Chatsworth, CA) equilibrated with buffer A and wash the column with 10 ml of buffer A.
4. Elute the translated caspase-3 with buffer A containing 250 m*M* imidazole.

---

[11] X. Wang, J.-T. Pai, E. A. Wiedenfeld, J. C. Medina, C. A. Slaughter, J. L. Goldstein, and M. S. Brown, *J. Biol. Chem.* **270,** 18044 (1995).

Assay for dATP-Dependent Activation of Caspase-3

1. Incubate an aliquot of 3 μl of the *in vitro*-translated caspase-3 with 50 μg of HeLa cell S-100, or with 50 μg of HeLa cell S-cytosol plus 0.2 μg of cytochrome c, in the presence of 1 mM dATP, and 1 mM additional MgCl$_2$ at 30° for 1 hr in a final volume of 20 μl of buffer A.

2. Add 7 μl of 4× SDS sample buffer to each reaction.

3. Boil each sample for 3 min and load onto an SDS–15% (w/v) polyacrylamide gel.

4. Transfer the gel to a nitrocellulose filter and expose the filter to Kodak (Rochester, NY) X-Omat AR X-ray film for 16 hr at room temperature. See Fig. 1 for an example.

Comments: To see the results quickly, expose the filter to a phosphorimager plate for 30 min and visualize the results through a phosphorimager machine. Such a protocol has also been used to assay caspase-3 activation in S-100 prepared from human embryonic kidney 293 cells, human leukemia HL60 cells, and human germ cell tumor Tara-2 cells.

Assay for DNA Fragmentation Factor

*Expression and Purification of Recombinant His$_6$-Tagged
Activated Caspase-3*

1. Transform the plasmid carrying a gene encoding a fusion protein of six histidine and hamster caspase-3 (amino acids 29–277) as described above into DE3 competent cells (Novagen).

2. Grow a 1-liter bacterial culture at 37° until the density reaches an OD$_{600}$ reading of 0.6.

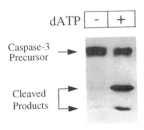

FIG. 1. *In vitro* assay for dATP-dependent caspase-3 activation. A 3-μl aliquot of *in vitro*-translated [^{35}S]methionine-labeled caspase-3 was incubated with 50 μg of HeLa cell S-100 in the absence or presence of 1 mM dATP at 37° for 1 hr in a final volume of 20 μl of buffer A. The samples were subjected to SDS–PAGE and transferred to a nitrocellulose filter. The filter was exposed to Kodak X-Omat AR X-ray film for 16 hr at room temperature.

3. Add isopropyl-β-D-thiogalactopyranoside (IPTG) to the bacterial culture to a final concentration of 2 mM.

4. Pellet the bacteria after 1 hr of induction by centrifugation at 1000g for 20 min at 4°.

5. Lyse the bacteria in buffer A by sonication.

6. Centrifuge at 15,000 rpm at 4° in as SS-34 rotor (Sorvall, Newtown, CT).

7. Load the supernatant onto a 3-ml nickel–agarose (Qiagen) column equilibrated with buffer A.

8. Wash the column with 10 ml of buffer A, followed by 10 ml of buffer A containing 500 mM NaCl, and again with 10 ml of buffer A.

9. Elute the fusion protein with buffer A containing 250 mM imidazole in 1-ml fractions.

10. Pool the protein peak fractions and load onto a fast protein liquid chromatography (FPLC) Superdex-200 16/30 column equilibrated with buffer A and collect 1-ml fractions.

11. Analyze the purity of recombinant caspase-3 enzyme by silver staining.

Comments: Expression of caspase-3 in bacteria results in the active form of caspase-3. The recombinant caspase-3 purified through a gel-filtration column may still contain several other bacterial proteins. To obtain caspase-3 in a purer form, pass the nickel column eluate through a MonoS 5/5 column and check the purity of caspase-3 by silver staining.

Preparation of Hamster Liver Nuclei

1. Rinse livers from four male golden Syrian hamsters (Sasco, Omaha, NE) with ice-cold PBS.

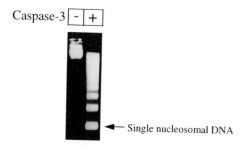

Caspase-3 | - | + |

← Single nucleosomal DNA

FIG. 2. *In vitro* assay for DNA fragmentation. An aliquot of 250 μg of HeLa cell S-100 was incubated with 6 μl of hamster liver nuclei in the absence or presence of 150 μg of active caspase-3 at 37° for 2 hr in a final volume of 60 μl of buffer A. The genomic DNA from these samples was extracted, analyzed, and visualized as described in the DNA fragmentation assay.

2. Mix the livers with buffer D in the ratio of 0.25 g/ml and homogenize them by three strokes of a motor-driven homogenizer.

3. Centrifuge the homogenates through a 10-ml cushion of buffer C at 25,000 rpm for 1 hr in a Beckman SW 28 rotor at 4°.

4. Resuspend the nuclei pellet in buffer E at 8.5×10^7 nuclei/ml and store them at $-80°$ in multiple aliquots.

5. Check the appearance of the nuclei by staining them with 4,6'-diamino-2-phenylindole (DAPI) and observing them under a fluorescence microscope. Each nucleus should have a round shape, with DNA uniformly distributed throughout the entire nucleus.

DNA Fragmentation Assay

1. Incubate 150 μg of active caspase-3 with 6 μl of hamster liver nuclei and an aliquot of 250 μg of HeLa cell S-100 at 37° for 2 hr in a final volume of 60 μl of buffer A.

2. Add an aliquot of 330 μl of buffer F to each reaction and incubate at 37° overnight.

3. Add NaCl to each reaction to a final concentration of 1.5 M.

4. Centrifuge for 15 min in a microcentrifuge at room temperature, to eliminate nuclear debris.

5. Add an equal volume of 100% ethanol to the supernatant and centrifuge at 4° for 30 min at full speed in a microcentrifuge.

6. Wash the DNA precipitate once with 500 μl of 70% ethanol and centrifuge for 15 min at 4°.

7. Resuspend the DNA in 40 μl of buffer G and incubate for 1 hr at 37°.

8. Load the DNA onto a 2% (w/v) agarose gel and run the gel at 50 V for 2 hr in 0.5× TBE buffer.

9. Stain the gel with ethidium bromide (2 μg/ml) for 15 min and destain it with water for 1 hr, and visualize under UV light. See Fig. 2 for example.

Comments: To assay DNA fragmentation in cells undergoing apoptosis, collect the cells and digest them with 330 μl of buffer F at 37° overnight, then pass the samples through a 22-gauge needle 15 times and start at step 3 as described above.

[16] Cell-Free Apoptosis in *Xenopus laevis* Egg Extracts

By Oliver von Ahsen and Donald D. Newmeyer

Introduction

Apoptotic cell death is thought to be mediated by conserved biochemical mechanisms that culminate in the orderly destruction of cellular components. The pathways leading to activation of the effector molecules that direct and carry out the final dismantling of the cell are now partly understood. Central to this process is the caspase family of apoptotic proteases (reviewed, e.g., in Refs. 1–3). Caspases can cleave key cellular proteins, thereby either destroying their normal cellular maintenance functions or activating them as death effectors. Caspases can also process and activate each other; thus, one of the key commitment steps of the apoptotic process is the activation of the most upstream member of a caspase cascade. This can occur in two main ways. The first of these involves the binding of ligands to death receptors at the cell surface, leading to the oligomerization and self-activation of proximal caspases (e.g., caspase-8/FLICE), as occurs in cell death mediated by CD95 (Fas/Apo-1) and the tumor necrosis factor (TNF) receptor (reviewed, e.g., in Refs. 4–8). The second mode of caspase activation involves mitochondria as a point of control. These organelles can mediate cell death through the release of proapoptotic proteins, particularly cytochrome *c*, from the mitochondrial intermembrane space.[9] Once translocated to the cytosol, cytochrome *c* binds to the Apaf-1 protein, which then triggers the activation of caspase-9, in turn leading to the processing of downstream caspases in cascade fashion.[10,11] The release of cytochrome *c* from mitochondria can be regulated by proteins belonging to the Bcl-2

---

[1] S. J. Martin and D. R. Green, *Cell* **82**, 349 (1995).
[2] N. A. Thornberry, *Br. Med. Bull.* **53**, 478 (1997).
[3] P. Villa, S. H. Kaufmann, and W. C. Earnshaw, *Trends Biochem. Sci.* **22**, 388 (1997).
[4] R. H. Arch, R. W. Gedrich, and C. B. Thompson, *Genes Dev.* **12**, 2821 (1998).
[5] S. J. Baker and E. P. Reddy, *Oncogene* **12**, 1 (1996).
[6] A. M. Chinnaiyan and V. M. Dixit, *Semin. Immunol.* **9**, 69 (1997).
[7] S. Nagata, *Intern. Med.* **37**, 179 (1998).
[8] M. E. Peter and P. H. Krammer, *Curr. Opin. Immunol.* **10**, 545 (1998).
[9] X. Liu, C. N. Kim, J. Yang, R. Jemmerson, and X. Wang, *Cell* **86**, 147 (1996).
[10] P. Li, D. Nijhawan, I. Budihardjo, S. M. Srinivasula, M. Ahmad, E. S. Alnemri, and X. Wang, *Cell* **91**, 479 (1997).
[11] H. Zou, W. J. Henzel, X. Liu, A. Lutschg, and X. Wang, *Cell* **90**, 405 (1997).

family,[12-18] some of which are proapoptotic while others are antiapoptotic. Bcl-2 relatives may also participate in cross-talk between death receptor- and mitochondria-mediated pathways of cell death.[16,19-21]

Although Bcl-2 and its relatives have been shown to regulate cyto-chrome c release, to mitigate the effects of cellular oxidants[22,23] and to regulate the partitioning of intracellular calcium,[24] these effects are proba-bly all indirect, and the precise functions of Bcl-2 family members are not clearly understood. Bcl-2, Bcl-x_L, and Bax can form channels in synthetic lipid bilayers[25-27]; however, it has yet to be proved that these proteins regulate apoptosis via channel formation in intracellular membranes.

Because of the gaps in our understanding of the nature and sequence of biochemical events in apoptosis, it is desirable to have at our disposal cell-free systems that permit the biochemical dissection of apoptotic pathways. One such system was described by Lazebnik et al.,[28] who utilized cytosolic extracts from doubly synchronized chicken hepatoma cells

[12] R. Eskes, B. Antonsson, A. Osen-Sand, S. Montessuit, C. Richter, et al., J. Cell Biol. **143**, 217 (1998).

[13] J. M. Jurgensmeier, Z. Xie, Q. Deveraux, L. Ellerby, D. Bredesen, and J. C. Reed, Proc. Natl. Acad. Sci. U.S.A. **95**, 4997 (1998).

[14] C. N. Kim, X. Wang, Y. Huang, A. M. Ibrado, L. Liu, G. Fang, and K. Bhalla, Cancer Res. **57**, 3115 (1997).

[15] R. M. Kluck, E. Bossy-Wetzel, D. R. Green, and D. D. Newmeyer, Science **275**, 1132 (1997).

[16] X. Luo, I. Budihardjo, H. Zou, C. Slaughter, and X. Wang, Cell **94**, 481 (1998).

[17] T. Rosse, R. Olivier, L. Monney, M. Rager, S. Conus, I. Fellay, B. Jansen, and C. Borner, Nature (London) **391**, 496 (1998).

[18] J. Yang, X. Liu, K. Bhalla, C. N. Kim, A. M. Ibrado, J. Cai, T. I. Peng, D. P. Jones, and X. Wang, Science **275**, 1129 (1997).

[19] T. Kuwana, J. J. Smith, M. Muzio, V. Dixit, D. D. Newmeyer, and S. Kornbluth, J. Biol. Chem. **273**, 16589 (1998).

[20] H. Li, H. Zhu, C. J. Xu, and J. Yuan, Cell **94**, 491 (1998).

[21] C. Scaffidi, S. Fulda, A. Srinivasan, C. Friesen, F. Li, K. J. Tomaselli, K. M. Debatin, P. H. Krammer, and M. E. Peter, EMBO J. **17**, 1675 (1998).

[22] D. M. Hockenbery, Z. N. Oltvai, X. M. Yin, C. L. Milliman, and S. J. Korsmeyer, Cell **75**, 241 (1993).

[23] D. J. Kane, T. A. Sarafian, R. Anton, H. Hahn, E. Butler Gralla, J. Selverstone Valentine, T. Oerd, and D. E. Bredesen, Science **262**, 1274 (1993).

[24] G. Baffy, T. Miyashita, J. R. Williamson, and J. C. Reed, J. Biol. Chem. **268**, 6511 (1993).

[25] B. Antonsson, F. Conti, A. Ciavatta, S. Montessuit, S. Lewis, I. Martinou, L. Bernasconi, A. Bernard, J. J. Mermod, G. Mazzei, K. Maundrell, F. Gambale, R. Sadoul, and J. C. Martinou, Science **277**, 370 (1997).

[26] A. J. Minn, P. Velez, S. L. Schendel, H. Liang, S. W. Muchmore, S. W. Fesik, M. Fill, and C. B. Thompson, Nature (London) **385**, 353 (1997).

[27] S. L. Schendel, Z. Xie, M. O. Montal, S. Matsuyama, M. Montal, and J. C. Reed, Proc. Natl. Acad. Sci. U.S.A. **94**, 5113 (1997).

[28] Y. A. Lazebnik, S. Cole, C. A. Cooke, W. G. Nelson, and W. C. Earnshaw, J. Cell Biol. **123**, 7 (1993).

("S/M extracts"). Exogenous nuclei placed in such extracts undergo the morphological changes typical of nuclei in apoptotic cells. This process was shown to depend on caspase activity.[29] Other investigators have employed apoptotic cell-free systems prepared from mammalian cells.[9,30–33]

We developed a different apoptotic cell-free system, based on extracts from *Xenopus* eggs.[34] When nuclei from various sources are placed in "apoptotic" egg extracts, they in time display the morphological changes typical of nuclei in apoptotic cells. Furthermore, these events are inhibited by the addition of exogenous Bcl-2 protein. *Xenopus* egg extracts can be used to study the roles of subcellular organelles, as the particulate components can easily be separated into two fractions: light membranes (LMs), largely composed of endoplasmic reticulum and nuclear membrane precursor vesicles, and heavy membranes (HMs), which are highly enriched in mitochondria. The LM fraction is required for nuclear membrane formation. In contrast, the HM fraction is dispensable for nuclear membrane formation but essential for the apoptotic activity of the extracts. The *Xenopus* system has proved to be useful in elucidation of the mechanisms of apoptosis and its regulation by Bcl-2 proteins.[15,19,34–41] Here we describe in detail the methods needed to prepare and use apoptotic *Xenopus* egg extracts.

Basic Features of *Xenopus* Egg Extracts

Because of the developmental strategy of certain amphibians, extracts from *Xenopus laevis* eggs have proved to be a versatile experimental resource for studies of the function of the cell nucleus (references cited in Newmeyer and Wilson[42]). The eggs of these organisms are large cells, more

[29] Y. A. Lazebnik, S. H. Kaufmann, S. Desnoyers, G. G. Poirier, and W. C. Earnshaw, *Nature (London)* **371,** 346 (1994).
[30] H. M. Ellerby, S. J. Martin, L. M. Ellerby, S. S. Naiem, S. Rabizadeh, G. S. Salvesen, C. A. Casiano, N. R. Cashman, D. R. Green, and D. E. Bredesen, *J. Neurosci.* **17,** 6165 (1997).
[31] M. Enari, R. V. Talanian, W. W. Wong, and S. Nagata, *Nature (London)* **380,** 723 (1996).
[32] S. J. Martin, D. D. Newmeyer, S. Mathias, D. M. Farschon, H.-G. Wang, J. C. Reed, R. N. Kolesnick, and D. R. Green, *EMBO J.* **14,** 5191 (1995).
[33] M. Muzio, G. S. Salvesen, and V. M. Dixit, *J. Biol. Chem.* **272,** 2952 (1997).
[34] D. D. Newmeyer, D. M. Farschon, and J. C. Reed, *Cell* **79,** 353 (1994).
[35] S. C. Cosulich, S. Green, and P. R. Clarke, *Curr. Biol.* **6,** 997 (1996).
[36] S. C. Cosulich, V. Worrall, P. J. Hedge, S. Green, and P. R. Clarke, *Curr. Biol.* **7,** 913 (1997).
[37] E. K. Evans, T. Kuwana, S. L. Strum, J. J. Smith, D. D. Newmeyer, and S. Kornbluth, *EMBO J.* **16,** 7372 (1997).
[38] E. K. Evans, W. Lu, S. L. Strum, B. J. Mayer, and S. Kornbluth, *EMBO J.* **16,** 230 (1997).
[39] D. M. Farschon, C. Couture, T. Mustelin, and D. D. Newmeyer, *J. Cell Biol.* **137,** 1117 (1996).
[40] S. Faure, S. Vigneron, M. Doree, and N. Morin, *EMBO J.* **16,** 5550 (1997).
[41] R. M. Kluck, S. J. Martin, B. M. Hoffman, J. S. Zhou, D. R. Green, and D. D. Newmeyer, *EMBO J.* **16,** 4639 (1997).
[42] D. D. Newmeyer and K. L. Wilson, *Methods Cell Biol.* **36,** 607 (1991).

than 1 mm in diameter, and contain a huge stockpile of all the materials required for the biogenesis of nuclei and other organelles, except for DNA.[43] These stores are mobilized during early development, a period in which the embryos undergo several rounds of cell division without significant new gene expression. The *Xenopus* egg, moreover, is arrested in metaphase of meiosis II, which means that the components of the nucleus (including lamins, pore complex proteins, and nuclear membrane precursor vesicles) are present in their disassembled mitotic form. When the egg is fertilized or artificially activated, endogenous Mos protein and cyclin B are degraded via a Ca^{2+}-dependent mechanism, and the metaphase block is released.[44–47] As a result, the stockpiled nuclear components become available for nuclear assembly, as long as a suitable chromatin substrate is introduced into the egg.

Extracts of *Xenopus* eggs can be prepared in either a metaphase or an interphase state, depending on the presence or absence, respectively, of Ca^{2+} chelators and phosphatase inhibitors in the lysis buffer. Interphase extracts support nuclear assembly, whereas mitotic extracts disassemble nuclei. However, the interphase extracts have the capacity to reenter mitosis, and when prepared carefully, can proceed through several cycles of M and S phases.[48] This is useful for cell cycle investigations, but a nuisance when the functions of intact nuclei are being studied. To investigate apoptotic events, especially nuclear condensation and fragmentation, the extracts are generally arrested in interphase by including cycloheximide in the egg lysis buffer, thereby preventing the resynthesis of cyclin B and the subsequent reentry into mitosis.[49] However, mitotic extracts also appear to undergo apoptosis (D. D. Newmeyer, unpublished observation).

Xenopus eggs provide a convenient source of reasonably large amounts of homogeneous and naturally synchronized apoptotic cell extracts suitable for biochemical subfractionation. Thus, this system is well suited for studying subcellular events in apoptosis. Apoptotic extracts[34] are prepared in exactly the same way as extracts used for the study of nuclear assembly and protein import.[42] We formerly believed it important to give the frogs an altered schedule of hormone injections to induce a preapoptotic state;

[43] R. A. Laskey, J. B. Gurdon, and M. Trendelenburg, *Br. Soc. Dev. Biol. Symp.* **4**, 65 (1979).
[44] T. Lorca, F. H. Cruzalegul, D. Fesquet, J. C. Cavadore, J. Mery, A. Means, and M. Doree, *Nature (London)* **366**, 270 (1993).
[45] J. Minshull, J. J. Blow, and T. Hunt, *Cell* **56**, 947 (1989).
[46] N. Watanabe, T. Hunt, Y. Ikawa, and N. Sagata, *Nature (London)* **352**, 247 (1991).
[47] N. Watanabe, G. F. Vande Woude, Y. Ikawa, and N. Sagata, *Nature (London)* **342**, 505 (1989).
[48] A. W. Murray, *Methods Cell Biol.* **36**, 581 (1991).
[49] J. Newport, *Cell* **48**, 205 (1987).

however, experience in our laboratory has shown that most extracts (as long as they contain mitochondria) display apoptotic behavior, regardless of the hormone schedules used. The apoptotic character of egg extracts may reflect normal atretic processes *in vivo*, which, at least in mammals, are known to be apoptotic.[50] For the care and maintenance of *Xenopus*, see Wu and Gerhart.[51]

Preparation of Apoptotic Extracts from *Xenopus* Eggs

Interphase Extract

This procedure is modified from Newport[49] and Wilson and Newport.[52] All steps are done at room temperature unless otherwise noted.

1. Inject female frogs [available in the United States from Nasco (Ft. Atkinson, WI), *Xenopus* I, (Ann Arbor, MI), or *Xenopus* Express (Homosassa, FL)] 14–28 days prior to egg laying with 100 U of pregnant mare serum gonadotropin (PMSG; Calbiochem, La Jolla, CA) in 0.5 ml of sterile phosphate-buffered saline (PBS). Frogs are injected subcutaneously into their dorsal lymph sac.

2. The evening (16 hr) before egg collection: Inject the frogs with 1000 U of human chorionic gonadotropin (hCG; Sigma, St. Louis, MO) in 1 ml of sterile PBS and place the frogs in individual containers (preferably clear or translucent nonporous plastic) containing 8 liters of salt water (5 g of food-grade NaCl per liter). Often, the frogs will not have finished laying all their eggs by morning, and an additional quantity can be harvested by squeezing the animal firmly (but not roughly) while holding it in the salt water. Transfer the frogs to fresh water, and allow the eggs to settle to the bottom of the containers. Pour off most of the salt water, and pour the eggs into 500-ml beakers. Allow the eggs to settle out, and pour off most of the remaining salt water.

3. Prepare 4% (w/v) cysteine solution (ambient temperature), about 100 ml per batch of eggs, and adjust to pH 7.9 with NaOH. Add an equal volume of cysteine solution to the egg suspension, bringing cysteine to 2% (w/v). Wait 5–10 min, periodically swirling the eggs gently; when the eggs pack tightly with no space between them, the jelly coats have been dissolved. Do not allow the eggs to sit in cysteine solution any longer than necessary.

4. Rinse the eggs several times with MMR (buffers and reagents are listed at the end of the chapter; prepare about 500 ml of MMR per frog).

[50] A. Kaipia and A. J. Hsueh, *Annu. Rev. Physiol.* **59**, 349 (1997).
[51] M. Wu and J. Gerhart, *Methods Cell Biol.* **36**, 3 (1991).
[52] K. L. Wilson and J. Newport, *J. Cell Biol.* **107**, 57 (1988).

Be gentle and thorough with all rinses and avoid jarring the eggs. Use a Pasteur pipette to remove as much buffer as possible between rinses.

5. Pour each batch of eggs into a glass petri dish large enough to spread the eggs in a monolayer, and examine the eggs under a dissecting microscope. Because of the variability that tends to occur between extracts, it is a good idea to keep batches of eggs separate, if possible. Remove debris and all the bad eggs with a glass Pasteur pipette. If a particular batch of eggs contains more than 10% mottled or lysed (white) eggs, it is probably not worth saving. Eggs that are furrowed (pseudocleaved) have been activated by mechanical damage to the outer membrane, but in small numbers are usually not detrimental to the preparation of S-phase extracts. For some reason, batches of eggs that come out of the frog in strings of jelly are usually bad. Occasionally, frogs will lay a small proportion of unmatured oocytes, lacking the white spot indicative of nuclear breakdown. Remove these if M-phase extracts are being prepared.

6. Rinse three times in egg lysis buffer (ELB) (prepare about 300 ml of ELB per batch of eggs). Remove excess buffer. Add the protease inhibitors, aprotinin and leupeptin (A/L), to 10-μg/ml final concentration. Transfer the eggs gently to 15-ml centrifuge tubes (Falcon #2059; Becton Dickinson Labware, Lincoln Park, NJ). Do not fill the tubes above 12 ml. To estimate the yield of extract, measure the volume of good eggs. Experience has shown that if the volume of dejellied eggs is $\sim V$, then the volume of crude cytoplasmic extract is $\sim V/3$, and the approximate volume of cytosol fraction from a 200,000g spin (see below) comes to $V/10$. Remove excess buffer. Pack the eggs by centrifuging for 2 min at 2000 rpm in a Sorvall (Newtown, CT) centrifuge, with an HB-4 (swinging bucket) rotor or the equivalent. *Important:* Again, remove excess buffer. The idea is to make the extracts as concentrated as possible.

7. Centrifuge the packed eggs at 10,000g (10,000 rpm in an HB-4 rotor) for 12 min at 4°. The preparation from this point is shown schematically in Fig. 1. Centrifugal forces lyse the eggs and stratify the egg contents. Dense yolk and pigment granules go to the bottom, and the crude cytoplasmic extract floats above. [*Note:* For small volumes of eggs, lyse by centrifuging in a 1.5-ml Eppendorf tube for 5 min at maximum speed in a Beckman (Fullerton, CA) Microfuge or an Eppendorf (Hamburg, Germany) centrifuge kept at 4°, using a horizontal rotor.] After lysis, keep extracts on ice. The cytoplasmic extract is withdrawn from the side of the tube, using a 3-ml syringe and 21-gauge needle. Insert the needle, bevel up, just above the pigment layer (see Fig. 1). If the first needle becomes clogged, use a second syringe with fresh needle and insert quickly (to prevent leakage) into the hole made by the first one. Transfer the cytoplasm to a clean tube on ice.

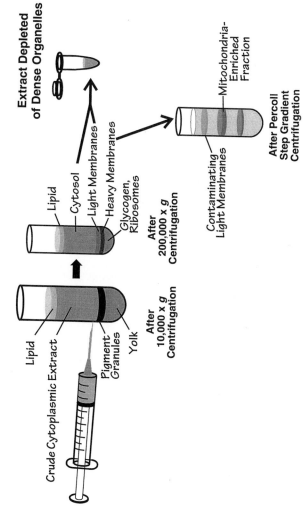

FIG. 1. Schematic illustration of egg extract fractionation procedure. Eggs are crushed by centrifugation at 10,000g; the crude cytoplasmic extract is then removed from the side of the tube, using a syringe. This material may be fractionated further by high-speed centrifugation, yielding cytosol and light and heavy membrane fractions. The cytosol and light membrane fractions are recombined to produce a nuclear assembly extract depleted of dense organelles, and lacking apoptotic activity. The HM fraction is enriched further by Percoll gradient centrifugation and added back to the HM-depleted extract to restore apoptotic activity. Adapted from D. D. Newmeyer and K. L. Wilson, *Meth. Cell Biol.* **36**, 607 (1991).

8. Repeat the centrifugation to remove remaining yolk, pigment, and lipid. [If preparing small volumes of extract, for this second centrifugation it is best to use long, narrow 0.4-ml polyethylene microcentrifuge tubes (Beckman); after centrifuging for 5 min at top speed, lay the tube on its side, cut off the top of the tube just below the lipid layer with a clean razor blade, and use a Pasteur pipette to remove the cytoplasmic material.] The resulting crude extract is adequate for most kinds of experiments that do not need separate incubations of cytosol and membranous material. Crude extract must be used fresh.

Mitotic Extract

This procedure is from Newport and Spann.[53]

For optimal separation of light and heavy membrane components and especially the purification of light membranes that can be used for nuclear reassembly, the cell extract must be arrested in a mitotic state. With S-phase extracts, some of the nuclear membrane vesicles are recovered in the heavy membrane fraction,[52] presumably because they begin to fuse or associate with other extract components after the eggs are crushed. Thus, for experiments that combine cytosol and fractionated membrane components, it is normally necessary to prepare two extracts: an S-phase extract, for cytosol, and an M-phase extract, for membrane components.

The only methodological difference between mitotic and interphase extracts is the buffer in which the eggs are lysed. M-phase egg lysis buffer (MELB) has two ingredients that preserve the meiotic/mitotic state by blocking the calcium-dependent degradation of the Mos protein and cyclin B: β-glycerophosphate, a competitive inhibitor of phosphatase activity; and EGTA, which chelates Ca^{2+}. Alternatively, S-phase ELB containing at least 5 mM EGTA could be used.[54] For details of S-phase extract preparation, refer to the previous section.

It has been shown that the addition of staurosporine (3–5 μM) can convert a mitotic extract (prepared with 5 mM EGTA in ELB, rather than MELB) into a pseudo-interphase extract that assembles nuclei but cannot replicate DNA.[54] Hence, if the particular processes being studied are unaffected by the presence of staurosporine (as is the case for apoptotic egg extracts[34]), it is possible to do fractionation experiments with a single M-phase extract.

---

[53] J. Newport and T. Spann, *Cell* **48**, 219 (1987).
[54] Y. Kubota and H. Takisawa, *J. Cell Biol.* **123**, 1321 (1993).

Subcellular Fractionation of Crude Egg Extracts

Ultracentrifugation separates the membranes and soluble components. Each component can be frozen and stored for at least 6 months; however, the light membranes lose some of their activity on each freeze–thaw cycle and each time they are pelleted.

1. Transfer the crude extract into an ultraclear ultracentrifuge tube, on ice. Centrifuge at 200,000*g* for 1 hr at 4°, in a swinging bucket rotor. *Best*: Use a Beckman tabletop TL-100 ultracentrifuge with a TLS-55 swinging bucket rotor at 55,000 rpm. Other rotors may not achieve adequate separation in 1 hr.

2. Recover the cytosol fraction from the side, using syringe and needle as before, being careful not to disturb the underlying membrane layers (Figure 1). To obtain very pure cytosol, centrifuge again at 200,000*g* for 20 min at 4° to remove most of the residual membranes. For the best separation of cytosol and membranes add ATP/phosphocreatine (PC) and creatine kinase before recentrifugation. Freeze aliquots in liquid nitrogen and store them at −80°.

3. To obtain the light membrane fraction, remove and discard the remaining cytosol and lipid remaining above the membrane layers.

4. Using a wide-bore 200-μl micropipette tip (the end chopped off with a razor blade), collect most of the pale yellow light membrane layer, about 50 μl at a time. With care, it is possible to avoid contaminating this material with the brown heavy membrane layer underneath.

5. Resuspend in at least 20 volumes of cold membrane wash buffer (MWB). Pellet the membranes by centrifugation at 40,000*g* for 15 min at 4° (25,000 rpm in the TLS-55 rotor). Remove and discard the supernatant. Recover the membranes in the smallest possible volume of MWB, preferably less than 1/10 of the volume of the cytosol fraction.

6. If long-term storage is desired, supplement the light membranes with sucrose to at least 0.5 *M* final concentration. Freeze small aliquots in liquid nitrogen and place them at −80°.

7. Heavy membrane fraction[34]: After removal of most of the LM fraction, the crude HM material (contaminated with a small amount of LM material) is removed with a micropipettor. To isolate a mitochondria-enriched fraction uncontaminated with LMs, centrifuge the crude HM material over a Percoll step gradient (modified from Reinhart *et al.*[55]).

8. Dilute the crude HM with an equal volume of mitochondrial isolation buffer (MIB). Layer up to 100 μl of the diluted HM material over a

[55] P. H. Reinhart, W. M. Taylor, and F. L. Bygrave, *Biochem. J.* **204,** 731 (1982).

1.8-ml Percoll step gradient consisting of four 0.45-ml layers with Percoll concentrations of 25, 30, 37, and 42%, in MIB. Centrifuge at 4° for 20 min in a TLS-55 rotor, at 25,000 rpm, using a Beckman TL-100 centrifuge, with the brake set at 5.

9. With a Pasteur pipette inserted into the gradient from above, remove the brown material forming the major band at the interface between 30 and 37% Percoll. Try to recover it in as small a volume as possible.

10. To wash out residual Percoll, resuspend the heavy membranes in at least 100 vol of MIB, centrifuge at 1600g for 10 min at 4°, and resuspend the pellet in an equal volume of MIB. The resulting enriched HM fraction can be frozen in small aliquots in liquid nitrogen and then stored at −80°.

Preparation of Demembranated Sperm Chromatin

This procedure is slightly modified from Lohka and Masui.[56] All steps are done at room temperature unless otherwise noted. Approximate buffer volumes needed per frog: 4.0 ml of sperm buffer (SB), 1.0 ml of SB with 0.05% (w/v) lysolecithin, and 5.0 ml of SB with 3% (w/v) bovine serum albumin (BSA).

1. Anesthetize male frogs in 1% (w/v) tricaine (3-aminobenzoic acid ethyl ester; Sigma) for 15–20 min. When the frogs no longer move and when they are turned belly-up, they are ready.

2. Make an abdominal incision and remove the testes, being careful not to puncture them. Incubate them overnight at 18–22° in MMR solution containing hCG (10 U/ml) and gentamicin (250 μg/ml).

3. Trim away fat and connective tissue and rinse in SB. Release the sperm by crushing the testes in a 1.5-ml Eppendorf tube containing SB (1 ml per pair of testes). Use broad forceps or, better, a plastic homogenizer pestle designed to fit snugly into the conical bottoms of 1.5-ml Eppendorf tubes (pellet pestle mixer; Kontes, Vineland, NJ). To avoid contamination with somatic cells, take care not to disintegrate the outer tissue of the testis.

4. Clear debris by centrifugation for 10 sec at setting #3 in a clinical centrifuge or at 1000 rpm in an Eppendorf centrifuge. Transfer the supernatant to a new Eppendorf tube. Add 500 μl of SB to the pelleted tissue, mix gently, and recentrifuge. Combine the two supernatants.

5. Pellet the sperm by centrifugation for 2 min at 6000 rpm in an Eppendorf centrifuge. After discarding the supernatant, carefully resuspend the white part of the pellet in SB, 50–100 μl at a time, for a total of ~500 μl. Avoid resuspending the red bottom of the pellet, which contains

[56] M. J. Lohka and Y. Masui, *Science* **220**, 719 (1983).

unwanted erythrocytes and somatic cells. Repeat the pelleting/resuspension procedure once more to eliminate more somatic cells.

6. Pellet the sperm a third time, and resuspend in 100 μl of SB. Add 1 ml of SB/0.05% (w/v) lysolecithin. Lysolecithin removes the plasma and nuclear membranes. After a 5-min incubation at room temperature, pipette the suspension into 3 ml of SB/3% (w/v) BSA and mix gently. The BSA aids in the removal of free lysolecithin.

7. Pellet the demembranated sperm nuclei for 8–10 min at 2000 rpm in a Sorvall centrifuge. Resuspend the pellets in 200 μl each of SB/3% (w/v) BSA, pool into one tube, and add 1 ml of SB/3% (w/v) BSA.

8. Pellet the sperm nuclei for 8–10 min at 2000 rpm, and resuspend in 50 μl of ice-cold SB. Count the sperm, using a hemacytometer.

9. Dilute with SB to obtain a final concentration of $4 \times 10^5/\mu$l. Freeze 10-μl aliquots in liquid nitrogen, and store at $-80°$.

Isolation of Rat Liver Nuclei

This procedure is from Blobel and Potter,[57] with crucial buffer modifications as per Newport and Spann.[53]

1. Make all solutions in advance and cool to 4°. Precool ultracentrifuge, SW28 rotor, and buckets to 4°.

2. Kill rats and remove livers, placing them in a petri dish on ice. Cut away the tough connective tissue between lobes. Record the weight (*W*) of the liver tissue.

3. Transfer the tissue to a clean glass plate on ice; mince with razors. Scrape the minced tissue into a chilled homogenizer chamber (~30 ml in size). Add ($2 \times W$) ml of chilled solution A. Homogenize with 12–14 strokes (by Dounce or motorized homogenizer).

4. Filter the homogenate through four layers of cheesecloth in a funnel.

5. Add ($2 \times W$) ml of solution B; mix thoroughly by inversion. Pipette into SW28 centrifuge tubes (ultraclear; ~30 ml/tube). Drip 5 ml of solution B into each tube to form a cushion at the bottom. Balance the tubes carefully, and centrifuge for 1.5 hr at 22,000 rpm at 4°, with the brake on.

6. The nuclei form a white pellet. Aspirate and remove all supernatant debris. Use a lint-free paper towel to clean the inside tube wall. Resuspend each pellet gently in 200 μl of solution A.

7. Count the nuclei, using a hemacytometer. Adjust the concentration to $1.6–2.0 \times 10^5$ nuclei/μl. Freeze small (5- to 20-μl) aliquots in liquid nitrogen; store at $-80°$.

[57] G. Blobel and V. R. Potter, *Science* **154**, 1662 (1966).

Assembly of Nuclei around Demembranated Sperm Chromatin

Each egg contains enough nuclear components to construct about 4000 embryonic nuclei.[58] The amount of extract obtained from one egg is about 0.7 μl. A typical nuclear assembly reaction contains the following:

1. 100 μl of soluble fraction (~130 egg equivalents)
2. 10 μl of light membranes (LMs) and 2.5 μl of heavy membranes (HMs; necessary for apoptotic extracts but not for nuclear assembly)
3. 0.7 μl of sperm chromatin (at 400,000 sperm/μl; final sperm concentration, ~4000 per egg equivalent)
4. 7 μl of ATP/phosphocreatine (ATP/PC)
5. 3 μl of creatine phosphokinase stock solution (CK)

Nuclear assembly reactions reconstituted from frozen components typically yield decent nuclear envelopes and decondensed chromatin within 1 hr. An assay for nuclear integrity is described below.

Reconstitution and Analysis of Apoptotic Events Using Fractionated Extracts

Cell-free fractionated extracts prepared from *Xenopus* eggs offer the advantage that different partial reactions or temporal phases occurring during apoptosis can be studied separately. For example, one can study the activation of caspases by exogenous purified cytochrome *c*, the response of nuclei to prior events in the cytoplasmic extract, or mitochondrial changes resulting from various treatments. In reconstituted extracts, light membranes are required only if nuclei or chromatin will be added. As most apoptotic events in this system do not require the presence of nuclei, most studies will require only a minimal cell-free system consisting of cytosol and mitochondria.[15]

1. Keep all isolated components on ice before use, and mix gently to avoid damaging mitochondria or nuclei when these are used.
2. Add 3 μl of CK and 7 μl of ATP/PC to 100 μl of clear cytosol. An ATP-regenerating system helps provide energy for prolonged apoptotic reactions.
3. For investigations concerning the role of mitochondria, heavy membranes should be added to the reaction mix to a final concentration of ~2–3% (v/v; in this calculation we refer to the volume of the packed HM pellet *before* dilution 1 : 1 with buffer). Nuclei should be used at concentrations of about 1000 nuclei/μl.

58 J. Newport and D. J. Forbes, *Annu. Rev. Biochem.* **56,** 535 (1987).

4. Start the experiment by shifting the samples to room temperature.

5. Take samples at different time points to follow cytochrome *c* release from mitochondria, caspase activation, and nuclear fragmentation.

Caspase Activity

Because caspases are the central executioners involved in apoptosis, their activity is a hallmark of this form of cell death. Caspase activation by the Apaf-1/caspase-9 "apoptosome" complex can also be used as a convenient indirect assay for cytochrome *c* release from mitochondria. Detection of caspase activity is conveniently done with chromogenic substrates such as DEVD-pNA or its more sensitive fluorogenic counterpart, DEVD-AFC. Because of the high caspase concentrations in crude extracts or cytosol preparations, it is possible to perform activity assays in a conveniently small sample volume.

1. Dilute DEVD-pNA or DEVD-AFC into DEVDase assay buffer to 40 μM final concentration. (Other tetrapeptide substrates can also be used to assay caspases with different specificities.)

2. To a 2-μl sample of the extract that is to be assayed, add 200 μl of assay mix.

3. For DEVD-pNA, measure the increase in absorbance at 405 nm for 30 min at room temperature; for DEVD-AFC, measure light emission at 505 nm with excitation at 400 nm.

4. Especially for long time courses and large experiments, it is convenient to freeze the 2-μl samples in a 96-well plate, placed on dry ice. Cover the plate to avoid condensation. Add the substrate solution after thawing the plate.

Mitochondrial Cytochrome c Release

The release of cytochrome *c* from mitochondria has been shown to be a key event during the commitment phase of at least some forms of apoptosis, and this release can be regulated by Bcl-2 family members.[9–17,19,20,30,41,59–66] The mitochondrial events leading to cytochrome *c*

[59] E. Bossy-Wetzel, D. D. Newmeyer, and D. R. Green, *EMBO J.* **17**, 37 (1998).

[60] H. Kojima, K. Endo, H. Moriyama, Y. Tanaka, E. S. Alnemri, C. A. Slapak, B. Teicher, D. Kufe, and R. Datta, *J. Biol. Chem.* **273**, 16647 (1998).

[61] A. Krippner, A. Matsuno-Yagi, R. A. Gottlieb, and B. M. Babior, *J. Biol. Chem.* **271**, 21629 (1996).

[62] F. Li, A. Srinivasan, Y. Wang, R. C. Armstrong, K. J. Tomaselli, and L. C. Fritz, *J. Biol. Chem.* **272**, 30299 (1997).

[63] S. Manon, B. Chaudhuri, and M. Guerin, *FEBS Lett.* **415**, 29 (1997).

[64] H. Stridh, M. Kimland, D. P. Jones, S. Orrenius, and M. B. Hampton, *FEBS Lett.* **429**, 351 (1998).

release are therefore an important subject of current research. To measure cytochrome c release in the *Xenopus* cell-free system:

1. Take a 10-μl sample of extract and pellet particulate components, including mitochondria, by microcentrifugation for 5 min at 14,000 rpm.

2. Separate the supernatant from the pellet and assay for cytochrome c in both fractions by SDS–PAGE and Western blotting. A anti-cytochrome c antibody cross-reacting with the *Xenopus* protein can be purchased from PharMingen (San Diego, CA; clone 7H8.2C12).

If mitochondria are incubated in buffer instead of cytosol, make sure that the buffer contains at least 80 mM KCl; otherwise, cytochrome c remains associated with the mitochondria even when the outer membrane is permeabilized (R. M. Kluck and D. D. Newmeyer, unpublished observation).

Assay of Nuclear Integrity by Fluorescence Microscopy

1. To visualize DNA, mix about 4 μl of the nuclear assembly reaction with 0.5 μl of formaldehyde/Hoechst mixture: Hoechst 33258 (100 μg/ml) in 37% (v/v) formaldehyde. View with a fluorescence microscope equipped with filters appropriate for Hoechst or 4',6-diamidino-2-phenylindole (DAPI) fluorescence.

2. To visualize membranes, use phase-contrast microscopy or use the membrane dye 3,3'-dihexyloxacarbocyanine (DHCC) and view in the fluorescence microscope, using a filter set appropriate for fluorescein. DHCC is light sensitive; the following working solution should be made fresh each day. Dilute the stock DHCC solution, 10 mg/ml in DMSO, into ~50 vol of Hoechst buffer, centrifuge for 2 min in a microcentrifuge to pellet insoluble chunks of DHCC, and remove the supernatant as working solution. To reduce the background a membrane vesicles, mix on the microscope slide 1 μl of extract and 4 μl of working DHCC solution.

Buffers and Reagents

PBS (10×): 74.2 g of NaCl, 2 g of KCl, 2 g of KH_2PO_4 and 21.7 g of $Na_2HPO_4 \cdot 7H_2O$ in 1 liter of H_2O, adjusted to pH 7.4 with HCl

[65] M. G. Vander Heiden, N. S. Chandel, E. K. Williamson, P. T. Schumacker, and C. B. Thompson, *Cell* **91,** 627 (1997).

[66] B. Zhivotovsky, S. Orrenius, O. T. Brustugun, and S. O. Doskeland, *Nature* (*London*) **391,** 449 (1998).

Dejellying solution: 4% (w/v) cysteine, adjusted to pH 7.9 with NaOH (prepare freshly before use)

MMR (10×): 1 M NaCl, 20 mM KCl, 20 mM CaCl$_2$, 10 mM MgCl$_2$, 50 mM HEPES-KOH, pH 7.4. Autoclave and store at room temperature. If kept for short periods of time, the 1× solution does not need to be sterilized

Egg lysis buffer (ELB): 250 mM sucrose, 50 mM KCl, 2.5 mM MgCl$_2$, 20 mM HEPES-KOH (pH 7.5), 1 mM dithiothreitol (DTT), cyclo-heximide (50 μg/ml), cytochalasin B (5 μg/ml), and aprotinin and leupeptin (10 μg/ml each). Add cycloheximide, aprotinin, and leu-peptin fresh from 10-mg/ml stocks in water, cytochalasin B from a 10-mg/ml stock in dimethyl sulfoxide (DMSO)

M-phase egg lysis buffer (MELB): 240 mM β-glycerophosphate (pH 7.4), 60 mM EGTA, 45 mM MgCl$_2$, 1 mM DTT

Mitochondrial isolation buffer (MIB): 210 mM mannitol, 60 mM su-crose, 10 mM KCl, 10 mM succinic acid, 5 mM EGTA, 1 mM ADP, 0.5 mM DTT, 20 mM HEPES-KOH (pH 7.5)

Membrane wash buffer (MWB): 250 mM sucrose, 50 mM KCl, 2.5 mM MgCl$_2$, 20 mM HEPES-NaOH (pH 8.0), 1 mM DTT, 1 mM ATP, leupeptin (1 μg/ml), aprotinin (1 μg/ml). Store aliquots at $-20°$. Add DTT, ATP, and protease inhibitors just before use (ATP interferes with the BCA protein assay)

A/L: The protease inhibitors aprotinin and leupeptin are used at 10 mg/ml in water as a 1000× stock. Store in aliquots at $-20°$

CK: Creatine kinase (5 mg/ml) in 50% (v/v) glycerol, 20 mM HEPES-KOH (pH 7.5)

ATP/PC: 100 mM ATP, 200 mM phosphocreatine in 20 mM HEPES-KOH, pH 7.5

DEVDase substrate stock: 20 mM DEVD-pNA in DMSO

DEVDase assay buffer: 250 mM sucrose, 50 mM KCl, 20 mM HEPES-KOH (pH 7.5), 2.5 mM MgCl$_2$

Hoechst fixative: 100 μg of Hoechst 33258 (bisbenzimide) in 37% (v/v) formaldehyde. A 100× H33258 stock (10 mg/ml in DMSO) can be used to prepare it

Hoechst buffer: 200 mM sucrose, 5 mM MgCl$_2$, 1× buffer A salts, Hoechst 33258 (10 μg/ml). *Option*: Add formaldehyde to 3.7% (v/v). Store in the dark at 4°

Buffer A salts (10×): 800 mM KCl, 150 mM NaCl, 50 mM EDTA, 150 mM PIPES-NaOH, pH 7.4

Solution SB: 1× buffer A salts, 0.2 M sucrose, 7 mM MgCl$_2$

Solution A: 250 mM sucrose, 0.5 mM spermidine, 0.2 mM spermine, 1 mM DTT and 1 mM phenylmethylsulfonyl fluoride (PMSF) in 1×

buffer A salts. (Add DTT and PMSF freshly, PMSF from a 100 mM stock in ethanol)

Solution B: 2.3 M sucrose, 0.5 mM spermidine, 0.2 mM spermine, 1 mM DTT, and 1 mM PMSF in 1× buffer A salts

[17] Cytofluorometric Quantitation of Nuclear Apoptosis Induced in a Cell-Free System

By Hans K. Lorenzo, Santos A. Susin, and Guido Kroemer

Introduction

Cytoplasmic extracts and mitochondrial intermembrane proteins contain factors that can induce apoptotic changes including chromatin condensation and oligonucleosomal DNA fragmentation in isolated nuclei *in vitro*. These changes can be revealed by agarose gel electrophoresis of DNA or by staining of nuclei with 4′,6-diamidino-2-phenylindole dihydrochloride (DAPI) or other DNA-intercalating dyes, followed by visual inspection of nuclei in a fluorescence microscope.[1-7] We have developed a simple method for the quantitation of nuclear apoptosis by cytofluorometric analysis of propidium iodine-stained HeLa nuclei.[8,9] Nuclei whose DNA is fragmented contain less DNA than control nuclei and thus give rise to a sub-$G_{0/1}$ peak. The frequency of hypoploid nuclei strictly correlates with that determined by morphological criteria.

Required Materials

Thomas Potter homogenizes with Teflon inlet

Corex 15-ml tubes

HNB buffer: 10 mM KCl, 10 μM cytochalasin B, 2 mM MgCl$_2$, 1 mM dithiothreitol (DTT), 0.1 mM phenylmethylsulfonyl fluoride

[1] Y. A. Lazebnik, S. Cole, C. A. Cooke, W. G. Nelson, and W. C. Earnshaw, *J. Cell. Biol.* **123**, 7 (1993).
[2] S. J. Martin, D. D. Newmeyer, S. Mathisa, *et al.*, *EMBO J.* **14**, 5191 (1995).
[3] Y. A. Lazebnik, A. Takahashi, G. G. Poirier, S. H. Kaufman, and W. C. Earnshaw, *J. Cell Sci.* **S19**, 41 (1995).
[4] M. Enari, H. Hug, and S. Nagata, *Nature (London)* **375**, 78 (1995).
[5] N. Zamzami, S. A. Susin, P. Marchetti, *et al.*, *J. Exp. Med.* **183**, 1533 (1996).
[6] S. A. Susin, N. Zamzami, M. Castedo, *et al.*, *J. Exp. Med.* **184**, 1331 (1996).
[7] X. Liu, C. N. Kim, J. Yang, R. Jemmerson, and X. Wang, *Cell* **86**, 147 (1996).
[8] S. A. Susin, N. Zamzami, M. Castedo, *et al.*, *J. Exp. Med.* **186**, 25 (1997).
[9] S. A. Susin, N. Zamzami, N. Larochette, *et al.*, *Exp. Cell Res.* **236**, 397 (1997).

(PMSF), 10 mM piperazine-N,N'-bis(2-ethanesulfonic acid) (PIPES), pH 7.4

Sucrose solution: Sucrose (30%, w/v) diluted 1:1 (v/v) in HNB buffer

HNB freezing buffer: 80 mM KCl, 20 mM NaCl, 250 mM sucrose, 5 mM EGTA, 1 mM DTT, 0.1 mM PMSF, 10 mM PIPES (pH 7.4) plus 50% (v/v) glycerol

CFS buffer: 220 mM mannitol, 68 mM sucrose, 2 mM NaCl, 2.5 mM KH$_2$PO$_4$, 0.5 mM EGTA, 2 mM MgCl$_2$, 5 mM pyruvate, 0.1 mM PMSF, leupeptin (1 μg/ml), pepstatin A (1 μg/ml), antipain (50 μg/ml), chymopapain (10 μg/ml), 1 mM dithiothreitol, 10 mM HEPES-NaOH (pH 7.2). This buffer is stored at $-20°$ and all protease inhibitors, as well as dithiothreitol, are added shortly before the experiment

Propidium iodine (PI) stock solution: 1 mg/ml in distilled water

Cytofluorometer

Protocol

Purification and Storage of Nuclei. HeLa cells (or other cell lines) are washed twice in phosphate-buffered saline (PBS) (600g, 10 min), resuspended in HNB buffer, incubated for 45 min on wet ice, lysed in a Thomas Potter homogenizer (20 strokes, $\sim$500 rpm), and layered over HNB buffer plus 30% (w/v) sucrose (1:1, v/v). The nuclei (in the interface) are then washed twice in HNB buffer (800g, 10 min) and resuspended at a final concentration of 10^7/ml. Nuclei are conserved at $-20°$ in HNB freezing buffer for up to 15 days.

Preparation of Supernatants from Mitochondria. Mitochondria are purified and resuspended in CFS buffer (10 mg of protein per milliliter), as described in [18] in this volume.[10] They are then treated for 30 min with freshly purified atractyloside (10 mM, final concentration), followed by two rounds of centrifugation (first 7000g, 10 min, 4°, then 1.5 × 10^5g, 1 hr, 4°), to recover the supernatant. This supernatant can be snap-frozen and stored at $-80°$. It contains apoptosis-inducing factor (AIF) and can serve as positive control of the cell-free system.

Cell-Free System of Apoptosis. Nuclei (10^3 nuclei/μl) are washed in HNB buffer (1200g, 10 min, 4°; two washes), resuspended in CFS buffer, and cultured in the presence of CFS buffer only (negative control) or CFS buffer containing various concentrations of mitochondrial supernatant (positive control) or other preparations whose apoptogenic potential is to

---

[10] S. A. Susin, N. Larochette, M. Geuskens, and G. Kroemer, *Methods Enzymol.* **322**, [18], 2000 (this volume).

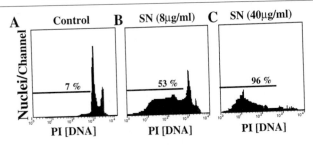

FIG. 1. Representative propidium iodine (PI)-staining profiles of isolated HeLa nuclei cultured for 90 min in the presence of CFS buffer only (A) or in the presence of various concentrations of mitochondrial supernatants (B and C).

be determined. The usual incubation time is 90 min at 37°, and the reaction is stopped by transferring the nuclei on wet ice.

Cytofluorometric Analysis of Nuclear Apoptosis. Nuclei are stained with PI [10 μg/ml (Sigma, St. Louis, MO), minimum 5 min at room temperature, analyzed in FL-3], followed by cytofluorometric analysis is an EPICS Profile II analyzer (Coulter, Hialeah, FL), while gating the forward and side scatters on single-nucleus events (Fig. 1). Fluorescence is excited with an argon laser. The lipophilic dye 5-methyl-bodipy-3-dodecanoic acid [100 nM (Molecular Probes, Eugene, OR), 15 min at room temperature, analyzed in FL-1] can be used to stain the nuclear membrane.[11] A minimum of 10^4 events is recorded for each data point.

Expected Results and Pitfalls

This method provides a highly reproducible and reliable method for the accurate and rapid quantitation of nuclear apoptosis induced in the cell-free system. Background values depend on the quality of HeLa cultures (spontaneous apoptosis) and on the duration of storage of nuclei (ideally shorter than 3 weeks). As in other biochemical assays, dose–response curves should be performed for each sample to allow for quantitative comparisons. This technique is superior to previously published protocols on the basis of visual inspection of isolated nuclei, because it provides quantitative data and allows for the analysis of hundreds of samples per day. It can be adapted for the study of different apoptogenic activities and their inhibitory profiles. Note that, as in other biological assays, dose–response curves should be performed on each sample to quantify its apoptogenic potential.

---

[11] A. Macho, Z. Mishal, and J. Uriel, *Cytometry* **23**, 166 (1996).

Results may be expressed as units, considering 1 U as the activity causing 50% nuclear apoptosis.

Acknowledgments

This work was supported by grants from the French Ministry of Science, ANRS, ARC, CNRS, FF, FRM, INSERM, and a special grant from LNC to G.K. S.A.S. receives an EC Marie Curie fellowship.

Section V

Mitochondria and Apoptosis

[18] Purification of Mitochondria for Apoptosis Assays

By SANTOS A. SUSIN, NATHANAEL LAROCHETTE, MAURICE GEUSKENS, and GUIDO KROEMER

Introduction

Mitochondria have been discovered to exert an important, perhaps essential, role in the process of apoptosis (for review see Refs. 1 and 2). It appears that one rate-limiting step of apoptosis is an increase in the permeability of the inner and/or outer mitochondrial membrane accompanied by the release of proteins normally confined to the intermembrane space. As a result, mitochondria or mitochondrial intermembrane proteins are essential for the induction of nuclear apoptosis *in vitro* in most, if not all, experimental systems of mammalian cell death.[3–6] Moreover, it appears that mitochondria are the principal site of action of Bcl-2-like oncoproteins.[3,5,7,8] As a result, the study of the regulation of mitochondrial function *in vitro,* using isolated mitochondria, has an increasing impact on apoptosis research.

Principle of Protocol

Tissues or cells are mechanically disrupted in a buffer compatible with mitochondrial integrity, followed by several rounds of differential centrifugation in order to separate mitochondria from other organelles and debris. For a better separation of mitochondria, particularly from endoplasmic reticulum and ribosomes, enriched mitochondria can be purified on a discontinuous density gradient (Fig. 1). This latter step can be omitted if a low degree of purity is acceptable.

[1] G. Kroemer, B. Dallaporta, and M. Resche-Rigon, *Annu. Rev. Physiol.* **60,** 619 (1998).
[2] S. A. Susin, N. Zamzami, and G. Kroemer, *Biochim. Biophys. Acta* (*Bioenerg.*) **1366,** 151 (1998).
[3] N. Zamzami, S. A. Susin, P. Marchetti *et al., J. Exp. Med.* **183,** 1533 (1996).
[4] X. Liu, C. N. Kim, J. Yang, R. Jemmerson, and X. Wang, *Cell* **86,** 147 (1996).
[5] S. A. Susin, N. Zamzami, M. Castedo *et al., J. Exp. Med.* **184,** 1331 (1996).
[6] S. A. Susin, N. Zamzami, N. Larochette *et al., Exp. Cell Res.* **236,** 397 (1997).
[7] J. Yang, X. Liu, K. Bhalla *et al., Science* **275,** 1129 (1997).
[8] R. M. Kluck, E. Bossy-Wetzel, D. R. Green, and D. D. Newmeyer, *Science* **275,** 1132 (1997).

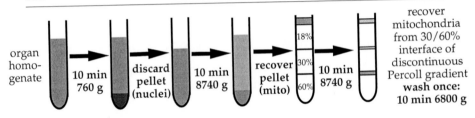

Fig. 1. Scheme for purification of liver mitochondria. For details consult text.

Required Equipment and Reagents

Thomas Potter homogenizer with Teflon inlet (Polylabo, Strasbourg, France)

Hamilton (Reno, NV) 500-μl syringe

Corex 15-ml tubes (Polylabo)

Cylinder made of cork loosely fitting into the Corex tube (height approximately 2 mm), attached in the center to nylon sewing cotton. This cork cylinder can be easily produced from the cork of a wine bottle

Homogenization (H) buffer: 300 mM saccharose, 5 mM N-tris(hydroxymethyl)methyl-2-aminoethanesulfonic acid (TES), 200 $\mu$$M$ EGTA, pH 7.2, to be stored at 4°

Percoll (P) buffer: 300 mM saccharose, 10 mM TES, 200 $\mu$$M$ EGTA, pH 6.9, to be stored at 4°

Solution A: 90 ml of P buffer plus 2.05 g of saccharose

Solution B: 90 ml of P buffer plus 3.96 g of saccharose

Solution C: 90 ml of P buffer plus 13.86 g of saccharose

Percoll stock solution (100%; Pharmacia, Piscataway, NJ)

Percoll (18%): 6.55 ml of solution A + 1.45 ml of Percoll solution (prepare fresh)

Percoll (30%): 5.6 ml of solution B + 2.4 ml of Percoll solution (prepare fresh)

Percoll (60%): 3.2 ml of solution C + 4.8 ml of Percoll solution (prepare fresh)

Protocol for Purification of Mitochondria from Mouse Liver

1. After cervical dislocation, remove livers rapidly, rinse with cold H buffer, and cut into small pieces with a pair of scissors. Note that all steps are performed at 4° or on wet ice.

2. Homogenize the tissue in the Potter-Thomas homogenizer (approximately 20 strokes, approximately 500 rpm).

3. Distribute the homogenate in two Corex tubes and centrifuge for 10 min at 760g.

4. Recover the supernatant. Resuspend the pellet in H buffer, centrifuge as above, and recover the supernatant. Combine the two supernatants.

5. Centrifuge for 10 min at 8740g. In the meantime, prepare the Percoll gradient. Pipette 4 ml of 60% Percoll solution into a Corex glass tube, then introduce the cork cylinder, which floats on the solution. Keep the tube vertical and add the 30% Percoll solution, then the 18% solution (4 ml each) directly onto the cork, using a 5-ml plastic pipette. Remove the cork by pulling the attached string.

6. Recover the pellet and resuspend in 1 ml of H buffer; carefully add the mitochondria-containing solution on top of the 60%/30%/18% Percoll gradient while inclining the tube at 45°.

7. Centrifuge for 10 min at 8740g.

8. Remove the lower interface (between 60 and 30% Percoll) with the Hamilton syringe. Dilute 10× in H buffer.

9. Centrifuge for 10 min at 6800g to remove Percoll, which is toxic for mitochondria.

10. Discard the supernatant and resuspend in the appropriate buffer, e.g., CFS buffer [220 mM mannitol, 68 mM sucrose, 2 mM NaCl, 2.5 mM H_2KPO_4, 0.5 mM EGTA, 2 mM $MgCl_2$, 5 mM pyruvate, 0.1 mM phenylmethylsulfonyl fluoride (PMSF), 1 mM dithiothreitol, 10 mM HEPES-NaOH, pH 7.4] for the introduction of mitochondria into cell-free systems.

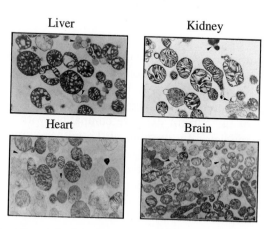

FIG. 2. Electron microscopic pictures of mitochondrial preparations obtained from mouse liver, kidney, heart, and brain. Note that each preparation is >90% pure and that the morphology of mitochondria depends on their source. Arrowheads exemplify nonmitochondrial vesicles.

Note that this protocol is optimized for mouse tissues. Check the purity of preparations (which should be higher than 90%) by electron microscopy (Fig. 2). For cell lines, an alternative protocol of disruption, i.e., nitrogen cavitation, can be used to liberate mitochondria from the cells. Cells (3×10^7 to 1×10^8) are suspended in H buffer, washed ($600g$, 10 min, room temperature) twice in this buffer, and exposed to a nitrogen decompression of 150 lb/in^2 for 30 min, using a "cell disruption bomb" (Parr Instrument, Moline, IL). Thereafter the cell lysate is subjected to differential centrifugation as described above. Note that the resuspension buffer of mitochondria depends on the intended use of the mitochondria (cell-free system, determination of large-amplitude swelling, etc.).

Acknowledgments

This work has been partially supported by the French Ministry of Science, ANRS, ARC, CNRS, INSERM, and a special grant from LNC (to G.K.). S.A.S. received a fellowship from the European Commission. M.G. is a senior research associate of the Belgian National Fund for Scientific Research.

[19] Quantitation of Mitochondrial Transmembrane Potential in Cells and in Isolated Mitochondria

By Naoufal Zamzami, Didier Métivier, and Guido Kroemer

Introduction

Accurate quantitation of the mitochondrial inner transmembrane potential ($\Delta\Psi_m$) using appropriate potentiometric fluorochromes reveals a $\Delta\Psi_m$ decrease in most if not all models of apoptosis. This $\Delta\Psi_m$ decrease is an early event and thus precedes the activation of downstream caspases and nucleases. Most of the current data are compatible with the notion that a $\Delta\Psi_m$ decrease mediated by opening of the mitochondrial permeability transition (PT) pore (also called megachannel) constitutes an irreversible event of the apoptotic process.[1,2] Therefore, the determination of the $\Delta\Psi_m$ allows for the detection of early apoptosis both *in vitro* and *ex vivo*. In cell-free systems of apoptosis, purified mitochondria can be exposed to

---

[1] N. Zamzami, C. Brenner, I. Marzo, S. A. Susin, and G. Kroemer, *Oncogene* **16**, 2265 (1998).
[2] S. A. Susin, N. Zamzami, and G. Kroemer, *Biochim. Biophys. Acta* (*Bioenerg.*) **1366**, 151 (1998).

apoptogenic molecules. The quantitation of the ΔΨ$_m$ of isolated mitochondria is useful because it detects changes in the mitochondrial inner membrane permeability that usually are accompanied by an increase in outer membrane permeability leading to the release of caspase- and nuclease-activating proteins normally stored in the intermembrane space.[3,4] Here we propose several protocols for the quantitation of the ΔΨ$_m$ in intact cells and in purified mitochondria.

Quantification of Mitochondrial Transmembrane Potential in Cells

Lipophilic cations accumulate in the mitochondrial matrix, driven by the electrochemical gradient following the Nernst equation, according to which every 61.5 mV increase in membrane potential (usually 120–170 mV) corresponds to a 10-fold increase in cation concentration in mitochondria. Therefore, the concentration of such cations is 2 to 3 logs higher in the mitochondrial matrix than in the cytosol. Several different cationic fluorochromes can be employed to measure mitochondrial transmembrane potentials. These markers include 3,3'-dihexyloxacarbocyanine iodide [DiOC$_6$(3)] (fluorescence in green),[5] chloromethyl-X-rosamine (CMXRos) (fluorescence in red),[6,7] and 5,5',6,6'-tetrachloro-1,1',3,3'-tetraethylbenzimidazolcarbocyanine iodide (JC-1) (fluorescence in red and green).[8] As compared with rhodamine 123 (Rh123), which we do not recommend for cytofluorometric analyses,[9] DiOC$_6$(3) offers the important advantage that it does not show major quenching effects. JC-1 incorporates into mitochondria, where it either forms monomers (fluorescence in green, 527 nm) or, at high transmembrane potentials, aggregates (fluorescence in red, 590 nm).[8] Thus, the quotient between green and red JC-1 fluorescence provides an estimate of ΔΨ$_m$ that is (relatively) independent of mitochondrial mass.

Materials

Stock solutions of fluorochromes: DiOC$_6$(3) should be diluted to 40 μM in dimethyl sulfoxide (DMSO), CMXRos to 1 mM in DMSO, and JC-1 to 10 mM in DMSO. All three fluorochromes can be

[3] N. Zamzami, S. A. Susin, P. Marchetti et al., J. Exp. Med. **183,** 1533 (1996).
[4] S. A. Susin, N. Zamzami, M. Castedo et al., J. Exp. Med. **184,** 1331 (1996).
[5] P. X. Petit, J. E. O'Connor, D. Grunwald, and S. C. Brown, Eur. J. Biochem. **220,** 389 (1990).
[6] M. Castedo, T. Hirsch, S. A. Susin et al., J. Immunol. **157,** 512 (1996).
[7] A. Macho, D. Decaudin, M. Castedo et al., Cytometry **25,** 333 (1996).
[8] S. T. Smiley, Proc. Natl. Acad. Sci. U.S.A. **88,** 3671 (1991).
[9] D. Metivier, B. Dallaporta, N. Zamzami et al., Immunol. Lett. **61,** 157 (1998).

purchased from Molecular Probes (Eugene, OR) and should be stored, once diluted, at $-20°$ in the dark

Working solutions: Dilute $DiOC_6(3)$ to 400 nM [10 μl of stock solution plus 1 ml of phosphate-buffered saline (PBS)], CMXRos to 10 μM nM (10 μl of stock solution plus 1 ml of PBS or mitochondrial resuspension buffer), and JC-1 to 20 μM (2 μl of stock solution plus 1 ml of PBS). These solutions should be prepared fresh for each series of stainings

Carbonyl cyanide m-chlorophenylhydrazone (CCCP) diluted in ethanol (stock at 20 mM): A protonophore required for control purposes ($\Delta\Psi_m$ disruption)

Cytofluorometer with appropriate filters

Staining Protocol

1. Cells (5–10×10^6 in 0.5 ml of PBS) should be kept on ice until staining. If necessary, cells can be labeled with specific antibodies conjugated to compatible fluorochromes [e.g., phycoerythrin for $DiOC_6(3)$, fluorescein isothiocyanate for CMXRos] before determination of mitochondrial potential.

2. For staining, add the following amounts of working solutions to 0.5 ml of cell suspension: 25 μl of $DiOC_6(3)$ (final concentration, 20 nM), 25 μl of CMXRos (final concentration, 100 nM), or 25 μl of JC-1 (final concentration, 1 μM) and transfer the tubes to a water bath kept at $37°$. After 15–20 min of incubation, return the cells to ice. Do not wash the cells. As a negative control, in each experiment aliquots of cells should be labeled in the presence of the protonophore CCCP (100 μM).

3. Perform cytofluorometric analysis within 10 min, while gating the forward and sideward scatters on viable, normal-sized cells. When large series of tubes are to be analyzed (>10 tubes), the interval between labeling and cytofluorometric analysis should be kept constant. When using an Epics Profile cytofluorometer (Coulter, Hialeah, FL), $DiOC_6(3)$ should be monitored in FL1 (Fig. 1), CMXRos in FL3 (excitation, 488 nm; emission, 599 nm), and JC-1 in FL1 versus FL3 (excitation, 488 nm; emission, 527 and 590 nm). The following compensations are recommended for JC-1: 10% of FL2 in FL1, and 21% of FL1 in FL2 (indicative values).

Quantification of Mitochondrial Transmembrane Potential in Isolated Mitochondria

The $\Delta\Psi_m$ of isolated mitochondria can be quantified by multiple methods. Here we propose two different methods. One is based on the cytoflu-

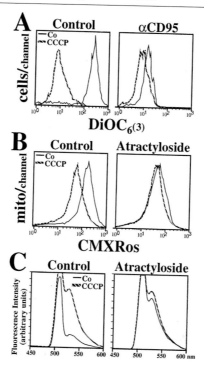

DiOC$_6$(3)

CMXRos

FIG. 1. Representative examples for $\Delta\Psi_m$ measurements. (A) $\Delta\Psi_m$ measurement in Jurkat cells that were either left untreated (control) or were treated for 3 hr with a CD95 cross-linking antibody (CH11; 1 μg/ml) for a period of 3 hr, followed by staining with DiOC$_6$(3) in the presence (solid line) or absence of CCCP (dashed line). (B) $\Delta\Psi_m$ measurement in isolated mouse liver mitochondria treated with 5 mM atractyloside, followed by staining with CMXRos in the presence or absence of CCCP and cytofluorometric evaluation of the CMXRos-dependent fluorescence. Note that higher fluorescence values imply a higher $\Delta\Psi_m$. (C) $\Delta\Psi_m$ measurement using Rh123. The same samples as in (A) were stained with Rh123 and evaluated in a fluorometer. Emission spectra for an excitation of 488 nm are shown. Note the inverse correlation between $\Delta\Psi_m$ and Rh123-dependent fluorescence.

orometric analysis of purified mitochondria on a per-mitochondrion basis. In this case, the incorporation of the dye CMXRos is measured and low levels of CMXRos incorporation indicate a low $\Delta\Psi_m$. The second protocol is performed as a bulk measurement, based on the quantitation of rhodamine 123 quenching. At a high $\Delta\Psi_m$ level, most of the rhodamine 123 is concentrated in the mitochondrial matrix and quenches. At lower $\Delta\Psi_m$ levels, rhodamine 123 is released, causing dequenching and an increase in rhodamine-123 fluorescence. Thus, a low $\Delta\Psi_m$ level corresponds to a higher value of rhodamine-123 fluorescence.

Required Materials

M buffer: 220 mM sucrose, 68 mM mannitol, 10 mM KCl, 5 mM KH$_2$PO$_4$, 2 mM MgCl$_2$, 500 μM EGTA, 5 mM succinate, 2 μM rotenone, 10 mM HEPES (pH 7.2)

Mitochondria: Purified as described,[10] kept on ice for a maximum of 4 hr, and resuspended in M buffer (or similar buffer)

CMXRos stock solution: Prepare as described above. Rh123 stock solution (10 mM in ethanol, to be kept at $-20°$, protected against light)

Advanced cytofluorometer capable of detecting isolated mitochondria

Spectrofluorometer

Protocol

Mitochondria are incubated for 30 min at 37° in the presence of the indicated reagent. Use 100 μM CCCP as a control. Determine the $\Delta\Psi_m$ using the potential-sensitive fluorochrome chloromethyl-X-rosamine (100 nM, 15 min, room temperature) and analyze in a FACSVantage cytofluorometer (Becton Dickinson, San Jose, CA) or similar, while gating on single-mitochondrion events in the forward and side scatters (Fig. 1B).

Alternatively, mitochondria (1 mg of protein per milliliter) are incubated in a buffer supplemented with 5 μM rhodamine 123 for 5 min, and the $\Delta\Psi_m$-dependent quenching of rhodamine fluorescence (excitation, 490 nm; emission, 535 nm) is measured[11] as shown in Fig. 1C in a fluorometer.

Anticipated Results and Pitfalls

The determination of the $\Delta\Psi_m$ in intact cells is a highly reproducible procedure. Note that, in each experiment, controls must be performed to assess background incorporation of fluorochromes in the presence of CCCP, a protonophore causing a complete disruption of the $\Delta\Psi_m$. Because the incorporation of these fluorochromes may be influenced by parameters not determined by mitochondria (cell size, plasma membrane permeability, efficacy of the multiple disease resistance pump, etc.), results can be interpreted only when the staining profiles obtained in different experimental conditions (e.g., controls versus apoptosis) are identical in the presence of CCCP (Fig. 1A). Working with isolated mitochondria requires that the mitochondrial preparations be optimal and fresh (<4 hr). Rhodamine 123

[10] S. A. Susin, N. Larochette, M. Geuskens, and G. Kroemer, *Methods Enzymol.* **322**, [18], 2000, this volume.
[11] S. Shimizu, Y. Eguchi, W. Kamiike *et al.*, *Proc. Natl. Acad. Sci. U.S.A.* **95**, 1455 (1998).

fluorescence measurements in a spectrofluorometer generate less problems than cytofluorometric measurements of isolated mitochondria.

Acknowledgments

Supported by the French Ministry of Science, ARC, ANRS, CNRS, FF, INSERM, and a special grant from LNC.

[20] Nitrogen Cavitation for Cell Disruption to Obtain Mitochondria from Cultured Cells

By ROBERTA A. GOTTLIEB and SOUICHI ADACHI

Background

Cell disruption by nitrogen cavitation is based on rapid decompression of a cell suspension from within a pressure vessel. This gentle method of cell disruption has gained wide popularity among investigators working with granule-laden cells such as neutrophils and platelets, because it allows efficient recovery of intact granules.[1-3] This method has been used for preparation of mitochondria from Chinese hamster ovary cells and from the AH130 hepatoma cell line.[4,5] We have applied this method to the recovery of mitochondria from cultured cell lines such as Jurkat T lymphoblasts and CEM cells.[6-8] Mitochondria prepared by this method possess an intact outer membrane and retain intermembrane space components, including cytochrome c and adenylate kinase.[8]

Nitrogen is dissolved in the cell suspension under high pressure within the cavitation chamber, or bomb. The cell suspension is then released dropwise through outflow tubing as the pressure is released. The nitrogen comes out of solution, forming bubbles that expand and rupture the cells. Shear stress also contributes to cell disruption, as bubbles stream through

[1] B. J. Del Buono, F. W. Luscinskas, and E. R. Simons, *J. Cell. Physiol.* **141**, 636 (1989).
[2] N. Borregaard, J. M. Heiple, E. R. Simons, and R. A. Clark, *J. Cell Biol.* **97**, 52 (1983).
[3] M. J. Broekman, *Methods Enzymol.* **215**, 21 (1992).
[4] L. Cezanne, L. Navarro, and M. Toho, *Biochim. Biophys. Acta* **1112**, 205 (1992).
[5] Y. Shinohara, I. Sagawa, J. Ichihara, K. Yamamoto, K. Terao, and H. Terada, *Biochim. Biophys. Acta* **1319**, 319 (1997).
[6] A. Krippner, A. Yagi, R. A. Gottlieb, and B. M. Babior, *J. Biol. Chem.* **271**, 21629 (1996).
[7] S. Adachi, A. R. Cross, B. M. Babior, and R. A. Gottlieb, *J. Biol. Chem.* **272**, 21878 (1997).
[8] S. Adachi, R. A. Gottlieb, and B. M. Babior, *J. Biol. Chem.* **273**, 19892 (1998).

the solution bathing the cells. The density of the bubbles is related to the amount of nitrogen dissolved in the solution, which is directly dependent on the pressure at which the chamber is equilibrated. For instance, disruption of human neutrophils and many other cultured cell lines is achieved at a pressure of 420 lb/in^2,* while certain other cell types or tissues may require higher pressures. Disruption of subcellular organelles may be achieved by increasing the pressure (>1000 lb/in^2) or by adding hypotonic stress.

Because each cell experiences the depressurization only once, there is no repetitive damage. In contrast, homogenization using either a glass–glass or Teflon–glass homogenizer is based on shear stress as cells stream between two tightly apposed surfaces. Organelles from cells disrupted early in the course of homogenization are subjected to repetitive shear stresses, which may result in their eventual rupture.

Overview

Cells are washed and resuspended at a concentration of 10^8/ml in cavitation buffer, and then placed in the cavitation vessel with a magnetic stir bar. The vessel is pressurized with nitrogen gas to 420 lb/in^2 and allowed to equilibrate for 5–20 min to allow time for the nitrogen to become fully dissolved in the buffer (larger volumes require the longer time). The valve for cell outflow is opened gradually, so that the suspension is released dropwise and collected in a tube. The lysate can then be processed as usual to recover nuclei and unbroken cells with a low-speed centrifugation, followed by a high-speed spin (10,000g) to recover mitochondria. The post-mitochondrial supernatant can be further processed by centrifugation (100,000g) to recover light membranes and cytosol.

Selection of Buffer for Cell Disruption

The cytosol recovered from this procedure will be diluted in this initial cavitation buffer, so it is recommended that the composition of this buffer be compatible with later experimental needs. Although the buffer we use is hypoosmolar (buffer B, below), similar results can be obtained with isotonic buffers. This set of buffers is adapted from a homogenization protocol described by Bourgeron et al.[9] for the preparation of mitochondria from B lymphoblastoid cell lines. Protease inhibitors may be included in this buffer, if desired. It is good practice to verify cell disruption by inspecting a

---

* 1 atm equals ~14.7 pounds per square inch (lb/in^2) at sea level.
[9] T. Bourgeron, D. Chretien, A. Rotig, A. Munnich, and P. Rustin, *Biochem. Biophys. Res. Commun.* **186**, 16 (1992).

sample stained with trypan blue. If necessary, the cavitation pressure can be increased or the osmolarity can be decreased to achieve efficient cell disruption. The lysate can also be subjected to a second round of cavitation. It may be necessary to establish these parameters for each cell type used, if the standard protocol is not adequate. For instance, we have found that frozen and revived lymphoblasts from patients with chronic lymphocytic leukemia are fairly difficult to disrupt, and therefore we employ a somewhat more hypotonic buffer for disruption of these cells.

Buffer Recipes

MA buffer: 100 mM sucrose, 1 mM EGTA, 20 mM 3-(N-morpholino) propanesulfonic acid (MOPS, pH 7.4), bovine serum albumin (BSA, 1 g/liter)

MB buffer: MA buffer plus 10 mM triethanolamine, 5% (w/v) Percoll, and an antiprotease mixture consisting of aprotinin, pepstatin A, and leupeptin, each at 10 μM, and 1 mM phenylmethylsulfonyl fluoride (PMSF)

MC buffer: 300 mM sucrose, 1 mM EGTA, 20 mM MOPS (pH 7.4), BSA (1 g/liter), and antiprotease mixture as described above

Respiration buffer: 250 mM sucrose, 10 mM $MgCl_2$, 5 mM KH_2PO_4, 10 mM K^+-HEPES (pH 7.4), BSA (1 g/liter)

Cavitation Procedure

Supplies

Nitrogen cavitation device (pressure bomb) (e.g., from Parr Instruments, Moline, IL; *parrinst.com*)

Cylinder of nitrogen gas (we use a cylinder capable of holding 304 ft^3 of gas)

High-pressure two-stage regulator: To allow easier pressure control, especially when releasing pressure sequentially from a multiwell bomb.

Sample chamber, if not part of the steel vessel itself

Conical 15-ml polypropylene tube for sample collection

A simplified diagram of a typical cavitation device is shown in Fig. 1.

We keep the apparatus in a cold room to ensure adequate sample chilling. The cell suspension is placed in the sample chamber of the prechilled vessel (which can be adapted to hold the sample in a plastic chamber with an outflow tube at the bottom of the tapered chamber), or added to a well in the base of the bomb itself, with an outflow tube at the base.

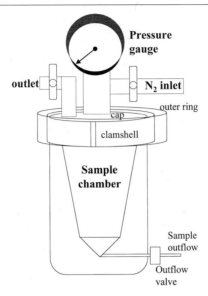

FIG. 1. Nitrogen cavitation device. The cell suspension is placed in the well of the *sample chamber,* along with a stir bar. The *cap* is positioned with an O-ring to provide a good seal. The two-component *clamshell* joins the cap to the base, and the clamshell is held in place with an *outer ring.* Nitrogen gas is introduced via the *N₂ inlet valve,* and after equilibration, the sample is depressurized by releasing it through the *outflow valve.* After the sample is collected, the remaining nitrogen gas is released through the *outlet* in the cap.

The outflow tube is connected to the outflow valve, which is closed. The sample chamber must also accommodate a magnetic stir bar, and the entire apparatus is placed on a magnetic stir plate. The stirring speed is adjusted to ensure adequate mixing without frothing or splashing the buffer. The cavitation chamber is then sealed.

A variety of designs have been introduced to ensure proper sealing to withstand the pressurization, but one of the simplest designs is that produced by Parr Instruments. The heavyweight stainless steel base rests on the stir plate, and has an O-ring between the base and the cap. The cap contains gas inlet and outlet valves and, in some cases, a pressure gauge. The cap is attached to the base by a strong steel "clamshell" (a two-component retaining ring, which in turn is secured by an outer ring held in place with a single screw). In other designs, the cap may be secured by a retaining ring that screws on and is tightened with bolts.

The tubing from the nitrogen tank is connected to the inlet valve of the bomb. All valves are closed. The main valve of the nitrogen cylinder is opened and the regulator pressure is set to 420 lb/in². The main outlet valve and sample outlet valves are double-checked to be sure that they are closed. Then the outlet valve of the regulator and the inlet valve of the bomb are opened, and nitrogen gas is introduced into the chamber. Nitrogen pressure is maintained at 420 lb/in^2 while the nitrogen dissolves into the sample and buffer. Five minutes should be adequate for small volumes of buffer (e.g., 1 ml), while larger volumes (>50 ml) may require longer times (~20 min). This should be determined empirically by inspecting a portion of the sample after staining with trypan blue.

After equilibration, the inlet valve is closed, as well as the outlet valve of the regulator and the main valve of the nitrogen cylinder. The collection tube is placed at the opening of the sample outlet, and the sample outflow valve is opened slowly. The suspension should be released dropwise. This step is critical, as opening it too rapidly results in shooting the sample out (possibly splashing it out of the collection tube) and also results in frothing and inefficient cell disruption. Some operators collect the sample in several fractions, to avoid blowing out the entire sample with a gust of nitrogen as the last bit of sample leaves the chamber. A high-quality mechanical valve is critical for control of sample outflow rate. Once the sample is collected (nearly total recovery should be expected, if the dead space for sample tubing is kept small), the remainder of the nitrogen can be released through the outflow valve in the cap. The filling tubing can be disconnected, and the bomb disassembled. The sample chamber and outflow tubing should be washed promptly to prevent the accumulation of debris.

Sample Processing for Mitochondria

The suspension of disrupted cells (2×10^8/ml in MB buffer) is recovered in a long test tube (e.g., a 15-ml conical tube) to avoid losing any of the material. If a hypoosmolar buffer was used, sucrose should be added to bring the solution back to isoosmolarity. Nuclei and unbroken cells are removed by two successive centrifugations (2500g, 5 min, 5°), and then the postnuclear supernatant is transferred to a microcentrifuge tube and centrifuged at 10,000g for 15 min. The crude mitochondrial pellet thus obtained is resuspended by trituration and can be washed twice more, or further purified by sucrose or Percoll gradient centrifugations. The postmitochondrial supernatant (representing 2×10^8 cell equivalents/ml) can be further separated into light membranes and cytosol by sedimentation at 100,000g for 30 min.

Evaluation of Cytochrome c Release or Dissociation

Because of the widespread interest in cytochrome c release from mitochondria during apoptosis, it is important to consider this topic in the context of outer membrane integrity.

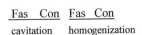

Fas   Con   Fas   Con
cavitation   homogenization

FIG. 2. Western blot of cytochrome c from cytosol of control and Fas-treated cells disrupted by nitrogen cavitation or homogenization. Jurkat cells were incubated with anti-Fas IgM or cultured in parallel for 2 hr, and then disrupted by cavitation or conventional homogenization. After sedimentation of nuclei and mitochondria, cytosol was obtained as the 100,000g supernatant, and resolved by SDS–PAGE and immunoblotting.

Western Blotting

Cytochrome c release has been demonstrated by homogenization of cells resuspended at 2×10^8 cells/ml in homogenization buffer [250 mM sucrose, 20 mM K^+-HEPES (pH 7.5), 10 mM KCl, 1.5 mM MgCl$_2$, 0.1 mM EDTA, 1 mM EGTA, 1 mM dithiothreitol, and 0.1 mM PMSF] and disrupted by homogenization with a Potter-Elvehjem homogenizer with a Teflon pestle.[10] Nuclei are removed by sedimentation at 2500g (5 min, done twice), and then mitochondria are collected by sedimentation at 10,000g (15 min). Cytosol is prepared by centrifugation at 100,000g for 30 min. A comparison of cytochrome c distribution from control and Fas-ligated cells disrupted by nitrogen cavitation or homogenization is shown in Fig. 2. The finding that cell disruption by nitrogen cavitation did not result in the appearance of cytochrome c in the cytosolic fraction suggests that cytochrome c release might be secondary to disruption of the outer mitochondrial membrane. A method for determining outer mitochondrial membrane integrity is presented below. It should be noted that cytochrome c can be more readily released from apoptotic mitochondria, implying that it is already dissociated from the electron transport chain, consistent with earlier findings.[6]

Oxygen Consumption

One indirect method for inferring cytochrome c release is to measure oxygen consumption with a Clarke oxygen electrode.[11] TMPD/ascorbate (1 mM ascorbate with 0.4 mM tetramethyl-p-phenylenediamine) is used as

[10] X. Liu, C. N. Kim, J. Yang, R. Jemmerson, and X. Wang, *Cell* **86**, 147 (1996).
[11] K. Schulze-Osthoff, A. C. Bakker, B. Vanhaesebroeck, R. Beyaert, W. A. Jacob, and W. Fiers, *J. Biol. Chem.* **267**, 5317 (1992).

a donor of electrons to cytochrome c. If whole cells (rather than mitochondria) are used, digitonin (0.005%, w/v) should be included to permit entry of TMPD/ascorbate.

Oxidation of Dihydrorhodamine

Another indirect method to detect cytochrome c dissociation is to measure oxidation of dihydrorhodamine 123 (DHR). DHR itself is nonfluorescent until oxidized to rhodamine 123. Although there is no clear evidence that DHR oxidation is directly dependent on electron transfer through cytochrome c, we have established conditions under which DHR oxidation is dependent on the availability of substrate (5 mM succinate plus 1 mM ADP) and is inhibitable by the addition of KCN (5 mM) (S. Adachi and R. A. Gottlieb, unpublished, 1999).

Intact cells or mitochondria from 2×10^6 cells are resuspended in respiration buffer containing 1 mM ADP and 5 mM succinate [include digitonin at 0.01% (w/v) for whole cells]. Cells or mitochondria are loaded at room temperature for 5 min with 0.1 μM dihydrorhodamine 123 (DHR; Molecular Probes, Eugene, OR). Flow cytometry is performed with excitation of the mitochondria at 488 nm and emission at 525 nm. The forward and side scatter histogram is generated with a 4-decade log scale for both parameters and all data are collected on a 4-decade log scale. Under these conditions, cells or mitochondria will oxidize dihydrorhodamine to an extent proportional to the amount of cytochrome c retained in mitochondria [when measured by Western blot after digitonin (0.06%, w/v) permeabilization and addition of 10 mM KCl].

Characterization of Mitochondrial Outer Membrane Integrity

Although electron transport and mitochondrial membrane potential are not impaired if the outer mitochondrial membrane is disrupted, investiga-

Ctrl	Bid 100ng	Bid 10ng	Dig 0.01%	Dig 0.03%	Dig 0.06%
0.056 (0.02)	0.091 (0.01)	0.076 (0.01)	0.058 (0.02)	0.111 (0.03)	0.172 (0.01)

FIG. 3. Cytochrome oxidase assay. Mitochondria isolated by nitrogen cavitation were incubated with the indicated concentrations of recombinant truncated Bid or digitonin. The decrease in absorbance at 550 nm (indicating cytochrome c oxidation) was monitored for 3 min. Rates are indicated within each frame, and reflect the mean (± 1 SD) of triplicates.

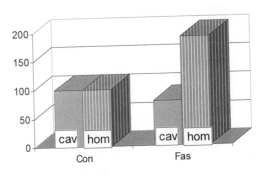

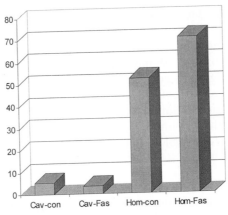

FIG. 4. Effect of cell disruption method on cytochrome *c* release and outer membrane permeability. Jurkat cells were incubated with anti-Fas IgM or cultured in parallel for 2 hr, then disrupted by cavitation or conventional homogenization. *Top:* Cytosolic cytochrome *c* was measured by immunoblotting and reported as a percent of the control (= 100%). *Bottom:* The lysate was tested for cytochrome oxidase activity, reported as a percentage of the maximal rate obtained with digitonin (= 100%).

tions of the intermembrane space of mitochondria depend critically on mitochondrial preparations that have intact outer membranes. This can be evaluated through the use of a cytochrome oxidase assay,[6,8] which we have modified to be suitable for use in a 96-well plate format.

Cytochrome *c* must be reduced and prepared freshly each time. Horse heart cytochrome *c* (70 mg, MW 12,384; Sigma Chemicals, St. Louis, MO) is dissolved in 565 μl of 2-(N-morpholino) ethanesulfonic acid (MES) buffer (0.1 M MES, pH 6.0) and 700 μl of 1 M ascorbate (pH 7.0) is added. The material is applied to a PD-10 desalting column equilibrated with 0.1 M MES. Collect 2 ml of the darkest pink solution. The final concentration of cytochrome c is approximately 0.1 mM.

Cytochrome oxidase buffer (0.1 M MES, pH 6.0, with 250 mM sucrose and 10 μM EGTA) is prepared and 150 or 160 μl is added to each well [depending on whether 10 μl of 1.25% (w/v) digitonin will be added to that well].

Ten microliters of cytochrome c is added to each well, and wells are checked in the plate reader for a stable baseline. If the baseline is unstable, the addition of 5 mM MgCl$_2$ may be helpful.

The reaction is initiated by the addition of 30 μl of mitochondria equals 2.3 × 10^7 cell equivalents/ml) (suspended in MC buffer at ~7 × 10^8 cell equivalents/ml). In the case of Jurkat cells, this corresponds to about 20 μg of mitochondrial protein. The rate of cytochrome c oxidation is monitored in a plate reading spectrophotometer (Molecular Devices, Palo Alto, CA) at 550 nm for a period of 3 min. Permeability is reported as the rate of cytochrome c oxidation in the sample as a percentage of the maximal rate obtained in the presence of digitonin [expected optical density (OD) change of 0.5 units over 3 min]. Raw data from a typical assay are shown in Fig. 3.

The integrity of the outer membrane depends critically on the method employed for cell disruption, as shown in Fig. 4. While nitrogen cavitation is associated with only a small degree of leakiness (corresponding to low numbers of leaky mitochondria), homogenization results in extensive disruption of the outer membrane. Although by nitrogen cavitation Fas ligation does not appear to significantly increase the leakiness of the outer membrane, there is a significant increase in leakiness of the apoptotic mitochondria prepared by homogenization. This suggests that the outer mitochondrial membrane may be more easily disrupted in apoptotic cells, perhaps because the mitochondria may be swollen or more rigid.

In summary, cell disruption by nitrogen cavitation yields mitochondria with an intact outer mitochondrial membrane. Under these conditions, cytochrome c and other intermembrane space components may be retained even in mitochondria from apoptotic cells.

Acknowledgments

The authors acknowledge Akemi Matsuno-Yagi for useful discussions. This work was supported in part by NIH R01 AG-13501.

[21] Apoptosis-Related Activities Measured with Isolated Mitochondria and Digitonin-Permeabilized Cells

By Gary Fiskum, Alicia J. Kowaltowksi, Alexander Y. Andreyev, Yulia E. Kushnareva, and Anatoly A. Starkov

Introduction

Mitochondrial alterations have been recognized for many years as contributors to cell death. Discoveries made within the last decade have further identified mitochondria as key players in the process of apoptosis.[1,2] In response to several different apoptotic stimuli, e.g., elevated intracellular Ca^{2+} and the subcellular redistribution of cell death proteins such as Bax, mitochondria release cytochrome *c*, procaspase 9, and apoptosis-initiating factor (AIF) into the cytosol. These proteins then form complexes with other factors, resulting in activation of the cell death protease (caspase) cascade that mediates the final stages of apoptosis, i.e., nuclear fragmentation and the disorganization of the cell into vesicular apoptotic bodies.

Although an appreciation of the roles that mitochondria play in apoptosis has developed largely from indirect measurements of mitochondrial structure, function, and permeability within cells, understanding the molecular mechanisms responsible for release of apoptotic proteins from mitochondria requires simpler experimental models that allow for better control over the biochemical milieu. These models include measurements performed with isolated mitochondria and with cells permeabilized so that the soluble components of the cytosol are in equilibrium with those of the medium. In addition to providing direct access to mitochondria of exogenous molecules and ions, these systems enable measurements of mitochondrial activities and characteristics, e.g., respiration and mitochondrial membrane potential, that are much more difficult to assess accurately *in situ* within intact cells and tissues. However, activities measured with mitochondria and permeabilized cells in the presence of artificial media, even in the presence of cytosolic extracts, are likely to be at least quantitatively if not qualitatively different from those that exist within the true intracellular environment. Therefore, comparisons of results and interpretations obtained with both cellular and subcellular paradigms help validate each approach.

---

[1] D. R. Green and J. C. Reed, *Science* **281**, 1309 (1998).
[2] A. N. Murphy, G. Fiskum, and M. F. Beal, *J. Cereb. Blood Flow Metab.* **19**, 231 (1999).

The following examples of evaluations of mitochondrial activities represent those being used by many laboratories that seek an understanding of both the mechanisms and regulation of mitochondrial roles in apoptosis and the impact that apoptotic stimuli have on normal mitochondrial energy-transducing activities.

Materials and Methods

Reagents and Chemicals

tert-Butyl hydroperoxide, safranine O, rotenone, succinate, malate, glutamate, digitonin, carbonyl cyanide *p*-trifluoromethoxyphenylhydrazone (FCCP), alamethicin, and protease inhibitor cocktail are obtained from Sigma (St. Louis, MO). Cyclosporin A is from Alexis (San Diego, CA). Tetraphenylphosphonium chloride is from Aldrich Chemicals (Milwaukee, WI). Precast gradient gels and polyvinylidene difluoride (PVDF) membrane-filter paper sandwiches are from Novex (San Diego, CA). X-Omat AR film is from Kodak (Rochester, NY). The source of other chemicals is provided in text.

Isolation of Mitochondria from Tissues and Cells

Many different procedures are established for the isolation of mitochondria from different tissues and cultured cells.[3-5] They all involve either homogenization or disruption of tissues and cells so that mitochondria can be separated from other cellular constituents. Separation normally involves either differential centrifugation or centrifugation through gradients of sucrose, Ficoll, or Percoll.[6,7] Some of these techniques provide for optimal purity of mitochondria while others were developed to obtain mitochondria with optimal energy-linked activities, e.g., oxygen consumption coupled to ATP synthesis. It should always be kept in mind that isolated mitochondria likely do not represent the entire population of mitochondria present within a particular tissue or collection of cells. In fact, some procedures are designed to separate normal subpopulations of mitochondria.[6,7] In addition, because many isolation procedures are intended to yield optimal bioenergetic mitochondrial activities, they often exclude "damaged" mitochondria

[3] *Methods Enzymol.* **10** (1967).
[4] C. P. Lee, *Biochim. Biophys. Acta* **1271,** 21 (1995).
[5] R. Moreadith and G. Fiskum, *Anal. Biochem.* **137,** 360 (1984).
[6] R. Hovius, H. Lambrechts, K. Nicolay, and B. Kruijff, *Biochim. Biophys. Acta* **1021,** 217 (1990).
[7] F. Dagani, F. Zanada, F. Marzatico, and G. Benzi, *J. Neurochem.* **45,** 653 (1985).

that are generally thought to be generated by the isolation process. However, such exclusion can also result in the loss of subsets of mitochondria that are normally less bioenergetically sound than that of other subsets or that are altered as a consequence of stress that occurs *in vivo*. Alternatively, procedures that strive to include all populations of mitochondria probably result in preparations that are "contaminated" with a subset of mitochondria that are altered because of damage suffered *ex vivo*.

While disruption of animal tissues, e.g., liver, brain, kidney, and heart, and the purification of mitochondria from the tissue homogenates is commonplace, isolation of mitochondria from cultured cells is more problematic because of the relative difficulty in disrupting suspensions of cells without disrupting the mitochondrial membranes. Cells can be made much more vulnerable to disruption with mechanical homogenizers by first selectively "tenderizing" the plasma membranes. This approach has involved the controlled use of proteolytic enzymes, e.g., Nagarse, or the extraction of cholesterol and other β-hydroxysterols by the addition of the steroid glycoside digitonin.[5,8] The specific affinity of digitonin for cholesterols allows it to selectively disrupt eukaryotic plasma membranes, which have a high cholesterol-to-phospholipid ratio, without disrupting intracellular membranes and, in particular, the mitochondrial inner membrane, which contains little cholesterol.[9] In one study comparing respiratory characteristics of mitochondria isolated from Ehrlich ascites tumor cells by the Nagarse method or the digitonin method, the digitonin method was shown to be superior.[5] However, as the mechanisms by which the mitochondrial outer membrane becomes permeable to large molecules such as cytochrome *c* are a current focus of research, digitonin should be used with caution. Although the mitochondrial inner membrane has low levels of cholesterol, the outer membrane contains cholesterol and is therefore subject to structural and functional alterations caused by digitonin, depending on the amount used in the mitochondrial isolation procedure. Thus it is important to use digitonin at the lowest level that will permeabilize the plasma membrane, which generally falls within the range of 0.005–0.020% (w/v).

Permeabilized Cells

In addition to the use of permeabilized cells for mitochondrial isolation procedures, this method has proved to be valuable for measuring the activities of mitochondria and other cellular organelles "*in situ*" without the need for isolation procedures that can cause functional alterations. These

[8] S. Kobayshi, B. Hagihara, M. Masuzumi, and K. Okunuki, *Biochim. Biophys. Acta* **113,** 421 (1966).
[9] G. Fiskum, *Cell Calcium* **6,** 25 (1985).

systems provide the added advantage of monitoring interactions between activities, e.g., Ca^{2+} transport, between different organelles.[10] Although the soluble cytosolic components are lost to the extracellular environment after cell permeabilization, the cytoskeleton remains relatively intact and retains structural associations with mitochondria and other cellular components.[11] Digitonin is the most commonly used agent for permeabilizing cells in mitochondrial studies[5,9-15]; however, other related compounds have been successfully employed, e.g., filipin and the heterogeneous mixture of digitonin-like compounds called saponins.[16-18]

An example of how permeabilized cells can be used to identify apparent alterations in mitochondrial function induced during a mitochondrial isolation procedure is described in Fig. 1.[18a,b] In this experiment, the basic ADP-stimulated (state 3) and nonphosphorylating (state 4) rates of respiration were measured with both digitonin-permeabilized GT1-7 neural cells and isolated GT1-7 mitochondria. Both Bcl-2-overexpressing transfectants (Bcl-2$^+$) and control transfectants (Bcl-2$^-$) were used to determine if overexpression of the antiapoptotic gene Bcl-2 results in altered electron transport and coupling to oxidative phosphorylation. The first recording of O_2 consumption demonstrates a simple technique for titrating digitonin with cells suspended in an O_2 electrode apparatus. In the presence of rotenone, an inhibitor of complex I of the electron transport chain,[19] endogenous cellular respiration is largely inhibited and limited by the permeability of the complex I-independent oxidizable substrate succinate across the plasma

[10] G. Becker, G. Fiskum, and A. L. Lehninger, J. Biol. Chem. **255**, 9009 (1980).

[11] G. Fiskum, S. W. Craig, G. Decker, and A. L. Lehninger, Proc. Natl. Acad. Sci. U.S.A. **77**, 3430 (1980).

[12] J. B. Hoek, J. L. Farber, A. P. Thomas, and X. Wang, Biochim. Biophys. Acta **1271**, 93 (1995).

[13] T. Jiang and D. Acosta, Jr., Toxicology **95**, 155 (1995).

[14] Y. V. Evtodienko, V. V. Teplova, J. Duszynski, K. Bogucka, and L. Wojtczak, Cell Calcium **15**, 439 (1994).

[15] A. E. Vercesi, C. F. Bernardes, M. E. Hoffmann, F. R. Gadelha, and R. Docampo, J. Biol. Chem. **266**, 14431 (1991).

[16] H. S. Gankema, E. Laanen, A. K. Groen, and J. M. Tager, Eur. J. Biochem. **119**, 409 (1981).

[17] H. Breitbart, R. Wehbie, and H. Lardy, Biochim. Biophys. Acta **1027**, 72 (1990).

[18] G. M. Burgess, J. S. McKinney, A. Fabiato, B. A. Leslie, and J. W. Putney, Jr., J. Biol. Chem. **258**, 15336 (1983).

[18a] A. N. Murphy, D. E. Bredesen, G. Cortopassi, E. Wang, and G. Fiskum, Proc. Natl. Acad. Sci. U.S.A. **93**, 9893 (1996).

[18b] A. N. Murphy, K. M. Myers, and G. Fiskum, in "Pharmacology of Cerebral Ischemia, 1996" (J. Krieglstein, ed.), pp. 163–172. Medpharm Scientific Publishers, Stuttgart, Germany, 1996.

[19] M. Degli Esposti, Biochim. Biophys. Acta **1364**, 222 (1998).

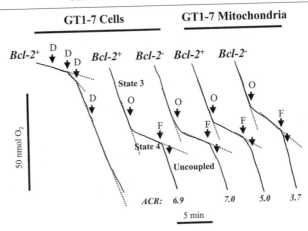

FIG. 1. Polarographic measurements of oxygen consumption by digitonin-permeabilized GT1-7 cells and isolated GT1-7 mitochondria. GT1-7-transformed murine hypothalamic cells that were transfected with and overexpressed the human *Bcl-2* gene (*Bcl-2*⁺) or that were transfected with the control construct that contained a reporter gene (*Bcl-2*⁻) were trypsinized from culture flasks and maintained in suspension in normal growth medium supplemented with 5 m*M* EGTA at room temperature for up to 4 hr. Immediately before each experiment, cells were centrifuged at 200*g* for 5 min and resuspended to 1 × 10⁷ cells/ml at 37° in 0.6 ml of respiration medium containing 130 m*M* KCl, 2 m*M* KH₂PO₄, 1 m*M* MgCl₂, 5 m*M* HEPES (pH 7.0), and different oxidizable substrates. For the first measurement, GT1-7 Bcl-2⁺ cells were suspended in medium containing 5 m*M* succinate, 2 μ*M* rotenone, and 0.25 m*M* ADP. Sequential aliquots of 0.005% (w/v) digitonin (D) from a 10% (w/v) stock solution in dimethyl sulfoxide were added and the oxygen concentration present within the closed chamber monitored with a Clark-type oxygen electrode. For the other two measurements performed with cells, the respiratory medium contained 5 m*M* malate, 5 m*M* glutamate, 0.25 m*M* ADP, and 0.015% (w/v) digitonin. Once a steady state 3 rate of respiration was reached, oligomycin (O, 2 μg/ml) was added. After a steady state 4 rate of respiration was reached, sequential additions of FCCP (F, 50 n*M*) were made to attain a fully uncoupled rate of respiration. For measurements of mitochondrial respiration, an aliquot of isolated GT1-7 mitochondria was added directly to the oxygen electrode chamber (final concentration, 0.5 mg of protein/ml) containing respiratory medium plus 5 m*M* malate, 5 m*M* glutamate, and 0.25 m*M* ADP. Oligomycin and FCCP were then added as for the measurements of cellular respiration. The acceptor control ratio (ACR) is defined as the ratio of state 3 to state 4 rates of respiration. See Refs. 18a and 18b for additional details.

membrane. With successive additions of digitonin, the plasmalemma becomes permeable to succinate and respiration is greatly stimulated. The minimum digitonin concentration necessary to provide maximal respiration is then used in other experiments.

The second and third measurements of O₂ consumption shown in Fig. 1 were used to determine the rates of state 3, state 4, and uncoupler-stimulated respiration for Bcl-2⁺ and Bcl-2⁻ cells. The acceptor control ratio, defined as the ratio of state 3 to state 4 respiration, was also calculated and provided in Fig. 1. The respiratory rates and acceptor control ratios were found not to be significantly different for permeabilized Bcl-2$^+$ and Bcl-2$^-$ GT1-7 cells. These same measurements were performed with mitochondria isolated from these cells, as described by the fourth and fifth respiratory measurements shown in Fig. 1. In addition to the acceptor control ratios being lower than those determined for the permeabilized cells, a significant difference in the ratios obtained for mitochondria isolated from Bcl-2$^+$ compared with Bcl-2$^-$ cells was observed. These results and others have led us to the conclusion that the overexpression of *Bcl-2* does not cause fundamental alterations in mitochondrial electron transport and oxidative phosphorylation and that the apparent differences observed with isolated mitochondria are primarily the result of Bcl-2 protein protecting against *ex vivo* damage to mitochondria that occurs during the isolation procedure. Therefore, the digitonin-permeabilized cell system may provide a more accurate representation of the true mitochondrial characteristics exhibited by the spectrum of mitochondria and cells present within a typical cell culture.

Safranine Measurements of Mitochondrial Membrane Potential

There are several ways that mitochondrial membrane potential ($\Delta\Psi$) can be monitored. One relatively easy method, applicable for use with isolated mitochondria and permeabilized cells, employs the cationic dye safranine.[20,21] Spectrophotometric determinations of mitochondrial $\Delta\Psi$ can be made with this agent because of the uptake of safranine into mitochondria and the metachromatic shift in its absorption spectrum that occurs on establishment of an electrical potential across the mitochondrial inner membrane. As with many other spectrophotometric methods applied to suspensions of mitochondria, dual-wavelength spectrophotometry should be used to correct for changes in light scattering or changes in absorption of other chromophores during the course of the experiment. Dual-wavelength spectrophotometers are relatively quite expensive, so other methods of measurement are often employed. Fortunately, safranine also fluoresces and its fluorescence is quenched by the accumulation of safranine in mitochondria in response to $\Delta\Psi$. Thus most spectrofluorometers equipped with a rapid stirring device that fully maintains mitochondria and cells in suspension are capable of monitoring mitochondrial $\Delta\Psi$ with safranine. Another

---

[20] K. E. Akerman, *Microsc. Acta* **81,** 147 (1978).
[21] A. Zanotti and G. F. Azzone, *Arch. Biochem. Biophys.* **201,** 255 (1980).

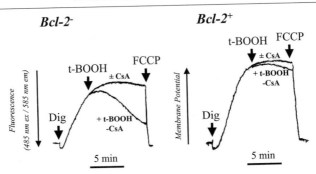

FIG. 2. Safranine O measurements of changes in mitochondrial membrane potential within digitonin-permeabilized PC12 cells. PC12 pheochromocytoma cells that were transfected with and overexpressed the human *Bcl-2* gene (Bcl-2$^+$) or that were transfected with the control construct that contained the reporter gene were trypsinized from culture flasks and maintained in suspension in normal growth medium supplemented with 5 mM EGTA at room temperature for up to 4 hr. Immediately before each experiment, 4×10^7 cells were centrifuged at 200g for 5 min and resuspended in 2.0 ml of medium containing 130 mM KCl, 5 mM succinate, 2 $\mu$$M$ rotenone, 5 mM HEPES (pH 7.0), 8.0 $\mu$$M$ CaCl$_2$, and 5 $\mu$$M$ safranine O at 30° in a quartz cuvette. The cuvette contained a magnetic stir bar and was placed in a spectrofluorom-eter with a thermostatically controlled cuvette holder and a magnetic stirring device. Fluores-cence measurements were made with excitation and emission wavelengths of 485 and 585 nm, respectively, and recorded on a strip-chart recorder. Digitonin (Dig) (0.01%, w/v) was added followed by 200 $\mu$$M$ *tert*-butyl hydroperoxide (*t*-BOOH) and 1 $\mu$$M$ FCCP.

useful fluorescent probe for monitoring mitochondrial $\Delta\Psi$ is tetrameth-ylrhodamine methyl ester (TMRM).[22]

Figure 2 describes spectrofluorometric safranine measurements of mito-chondrial $\Delta\Psi$ within digitonin-permeabilized PC12 neural cells. The goal of these measurements was to test the hypothesis that *Bcl-2* overexpression inhibits the drop in $\Delta\Psi$ induced by the prooxidant *tert*-butyl hydroperoxide (*t*-BOOH) in the presence of Ca^{2+}. The exposure of mitochondria to Ca^{2+} and a prooxidant is a commonly used technique for inducing the mitochon-drial permeability transition (MPT)[23,24] thought to be involved in the release of cytochrome c and other apoptogenic factors from mitochondria in many apoptosis paradigms.[25,26] The drop in $\Delta\Psi$ can represent an increase in the

[22] R. C. Scaduto, Jr. and L. W. Grotyohann, *Biophys. J.* **76,** 469 (1999).

[23] G. Fiskum and A. Pease, *Cancer Res.* **46,** 3459 (1986).

[24] A. J. Kowaltowski and A. E. Vercesi, *Free Radic. Biol. Med.* **26,** 463 (1999).

[25] J. J. Lemasters, A. L. Nieminen, T. Qian, L. C. Trost, S. P. Elmore, Y. Nishimura, R. A. Crowe, W. E. Cascio, C. A. Bradham, D. A. Brenner, and B. Herman, *Biochim. Biophys. Acta* **1366,** 177 (1998).

[26] G. Kroemer, B. Dallaporta, and M. Resche-Rigon, *Annu. Rev. Physiol.* **60,** 619 (1998).

permeability of the inner membrane to ions and small molecules that may result in sufficient osmotic swelling to rupture the mitochondrial outer membrane. Cytochrome c and other intermembrane proteins are then subject to being released into the extramitochondrial milieu.

As seen in Fig. 2, when digitonin is added to a suspension of Bcl-2$^-$ cells, the safranine fluorescence decreases because of the increased accessibility of safranine to mitochondria within the cell. In the absence of further additions, the decreased fluorescence remains stable until $\Delta\Psi$ is collapsed by the addition of the proton ionophore carbonyl cyanide p-trifluoromethoxyphenylhydrazone (FCCP). The collapse of $\Delta\Psi$ releases the accumulated safranine, thus eliminating the mitochondrial quenching of its fluorescence. In the presence of a low level of added Ca^{2+} (8 μM) that is insufficient to induce the MPT over the 15-min experimental period, the further addition of t-BOOH results in a substantial decline in $\Delta\Psi$ in Bcl-2$^-$ cells. Evidence that this decline in $\Delta\Psi$ is due to the MPT rather than to simple respiratory inhibition or uncoupling comes from the observation that the effect of t-BOOH is inhibited by the presence of the MPT inhibitor cyclosporin A (CsA).

The ability of t-BOOH plus Ca^{2+} to collapse mitochondrial $\Delta\Psi$ by induction of the MPT is also evidenced by the morphology of mitochondria present within the t-BOOH-treated permeabilized cells. Figure 3 provides transmission electron micrographs of digitonin-permeabilized PC12 Bcl-2$^-$ cells in the absence and presence of t-BOOH. Figure 3A and B provides low- and high-magnification images of cells permeabilized with digitonin and maintained under the conditions described for Fig. 2 in the absence of t-BOOH. As described previously for other cell types, the appearance of these cells is near normal and the mitochondria are intact and unswollen.[11] However, after a 10-min exposure to t-BOOH in the presence of 8 μM added Ca^{2+}, the mitochondria appear swollen (Fig. 3B), as expected from the induction of MPT and the ensuing net influx of ions, molecules, and H$_2$O into the mitochondrial matrix.

Multiparameter Measurements of Mitochondrial Ca^{2+} Transport, Membrane Potential, and Volume Changes

There is an increasing need to simultaneously monitor several different mitochondrial parameters to avoid misinterpretation of results obtained with any one measurement and to maximize the information obtained from a limited amount of biological material. For example, measurements of changes in mitochondrial $\Delta\Psi$ alone do not necessarily represent the activation or inhibition of permeability transitions, as many conditions, toxins, and pharmacological agents can either inhibit respiration or affect mito-

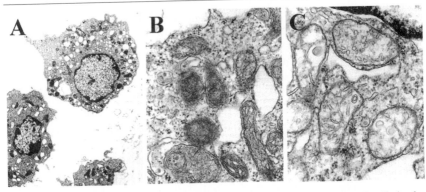

Fɪɢ. 3. Transmission electron microscopy of digitonin-permeabilized PC12 cells in the absence and presence of conditions leading to the mitochondrial membrane permeability transition. PC12 Bcl-2⁻ cells were digitonin permeabilized and incubated under the conditions described for Fig. 1. Samples of the cellular suspensions were centrifuged at 200g for 5 min and fixed overnight in a solution containing 4% (v/v) formaldehyde plus 1% (v/v) glutaralde-hyde, and postfixed in 1% (w/v) osmium tetroxide. Dehydration was performed in a series of ethanol and propylene oxide extractions, before sample embedding in Polibed. Sections were cut at 0.1 μm and stained with uranyl acetate plus lead citrate. (A) Low-magnification (×3500) photomicrograph of permeabilized cells in the absence of t-BOOH and FCCP. (B) High-magnification (×30,000) photomicrograph of permeabilized cells in the absence of t-BOOH and FCCP. (C) High-magnification (×30,000) photomicrograph of permeabilized cells after a 5-min exposure to 200 μM t-BOOH.

chondrial gradients of protons and other ions in ways that will lower $\Delta\Psi$ independent of any effects on permeability transition.

One method that is widely used to measure changes in mitochondrial volume due to alterations in the permeability of the inner membrane is light scattering. The scattering of light by vesicles that are osmotically responsive, e.g., mitochondria, is reduced as their volume decreases. The scattering of light by mitochondria can easily be monitored by measuring the absorbance of light at a wavelength between approximately 500 and 600 nm. At these high wavelengths, there is little (if any) interference due to light scattering by the relatively small absorbance changes that can be caused by mitochondrial chromophores, e.g., cytochromes. A single-beam spectrophotometer can be used to measure the light absorbance due to light scattering, although as usual, the cuvette holder must be equipped with a good stirring device. Alternatively, a thermostatically controlled chamber equipped with a red light-emitting diode (λ_{max}, 635–660 nm) and a light-sensitive photodiode or photocell (available at most electronics stores) is also satisfactory.

An example of experiments in which swelling of isolated mitochondria

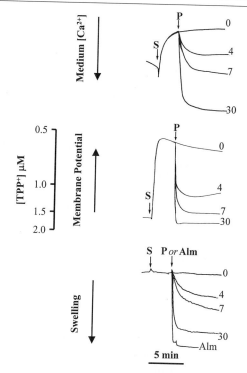

FIG. 4. Multiparameter measurements of mitochondrial Ca^{2+} transport, membrane potential, volume changes after exposure to a synthetic human cytochrome oxidase (COX) subunit IV signal peptide. Isolated rat forebrain mitochondria were suspended at a concentration of 0.4 mg of protein/ml in medium at 30° containing 210 mM mannitol, 70 mM sucrose, 3 mM KH_2PO_4, 10 mM HEPES (pH 7.4), 2 μM rotenone, and 2 μM tetraphenyl phosphonium (TPP$^+$) chloride. After the addition of mitochondria, 5 mM succinate (S) was added, followed by 4–30 μM COX IV$_{1-22}$ peptide or alamethicin (Alm, 30 μg/ml). Medium Ca^{2+} and TPP$^+$ concentrations were monitored with Ca^{2+}- and TPP$^+$-selective electrodes. Mitochondrial volume changes were monitored simultaneously by measuring the light scattering of suspensions via the absorbance of light at 660 nm.

is monitored by measuring absorbance of light at 560 nm, using a diode and photocell, is shown in Fig. 4. These measurements employed a multiparameter device additionally equipped with a Ca^{2+}-selective electrode for monitoring changes in the medium [Ca^{2+}], and an electrode that measures the medium concentration of tetraphenyl phosphonium ion (TPP$^+$) as an indicator of mitochondrial $\Delta\Psi$.[27] The medium [TPP$^+$] is inversely related

[27] Chamber constructed by B. Krasnikov (A. N. Belozersky Institute, Moscow State University, Russia).

to $\Delta\Psi$ because the negative-inside membrane potential allows for reversible accumulation of this lipophilic cation.[28,29] Although TPP$^+$ electrodes are not currently commercially available, advice regarding their assembly is available.[30] The experiment described in Fig. 4 demonstrates the ability of mitochondrial signal peptides to induce a permeability transition in the presence of a transmembrane electrical potential generated by respiration. The ability of these small, helical, cationic peptides to trigger a permeability transition and to open a multiple conductance channel has been described previously.[31] Figure 4 describes the effects of a mitochondrial signal peptide on the retention of accumulated Ca^{2+}, the $\Delta\Psi$, and the volume of isolated rat liver mitochondria. On addition of the oxidizable substrate succinate to the mitochondrial suspension containing 2 μM TPP$^+$ Cl$^-$, much of the TPP$^+$ is taken up by the mitochondria, reflecting the respiration-dependent generation of $\Delta\Psi$. At the same time, much of the contaminating Ca^{2+} present in the medium ($\sim$5 μM) is also accumulated. These events are accompanied by essentially no change in light scattering. On addition of micromolar concentrations of a synthetic signal sequence for the human cytochrome oxidase subunit IV protein, the $\Delta\Psi$ rapidly declines, accompanied by the release of accumulated and endogenous mitochondrial Ca^{2+} and a substantial decrease in light scattering, indicative of mitochondrial swelling. Indeed, when light scattering is observed after the addition of the pore-forming molecule alamethicin, the decrease is comparable to that obtained at the highest level of peptide. As several intracellular proteins, e.g., Bax and Bid, have been shown to induce the release of cytochrome c and other proteins from mitochondria,[32,33] multiparameter measurements such as those described in Fig. 4 should be helpful in assessing their effects on mitochondrial membranes and in understanding their mechanism of action.

Immunoblot Measurements of Mitochondrial Cytochrome c Release

In addition to measurements of mitochondrial characteristics, such as $\Delta\Psi$ and volume, which may change in response to apoptotic stimuli, direct measurements of the release of mitochondrial apoptogenic proteins, e.g.,

[28] E. A. Liberman, V. P. Topaly, L. M. Tsofina, A. A. Jasaitis, and V. P. Skulachev, *Nature* (*London*) **222**, 1076 (1969).
[29] H. Rottenberg, *J. Membr. Biol.* **81**, 127 (1984).
[30] Contact A. A. Starkov at *stark@izbe.ru*.
[31] Y. E. Kushnareva, M. L. Campo, K. W. Kinnally, and P. M. Sokolove, *Arch. Biochem. Biophys.* **366**, 107 (1999).
[32] J. M. Jürgensmeier, Z. Xie, Q. Deverzux, L. Ellerby, D. Bredesen, and J. C. Reed, *Proc. Natl. Acad. Sci. U.S.A.* **95**, 4997 (1998).
[33] E. Bossy-Wetzel and D. R. Green, *J. Biol. Chem.* **274**, 17484 (1999).

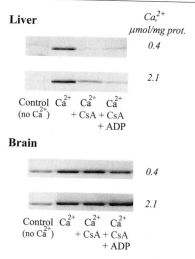

Fig. 5. Immunoblot determinations of cytochrome c released from suspensions of isolated liver and brain mitochondria in response to added Ca^{2+} in the absence and presence of inhibitors of the mitochondrial permeability transition. Determinations were made with aliquots of supernatants obtained after centrifugation of mitochondrial suspensions (0.5 mg of protein/ml) after 10-min incubations in medium maintained at 37° and containing 130 mM KCl, 2 mM K_2HPO_4, 3 mM ATP, 4 mM $MgCl_2$, 5 mM HEPES-KOH (pH 7.0), and 5 mM succinate plus 2 μM rotenone. When present, $CaCl_2$ was added at the indicated levels 2 min after the addition of mitochondria. When present, 1 μM cyclosporin A (CsA) ± 3 mM ADP was added before the addition of mitochondria. See Ref. 35 for additional details.

cytochrome c, are important in understanding the role of mitochondria in the apoptotic cascade. We have had success with Western immunoblot measurements of cytochrome c present in the medium after incubating mitochondria or permeabilized cells under conditions *in vitro* that may reflect the triggering of cytochrome c release within cells undergoing apoptosis.[34,35] The duration of incubation at temperatures of 30–37° is normally kept shorter than 20 min to avoid spontaneous mitochondrial degradation and permeabilization, which can occur easily at these elevated temperatures. While other incubation conditions are quite variable, the presence of sufficient oxidizable substrates and O_2 to avoid complications due to inhibition of electron transport should be maintained. At the end of an experimental period, the suspension of mitochondria or permeabilized cells is centrifuged in a microcentrifuge at 12,000–15,000g for 2–5 min. An aliquot of the supernatant is carefully removed, taking care not to disturb

[34] A. Y. Andreyev, B. Fahy, and G. Fiskum, *FEBS Lett.* **439,** 373 (1998).
[35] A. Y. Andreyev and G. Fiskum, *Cell Death Differ.* **6,** 825 (1999).

the pellet. The remainder of the supernatant is carefully aspirated and the pellet is resuspended to the volume of the original suspension. Protease inhibitors are added to these samples and they are frozen on dry ice and stored at $-70°$. Aliquots of carefully thawed samples are run on 4–20% Tris–glycine gradient gels. Proteins are electrotransferred to PVDF membranes that are rinsed and blocked overnight. Cytochrome c is immunostained with a primary antibody, e.g., the 7H8 mouse anti-cytochrome c antibody from PharMingen (San Diego, CA), and a secondary anti-mouse immunoglobulin bound to horseradish peroxidase (1 : 2000 dilution each; Amersham, Arlington Heights, IL). Peroxidase activity is detected with an Enhanced ChemiLuminescence detection kit (Amersham) and X-ray film. Band intensities are analyzed densitometrically.

Figure 5 describes the results of experiments designed to compare the characteristics of Ca^{2+}-induced cytochrome c release from isolated rat liver and rat forebrain mitochondria suspended in a K^+-based medium containing physiologically realistic concentrations of Mg^{2+} and ATP. The immunoblots indicate that although cytochrome c release from liver mitochondria was effectively inhibited by cyclosporin A (CsA) or CsA plus ADP, both inhibitors of the classical MPT, release of cytochrome c from brain mitochondria was unaffected or possibly enhanced by these MPT inhibitors. Results from measurements of the type described earlier provided additional evidence that while cytochrome c release from liver mitochondria is mediated by the MPT, cytochrome c release from brain mitochondria occurs by a different mechanism that does not necessitate mitochondrial swelling or disruption of the outer membrane.[34,35]

Acknowledgments

This work was supported by the National Institutes of Health (NS34152) and by a grant from the Bayer Corporation to G.F.

[22] Assays for Cytochrome *c* Release from Mitochondria during Apoptosis

By ELLA BOSSY-WETZEL and DOUGLAS R. GREEN

Introduction

In living cells, cytochrome *c* is confined to the mitochondrial intermembrane space, where it participates as an electron carrier in the respiratory chain. In apoptotic cells, cytochrome *c* leaves its normal mitochondrial localization and accumulates in the cytoplasm, where it binds to the adaptor molecule, Apaf-1, rendering it competent to bind and activate the initiator caspase, procaspase 9.[1–4] Caspase 9, in turn, activates the downstream effector caspases, inducing a caspase cascade. Thus, cytochrome *c* acts as an essential cofactor in the activation of the dormant killer proteases. Cytochrome *c* release from mitochondria is an early event in apoptosis. It occurs in most mammalian cells and by many apoptotic stimuli, thus increasing the demand for reliable techniques to monitor this critical event.[1–8]

In this chapter we describe three techniques to detect the translocation of cytochrome *c* from mitochondria to the cytoplasm. In the first method intact cells are separated into cytosolic and mitochondrial fractions and cytochrome *c* levels are quantified by immunoblotting in each fraction.[5] The second method describes an immunocytochemical procedure, permitting a qualitative assessment of cytochrome *c* localization in intact cells. This technique can also be combined with DNA- or mitochondria-specific dyes, which enables cytochrome *c* release to be related to other apoptotic events, such as chromatin condensation and loss of mitochondrial transmembrane potential. Immunocytochemistry also can be used to confirm data obtained by subcellular fractionation. Last, we describe an *in vitro* system to assay cytochrome *c* release from isolated mitochondria.[6–8] This is of particular

[1] X. Liu, L. N. Kim, J. Yang, R. Jemmerson, and X. Wang, *Cell* **81,** 147 (1996).
[2] R. M. Kluck, E. Bossy-Wetzel, D. R. Green, and D. D. Newmeyer, *Science* **275,** 1132 (1997).
[3] J. Yang, X. Liu, K. Bhalla, C. N. Kim, A. M. Ibrado, J. Cai, T.-I. Peng, D. P. Jones, and X. Wang, *Science* **275,** 1129 (1997).
[4] P. Li, D. Nijhawan, I. Budihardjo, S. M. Srinivasula, M. Ahmad, E. S. Alnemri, and X. Wang, *Cell* **91,** 476 (1997).
[5] E. Bossy-Wetzel, D. D. Newmeyer, and D. R. Green, *EMBO J.* **17,** 37 (1998).
[6] H. Li, H. Zhu, C. J. Xu, and J. Yuan, *Cell* **94,** 481 (1998).
[7] X. Luo, I. Budihardjo, H. Zou, C. Slaoughter, and X. Wang, *Cell* **94,** 481 (1998).
[8] M. Steemans, V. Goossens, M. Van de Craen, F. Van Herreweghe, K. Van Compernolle, K. DeVos, P. Vandenabeele, and J. Grooten, *J. Exp. Med.* **188,** 2193 (1998).

interest in identifying and characterizing molecules involved in the regulation of cytochrome c release.

Detection of Cytochrome c Release by Subcellular Fractionation and Immunoblotting

Reagents and Buffers

Phosphate-buffered saline (PBS, 2×) (GIBCO, Gathersburg, MD)

Buffer A (cytosolic extract): 250 mM sucrose, 20 mM HEPES-KOH (pH 7.4), 10 mM KCl, 1.5 mM Na-EGTA, 1.5 mM Na-EDTA, 1 mM MgCl$_2$, 1 mM dithiothreitol (DTT), and cocktail of protease inhibitors; sterilize by filtration and store at 4°

Buffer B (mitochondrial extract): 50 mM HEPES (pH 7.4), 1% (v/v) Nonidet P-40 (NP-40), 10% (v/v) glycerol, 1 mM EDTA, 2 mM DTT, cocktail of protease inhibitors; sterilize by filtration and store at 4°

Cocktail of protease inhibitors (Boehringer Mannheim, Indianapolis, IN)

Trypan blue solution: 0.4% (w/v) in PBS; store at room temperature

Sodium dodecyl sulfate (SDS) sample buffer (4×): 4% (w/v) SDS, 20% (v/v) glycerol, 200 mM DTT, 120 mM Tris-HCl (pH 6.8), 0.002% (w/v) bromphenol blue; store in aliquots at −20°

Transfer buffer: 20 mM Tris base, 150 mM glycine, 20% (v/v) methanol; prepare fresh for each experiment and precool on ice

Blocking solutions: 3% (w/v) nonfat milk powder, 3% (w/v) bovine serum albumin (BSA), 0.1% (v/v) Tween 20 in PBS; prepare fresh each time

Immunoblotting solution: 3% (w/v) nonfat milk powder, 0.1% (v/v) Tween 20 in PBS; prepare fresh each time

Anti-cytochrome c antibody (7H8.2C12) (PharMingen, San Diego, CA)

Anti-actin antibody (clone 14) (ICN Chemical Credential, Aurora, CA)

Anti-cytochrome c oxidase (subunit II) antibody (Molecular Probes, Eugene, OR)

Donkey anti-mouse horseradish peroxidase antibody (Amersham, Buckinghamshire, England)

Hybond-ECL nitrocellulose (0.45 μm) (Amersham, Buckinghamshire, England)

Super-signal chemiluminescence reagent (Pierce, Rockford, IL)

Stripping solution: 62.5 m*M* Tris-HCl (pH 6.8), 2% (w/v) SDS, 100 m*M* 2-mercaptoethanol; store at room temperature

Equipment

Glass Dounce homogenizer (2 ml) with a sandpaper-polished tight pestle (B-type)
Minielectrophoresis and transfer unit (Bio-Rad, Hercules, CA)

Procedures

Cytosolic and Mitochondrial Extract Preparation. For this protocol it is critical to disrupt the cells in buffer containing 250 m*M* sucrose, which leaves mitochondria intact during the homogenization process. The cells are gently broken with a glass Dounce. By this procedure no or little cytochrome *c* is found in cytosolic extracts of living cells. The time when cytochrome first can be detected in cytosol of apoptotic cells depends on the cell type and the apoptotic stimulus. It may range from 1 to 12 hr after exposure to the apoptotic agent.

Approximately 1×10^7 cells are required for each preparation. To harvest the cells, transfer the cell suspension into a 15-ml tube. When adherent cells are used, gently scrape the cells from the culture dish with a rubber policeman. The cells are pelleted by centrifugation at 200*g* for 5 min, using a tabletop centrifuge. The cell pellets are washed twice with 10 ml of cold PBS, pH 7.4. It is important to remove all PBS from the cell pellet, before resuspension with 500–700 μl of buffer A. Incubate the resuspended cells for 20 to 30 min on ice. During this time the tubes are tapped from time to time in order to resuspend the cell pellet.

Cells are then disrupted by 20 to 50 strokes with the glass Dounce and a tight pestle (B-type). After Dounce homogenization, the cell homogenates are transferred to an Eppendorf tube and nuclei, unbroken cells, and large debris are removed by centrifugation at 800*g* for 10 min at 4°. Supernatants containing mitochondria are transferred to a new Eppendorf tube and are further centrifuged at 22,000*g* for 15 min at 4°. The resulting supernatants are saved as cytosolic extracts at −70° until further analysis. The mitochondrial pellets are lysed with 100 μl of buffer B. Samples are vortexed from time to time during the 20-min incubation period on ice. Cellular debris are removed by centrifugation at 22,000*g* for 15 min at 4°. The supernatants containing mitochondrial proteins can be stored at −70°.

Note: The optimal conditions for cell permeabilization depend on the Dounce and the cell type used. Accordingly, one should define how many strokes are required to observe cytochrome *c* in apoptotic, but not in control, samples. To establish the optimum conditions for cell homogenization, the trypan blue exclusion assay, which discriminates between broken (stained) cells and intact cells (unstained), can be used. For the trypan blue exclusion test, a 0.4% (w/v) solution of trypan blue in PBS is diluted 1:10 with the cell suspension and is examined under the microscope.

Immunoblotting Using Anti-cytochrome c Antibodies. To analyze the cytochrome *c* level in cytosol and mitochondria, 10 to 50 μg of protein is mixed with 4× protein sample buffer. The samples are heated at 95° for 10 min and loaded onto a 15% SDS–polyacrylamide gel (minigel electrophoresis unit). After electrophoresis at 90 V for approximately 3 hr, proteins are transferred to a nitrocellulose filter in cold transfer buffer at 150 mA overnight (minitransfer unit).

Nonspecific protein binding of the filter is prevented by incubation in blocking buffer on a rocking table. After 3 hr of incubation at room temperature the nitrocellulose membrane is transferred to a plastic bag and 6 ml of monoclonal anti-cytochrome *c* antibody 7H8.2C12 (diluted 1:2000 in immunoblotting buffer) is added. The sealed plastic bag is then incubated overnight on a rocking plate at 4°.

The next day the filter is removed from the plastic bag and washed four times with 100 ml of immunoblotting solution for 10 min. Last, 25 ml of horseradish peroxidase-coupled anti-mouse antibody (diluted 1:1000 in immunoblotting buffer) is added and incubated on a rocking plate for 2 hr at room temperature. The filter is then washed four times with 100 ml of immunoblotting buffer for 10 min. The specific protein complexes are visualized with chemiluminescence reagent.

Note: To control for similar protein loading or contamination of cytosolic extracts with mitochondria, the immunoblots can be stripped and reprobed with anti-actin antibodies (diluted 1:3000) or anti-cytochrome *c* oxidase antibodies (diluted 1:300). To strip immunocomplexes, place the filter into a plastic bag and add 20 ml of stripping solution. Incubate the sealed plastic bag for 10 min in a 60° water bath. After incubation remove the filter from the bag and rinse extensively with distilled water. The filter is then incubated with blocking solution and treated as described above.

Detection of Cytochrome c Translocation by Immunocytochemistry

Reagents and Buffers

Glass coverslips (1 mm), sterilized by autoclaving

Paraformaldehyde (3%, w/v) in PBS: To prepare fixative, work under a fume hood. Six grams of paraformaldehyde (EM grade) is added to 100 ml of H_2O and the mix is heated under stirring to 60°. NaOH (1 *N*) is added drop by drop until solids are dissolved. The solution is placed on ice and 100 ml of 2× PBS, pH 7.4, is added. After filtration the fixative is ready to use. The fixative should be prepared fresh each time. Alternatively, aliquots can be stored at −20° until needed

Saponin (0.25%, w/v) in PBS, pH 7.4

Anti-cytochrome *c* antibody (6H2.B4) (PharMingen)

Goat anti-mouse antibody labeled with high-fluorescence fluorescein isothiocyanate (FITC) (Antibodies Incorporated, Davis, CA)

FluoroGuard antifade reagent (Bio-Rad)

Hoechst 33342 stock solution: 100 μg/ml in PBS

Propidium iodide stock solution: 500 μg/ml in PBS

Mitotracker Orange CMTMRos (Molecular Probes, Eugene, OR) stock solution: 30 μM in dimethyl sulfoxide (DMSO)

Equipment

Confocal laser scanning microscope or conventional fluorescence microscope

Procedures

The described protocol works best for adherent cell types. A glass coverslip is placed into a 30-mm culture dish, using sterile forceps. Approximately 5 × 10^4 cells are seeded into each dish and allowed to adhere for 24 hr under normal growth conditions. Cells are then treated with the apoptotic stimulus. After the desired time, the culture medium is aspirated and the cells are fixed with 2 ml of 3% (w/v) paraformaldehyde in PBS, pH 7.4, for 30 min at room temperature. It is convenient to leave the glass coverslips in the culture dish for fixation and subsequent wash steps. After fixation the cells are washed three times with PBS, pH 7.4, for 5 min. The cells are permeabilized by incubation with 0.25% (w/v) saponin in PBS for 5 min and subsequently washed three times with PBS, pH 7.4, for 5 min. To incubate the fixed cells with primary antibody, several moist paper towels are placed at the bottom of a plastic container and are covered with Parafilm. Approximately 100 μl of primary anti-cytochrome *c* antibody 6H2.B4 [10 μg/ml in PBS (pH 7.4), 3% (w/v) BSA] is mounted onto the Parafilm for each coverslip. The glass coverslips are carefully transferred with forceps on top of the antibody solution (cells facing bottom) and incubated for 3 hr or overnight at room temperature by covering the plastic container to avoid drying of the samples. To remove excess antibody binding the glass coverslips are transferred back to the culture dishes and are washed four times with PBS, pH 7.4, for 5 min. The secondary goat anti-mouse antibody (FITC labeled) is diluted 1 : 100 in PBS–3% (w/v) BSA and incubated at room temperature for 1 hr as described for the primary antibody. To prevent photobleaching of the FITC-conjugated secondary antibody, protect the glass coverslips from light. The coverslips are then rinsed three times with PBS, pH 7.4, for 5 min. Last, the coverslips are mounted onto microscope slides, using a small amount of antifade reagent. Excess mounting medium is removed with a clean tissue and the glass coverslips are sealed with clear nail polish and analyzed by confocal scanning microscopy or conventional fluorescence microscopy. The slides can be stored up to 4 weeks at 4°, if protected from light.

Note: To identify apoptotic cells, it is convenient to colabel the cells with DNA-specific dyes, such as Hoechst 33342 or propidium iodide. Apoptotic cells show condensed chromatin and fragmented nuclei, which can be visualized easily with these dyes. The fixed cells are incubated with the dyes after the secondary antibody incubation. To stain nuclei dilute the Hoechst 33342 stock solution 1 : 100 in PBS, or the propidium iodide dye stock solution 1 : 1000 in PBS, and incubate for 15 or 1 min, respectively. Wash the glass coverslips three times with PBS for 5 min and mount onto glass slides as described above.

It is possible to combine this technique also with mitochondria-specific dyes that are resistant to fixation, such as Mitotracker Orange or Red (Molecular Probes).[10] Mitotracker Orange or Red incorporates into the mitochondrial matrix, depending on the mitochondrial innermembrane potential. To label mitochondria, dilute Mitotracker stock solution 1 : 200 with prewarmed culture medium. Load the cells before fixation with the dye for 15–30 min under cell culture conditions. The cells are then fixed as described above. It is of note that during apoptosis the mitochondrial transmembrane potential dissipates, resulting in a loss of Mitotracker staining. To label mitochondria in dead cells Mitotracker Green (Molecular Probes), which stains mitochondria irrespective of the mitochondrial transmembrane potential, can be used.

Cytochrome c Release Assay from Isolated Liver Mitochondria in Vitro

Reagents and Buffers

Mitochondria isolation buffer (MIB): 220 mM mannitol, 68 mM sucrose, 10 mM KCl, 1 mM EDTA, 1 mM EGTA, 10 mM HEPES-KOH (pH 7.4), 0.1% (w/v) BSA, supplemented with cocktail of protease inhibitors

[9] E. Bossy-Wetzel and D. R. Green, J. Biol. Chem. 274, 17484 (1999).
[10] A. Macho, D. Decaudin, M. Castedo, T. Hirsch, S. A. Susin, N. Zamzami, and G. Kroemer, Cytometry 25, 333 (1996).

Mitochondria resuspension buffer (MRB): 200 mM mannitol, 50 mM sucrose, 10 mM succinate, 5 mM potassium phosphate (pH 7.4), 10 mM HEPES-KOH (pH 7.4), 0.1% (w/v) BSA

Cytosol extraction buffer: 220 mM mannitol, 68 mM sucrose, 20 mM HEPES-KOH (pH 7.0), 10 mM KCl, 1.5 mM MgCl$_2$, 1 mM Na-EDTA, 1 mM Na-EGTA, 1 mM DTT, supplemented with cocktail of protease inhibitors

Purified caspases (PharMingen)

Equipment

Glass Dounce (15 ml) with tight Teflon pestle
Glass Dounce (2 ml) with glass pestle (B type)

Procedures

Preparation of Mouse Liver Mitochondria. After sacrifice, the liver of a 2- to 6-month-old mouse is rapidly excised and placed into a culture dish on ice. All subsequent steps are carried out on ice or at 4°. Special care should be taken not to disrupt or include the gall bladder in the preparation. The tissue is briefly rinsed with cold PBS, pH 7.4, and placed into an empty culture dish, where it is minced into small pieces with fine scissors. Half of the chopped liver tissue is resuspended in 10 ml of cold MIB buffer, transferred to a glass Dounce with a tight Teflon pestle, and homogenized by up-and-down strokes, until no tissue pieces remain. The same is repeated for the other half of the tissue. The liver homogenates are distributed into two 15-ml polypropylene tubes (round bottom), and nuclei, unbroken cells, and large debris are pelleted by centrifugation at 600g for 10 min, using a Sorvall centrifuge and an HB-4 rotor. Supernatants containing mitochondria are transferred to new tubes without disturbing the pellet and are recentrifuged at 3500g for 15 min. After centrifugation, supernatants and floating lipid layers are aspirated and the mitochondrial pellet is resuspended in 15 ml of fresh MIB buffer and centrifuged at 1500g for 5 min at 4° to remove additional nuclei or unbroken cells. To pellet mitochondria, supernatants are recentrifuged at 5500g for 10 min. The last two steps are repeated twice. Last, the mitochondrial pellets are resuspended in 1 ml of MRB buffer, kept on ice, and used within 4 hr.

Preparation of Cytosolic Extracts. To obtain cytosolic extracts we use Jurkat, CEM, or HeLa cells. Cells are harvested by centrifugation at 200g for 10 min at 4°, washed twice with cold PBS, pH 7.4, and resuspended in 3 vol of cold cytosolic extraction buffer. After incubation on ice for 20 to 30 min, cells are disrupted by homogenization with a glass Dounce and a tight glass pestle, applying 50 strokes. Nuclei and unbroken cells are removed by centrifugation at 800g for 10 min at 4° in an Eppendorf centrifuge. Supernatants are transferred to new Eppendorf tubes and further centrifuged at 22,000g for 30 min. Supernatants are stored as cytosolic extracts at −70°, until needed for *in vitro* assays.

In Vitro Assay for Cytochrome c Release. Standard reactions are carried out in a 50-μl final reaction volume containing 20 to 50 μg of cytosolic extract, 3 to 5 μg of isolated liver mitochondria, and 50 nM caspase 3 or caspase 8 at 37°.[9] Cytochrome *c* can also be released from purified mitochondria without cytosolic extract when purified Bax or Bid protein is used.[11–14] Instead of caspases, Bax, or Bid, purified protein fractions or candidate molecules can be tested for cytochrome *c*-releasing activity by this assay. At the end of the incubation, the samples are centrifuged at 22,000g in an Eppendorf centrifuge to pellet mitochondria. After centrifugation, 45 μl of supernatant is transferred to a new Eppendorf tube and mixed with 12 μl of 4× protein sample buffer. The mitochondrial pellet is resuspended in 60 μl of 1× protein sample buffer. Samples are boiled for 10 min and separated on a 15% SDS–polyacrylamide gel electrophoresis and immunoblotting with anti-cytochrome *c* antibodies as described above.

Acknowledgments

This work was supported by National Institutes of Health Grants AI40646 and CA6938 to D.R.G. and by a fellowship of the Swiss National Science Foundation (823A-046638) to E.B.-W. We thank M. Pinkoski and Martin Schuler for their comments on this chapter.

[11] J. M. Jürgensmeier, Z. Xie, Q. Deveraux, L. Ellerby, D. Bredesen, and J. C. Reed, *Proc. Natl. Acad. Sci. U.S.A.* **95**, 4997 (1998).
[12] R. Eskes, B. Antonsson, A. Osen-Sand, S. Montessuit, C. Richter, R. Sadoul, G. Mazzei, A. Nichols, and J. C. Martinou, *J. Cell Biol.* **143**, 217 (1998).
[13] M. Narita, S. Shimizu, T. Ito, T. Chittenden, R. J. Lutz, H. Mat Suda, and Y. Tsujimoto, *Proc. Natl. Acad. Sci. U.S.A.* **95**, 14681 (1998).
[14] D. M. Finucane, E. Bossy-Wetzel, N. J. Waterhouse, T. G. Cotter, and D. R. Green, *J. Biol. Chem.* **274**, 2225 (1999).

[23] Purification and Liposomal Reconstitution of Permeability Transition Pore Complex

By Catherine Brenner, Isabel Marzo, Helena L. de Araujo Vieira, and Guido Kroemer

Introduction

The permeability transition (PT) pore, also called the mitochondrial megachannel or multiple conductance channel,[1–3] is a dynamic multiprotein complex located at the contact site between the inner and outer mitochondrial membranes, one of the critical sites of metabolic coordination between the cytosol, the mitochondrial intermembrane space, and the matrix. The PT pore participates in the regulation of matrix Ca^{2+}, pH, $\Delta\Psi_m$ (mitochondrial transmembrane potential), and volume and functions as a Ca^{2+}-, voltage-, pH-, and redox-gated channel with several levels of conductance and little if any ion selectivity.[1–6] In isolated mitochondria, opening of the PT pore causes matrix swelling with consequent distension and local disruption of the mitochondrial outer membrane (whose surface is smaller than that of the inner membrane), release of soluble products from the intermembrane space, dissipation of the mitochondrial inner transmembrane potential, and release of small molecules up to 1500 Da from the matrix, through the inner membrane.[1,2,7,8] Similar changes are found in apoptotic cells, perhaps with the exception of matrix swelling, which is observed only in a transient fashion, before cells shrink.[9–13] In several models of apoptosis, pharmaco-

[1] M. Zoratti and I. Szabò, *Biochim. Biophys. Acta Rev. Biomembr.* **1241,** 139 (1995).

[2] P. Bernardi, *Biochim. Biophys. Acta Bioenerg.* **1275,** 5 (1996).

[3] K. W. Kinnally, T. A. Lohret, M. L. Campo, and C. A. Mannella, *J. Bioenerg. Biomembr.* **28,** 115 (1996).

[4] G. Beutner, A. Rück, B. Riede, W. Welte, and D. Brdiczka, *FEBS Lett.* **396,** 189 (1996).

[5] I. Marzo, C. Brenner, N. Zamzami, S. A. Susin, G. Beutner, D. Brdiczker, R. Remy, Z.-H. Xie, J. C. Reed, and G. Kroemer, *J. Exp. Med.* **187,** 1261 (1998).

[6] F. Ichas, L. S. Jouavill, and J.-P. Mazat, *Cell* **89,** 1145 (1997).

[7] S. P. Kantrow and C. A. Piantadosi, *Biochem. Biophys. Res. Commun.* **232,** 669 (1997).

[8] P. X. Petit, M. Goubern, P. Diolez, S. A. Susin, N. Zamzami, and G. Kroemer, *FEBS Lett.* **426,** 111 (1998).

[9] G. Kroemer, P. X. Petit, N. Zamzami, J.-L. Vayssière, and B. Mignotte, *FASEB J.* **9,** 1277 (1995).

[10] X. Liu, C. N. Kim, J. Yang, R. Jemmerson, and X. Wang, *Cell* **86,** 147 (1996).

[11] G. Kroemer, N. Zamzami, and S. A. Susin, *Immunol. Today* **18,** 44 (1997).

[12] M. G. vander Heiden, N. S. Chandal, E. K. Williamson, P. T. Schumacker, and C. B. Thompson, *Cell* **91,** 627 (1997).

logical inhibition of the PT pore is cytoprotective, suggesting that opening of the PT pore can be rate-limiting for the death process.[13-15] Moreover, the cytoprotective oncoprotein Bcl-2 has been shown to function as an endogenous inhibitor of the PT pore.[15-17] It has been speculated that several proteins interacting in the contact sites between the inner and the outer mitochondrial membranes might build up the so-called permeability transition pore complex (PTPC).[1,4] Given the probable importance of the PT pore for apoptosis regulation, we have developed a protocol for the purification of PTPCs, their reconstitution in liposomes, and the quantitation of PT pore-regulated membrane permeability.[5]

Purification of Permeability Transition Pore Complexes and Their Reconstitution into Liposomes

Principle

Hexokinase type 1 associates with the outer mitochondrial membrane protein VDAC (voltage-dependent anion channel, also called porin), which, in turn, can interact with the inner mitochondrial membrane protein ANT (adenine nucleotide translocator, also called ATP/ADP carrier)[4,18] The method to purify PTPCs thus consists in tracing hexokinase activity copurifying with a protein complex that is water insoluble in brain homogenates, partitions into the Triton-soluble fraction, elutes from an anion-exchange fast protein liquid chromatography (FPLC) column at a relatively high salinity, and incorporates into phosphatidylcholine/cholesterol liposomes.[5] The overall principle of the method is outlined in Fig. 1.

Required Equipment

Potter homogenizer, 50 ml
DE52 resin (Whatman, Clifton, NJ), preswollen
FPLC system equipped with two pumps (Pharmacia Biotech, Uppsala, Sweden)

---

[13] G. Kroemer, B. Dallaporta, and M. Resche-Rigon, *Annu. Rev. Physiol.* **60,** 619 (1998).
[14] P. Marchetti, M. Castedo, S. A. Susin, N. Zamzami, T. Hirsch, A. Haeffner, F. Hirsch, M. Geuskens, and G. Kroemer, *J. Exp. Med.* **184,** 1155 (1996).
[15] N. Zamzami, S. A. Susin, P. Marchetti, T. Hirsch, J. Gomez-Monterrey, M. Castedo, and G. Kroemer, *J. Exp. Med.* **183,** 1533 (1996).
[16] S. A. Susin, N. Zamzami, M. Castedo, T. Hirsch, P. Merchetti, A. Macho, E. Daugas, M. Geuskens, and G. Kroemer, *J. Exp. Med.* **184,** 1331 (1996).
[17] G. Kroemer, *Nature Med.* **3,** 614 (1997).
[18] B. Gelb, V. Adams, S. Jones, L. Griffin, G. MacGregor, and E. McCabe, *Proc. Natl. Acad. Sci. U.S.A.* **89,** 202 (1992).

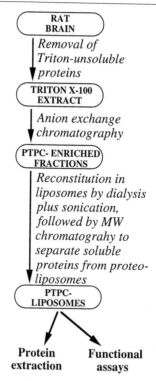

FIG. 1. Principle of PTPC purification and reconstitution in liposomes.

FPLC column (XK 16/20; Pharmacia Biotech)
Centrifuge and ultracentrifuge (Beckman, Fullerton, CA)
Spectrophotometer (366 nm, 1-ml cuvettes)
Sonicator (ultrasonic processor; Bioblock, Illkirch, France)
Membrane for dialysis (cut-off, 10 kDa; Pierce, Rockford, IL)
Column C16/20 packed with Sephadex G50 beads (Pharmacia)
Peristaltic pump (Pharmacia)
Glassware cleaned with chloroform

Reagents

Buffer 1: 1 mM α-monothioglycerol + 10 mM glucose, pH 8.0
Buffer 2: 1.5 mM Na$_2$HPO$_4$, 1.5 mM K$_2$HPO$_4$, 100 mM glucose, 1 mM dithioerythritol, pH 8.0, freshly prepared
Buffer 3: Like buffer 1, plus 50 mM KCl
Buffer 4: Like buffer 1, plus 400 mM KCl
Buffer 5: 125 mM sucrose, 10 mM HEPES (pH 7.4)
TraPB buffer: 100 mM triethanolamine, 10 mM EDTA, 16 mM MgSO$_4$·7H$_2$O, 150 mM KCl, pH 7.6 (store at −20°)

Glucose-6-phosphate dehydrogenase (Boehringer Mannheim, India-
napolis, IN)
NADPH: 20 mg/ml
Glucose: 0.1 M
ATP: 0.1. M (prepare fresh)
Phosphatidylcholine from egg yolk
Cholesterol
n-Octyl-β-D-glucopyranoside
Chloroform
Chloroform–methanol (2 : 1, v/v)
KCl: 1 M
Malate: 0.5 M
Acetone (80%, v/v) in distilled water
Triton X-100 (Boehringer Mannheim, special for membrane research)
Optional: [^{3}H]Glucose (625 GBq/mmol) and [^{14}C]cholesterol (925
GBq/mmol; Amersham, Arlington Heights, IL)
All products are of analytical grade and from Sigma (St. Louis, MO)
if unspecified. All buffers are sterilized through 0.45-μm pore size
filters.

Protocol

1. Wistar rats (3- to 4-month-old males) are killed by decapitation.
Brains are immediately removed and stored at $-80°$ for up to 3 months.
2. Four brains are cut down with scissors in ice-cold buffer 1. The pieces
(discard buffer) are homogenized in 40 ml of cold buffer 1 in a Potter
homogenizer and centrifuged twice (15 min, 12,000g, 4°). Supernatants
are discarded.
3. The pellet is resuspended in 40 ml of buffer 1 containing 2 ml of 10%
(v/v) Triton X-100 (Boehringer Mannheim) for 30 min at room temperature
while stirring, in the dark. After centrifugation (40 min, 50,000g, 10°), the
pellet is discarded.
4. The supernatant is mixed with 17 g of DE52 resin (Whatman) previously equilibrated with buffer 2. These beads are packed into an FPLC
column (XK16/20; Pharmacia Biotech) and eluted with buffer 2 supplemented with 50 mM KCl (buffer 3) and 400 mM KCl (buffer 4). Elution
conditions (0.8 ml min^{-1}): 0 to 7.5 min, 100% buffer 3; 7.5 to 250 min, linear
gradient to 100% buffer 4; 250 to 252 min, linear gradient to 0% buffer 4;
and 252 to 260 min, 0% buffer 4. Fractions of 2.4 ml are collected and kept
on ice, and their hexokinase activity is determined. The three fractions with
maximum hexokinase activity are employed for incorporation into liposomes.
5. Lipid vesicles are prepared by mixing 300 mg of phosphatidylcholine
and 60 mg of cholesterol in 1 ml of chloroform. At this step, 150 μl of

[^{14}C]cholesterol (5 μCi; Amersham) in toluene can be added optionally. The solvent is evaporated under nitrogen and the tubes are incubated for 1 hr at 37° to eliminate all traces of chloroform. The dried lipids are resuspended in 3 ml of buffer 5 containing 0.3% n-octyl-β-D-pyranoside by vigorous vortexing until the mixture becomes homogeneously white (usually 60 min, room temperature). The vesicles are incubated with an equal volume of PTPC-containing solution for 20 min at room temperature and then dialyzed overnight (4°) against 5 liters of buffer 5. External proteins (e.g., recombinant Bcl-2) can be added before the dialysis step.

6. Liposomes (minimum volume, 2 ml) recovered from dialysis are ultrasonicated on ice (120 W) for 7 sec in buffer 5 supplemented with 5 mM malate and 10 mM KCl or a radiolabeled compound ([^{3}H]glucose).

7. The proteoliposomes are charged on Sephadex G50 columns (C16/20; Pharmacia), and eluted with buffer 5 at a rate of 0.8 ml min^{-1} with a peristaltic pump. Fractions (2 ml/fraction) are recovered and essayed for hexokinase activity. Only the two fractions with maximum hexokinase activity are used for functional experiments.

Method for Hexokinase Determination

1. Mix in a 1 ml-cuvette:

H$_2$O	500 μl
TrapB buffer	500 μl
Glucose (0.1 M)	20 μl
NADP (20 mg/ml)	20 μl
Glucose-6-phosphate-dehydrogenase (Boehringer; dilute fivefold in water immediately before use)	5 μl
H$_2$O	500 μl
Sample	20 μl

2. Add 20 μl of ATP (0.1 M) to start the reaction and mix. After 5 min, read the absorbance at 360 nm in a spectrophotometer. The slope of the absorbance can be calculated in mmol NADPH min^{-1} mg^{-1} protein), considering that the specific $E_{\text{NADH}} = 3.3$ mM^{-1} cm^{-1}.

Remarks

The timing must be carefully respected to avoid a loss of activity, notably until the chromatographic step. Maximum hexokinase activity is usually found in fractions 28–31 of the first chromatography step and is found in the second tube containing liposomes of the second chromatography step. This activity is sensitive to storage (~50% of loss overnight at 4°) and inhibited by 250 mM NaCl and by metals such as Cu, Zn, and Ni. Proteins may be extracted from the liposome preparation (1 ml) by mixing with

2 ml of KCl, and 6 ml of chloroform–methanol (2:1, v/v) and are recovered from the interface after centrifugation,[19] followed by resuspension
in 0.1% (w/v) sodium dodecyl sulfate (SDS). They can then be precipitated
with 80% (v/v) acetone for further analysis. Protein content is quantified
by the Bradford method, using bovine serum albumin (BSA) as standard.
Alternatively, the liposomes can be subjected to functional assays, for instance to measure the permeability of the liposomal membrane. Two techniques for the assessment of liposomal permeability are outlined below.

Quantitation of Permeability Transition Pore Complex Opening by Cytofluorometry

Principle

PTPC-containing proteoliposomes can be exposed to different inducers
of PT pore opening, followed by quantitation of membrane permeability.
The proteoliposomes are equilibrated with the amphophilic cationic fluorochrome 3,3′-dihexyloxacarbocyanine iodide [$DiOC_6(3)$, MW 573]. The
retention of $DiOC_6(3)$ fluorescence is then monitored in a cytofluorometer
[i.e., in a flow in which liposomes are diluted and the external $DiOC_6(3)$
concentration approaches zero], while gating on a population of liposomes
with defined forward and side scatter characteristics (estimated diameter,
0.15 to 0.3 μm).

Equipment

FACS-Vantage cytofluorometer (Becton Dickinson, San Jose, CA)

Reagents

Buffer 5: As described above
$DiOC_6(3)$ (Molecular Probes, Eugene, OR) stock solution: 10 mM in
  ethanol; keep at −20°
PT inducers or inhibitors prepared in buffer 5
Atractyloside (Sigma) prepared fresh at 100 mM in H_2O
SDS [10% (w/v) stock solution]

Protocol

1. Ten-microliter aliquots of liposomes obtained as described above
are incubated for a minimum of 30 min at room temperature in buffer 5
supplemented with PT pore modulators. As a positive control for PT pore
opening, 100 μM freshly prepared atractyloside can be used.

---

[19] J. Folch, M. Lees, and G. M. S. Stanley, *J. Biol. Chem.* **226**, 447 (1957).

2. Diluted (1 ml) liposomes are incubated with 2 μl of 3,3'-dihexyloxa-carbocyanine iodide $DiOC_6(3)$, 40 nM, 20–30 min at room temperature; Molecular Probes).

3. $DiOC_6(3)$ retention is determined in a FACS-Vantage cytofluo-rometer (Becton Dickinson). The forward scatter threshold is set at 30 (16 A) and the flow rate at 1500 events/sec. The photomultiplier of the side scatter and FL1 are set at 700 and 700–800 mV, respectively. The fluorescence is excited with an argon laser (excitation wavelength, 488 nm) and analyzed in FL1 (wavelength, 530 ± 30 nm). The forward and side scatters are gated on the quantitatively most abundant population of lipo-somes while excluding background noise, that is, the signal obtained with a liposome-free buffer. Results of a typical experiment are shown in Fig. 2.

Remarks

Because of the limited stability of proteoliposomes (some hours), it is preferable to start the incubations just after the preparation of PTPC liposomes. If incubations are performed as a series, it is necessary to repeat the control (untreated liposomes). Note that between two tubes analyzed in the cytofluorometer, extensive washing is required (sheet pressure, 11 Ψ; duration, at least 5 min) to avoid spillover from one sample to another. Triplicates of 5 × 10⁴ liposomes should be analyzed for each data point. Results can be expressed as a percentage of the reduction in $DiOC_6(3)$ fluorescence (log scale, geometric mean), considering the reduction ob-tained with 0.25% (w/v) SDS (15 min, room temperature) in PTPC lipo-somes as 100% value.

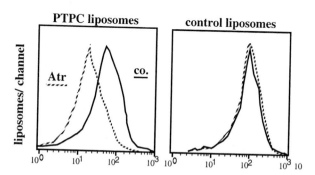

Fig. 2. Cytofluorometric profile of liposomes labeled with the fluorochrome $DiOC_6(3)$. Liposomes were reconstituted either in the presence of the hexokinase-containing fraction (PTPC liposomes) or in its absence (control liposomes), treated with atractyloside (Atr, 50 μM) or left untreated, then stained with $DiOC_6(3)$ and subjected to cytofluorometric analysis.

Quantitation of Permeability Transition Pore Complex Opening by
Means of Radioactive Compounds

Principle

In this assay, liposomal membranes are marked with [^{14}C]cholesterol,
whereas their hydrophilic lumen is marked with [^{3}H]glucose. After incubation with agents causing the PT pore to open, radioactive glucose is released
from the liposomes while cholesterol is retained. The separation of liposomes and soluble glucose is achieved by molecular sieve chromatography.

Equipment

Scintillation counter, for the determination of β irradiation, equipped
with a program allowing for the discrimination of ^{3}H and ^{14}C

Reagents

[^{3}H]Glucose (625 GBq/mmol) or [^{14}C]cholesterol (925 GBq/mmol;
Amersham)
Nick columns containing Sephadex G50 beads from Pharmacia
Liquid scintillation cocktail (Ultima Gold MV; Packard, Downers
Grove, IL)

Protocol

1. Proteoliposomes containing PTPCs are generated in the presence
of [^{14}C]cholesterol, as described above. [^{3}H]Glucose is added during the
sonication step, followed by chromatography on Sephadex G50 columns
(C16/20; Pharmacia), as described above.

2. Proteoliposomes containing maximum hexokinase activity (100 μl/
tube) are incubated at room temperature for a minimum of 60 min with
various PT pore-opening agents. Atractyloside (100 μM, prepared fresh)
can be used.

3. During this incubation step, Sephadex G50 nick columns are equilibrated with 5 ml of buffer 5, which is removed before addition of the liposomes.

4. Add liposomes to the columns. Allow them to enter the beads. Add
400 μl of buffer 5 and allow it to enter the beads. Then add 1.6 ml of buffer
5 and collect 5 drops per tube.

5. Add an appropriate scintillation liquid, vortex thoroughly, and count
radioactivity by using a program allowing for the discrimination of ^{14}C and
^{3}H. Tubes 1–4 usually contain liposomes (that is, all [^{14}C]cholesterol and
a variable portion of [^{3}H]glucose), whereas tubes 6–10 contain [^{3}H]glucose
released from the liposomes. The percentage of [^{3}H]glucose contained in

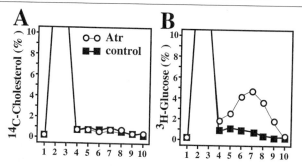

Fig. 3. Typical Sephadex S50 elution profiles of PTPC liposomes generated in the presence of [14C]cholesterol and loaded with [3H]glucose. PTPC liposomes containing both [14C]cholesterol and [3H]glucose were incubated for 60 min with 100 μM atractyloside (Atr) or left untreated (control), followed by separation of liposomes (fractions 1–4) and the buffer (fractions 5–10) on a Sephadex S50 column. Thereafter, 14C (A) and 3H (B) radioactivity was determined in each sample independently.

these latter fractions over the total radioactivity (tubes 1–10) reflects the percentage of release. Results from a typical experiment are shown in Fig. 3.

Remarks

The quantitation of PTPC opening by means of radioactive compounds has the advantage that it requires no cytofluorometer. Moreover, this technique should be useful to determine the molecular cutoff of the PTPCs, using various tritiated hydrophilic molecules. However, the number of samples that can be handled is lower than with the cytofluorometric method described above.

Expected Results and Perspectives

The functional data indicate that PTPC liposomes regulate membrane permeability in a fashion resembling that of the PT pore found in mitochondria. Thus, by using a number of different inducers and inhibitors of PT, we found an approximate functional equivalence between the natural (mitochondrial) PTPCs and the reconstituted (liposomal) PTPCs. Both in mitochondria and in PTPC liposomes, a similar panel of agents acts to permeabilize membranes (Ca^{2+}, Atr, prooxidants, and diamide, recombinant caspases) or to stabilize membrane function (cyclosporin A, monochlorobimane, bongkrekic acid, recombinant Bcl-2 or Bcl-X_L).[1,5,15,20,21] This may

[20] N. Zamzami, I. Marzo, S. A. Susin *et al., Oncogene* **16,** 1055 (1998).
[21] N. Zamzami, C. Brenner, I. Marzo, S. A. Susin, and G. Kroemer, *Oncogene* **16,** in press (1998).

indicate that the PT pore has indeed a major function of integration of pro- and antiapoptotic pathways. In rat brain, we have found that the PTPC copurifies with some apoptosis-regulatory proteins such as Bax and Bag-1.[5] However, the composition of PTPC may vary in different tissues, because some proteins are not ubiquitous (e.g. Bax, hexokinase 1), and isoforms of different PTPC pore components (adenine nucleotide translocator, hexokinase) are expressed in a tissue-specific fashion. The protocol for PTPC enrichment and incorporation into liposomes yields a reduced experimental system in which PTPC function can be analyzed without interference by other mitochondrial structures. We believe that the technique outlined above may help to elucidate the structure–function relationship between different components of the PT pore.

Acknowledgments

We thank Dr. Dieter Brdiczka (University of Konstanz, Konstanz, Germany) for helping us set up the system of PTPC purification. This work was supported by grants from the ANRS, ARC, CNRS, FF, FRM, INSERM, and a special grant from LNC (to G.K.). I.M. received a fellowship from the Spanish Ministry of Science, and E.L. de A.V. received an EC Leonardo da Vinci fellowship.

Section VI

Bcl-2 Family Proteins

[24] Monitoring Interactions of Bcl-2 Family Proteins in 96-Well Plate Assays

By José-Luis Diaz, Tilman Oltersdorf, and Lawrence C. Fritz

Introduction

The *bcl-2* gene family encodes a series of proteins involved in the regulation of programmed cell death. Members of the family can be grouped into two distinct sets with antagonistic functions.[1-5] For example, Bcl-2, Bcl-x_L, and Bcl-w have antiapoptotic, protective functions, and prevent the activation of downstream death-effector caspase proteases. In contrast, proteins such as Bax, Bak, Bad, and Bid have proapoptotic roles and can antagonize the cell-protective functions of Bcl-2.[6-9] A large number of studies have indicated that Bcl-2 family members can dimerize, forming homodimers and/or heterodimers, and that these interactions are important for their function.[10] Yang and Korsmeyer[11] have proposed a "life–death rheostat" model, in which the response of a cell to an apoptotic signal is determined by the ratio of antiapoptotic to proapoptotic Bcl-2 family proteins. In this model, apoptosis is regulated by the competitive dimerization between various pairs of family members.

Bcl-2 family members share stretches of sequence homology known as BH1, BH2, BH3, and BH4 domains.[12,13] The antiapoptotic family members

[1] Y. Tsujimoto, J. Cossman, E. Jaffe, and C. Croce, *Science* **228,** 1440 (1985).
[2] D. M. Hockenbery, G. Nunez, C. Milliman, R. D. Screiber, and S. J. Korsmeyer, *Nature (London)* **348,** 334 (1990).
[3] D. L. Vaux, I. L. Weissman, and S. K. Kim, *Science* **258,** 1955 (1992).
[4] T. Miyashita and J. C. Reed, *Blood* **81,** 151 (1993).
[5] L. H. Boise, M. Gonzalez-Garcia, C. E. Postema, L. Ding, T. Lindsten, L. Turka, X. Mao, G. Nunez, and C. B. Thompson, *Cell* **74,** 597 (1993).
[6] Z. Oltvai, C. Milliman, and S. J. Korsmeyer, *Cell* **74,** 609 (1993).
[7] S. N. Farrow, J. H. M. White, I. Martinou, T. Raven, K.-T. Pun, C. J. Grinham, J.-C. Martinou, and R. Brown, *Nature (London)* **374,** 731 (1995).
[8] T. Chittenden, E. A. Harrington, R. O'Connor, C. Flemington, R. J. Lutz, G. I. Evan, and B. C. Guild, *Nature (London)* **374,** 733 (1995).
[9] M. C. Kiefer, M. J. Brauer, V. C. Powers, J. J. Wu, S. R. Umansky, L. D. Tomei, and P. J. Barr, *Nature (London)* **374,** 736 (1995).
[10] X. M. Yin, Z. N. Oltvai, and S. J. Korsmeyer, *Nature (London)* **369,** 321 (1994).
[11] E. Yang and S. J. Korsmeyer, *Blood* **88,** 386 (1996).
[12] T. Chittenden, C. Flemington, A. B. Houghton, R. G. Ebb, G. J. Gallo, B. Elangovan, G. Chinnadurai, and R. J. Lutz, *EMBO J.* **14,** 5589 (1995).
[13] H. Zha, C. Aime-Sempe, T. Sato, and J. C. Reed, *J. Biol. Chem.* **271,** 7440 (1996).

generally possess all four homology domains. The three-dimensional structure of Bcl-x_L has been solved and demonstrates that the homology domains fold into a compact structure, forming a prominent hydrophobic groove on the surface of the protein.[14] In contrast, the proapoptotic family members contain only a subset of the homology domains; some contain domains BH1–3[12,13,15] but others possess only the BH3 domain.[16,17]

Although the proapoptotic family members vary in structure, mutagenesis experiments suggest that the BH3 domain is both necessary and sufficient for these proteins to heterodimerize with Bcl-2 and Bcl-x_L and to induce apoptosis.[12,13,18] Structural studies have now shown that the BH3 domain of Bak forms an α helix that binds into the hydrophobic groove of Bcl-x_L.[19] Indeed, the formation of both homodimers and heterodimers is dependent on BH3 domain binding, because BH3-derived peptides can fully inhibit the interactions between the full-length proteins.[20]

The mechanism by which the antiapoptotic Bcl-2 family members inhibit cell death remains unknown. However, mutagenesis studies suggest that these antiapoptotic proteins have intrinsic cell survival activity,[21] implying that this activity is inhibited when proapoptotic, BH3-containing proteins bind into the hydrophobic groove. Thus, the groove can be viewed as a pharmacological binding site for modulating apoptosis in cells. For example, compounds that bind the groove of Bcl-2 might mimic the proapoptotic proteins and inhibit cell survival function. Alternatively, compounds might bind into the groove in such a manner as to block the binding of proapoptotic proteins, but not inhibit the intrinsic activity of Bcl-2. Such a compound would be predicted to disinhibit Bcl-2, promoting survival. In

[14] S. W. Muchmore, M. Sattler, H. Liang, R. P. Meadows, J. E. Harlan, H. S. Yoon, D. Nettesheim, B. S. Changs, C. B. Thompson, S. Wong, S. Ng, and S. W. Fesik, *Nature* (*London*) **381**, 335 (1996).

[15] S. Ottilie, J.-L. Diaz, W. Horne, J. Chang, Y. Wang, G. Wilson, S. Chang, S. Weeks, L. C. Fritz, and T. Oltersdorf, *J. Biol. Chem.* **272**, 30866 (1997).

[16] J. M. Boyd, G. J. Gallo, B. Elangovan, A. B. Houghton, S. Malstrom, B. J. Avery, R. G. Ebb, T. Subramanian, T. Chittenden, R. J. Lutz, and G. Chinnadurai, *Oncogene* **11**, 1921 (1995).

[17] K. Wang, X.-M. Yin, D. T. Chao, C. L. Milliman, and S. J. Korsmeyer, *Genes Dev.* **10**, 2859 (1996).

[18] S. Ottilie, J.-L. Diaz, J. Chang, G. Wilson, K. M. Tuffo, S. Weeks, M. McConnell, Y. Wang, T. Oltersdorf, and L. C. Fritz, *J. Biol. Chem.* **272**, 16955 (1997).

[19] M. Sattler, H. Liang, D. Nettesheim, R. P. Meadows, J. E. Harlan, M. Eberstadt, H. S. Yoon, S. B. Shuker, B. S. Chang, A. J. Minn, C. B. Thompson, and S. W. Fesik, *Science* **275**, 983 (1997).

[20] T. W. Sedlak, Z. N. Oltvai, E. Yang, K. Wang, L. H. Boise, C. B. Thompson, and S. J. Korsmeyer, *Proc. Natl. Acad. Sci. U.S.A.* **92**, 7834 (1995).

[21] E. H.-Y. Cheng, B. Levine, L. H. Boise, C. B. Thompson, and J. M. Hardwick, *Nature* (*London*) **379**, 554 (1996).

either case, such compounds would be expected to block the binding of a BH3-containing protein to the groove-containing protein. Therefore, binding assays that measure the interaction of Bcl-2 family proteins may be useful in identifying compounds that promote or inhibit apoptosis.

Principle of Assay

Interactions between Bcl-2 family proteins have generally been studied by immunoprecipitation analyses and yeast two-hybrid systems.[10,12,13,18] However, these methods are not optimal for establishing the quantitative high-throughput assays necessary for the screening of large chemical libraries. The method described below is a quantitative binding assay that can be carried out in a 96-well format, using any combination of recombinant Bcl-2 family protein pairs. The assay is a solid-phase binding assay with an enzyme-linked immunosorbent assay (ELISA) readout, where one member of the protein pair is attached to the surface of a polystyrene 96-well plate in a nonspecific manner. The second protein is incubated and allowed to bind to the first protein, and then is specifically detected by a mouse monoclonal antibody. The readout for the assay is the colored product of a substrate converted by alkaline phosphatase enzyme, covalently conjugated to a second antibody.

Assay Methods

Bacterial Expression Plasmids

Human Bcl-2, Bcl-x_L, Bax, and Bad proteins are expressed in bacteria. Because soluble binding partners are desired, the C-terminal membrane-spanning domains of Bcl-2, Bcl-x_L, and Bax are deleted. Bcl-2 and Bcl-x_L are cloned into the pGEX-4T-1 (Pharmacia, Piscataway, NJ) vector, and are expressed as fusion proteins with N-terminal glutathione S-transferase (GST) tags. The construct for expression of GST-Bcl-2, encoding a fragment containing amino acids 1–218 of human Bcl-2, is a generous gift of J. Reed (Burnham Institute, La Jolla, CA). The construct for expression of GST-Bcl-x_L, encoding a fragment including amino acids 1–211, is obtained from a full-length Bcl-x_L cDNA template by polymerase chain reaction (PCR) with Pfu polymerase (Stratagene, La Jolla, CA). The primer sequences are 5′ AGT ATC GAA TTC ATG TCT CAG AGC AAC CGG 3′ and 5′ TAC AGT CTC GAG CTA GTT GAA GCG TTC CTG GCC CT 3′ with EcoRI as a 5′ cloning site and XhoI as a 3′ cloning site.

Proteins with N-terminal six-histidine (6H) tags are expressed using the pET-15b (Novagen, Madison, WI) vector. The construct for expression of

6-histidine-Bax, encoding amino acids 1–170, is obtained by PCR with the following primers: 5' ACG TAC CAT ATG GAC GGG TCC GGG GAG 3' and 5' TAC AGT CTC GAG CTA CCA CGT GGG CGT CCC AAA 3'. The fragment is cloned into the 5' NdeI and 3' XhoI cloning sites of pET-15b. Constructs encoding 6H-Bcl-x$_L$ (amino acids 1–211), 6H-Bcl-2 (amino acids 1–218), and 6H-Bad (amino acids 1–168) are obtained in an analogous fashion.

Protein Expression

For expression, pGEX-4-based plasmids are transformed into *Escherichia coli* XL-1 blue (Stratagene). Bacteria are grown in 3-liter batch cultures in shaker flasks to an absorbance of 0.85 at 600 nm at 37°. Cultures are cooled to 25 or 30° and isopropyl-β-D-thiopyranoside is added to induce protein expression. Exact conditions vary and are optimized for each of the various proteins. Cultures are grown until they reach an optical density of 3 units at 600 nm, or, for a maximum of 3 hr. Bacteria are harvested by pelleting at 4000g for 10 min. Bacterial pellets are resuspended in 50 ml of the appropriate binding buffer for further purification. Bacteria are lysed by sonication followed by processing through a microfluidizer (Microfluidics International, Newton, MA). Lysates are pelleted at 20,000g for 20 min. For GST fusion proteins only the soluble fraction is processed further. In the case of Bcl-x$_L$ and Bax the bulk of the expressed protein is contained in this soluble fraction, whereas for Bcl-2, a significant amount is contained in the insoluble pellet.

GST fusion proteins are purified by one-step affinity purification, using glutathione–Sepharose 4B (Pharmacia Biotech, Uppsala, Sweden) on a fast protein liquid chromatography (FPLC) system, essentially following the manufacturer instructions. Proteins with N-terminal 6H tags are expressed in the *E. coli* host strain BL-21 or BL-21-plysS and processed essentially like GST fusion proteins. Affinity chromatography is performed with Ni-NTA Superflow (Qiagen, Chatsworth, CA), following the manufacturer recommendations. When the bulk of the expressed protein is sequestered in inclusion bodies, the insoluble fraction of the bacterial extracts is solubilized in the appropriate loading buffer with 7 M urea. Ni-NTA chromatography is performed in the presence of urea. Chromatography is followed by a buffer exchange step using disposable PD-10 gel-filtration columns (Pharmacia Biotech). In some cases the final storage buffer and conditions must be optimized to avoid solubility problems.

The purity of proteins obtained by these methods is assessed by sodium dodecyl sulfate–polyacrylamide gel electrophoresis (SDS–PAGE). Protein concentrations are determined routinely by a bicinchoninic acid method

(Pierce, Rockford, IL) with a bovine serum albumin (BSA) standard. Expressed proteins are considered sufficiently pure when no contaminating bands can be detected in Coomassie-stained gels loaded with approximately 1 μg of protein per lane.

In Vitro Protein–Protein Binding Assay Procedure

1. *Coating of plate:* Dilute purified 6H-Bax to 4 μg/ml in phosphate-buffered saline (PBS) and coat onto 96-well microtiter plates (50 μl/well) (Immunosorb; Nunc, Roskilde, Denmark) for 18 hr at 4°.

2. *Blocking of plate:* Wash the plates two times with PBS containing 0.05% (v/v) Tween 20 (PBS-T) and then block with 150 μl of 2% (w/v) BSA in PBS for 2 hr at room temperature.

3. *Second protein:* Wash the plates two times with PBS-T and incubate with a range of concentrations (20 to 0.0001 μM) of GST-Bcl-2 in PBS-T containing 0.5% (w/v) BSA (50 μl/well). Allow the interaction to proceed for 2 hr at room temperature before washing the plates five times with PBS-T.

4. *Primary antibody:* The wells are then incubated for 1 hr at room temperature with 50 μl of the primary antibody, a mouse anti-GST monoclonal antibody used at 1 ng/ml in PBS-T plus BSA, before being washed five times with PBS-T. In our case, we use antibody 7E5A6 (a kind gift of J. Reed), but an anti-GST monoclonal antibody is commercially available from Santa Cruz Biotechnology (Santa Cruz, CA).

5. *Secondary conjugated antibody:* Add alkaline phosphatase-conjugated goat anti-mouse antibody (50 μl/well; Jackson ImmunoResearch Laboratories, West Grove, PA) at a concentration of 1 μg/ml and incubate for 1 hr at room temperature, and then wash the plates five times.

6. *Substrate:* Add 50 μl of enzyme substrate, *p*-nitrophenyl phosphate (Kirkegaard & Perry, Gaithersburg, MD) at 4 mg/ml in 10 mM diethanolamine (pH 9.5) containing 0.5 mM MgCl$_2$, allow the reaction to progress for 15 min at room temperature, and then stop the reaction by the addition of 0.4 M NaOH (50 μl/well). Read the optical density of the wells at 405 nm in a spectrophotometer (Molecular Devices, Palo Alto, CA).

Use of Assay to Map Binding Site

The solid-phase protein–protein binding assay may be used to determine protein sequences important for dimer formation. This may be done by running competition assays to test the ability of whole proteins and peptides to inhibit the dimerization of the protein pairs. For competition assays, the liquid-phase GST-tagged protein is used at a constant concentration. In the case of GST-Bcl-2, we use a concentration of 80 nM throughout. This concentration was determined in preliminary experiments to be on the rate-limiting part of the binding curve.

The coating and blocking steps are carried out as described above, after which the appropriate peptides (50 μl/well) are added to the wells in increasing concentrations [0.156–80 μM in PBS-T plus 0.5% (w/v) BSA]. The plates are incubated for 1 hr at room temperature and then the second protein is added to the wells without removing the test compounds.

The subsequent steps are exactly as described in the procedure above from step 3 onward. The controls usually run on each plate include a vehicle control and a negative control peptide containing a single amino acid change.

Method for Screening Compounds

An adaptation of the procedure described above may be used to screen for compounds that inhibit the dimerization of the protein pairs. For screening, the liquid-phase GST-tagged protein is also used at a constant concentration throughout. The coating and blocking steps are carried out as described above, after which the compounds are added to the wells in duplicate at a concentration of 20 μM (50 μl/well) in PBS-T plus 0.5% (w/v) BSA. The plates are incubated for 1 hr at room temperature and then the second protein is added to the wells still containing the test compounds. The subsequent steps are exactly as described in the procedure above. The controls usually run on each plate include a vehicle control and a known inhibitor at a concentration titrated to give a constant 50% inhibition.

Site of Action of Inhibitors

The binding assay can be modified to identify to which member of a dimer pair an inhibitory molecule is binding. After the usual coating and blocking steps using 6H-Bax, increasing concentrations of inhibitory compound (0.01 to 100 μM) are added to the wells and allowed to incubate for 2 hr at room temperature. The plates are then washed five times with PBS-T before addition of the second protein, Bcl-x_L, in liquid phase. An alternative assay is then set up, in which the plates are coated with Bcl-x_L and incubated with increasing concentrations of inhibitory compound. The compounds are washed off and then GST-Bax is added as the second protein in liquid phase. In this manner, only the protein on the solid phase is exposed to the inhibitory compound. Comparing the median inhibitory concentration (IC$_{50}$) values for the compound incubated singly with either the Bax or the Bcl-x_L protein gives an indication as to which of the dimerizing proteins is the target for the compound. In Fig. 1, for example, Bcl-x_L heterodimer formation is blocked by a peptide derived from the BH3 domain of Bax when preincubated with the Bcl-x_L (IC_{50} of 3.5 μM), but not when the peptide is preincubated with Bax ($IC_{50} > 100 \mu M$).

Reversible and Irreversible Inhibitors

A modification of the binding assay can also be used to help decipher the mode of action of the inhibitor molecule. A constant concentration of the GST-tagged protein in liquid phase is preincubated with increasing concentrations of inhibitory compound (0.01 to 100 μM) before addition to the 96-well plates coated with the appropriate proteins. Multiple dilution series of the compound are set up, and the preincubation time of the compound with the liquid-phase protein is varied from 0 to 24 hr. The results are plotted, and when the resulting IC_{50} values are compared, we see that there is a category of compounds whose IC_{50} values do not vary with time once they have reached equilibrium; these are reversible inhibitors. An

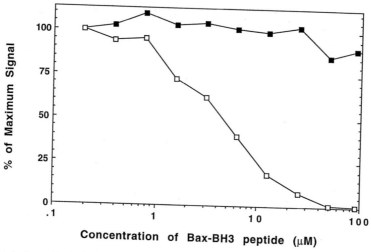

FIG. 1. A peptide derived from the BH3 domain of Bax blocks Bcl-x_L heterodimer formation by binding to the Bcl-x_L and not the Bax protein. Wells were coated with 6H-Bax (solid squares) or 6H-Bcl-x_L (open squares) and pretreated with increasing concentrations of the peptide Bax 52–72. The unbound peptide was washed off and the wells were incubated with a constant concentration of GST-Bcl-x_L (solid squares) or GST-Bax (open squares). Bound protein was quantitated in all cases by reaction with anti-GST antibody and alkaline phosphatase-coupled secondary antibody.

example of this is illustrated in Fig. 2, where a peptide derived from the BH3 domain of Bax (amino acids 52–72) inhibits Bcl-x_L heterodimer formation with similar IC_{50} values irrespective of time of preincubation. In contrast, IC_{50} values for irreversible inhibitors vary as a function of time of incubation.

Comments and Considerations

In the procedures described above, we have constantly referred to the Bax/Bcl-2 or Bax/Bcl-x_L dimerization as concrete examples, but this assay format may be used to investigate many of the different binding pairs in the Bcl-2 family of proteins. Some preliminary experimentation is needed to adapt this method to other binding pairs. In each case it is important to optimize the concentration of the coating protein, because the amount of protein adhering to the polystyrene surface will depend, among other things, on the size of the protein.

For the screening and competition assays, it is also important to create preliminary binding curves for each pair of dimerizing proteins to be used,

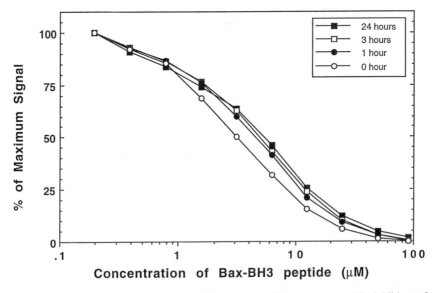

FIG. 2. A peptide derived from the BH3 domain of Bax is a reversible inhibitor of Bcl-x_L heterodimer formation. Wells were coated with 6H-Bax and incubated with a constant concentration of GST-Bcl-x_L that had been pretreated with increasing concentrations of competing peptide Bax 52–72 for 0 hr (open circles), 1 hr (closed circles), 3 hr (open squares), or 24 hr (closed squares). Bound protein was quantitated as in Fig. 1.

and to choose a concentration of the protein in liquid phase that lies on the steep logarithmic part of the binding curve.[22]

The binding assays can be carried out with purified proteins missing the transmembrane domain. Furthermore, in our hands the presence of tags, whether located on the N terminus or C terminus, does not generally interfere with dimerization. Thus, the tags do not need to be cleaved off. On the contrary, careful selection of the tags can be used to aid the assay. The formation of Bcl-2 homodimers, for example, can be detected by using 6H and GST tags on the solid–phase and liquid-phase partners, respectively. By expressing all potential liquid-phase partners with a uniform tag, such as GST or Glu–Glu–Phe (EEF), a single anti-GST or anti-EEF antibody can be used to monitor a wide range of dimerization pairs, without having to generate a specific antibody for each protein. An antibody directed against a terminal tag is also less likely to interfere with the dimerization of the proteins, or be affected by the action of inhibitory compounds, than an antibody directed against an epitope on the main sequence of the protein itself.

The assays can be run in various configurations with proteins tagged in various ways. However, in general, the assays work best when (1) the proteins to be coated onto the polystyrene plates are tagged with 6H, (2) the protein acting as a BH3 donor is used as the solid-phase partner, and (3) the partner with the hydrophobic binding cleft is used as the protein in liquid phase.

For each binding pair studied, preliminary experiments should be carried out to ensure the specificity of the binding observed. In our case, for each specific interaction, saturable binding was detected, whereas only low levels of background binding were seen when the control protein BSA was used as the solid-phase binding partner (Fig. 3). To confirm that these *in vitro* interactions reflect the interactions observed in cells, we assessed the ability of the G145A and G138A point mutants of Bcl-2 and Bcl-x_L, respectively, to bind to Bax; it has previously been shown by immunoprecipitation that these mutants fail to heterodimerize with Bax in cells. Recapitulating the cellular result, the mutants demonstrated a much-reduced binding to Bax in the plate-binding assay (Fig. 3).[22]

Advantages

This is an easy, reliable assay that can be used to investigate the binding properties of Bcl-2 family members with regard to homodimer and heterodimer formation. The assay can be performed with recombinant Bcl-2, Bcl-x_L, and Bax proteins and is highly reproducible (Table I). The format

[22] J.-L. Diaz, T. Oltersdorf, W. Horne, M. McConnell, G. Wilson, S. Weeks, T. Garcia, and L. C. Fritz, *J. Biol. Chem.* **272,** 11350 (1997).

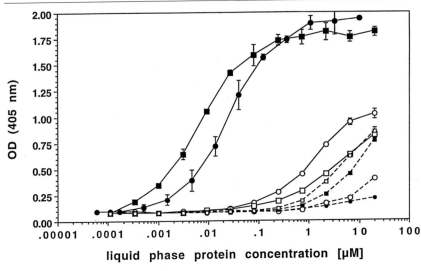

Fig. 3. Formation of Bcl-2 family heterodimers in solid-phase binding assays. Wells were coated with 6H-Bax (solid lines) or the control protein BSA (dashed lines) and incubated with increasing concentrations of GST-Bcl-2 (filled circles), GST-Bcl-x_L (filled squares), GST-Bcl-2 (G145A) (open circles), or GST-Bcl-x_L (G138A) (open squares). Bound protein was quantitated in all cases by reaction with anti-GST antibody and alkaline phosphatase-coupled secondary antibody. [Taken from Diaz et al.[22]]

of the assay makes it amenable to high-throughput screening, and the format has certain advantages compared with other assay formats designed to measure protein–protein interactions. For example, yeast two-hybrid assays can be used to monitor such interactions, but with the added variable of compound penetration through the yeast cell wall.

TABLE I
EC$_{50}$ VALUES FOR HETERODIMER AND HOMODIMER FORMATION[a]

Solid phase	Liquid phase	EC$_{50}$ (nM)	n
6H-Bax	GST-Bcl-2	26.1 ± 10.4	19
6H-Bax	GST-Bcl-x_L	14.7 ± 5.3	17
6H-Bcl-2	GST-Bcl-2	66.2 ± 36.0	9
6H-Bcl-x_L	GST-Bcl-x_L	109.7 ± 26.0	11

[a] 6H-Bax, 6H-Bcl-2, and 6H-Bcl-x_L were coated on microtiter planes and binding curves were established with increasing concentrations of GST-Bcl-2 or GST-Bcl-X_L. EC$_{50}$ values listed are the mean concentration of liquid-phase protein that yielded half-maximal binding ± standard deviation for n independent measurements. (Taken from Diaz et al.[22])

The assay can also be used in a quantitative manner. Indeed, we used these solid-phase binding assays to generate quantitative binding data that provided the first estimate for the strength of these protein–protein interactions.[22] The concentration of liquid-phase protein necessary for half-maximal binding for each reaction is presented in Table I.

Assumptions and Limitations

Although the protein–protein binding assay described above has been used to compare the affinity of the interactions between different Bcl-2 family proteins, the binding data generated are not true K_d values. There are two main reasons why we cannot generate K_d values in these assays: first, protein binding in this method does not occur in a homogeneous liquid phase; second, this procedure involves a separation step, and therefore the assay does not measure equilibrium binding. The binding assay involves extensive washing, and therefore one would expect that binding would be detected by this procedure only for interactions where dissociation rates were comparatively low. However, the assay yields little information on the kinetics of dissociation of the dimers. There may also be an avidity effect due to the local concentration of protein bound to the solid phase.

Experiments have been carried out in collaboration with Igen (Gaithersburg, MD), using their ORIGEN assay system. This is basically a method to study protein–protein interactions by electrochemiluminescence (ECL). Protein binding by this method occurs in a homogeneous liquid phase, using equilibrium binding procedures. Using ECL, we determined the K_d for the Bax/Bcl-2 interaction to be 8.4 nM (data not shown), which is reasonably close to the apparent K_d of the interaction in our plate assay (Table I).

Although dimerization data generated by these *in vitro* assays appear consistent with observations made by other methods such as immunoprecipitation, there are two major points of variance. It has been reported that Bax can form homodimers, yet we have been unable to observe Bax homodimer formation in solid-phase binding assays.[6,13,20] This could be highlighting a problem in our assays: for example, it is possible that Bax adopts a distinct and nonphysiological conformation when binding to the solid phase. However, published studies indicate that human Bax also fails to homodimerize in a yeast two-hybrid assay.[23] Furthermore, one study reported that Bax homodimerization, as monitored by immunoprecipitation, was found to be

---

[23] H. Zhang, B. Saeed, and S.-C. Ng, *Biochem. Biophys. Res. Commun.* **208,** 950 (1995).

dependent on the presence of detergent [1% (v/v) Triton X-100, as opposed to 0.05% (v/v) Tween 20 in our assays], but was not observed under more physiological conditions.[24] In addition, Bax has been shown to exist in the cytoplasm, but in that compartment it fails to form homodimers.[24]

Interestingly, Bcl-x_L readily formed homodimers in our assay, whereas attempts by others to observe this interaction in cells or with isolated protein have failed.[14,25] While we do not understand the basis for these differences, it is possible that a specific conformation of Bcl-x_L is favored when bound to plastic, and that this conformation facilitates homodimerization. For example, this conformation may alter the positioning of the α_2 helix (encompassing the BH3 domain), facilitating its interaction with a second molecule of Bcl-x_L.

Acknowledgments

We thank Dr. John Reed (Burnham Institute) for the GST-Bcl-2 expression construct and anti-GST antibody, Suzanne Weeks and Gary Wilson (Idun Pharmaceuticals) for the expression and purification of the Bcl-2 family proteins, and Steven Chang for refining and implementing the methods outlined above.

[24] Y.-T. Hsu and R. J. Youle, *J. Biol. Chem.* **272**, 13829 (1997).
[25] A. J. Minn, L. H. Boise, and C. B. Thompson, *J. Biol. Chem.* **271**, 6306 (1996).

[25] Analysis of Dimerization of Bcl-2 Family Proteins by Surface Plasmon Resonance

By Zhihua Xie and John C. Reed

Introduction and Background

Bcl-2 family proteins are important regulators of programmed cell death and apoptosis.[1,2] These proteins either inhibit or induce cell death, with the ratios of antiapoptotic relative to proapoptotic members of the Bcl-2 family representing a critical determinant of the ultimate sensitivity or resistance of mammalian cells to various apoptotic stimuli. Many Bcl-2 family proteins can physically interact with themselves and each other,

[1] J. Reed, *Oncogene* **17**, 3225 (1998).
[2] A. Gross, J. McDonnell, and S. Korsmeyer, *Genes Dev.* **13**, 1899 (1999).

forming homo- or heterodimers.[3] In some instances, these dimerization events appear to play important roles in the regulatory or effector functions of these proteins. Thus, a need exists to understand more about how dimerization among the Bcl-2 family proteins is controlled.

Although the concepts for the surface plasmon resonance (SPR) technique were developed more than 30 years ago, the SPR method for monitoring protein interactions became practical only in the early 1990s, when advances in instrumentation were introduced.[4] Because it provides real-time information on association and dissociation of biological macromolecules, SPR has found many applications in a broad variety of fields of biomedical research. Among the strengths of SPR are its capability to (1) monitor the sequential addition of components during formation of multiprotein complexes, (2) separate on rates from off rates; and (3) reconstitute protein interactions with artificial membranes. The SPR method detects changes in the optical properties at the surface of a thin gold film on a glass support, which is in contact with a solution through which light waves are reflected. Molecules that interact with the chip surface create a change in refractive index that is proportional to mass, thus forming the basis for a biomolecular interaction assay. After attachment of a desired protein to the chip surface, a second protein is passed over the sensor surface in a solution (mobile phase). This is typically done with a series of concentrations of the protein in the mobile solution. The data obtained from macromolecule binding to SPR chips are known as *sensorgrams* and contain information about the on and off rates of the mobile-phase protein as it interacts with immobilized protein on the chip surface. The sensorgram curves move upward as binding is initiated, and subsequently flatten and reach a plateau as equilibrium is reached. Solution lacking the mobile-phase protein is then flowed across the chip and the dissociation of the mobile protein is observed as a decline in the sensorgram reading.

SPR, however, is not without its difficulties. SPR is prone to certain potential artifacts, caused, for example, by immobilization of test proteins in multiple conformations, steric interference with interaction surfaces, and issues related to mass transport as mobile-phase proteins penetrate the 100–200 mm of dextran matrix. For detailed discussion, the reader is referred to reviews.[4–6]

In this chapter, we describe SPR-based procedures for monitoring real-time interactions between Bcl-2 family proteins, including the antiapoptotic

[3] A. Kelekar and C. B. Thompson, *Trends Cell Biol.* **8**, 324 (1998).
[4] R. J. Fisher and M. Fivash, *Curr. Opin. Biotechnol.* **5**, 389 (1994).
[5] Z. Salamon, H. A. Macleod, and G. Tollin, *Biochim. Biophys. Acta* **1331**, 117 (1997).
[6] P. Schuck, *Annu. Rev. Biophys. Struct.* **26**, 541 (1997).

proteins Bcl-2 and Bcl-X$_L$, the proapoptotic protein Bax, and the proapoptotic BH3-only protein Bid.[7]

General Considerations

Immobilization and Chips

In general, there are two approaches to effectively immobilize proteins of interest on SPR chips: covalent coupling and noncovalent capturing. The more straightforward of these is to immobilize the molecule of interest covalently on the chip surface through chemical coupling. The disadvantage of this method is that it may introduce modifications to the protein, generating subsets of immobilized target proteins that differ in terms of the amino acid residues, through which chip attachment occurs, or leading to severe steric hindrance problems if a small target protein has been chosen for immobilization. These problems may result in complete or partial blocking of the interaction between immobilized and solution-phase proteins. However, the advantage of covalent coupling is that it is generally relatively simple to regenerate the bioactive surface, allowing multiple experiments to be performed with a single chip. Particularly when small proteins are used, the regeneration can be accomplished through denaturation and subsequent refolding of the immobilized protein. Examples of noncovalent capture include (1) capturing molecules of interest using immobilized antibodies, (2) capturing His$_6$ fusion proteins by nickel surface (NTA chip), (3) capturing biotinylated protein by streptavidin (SA chip), or (4) capturing integral membrane proteins by insertion into artificial lipid monolayers attached to the chip surfaces (HPA chip). The disadvantage of noncovalent coupling is that loss of the immobilized proteins from the chip surface may occur during experiments, thus complicating interpretation of results. Moreover, harsh regeneration conditions typically cannot be employed, thus limiting reuse of bioactive surfaces. Finally, conditions that have been optimized for capturing a protein may not be suitable for subsequently performing interaction studies.

We primarily employ CM5 chips (Biacore AB, Uppsala, Sweden). This chip has a carboxymethyl dextran-modified gold surface. The attachment of the bound reactant is via the formation of N-hydroxysuccinimide esters from a subpopulation of available carboxyl groups on the chip. This is achieved by the reaction of N-hydroxysuccinimide (NHS) and N-ethyl-N'-(dimethylaminopropyl)carbodiimide hydrochloride (EDC). Coupling in this case is through the primary amine groups of the protein. The protein

[7] Z. Xie, S. Schendel, S. Matsuyama, and J. C. Reed, *Biochemistry* **37**, 6410 (1998).

to be immobilized is passed over the activated surface in a solution of low ionic strength at a pH value that is below the pI of the protein. The remaining esters on the chip are quenched with ethanolamine.

Regeneration

For many SPR studies, finding optimal conditions for regenerating bioactive surfaces can be a major challenge. For our studies, we take advantage of the fact that Bcl-2 family proteins are relatively small proteins ($\approx$20–25 kDa) that are easily refolded after denaturation. After each injection, chips containing covalently immobilized Bcl-2 family proteins are regenerated by injection of 5 μl of 4 M guanidinium-HCl (Gu-HCl). We observed that such chips regenerated with 4 M Gu-HCl can remain bioactive even after 50 cycles. Other conditions we have tested, including extremes of pH, high salt, and sodium dodecyl sulfate (SDS), do not work as well as Gu-HCl for Bcl-2 family proteins, either because chips are not regenerated completely (i.e., reactants remain bound) or chip surfaces lose activity after a few cycles.

Proteins

Bcl-2 family proteins are produced in and purified from bacteria as described in [24] in this volume,[8] as either glutathione S-transferase (GST) fusion or His$_6$-tagged proteins. When using GST fusion proteins, we typically remove the GST portion by thrombin-mediated proteolysis, rather than risk difficulties in interpreting results caused by dimerization of GST. However, GST fusion protein may be used for immobilization, although it is not recommended for use in the mobile phase. With respect to His$_6$-proteins, these should not be employed for the mobile phase if low pH conditions will be employed, because nonspecific interactions with chip surfaces are typically observed. In contrast, His$_6$ tag proteins are efficiently immobilized on standard chip surfaces by the protocol described below. Another potential application of the His$_6$ tag is that it promotes association with negatively charged artificial membranes on chip surfaces, mimicking to some extent the transmembrane-anchoring domain normally excluded from these recombinant proteins because of solubility considerations. In this regard, most investigators produce recombinant Bcl-2, Bcl-X$_L$, Bax, and Bak as truncated proteins that lack the C-terminal ~20-amino acid hydrophobic domain responsible for anchoring these proteins in membranes *in vivo*. Some Bcl-2 family proteins such as Bid lack such transmem-

[8] J. L. Diaz, T. Oltersdorf, and L. C. Fritz, *Methods Enzymol.* **322,** [24], 2000 (this volume).

brane domains and thus can be generated readily in bacteria as full-length proteins.

Solution Conditions

It is recommended that the proteins used in mobile phase be largely detergent free. In the case of Bcl-2, we observed that 0.01% (v/v) Tween 20, when injected across chips containing immobilized Bcl-2 family proteins, will generate 50–100 RU (resonance units) depending on the particular protein, presumably because of the high refractive index of detergent. HB buffer [10 mM HEPES (pH 7.4), 150 mM NaCl, 3.4 mM EDTA] is recommended for the running buffer, although other solutions may be equally appropriate. Importantly, the protein samples must be dialyzed against the chosen running buffer extensively to reduce the solution spike that occurs on sample injection into the SPR instrument.

Procedures

Kinetic measurements are performed with a BIAcore-II instrument with CM5 sensor chips and an amine coupling kit (Pharmacia Biosensor AB, Uppsala, Sweden), although comparable instruments should be equally appropriate. For immobilization of proteins, the sensor chip is equilibrated with HB buffer at 5 μl/min, and then activated by injecting 17 μl of 0.2 M EDC and 0.05 M NHS followed by a 35-μl injection of Bcl-2 family or control proteins at 50–100 μg/ml in 10 mM acetate, pH 3.5–4.8. Excess NHS esters on the surface are blocked with 17 μl of 1 M ethanolamine hydrochloride (pH 8.5). After immobilization, 10 μl of regeneration buffer [50 mM phosphate (pH 6.8), 4 M Gu-HCl] is injected to remove noncovalently bound material.

A total of 5000 or 1500–2000 RU produced by immobilized protein after regeneration is suitable for Bcl-2 family protein SPR analysis when carried out at neutral and acidic pH, respectively. When immobilizing Bcl-2 family proteins on CM5 chips, His$_6$-tagged proteins may be preferred because polyhistidine at low pH results in effective preconcentration, but differences in His$_6$-tagged or untagged proteins are not substantially different.

For studying protein–protein interactions, kinetic data are obtained by diluting the samples in interaction buffer (HB buffer) and injecting 20 μl at 10 μl/min across the prepared surface. The surface is regenerated after each injection with 5 μl of 50 mM phosphate (pH 6.8)–4 M Gu-HCl. A

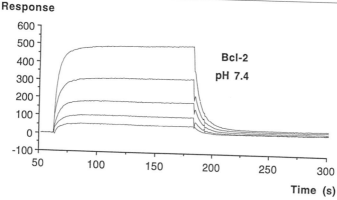

FIG. 1. Example of SPR analysis of Bcl-2 family protein interactions. Sensorgrams are presented, showing the binding of different concentrations of Bcl-X_L to and dissociation from immobilized Bcl-2 at pH 7.4. Bcl-X_L solution [ranging from 0.125 to 2 μM in 10 mM HEPES (pH 7.4), 150 mM NaCl, 3.4 mM EDTA] was injected (20 ml) at 10 μl/min across the immobilized Bcl-2 surface. The binding and dissociation were recorded. After each injection, the chip surface was regenerated with 5 μl of 50 mM phosphate (pH 6.8)–4 M Gu-HCl.

typical sensorgram is presented in Fig. 1, showing interaction between various concentrations of Bcl-X_L and immobilized Bcl-2 at neutral pH. Note that rapid binding occurs until a steady state equilibrium is reached, followed by rapid dissociation when the Bcl-X_L-containing mobile phase was replaced with flow buffer. The resonance units (RU) plateau obtained increases in a concentration-dependent manner.

Under acidic conditions *in vitro,* several Bcl-2 family proteins, including Bcl-2, Bax, Bcl-X_L and Bid, insert into lipid membranes, forming ion channels.[9–13] Presumably, the dependence on low pH for this phenomenon reflects the need for a conformational change that allows the putative pore-forming pair of α helices within these proteins to penetrate the lipid bilayer (reviewed in Ref. 14).

Although low pH can denature some proteins, far-UV circular dichroism (CD) spectra for Bcl-X_L, Bcl-2, Bax, and Bid are nearly superimposable at pH 4.0 and 7.0, indicating that the secondary structure of these proteins

[9] A. J. Minn, *et al., Nature (London)* **385,** 353 (1997).
[10] B. Antonsson, *et al., Science* **277,** 370 (1997).
[11] P. Schlesinger, *et al., Proc. Natl. Acad. Sci. U.S.A.* **94,** 11357 (1997).
[12] S. Schendel, R. Azimov, K. Pawlowski, A. Godzik, B. Kagan, and J. C. Reed, *J. Biol. Chem.* **274,** 21932.
[13] S. L. Schendel, Z. Xie, M. O. Montal, S. Matsuyama, M. Montal, and J. C. Reed, *Proc. Natl. Acad. Sci. U.S.A.* **94,** 5113.
[14] S. Schendel, M. Montal, and J. C. Reed, *Cell Death Differ.* **5,** 372 (1998).

is preserved at low pH. To examine how low pH affects heterodimerization among Bcl-2 family proteins, we examined the kinetics and magnitude of binding of Bcl-X_L with immobilized Bcl-2, Bax, Bid, and Bcl-X_L by SPR over a range of pH values (2.5–8.0).[7] The amount of Bcl-X_L interacting with chips containing immobilized Bcl-2 family proteins increases inversely with the pH of the flow solution, and is highest at approximately pH 4.0. Comparisons of measurements performed at pH 4.0 reveals that the association-phase kinetics for these binding events are not substantially different from those obtained at pH 7.4. In all cases, however, the dissociation rates are markedly slower at pH 4.0.

The rate constants k_{ass} and k_{diss} were calculated from the association and dissociation phases of sensorgram, and the apparent affinity constants K_D were calculated directly from the parameter k_{diss}/k_{ass}, using the BIAcore software package (Pharmacia). The affinity constants (K_D) of interactions among Bcl-2 family proteins at pH 4.0 in 150 mM NaCl ranged between 0.03 and 0.44 μM.[7]

SPR is generally not used at low pH, mainly because the CM5 chip surface is made from dextran. At low pH, it is highly negatively charged. Most proteins will be positively charged at that pH (<pI), and will nonspecifically bind to the chip surface. Indeed, this property has been used to enhance the immobilization of proteins to the chip surface when protein samples are diluted ("preconcentration"). However, when low pH is desirable for monitoring interactions between macromolecules, the main concern is to reduce nonspecific interactions of mobile-phase proteins with the chip surface. Among the Bcl-2 family proteins we have tested, all but Bcl-X_L exhibit nonspecific interactions with the chip at low pH. Thus, Bcl-X_L was employed in the mobile phase for all SPR experiments, with Bcl-2, Bax, and Bid each immobilized on the chip surface.[7] All other GST-tagged proteins exhibit unacceptable background interactions when injected over deactivated blank chip surfaces at low pH.

Control Experiments

Previous studies have demonstrated that dimerization among Bcl-2 family proteins depends on insertion of the BH3 domain (which is composed of an amphipathic α helix) of one partner into a surface hydrophobic pocket on the other.[15] To determine whether the interactions measured by SPR are specific, therefore, synthetic peptides representing BH3 domains can be employed as competitive antagonists. The minimum effective length of nonconstrained linear BH3 peptides that has proved successful is 16 amino

---

[15] M. Sattler, et al., Science **275**, 983 (1997).

acids.[7,15–17] Median inhibitory concentration (IC_{50}) values are typically ~25 μM. Unrelated non-BH3 peptides or BH3 peptides containing mutations should be included in all experiments as a control. Another excellent control for SPR experiments is to employ mutant Bcl-2 family proteins that are designed not to heterodimerize with other family members.[18–21]

Calculations

The basic Biacore rate equation[22] for the first-order reaction in association phase is

$$dR/dt = k_a C(R_{max} - R_t) - k_d R_t$$

where C is the concentration of mobile-phase protein, R_{max} is the total amount of binding sites of the immobilized protein (expressed as the SPR response in resonance units), R_t is the SPR response in resonance units at time t, k_a is the association rate constant, and k_d is the dissociation rate constant.

Thus, $(R_{max} - R_t)$ is the amount of the remaining free binding sites at time t, expressed as the SPR response in resonance units.

Rearranging,

$$dR/dt = k_a C R_{max} - (k_a C + k_d)R_t$$

To determine k_a, one must first determine the slope (k_s) of a plot of dR/dt (y axis) versus R_t (x axis), using either manual plotting or curve-fitting software. Using a minimum of three concentrations, a plot of k_s (y axis) against C then yields a line, the slope of which equals k_a.

The dissociation rate constant, k_d, can be obtained from the SPR dissociation phase data, base on the rate equation

$$dR/dt = -k_d R_t$$

either by manually plotting dR/dt against R_t or through software fitting. The equilibrium constant K_D is calculated using k_d and k_a ($K_D = k_d/k_a$). The rapid dissociation rates observed with some Bcl-2 family protein interactions at neutral pH make it difficult to accurately determine kinetic

[16] J.-L. Diaz, *et al.*, *J. Biol. Chem.* **272**, 11350 (1997).

[17] S. Ottilie, *et al.*, *J. Biol. Chem.* **272**, 30866 (1997).

[18] X. M. Yin, Z. N. Oltvai, and S. J. Korsmeyer, *Nature (London)* **369**, 321 (1994).

[19] A. Kelekar, B. Chang, J. Harlan, S. Fesik, and C. Thompson, *Mol. Cell. Biol.* **17**, 7040 (1997).

[20] H. Zha and J. C. Reed, *J. Biol. Chem.* **272**, 31482 (1997).

[21] E. H.-Y. Cheng, B. Levine, L. H. Boise, C. B. Thompson, and J. M. Hardwick, *Nature (London)* **379**, 554.

[22] Biacore, "Biacore Method Manual." Biacore AB, Uppsala, Sweden.

data (k_d). In those instances, it may be useful to employ a Scatchard analysis of the data as follows.

The BIAcore rate equation as equilibrium is reached is

$$dR/dt = 0 = k_a C(R_{max} - R_{eq}) - k_d R_{eq}$$

where R_{eq} is the SPR response in resonance units at equilibrium.

Substituting K_D for k_d/k_a, and rearranging,

$$R_{eq}/C = (R_{max} - R_{eq})/K_D$$

by plotting R_{eq}/C versus R_{eq} for each concentration of protein used (e.g., Bcl-X$_L$) in the mobile phase, a line is generated with the slope $1/K_D$. The apparent K_D for protein–protein interactions is thus derived from the slope of the linear portions of the curves. On the basis of this treatment of the SPR data, the affinity constant K_D of interactions among Bcl-2 family proteins at neutral pH ranges between 1.9 and 5.7 μM.[7]

Conclusions

SPR permits real-time analysis of dimerization among Bcl-2 family proteins. This technique can be useful for a variety of purposes, including structure–function studies of mutants, comparisons of potency and specify of BH3 domain peptide antagonists, and analysis of small-molecule compounds that target Bcl-2 family proteins.

[26] Measuring Pore Formation by Bcl-2 Family Proteins

By SHARON L. SCHENDEL and JOHN C. REED

Introduction

The Bcl-2 protein family, whose members include Bcl-2, Bcl-X$_L$, and Bax, controls a distal step in an evolutionarily conserved pathway for programmed cell death. Some family members act to sustain cell survival (including Bcl-2 and Bcl-X$_L$) while others (including Bax, Bid, and Bak) promote cell death.[1]

Bcl-2, Bcl-X$_L$, and Bax localize to several intracellular membranes, including the outer mitochondrial membrane, nuclear envelope, and parts of the endoplasmic reticulum. This localization is afforded by a stretch of

[1] J. C. Reed, *Oncogene* **17**, 3225 (1998).

~20 hydrophobic amino acids located at their extreme C termini.[2-4] Despite their known cellular localizations, the exact mechanisms by which the Bcl-2 family proteins control apoptosis remain unclear. The intracellular localization of these proteins, particularly at mitochondria, led to suggestions that they may play roles in (1) release of caspase-activating proteins from mitochondria,[5,6] (2) regulating mitochondrial membrane potential,[7] or (3) controlling Ca^{2+} fluxes.[8]

The determination of the three-dimensional structure of Bcl-X_L[9] showed that this protein had a silhouette strikingly similar to those of the previously determined structures of the pore-forming domains of the bacterial toxins diphtheria[10] and colicins A, E1, and Ia.[11-13] Subsequent determination of the three-dimensional structure of the proapoptotic protein Bid again revealed sriking similarity to bacterial toxin pore-forming domains.[14,15] The bacterial toxin proteins have a unique ability to exist either as soluble or membrane-inserted proteins. They are able to achieve this profound change in character by burying two, long, predominantly hydrophobic α helices within a shell of amphipathic helices. Under appropriate conditions, the outer amphipathic α helices splay away, freeing the central hydrophobic α-helical hairpin to insert into the membrane bilayer, which initiates channel formation.

Bcl-X_L consists of a bundle of seven α helices, arranged in three layers. The outer two layers of amphipathic helices shield between them two long central α helices, each ~20 amino acids in length with a bias toward

[2] S. Krajewski, S. Tanaka, S. Takayama, J. J. Schibler, W. Fenton, and J. C. Reed, *Cancer Res.* **53**, 4701 (1993).

[3] M. Gonzalez-Garcia, R. P. Perez-Ballestero, L. Ling, L. Duan, L. H. Boise, C. B. Thompson, and G. Nunez, *Development* **120**, 3033 (1994).

[4] H. Zha, H. A. Fisk, M. P. Yaffe, N. Mahajan, B. Herman, and J. C. Reed, *Mol. Cell Biol.* **16**, 6494 (1996).

[5] J. C. Reed, *Nature (London)* **387**, 773 (1997).

[6] J. C. Reed, *Cell* **91**, 559 (1997).

[7] D. R. Green and J. C. Reed, *Science* **281**, 1309 (1998).

[8] N. Namzani, C. Brenner, I. Marzo, S. A. Susin, and G. Koremer, *Oncogene* **16**, 2265 (1998).

[9] S. W. Muchmore, M. Sattler, H. Liang, R. P. Meadows, J. E. Harlan, H. S. Yoon, D. Nettesheim, B. S. Chang, C. B. Thompson, S.-L. Wong, S.-C. Ng, and S. W. Fesik, *Nature (London)* **381**, 335 (1996).

[10] S. Choe, M. J. Bennett, G. Fujii, P. M. G. Curmi, K. A. Kantardjieff, R. J. Collier, and D. Eisenberg, *Nature (London)* **357**, 216 (1992).

[11] M. W. Parker, J. P. M. Postma, F. Pattus, A. D. Tucker, and D. Tsernoglou, *J. Mol Biol.* **224**, 639 (1992).

[12] P. Elkins, A. Bunker, W. A. Cramer, and C. V. Stauffacher, *Structure* **5**, 443 (1997).

[13] M. Weiner, S. Freymann, P. Ghosh, and R. M. Stroud, *Nature (London)* **385**, 461 (1997).

[14] J. Chou, H. Li, G. S. Salvesen, J. Yuan, and G. Wagner, *Cell* **96**, 615 (1999).

[15] J. M. McDonnell, D. Fushman, C. L. Milliman, S. J. Korsmeyer, and D. Cowburn, *Cell* **96**, 625 (1999).

hydrophobic residues.[9] Bid has a similar structure, containing an additional α helix.[14,15] The two central α helices of Bcl-X_L and Bid are of sufficient length to span the hydrophobic cross-section of a membrane bilayer. The structural similarity between Bcl-X_L, Bid, and the pore-forming domains of the bacterial toxins suggests that the Bcl-2 protein family may possess pore-forming potential as well. Several studies have shown that Bcl-2, Bcl-X_L, Bax, and Bid all exhibit channel-forming activity *in vitro*.[16–20]

This chapter describes methods for examining, at a macroscopic level, channel formation by Bcl-2 family proteins. Macroscopic-level measurements have the advantage that, for channel activity to be detected, a majority of the protein molecules must participate in channel formation. However, the methods described do not yield information about channel formation on a single-channel basis, particularly in terms of conductivity and ion selectivity. For quantifying these characteristics, planar bilayer measurements are more appropriate. Explanation of this technique may be found in several reviews.[21,22]

The two methods of macroscopic channel formation described here both involve measurement of solute efflux from liposomes. The first method describes measurement of ion efflux with ion-selective electrodes, and the second method measures the efflux of encapsulated fluorescent dyes.

Liposome Preparation

Lipids

Two preferred lipid sources are Avanti Polar Lipids (Birmingham, AL) and Sigma (St. Louis, MO). The lipids are supplied either as powders or chloroform suspensions. In either case, the contact of these lipid stocks with air should be kept to a minimum to prevent lipid oxidation.

[16] S. L. Schendel, Z. Xie, M. O. Montal, S. Matsuyama, M. Montal, and J. C. Reed, *Proc. Natl. Acad. Sci. U.S.A.* **94**, 511 (1997).

[17] A. J. Minn, P. Velez, S. L. Schendel, H. Liang, S. W. Muchmore, S. W. Fesik, M. Fill, and C. B. Thompson, *Nature (London)* **285**, 353 (1997).

[18] B. Antonsson, F. Conti, A. Ciavatta, S. Montessuit, A. S. Lewis, I. Martinou, L. Bernasconi, A. Bernard, J. J. Mermod, G. Mazzei, K. Maundrell, F. Gambale, R. Sadoul, and J. C. Martinou, *Science* **277**, 370 (1997).

[19] P. H. Schlesinger, A. Gross, X.-M. Yin, K. Yamamot, M. Saito, G. Waksman, and S. J. Korsmeyer, *Proc. Natl. Acad. Sci U.S.A.* **94**, 11357 (1997).

[20] S. L. Schendel, R. Azimov, K. Palowski, A. Godzik, B. L. Kagan, and J. C. Reed, *J. Biol. Chem.* **274**, 21932 (1999).

[21] M. Montal and P. Mueller, *Proc. Natl. Acad. Sci. U.S.A.* **69**, 3561 (1972).

[22] B. L. Kagan and Y. Sokolov, *Methods Enzymol.* **235**, 691 (1994).

The lipids may be either synthetic or purified from a natural source, such as the soybean lipid, asolectin. The latter is a mosaic of lipids, and in the case of colicin E1, higher activity is often observed relative to liposomes produced from synthetic lipids.[23] The synthetic lipids allow greater control over vesicle composition, particularly the ratio of charged to uncharged lipids.

Studies of both bacterial toxins and the Bcl-2 family proteins have utilized lipid vesicles that contain at least 10 mol% negatively charged lipids. The negatively charged lipid most commonly used has a glycerol moiety at its head group, although phosphatidylserine carries a negative charge as well. In the case of colicin E1, the net negative charge of the vesicle is important for the tight binding of the protein to the membrane surface.[24] Presumably, this requirement may be applied to the Bcl-2 family as well.

Lipid Purification

Synthetic lipid stocks are of sufficient purity to be used as supplied. Further purification of asolectin is often performed. The following purification procedure is a modification of that described previously.[25]

1. Asolectin (10 g) is added to 100 ml of acetone and stirred overnight.
2. The mixture is centrifuged at 5000*g* for 10 min at room temperature.
3. The supernatant is discarded, the pellet is dissolved in 40 ml of anhydrous ether, and the mixture is centrifuged for 10 min at 5000*g*, at room temperature.
4. The ether is evaporated on a rotary evaporator device.
5. The dried lipid is dissolved in ~80 ml of chloroform.
6. The lipid concentration is determined by phosphate analysis.[26]

Liposome Production

For solute efflux to be measured, the liposomes must be large (>0.1 μm in diameter) and unilamellar, i.e., having only a single membrane bilayer. To produce such vesicles, the following procedure, representing a modification of that outlined by Peterson and Cramer,[27] is employed.

1. The appropriate amount of lipid is measured. Typically, a liposome suspension having the final lipid concentration of 10 mg/ml is used, of

[23] S. L. Schendel and W. A. Cramer, unpublished results (1999).
[24] S. D. Zakharov, J. B. Heymann, Y.-L. Zhang, and W. A. Cramer, *Biophys. J.* **70,** 2774 (1996).
[25] Y. Kawaga and E. Racker, *J. Biol. Chem.* **246,** 5477 (1971).
[26] G. R. J. Bartlett, *J. Biol Chem.* **234,** 466 (1959).
[27] A. A. Peterson and W. A. Cramer, *J. Mebr. Biol.* **99,** 197 (1987).

which ~80 and 20% are neutral and negatively charged lipids, respectively. This ratio can be varied. If powdered lipid stocks are used, the lipid powder is first dissolved in chloroform. The chloroform lipid suspension should be placed in a Pyrex tube.

2. The chloroform is evaporated under a stream of either nitrogen or argon while vortexing. Use of a vortex ensures an even distribution of the lipid film. Be certain that no pockets of chloroform are entrapped under a thin layer of dried lipid, as inadequate removal of organic solvents will affect liposome formation. The lipid film is further dried under vacuum for at least 2 hr to remove any remaining trace of organic solvent.

3. The lipid is then resuspended in buffer to the desired final concentration (typically 10 mg/ml). The buffer utilized in this laboratory is 10 mM dimethyl glutaric acid (DMG), 100 mM KCl, 2 mM Ca(NO$_3$)$_2$, pH 5.0, although HEPES or phosphate buffers can be employed. The suspension is vortexed under a stream of N$_2$ for at least 5 min to flush out any air from the tube. The tube is then sealed with at least two layers of Parafilm, the top layer being somewhat loose.

4. The tube is placed in a sonicator bath (Branson, Danbury, CT). For best results, the bath should be cleaned well and allowed to run for 30 min before lipid sonication. The tube should be placed in a visible "node" with the liquid meniscus even with the water level and adjusted until the liquid inside the tube splashes somewhat. Continue sonication until the solution has been converted from opaque to completely clear. The lipid suspension is now composed of small unilamellar vesicles (SUVs).

5. To convert the SUVs to large unilamellar vesicles (LUVs) several freeze–thaw cycles are employed. The suspension is frozen as rapidly as possible in a dry ice–ethanol bath or liquid nitrogen. The frozen solution is allowed to thaw slowly to room temperature. Repeat with at least four additional freeze–thaw cycles. The solution should again be opaque.

Dye-Encapsulating Liposomes

Two dyes have been used in studies of Bax and colicin E1 channel formation: carboxyfluorescein and 6-methoxy-*N*-(3-sulfopropyl)quinolinium (SPQ).[18,28]

Carboxyfluorescein has a fluorescence that is self-quenched by fluorescein-to-fluorescein excited state energy transfer. Thus, when this dye is packed within the liposome its fluorescence is decreased, and on leakage out of the vesicle this self-quenching is relieved and the quantum yield of the dye is increased. A disadvantage exists with the use of this probe

[28] B. A. Steer and A. R. Merrill, *Biochemistry* **33**, 1108 (1994).

to follow Bcl-2 protein family-induced solute efflux from liposomes: its fluorescence is highly pH dependent, with significantly decreased fluorescence at pH <6.0, and because Bcl-2 protein channel activity *in vitro* is largely undetectable above pH 5.0, the pH must be adjusted to pH >6.5 at the end of the measurement and the fluorescence recorded, precluding a continuous measurement of channel activity at acidic pH.

SPQ is an alternative to carboxyfluorescein. Its fluorescence is collisionally quenched by chloride, resulting in an ion concentration-dependent decrease in fluorescence. The fluorescence of SPQ is decreased by more than 50% at chloride concentrations above 10 mM.[28] As with carboxyfluorescein, SPQ fluorescence is low within the confines of the liposome, where [Cl$^-$] far exceeds 10 mM, but as the chloride exits the vesicle through the channel formed by a Bcl-2 family protein, the chloride quenching is relieved. In the case of colicin E1, the SPQ dye is probably of sufficient size so as to be excluded from the channel lumen (diameter, 8–9 Å), and therefore increases in SPQ fluorescence can be largely attributed to chloride efflux. In the case of Bcl-2 protein family channels the pore size is not yet defined, and so any decreased SPQ fluorescence cannot be assigned solely to chloride efflux. Although SPQ fluorescence is not significantly affected by changes in pH, and the dye has a high solubility in aqueous solution, its solubility is decreased below pH 6.0,[29] and channel activity measurements for colicin E1 were thus made at pH 6.0.[28] If channel formation is to be followed by changes in dye fluorescence, several modifications to the above procedure for liposome production are made.

1. Liposomes are prepared at a higher pH in the absence of Ca^{2+}. The lipid film may be resuspended in the following buffer: 10 mM DMG, 100 mM KCl, pH 6.0. An appropriate dye concentration is between 15 and 20 mM.

2. Unencapsulated dye must be removed and this may be achieved by several methods: (1) dialysis of the liposome suspension in the same buffer used for vesicle preparation, which lacks dye; (2) ultracentrifugation to pellet the liposomes and resuspension in dye-free buffer; and (3) passage of the vesicles over a G-25 or G-50 Sephadex column.

Channel Activity Measurements

Electrode Set-Up. Changes in chloride concentration are monitored with a chloride-specific electrode (94-17B; Orion, Beverly, MA) coupled to a double-junction reference electrode (90-02; Orion). The electrode signal is

[29] A. R. Merrill, personal communication (1999).

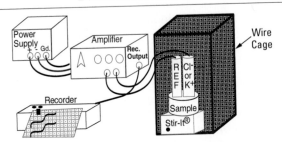

Fig. 1. Schematic diagram of ion-selective electrode set-up. The ion-selective electrode, coupled to a double-junction reference electrode (Ref) is housed inside a copper wire cage, which reduces extraneous signals when measurements are being conducted at high sensitivity. The electrode signal is out put to an amplifier, which feeds its signal to a strip chart recorder.

out put to a pH meter or other suitable voltmeter attached to a strip chart recorder. The meter utilized in this laboratory was home built and allows for expansion of the scale and readings of small concentration changes. A schematic diagram of the electrode set-up is shown in Fig. 1. In addition to chloride efflux, channels formed by the Bcl-2 protein family will also permit passage of K^+, and a K^+-sensitive electrode (93-19; Orion) may be easily substituted for the chloride electrode.

Extravesicular Buffer and Protein Buffers. The buffer commonly used for chloride efflux measurements in this laboratory is that described in Peterson and Cramer.[27] The buffer composition is 10 mM DMG, 100 mM choline nitrate, 2 mM Ca(NO$_3$)$_2$. Choline nitrate is formed choline bicarbonate (Sigma) and nitric acid. The buffer is degassed for at least 1 hr and titrated to the appropriate pH with NaOH. Other chloride-free buffers are also suitable for measurement, provided their ionic strength is similar to that used for vesicle preparation.

The protein sample should be in a buffer that is chloride free, if possible, and should also lack any compounds that would interfere with electrode response. 2-Mercaptoethanol has been found to especially perturb electrode response.[30]

Ion Efflux Measurements. Extravesicular buffer (15 ml) is placed in a disposable plastic beaker (Fisher, Pittsburgh, PA). Use of such inexpensive beakers removes the need for tedious washing between trials. Approximately 75–100 μl of the liposome suspension is pipetted into the beaker and the electrode allowed to equilibrate until a stable baseline is achieved. The solution should be stirred at medium speed at all times and no bubbles should adhere to the electrode tips. Once the baseline has stabilized, 2 μl

---

[30] S. L. Schendel and J. C. Reed, unpublished results (1999).

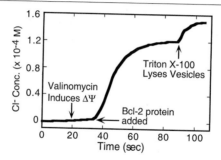

FIG. 2. Example of ion efflux from liposomes induced by Bcl-2. A stable baseline is achieved before the addition of a nanomolar quantity of the K^+ ionophore valinomycin, which escorts a few hundred K^+ ions out of each vesicle, resulting in a negative transmembrane potential. The Bcl-2 protein is then added and chloride ions immediately begin to efflux. As the Bcl-2-induced efflux tapers off, an excess of detergent is added to lyse the vesicles, allowing the determination of the fraction of encapsulated chloride released due to Bcl-2 action.

of a 70 μM methanol stock of the K^+-specific ionophore valinomycin is added to the beaker. Valinomycin induces the formation of a Nernst diffusion potential of ~135 mV, negative inside. The ionophore may be omitted if no membrane potential is desired. The electrode is again allowed to equilibrate after valinomycin addition. The Bcl-2 family protein is then added to a concentration that will induce release of 50–80% of the total encapsulated Cl^- and induce a response having a slope between 30 and 60° which facilitates accurate calculation of the efflux rate. After the protein-induced efflux has leveled off, Triton X-100 is added to the beaker to a final concentration of 0.1% (v/v). This detergent lyses the vesicles and releases any residual chloride, allowing the total amount of chloride encapsulated to be determined. A representative trace of chloride efflux by Bcl-2 is shown in Fig. 2.

At the end of the measurements, a calibration curve is produced by additions of known amounts of KCl. The KCl is added in the presence of an aliquot of liposomes, so that the background Cl^- concentration is similar. KCl is added in progressively increasing amounts until the chart recorder deflection equals the maximum observed by Triton X-100 lysis of the vesicles.

For colicin molecules, where the molecularity of the channel is known to be one,[31] a specific Cl^- efflux rate can be calculated by determining the ratio of colicin-induced Cl^- release to the total encapsulated Cl^-. This value is then divided by the number of colicin molecules present (assuming that

[31] W. A. Cramer, J. B. Heymann, S. L. Schendel, R. N. Dieriy, F. S. Coehn, P. A. Elkins, and C. V. Stauffacher, *Annu. Rev. Biophys. Biomol. Struct.* **24,** 611 (1995).

each molecule forms one channel) and the quotient divided by the time taken for the efflux to occur. Because the molecularity of the Bcl-2 protein family channels is still unclear, this rate calculation is not as straightforward. For purposes of comparison between trials, this laboratory simply determines the slope of the initial response.

Measurement of Solute Efflux from Dye-Encapsulating Liposomes

The fluorimeter parameters are set to reflect the optimum excitation and emission wavelengths of the dye in use. For carboxyfluorescein these wavelengths are 490 and 520 nm for excitation and emission, respectively. The slit widths are set between 4 and 8 nm for both excitation and emission. For SPQ the excitation and emission wavelengths are 347 and 445 nm, respectively, and the slit widths are between 4 and 8 nm, dependent on the intensity of the signal.

As with the ionselective electrodes, an aliquot of vesicles is added to the buffer in the cuvette and the suspension is constantly stirred. Typically, the vesicles will be diluted 50- to 100-fold. The suspension is allowed to equilibrate and a background fluorescence is monitored to ensure that no dye leakage occurs during the time of measurement. The valinomycin is added, followed by the Bcl-2 family protein, and the increase in fluorescence intensity is monitored (provided the pH is above 6.0) until the protein-induced efflux tapers. Triton X-100 is then added to note the total fluorescence.

Summary

Two methods for assaying Bcl-2 protein family-induced solute efflux from liposomes have been outlined. They utilize either ion-selective electrodes to follow ion efflux or fluorescence to monitor changes in fluorescence of the liposome-encapsulated dye SPQ or carboxyfluorescein. Both methods provide a simple means of determining protein activity. These methods do not have the capability to detect either single-channel conductivity or ion selectivity, but they indicate whether the bulk of the protein population is inducing solute efflux. Although in in vivo significance of Bcl-2 protein family pore formation remains to be determined, in vitro measurements of channel activity should provide a means to determine whether a given protein preparation has activity and whether mutations have an adverse effect on channel formation.

[27] Assays for Studying Bax-Induced Lethality in the Yeast *Saccharomyces cerevisiae*

By Qunli Xu, Ning Ke, Shigemi Matsuyama, and John C. Reed

Introduction

Bcl-2 family proteins play evolutionarily conserved roles in regulating the life and death of the cell (reviewed in Refs. 1–3). A member of this family, Bax, can promote apoptosis in both tissue culture cells and in animals.[4,5] In addition to functioning as a proapoptotic protein in mammalian cells, overexpression of Bax in some heterologous organisms also causes cell death. For example, expression of Bax in both the budding yeast *Saccharomyces cerevisiae* and the fission yeast *Schizosaccharomyces pombe* confers a lethal phenotype.[6–12] There are also indications that Bax can even induce lethality in certain bacteria.[13] Yeast cell death induced by Bax expression shares some commonalities with programmed cell death in mammalian cells. First, the lethal effect of Bax can be reverted by coexpression of certain antiapoptotic members of the Bcl-2 family, including Bcl-2, Bcl-X$_L$, and Mcl-1.[6] Second, mitochondria are implicated in yeast cell death induced by Bax overexpression. Bax is targeted to mitochondria in both yeast and mammalian cells; deletion of the transmembrane domain of Bax, which prevents Bax from localizing to mitochondria, renders Bax incapable of inducing lethality in yeast. Furthermore, replacing the transmembrane domain of Bax with the mitochondria-targeting sequence from

[1] J. C. Reed, *J. Cell Biol.* **124,** (1994).
[2] E. Yang and S. J. Korsmeyer, *Blood* **88,** 386 (1996).
[3] G. Kroemer, *Nature Med.* **3,** 614 (1997).
[4] Z. N. Oltvai, C. L. Milliman, and S. J. Korsmeyer, *Cell* **74,** 609 (1993).
[5] C. Yin, C. M. Knudson, S. J. Korsmeyer, and T. Van Dyke, *Nature (London)* **385,** 637 (1997).
[6] T. Sato, M. Hanada, S. Bodrug, S. Irie, N. Iwama, L. Boise, C. Thompson, L. Fong, H.-G. Wang, and J. C. Reed, *Proc. Natl. Acad. Sci. U.S.A.* **91,** 9238 (1994).
[7] M. Hanada, C. Aimé-Sempé, T. Sato, and J. C. Reed, *J. Biol. Chem.* **270,** 11962 (1995).
[8] H. Zha, H. A. Fisk, M. P. Yaffe, N. Mahajan, B. Herman, and J. C. Reed, *Mol. Cell. Biol.* **16,** 6494 (1996).
[9] W. Greenhalf, C. Stephen, and B. Chaudhuri, *FEBS Lett.* **380,** 169 (1996).
[10] J. M. Jürgensmeier, S. Krajewski, R. Armstrong, G. M. Wilson, T. Oltersdorf, L. C. Fritz, J. C. Reed, and S. Ottilie, *Mol. Biol. Cell* **8,** 325 (1997).
[11] B. Ink, M. Zornig, B. Baum, N. Hajibagheri, C. James, T. Chittenden, and G. Evan, *Mol. Cell. Biol.* **17,** 2468 (1997).
[12] W. Tao, C. Kurschner, and J. I. Morgan, *J. Biol. Chem.* **272,** 15547 (1997).
[13] S. Asoh, K. Nishimaki, R. Nanbu-Wakao, and S. Ohta, *J. Biol. Chem.* **273,** 11384 (1998).

a mitochondria protein, Msp70, did not affect the lethal effect of Bax in yeast, suggesting that mitochondria targeting of Bax is both necessary and sufficient to induce cell death in yeast.[8] Expression of Bax in yeast also induces cytochrome c release,[14] a phenomenon observed in mammalian cells on apoptosis stimulation, suggesting that the primary effect of Bax on mitochondria may be similar in both yeast and mammalian cells. Third, the same functional domain of Bax, i.e., the BH3 domain, is critical for Bax to kill both yeast and mammalian cells.[8] Taken together, these observations suggest that Bax utilizes the same activity to kill both yeast and mammalian cells. Yeast death caused by overexpression of Bax does not show typical apoptotic morphology; there is no obvious nuclear fragmentation and oligo-nucleosomal degradation. This lack of apoptotic phenotype is most likely due to the absence of caspases or caspase activity in yeast.[10]

A search of the complete S. cerevisiae genome did not reveal any poten-tial Bcl-2 homolog. The absence of a Bcl-2 homolog renders yeast an attractive system with which to study the function of individual Bcl-2 family proteins without the interference conferred by other family members. In addition, Bax-induced cell death in S. cerevisiae creates an equivocal oppor-tunity to apply yeast genetics and functional cloning approaches to the identification of other components of the Bax-dependent apoptotic path-way. Using Bax as an example, we describe methods used in our laboratory to study the function of death-inducing proteins in the budding yeast S. cere-visiae.

Methods

Expression of Bax in Yeast

When expressing mammalian genes in the yeast S. cerevisiae, cDNAs should be used instead of genomic DNAs, because promoter elements and splicing factors are generally not well conserved from yeast to mammals. A variety of expression vectors are available to express heterologous genes in S. cerevisiae in either a constitutive or a regulatable manner. Among the strong constitutive promoters, the promoters of the alcohol dehydrogenase (ADH1) and glyceraldehyde-3-phosphate dehydrogenase (GPD or GAP) genes are commonly employed. When a constitutive promoter is utilized to drive bax expression, the diminished number of transformants formed after introducing Bax expression plasmid into yeast can be used as a measure of lethality caused by Bax. A parallel transformation with an "empty" vector should be included as a control. However, this assay has its limita-

[14] S. Manon, B. Chaudhuri, and M. Guérin, FEBS Lett. 415, 29 (1997).

tions, because an inability to form transformants can result from either cell cycle arrest or from cell death. For this reason, we favor a regulatable system over a constitutive one. Several conditional promoters have been successfully employed to express heterologous genes in S. *cerevisiae;* the most commonly used is the *GAL1* and *GAL10* promoter. Transcription from the *GAL* promoter is controlled by the availability of carbon sources in the culture medium. The *GAL* promoter represses transcription in the presence of glucose and derepresses transcription in the presence of raffinose, but highly induces transcription in the presence of galactose and in the absence of glucose. Expression of Bax from the *GAL10* promoter-containing expression vector YEp51 efficiently kills yeast when cells are cultured in galactose-containing medium.[8]

Bax can also be expressed as a fusion protein, using some of the vectors employed in the yeast two-hybrid assay. For example, when Bax is fused next to the LexA DNA-binding domain in the vector pGilda (a *GAL1* promoter-containing vector), expression of LexA–Bax fusion protein strongly induced lethality in S. *cerevisiae* when cells are cultured in galactose-containing medium; in some cases an even stronger death-inducing activity was conferred by LexA–Bax fusion protein compared with Bax alone, presumably because of the higher stability of the fusion protein (our unpublished observations, 1999).

Bax needs to be expressed to high levels to induce cell death in yeast. By employing a strong yeast promoter, transcription of the heterologous gene should proceed efficiently. The next factor to consider is translation. In yeast, the distance between the promoter and the start codon eminently affects translation initiation in yeast AUG of the mRNA. Potential secondary structures in the 5'-untranslated leader sequence will likely inhibit translation initiation. When cloning a gene of interest into a yeast expression vector, it is advisable to use the first two restriction enzyme recognition sites in the multiple cloning sites of the vector and thus keep minimum spacing between the promoter and the start codon ATG of the cDNA.

After choosing a suitable promoter, a second factor to consider is the selectable marker carried by the vector, which needs to match the auxotrophic mutation of the yeast strain. Four selectable markers are commonly found in S. *cerevisiae* vectors; these are the *LEU2, HIS3, TRP1,* and *URA3* genes. Yeast vectors containing different promoters and selectable markers are available from the American Type Culture Collection (ATCC, Rockville, MD).

Strains and Media

Laboratory S. *cerevisiae* strains are of diverse origins and thus strains from different sources can be distantly related with respect to genetic background. We have examined the cytotoxic effect of Bax expression in a number of yeast strains. In all the strains tested, expression of Bax caused cell death; however, the efficacy of Bax killing appeared to be strain dependent; some strains are more easily killed by Bax, whereas some others exhibited a certain degree of resistance. Thus, it is advisable to test strains for their sensitivity to Bax or any other death-inducing genes before further analysis.

We routinely grow yeast in the rich YPD (yeast peptone dextrose) medium. Once a Bax expression plasmid is transformed into yeast, cells should be cultured in synthetic dropout (SD) medium lacking the particular amino acid selected for. For example, yeast containing pGilda-Bax can be maintained in synthetic medium lacking tryptophan (SD-Trp). It is also of note that in our experience the amount of nutrient in the medium also affects sensitivity of yeast to Bax; the lethality caused by Bax is most evident when yeast is cultured in media with minimum nutrient, such as the Burkholder's minimum medium (BMM).[15] Recipes for media used by us to assay for death-inducing activity of Bax in yeast are listed in Table I.[15,15a]

Introduction of Bax Expression Plasmid into Yeast by Transformation

After cloning *bax* cDNA into the pGilda vector (a *GAL1* promoter, and a LexA DNA-binding domain-containing vector), the resulting plasmid pGilda-bax can be introduced into the yeast strain, for example, EGY48 (genotype: *MATa leu2-3 112 his3-11,15 trp1-1 ura3-1 6LexAop-LEU2*) by transformation. We routinely used the lithium acetate transformation procedure and when the protocol below (adapted from Ito *et al.*[16] and Ausubel *et al.*[17]) is followed, transformation efficiency is about 10^4 transformants/ μg DNA. Factors such as yeast strain background, starting cell density, and quality of carrier DNA and plasmid DNA can all influence transformation efficiency.

Transformation Protocol

1. Inoculate a single colony of yeast into 50 ml of YPD medium in a 250-ml Erlenmeyer flask and grow the cells overnight in a 30° incubator–shaker (Lab-Line Instruments, Melrose Park, IL) set at 250 rpm.

[15] A. Toh-e, Y. Ueda, S.-I. Kakimoto, and Y. Oshima, *J. Bacteriol.* **113**, 727 (1973).

[15a] F. Sherman, *Methods Enzymol.* **194**, 3 (1991).

[16] H. Ito, Y. Fududa, K. Murata, and A. Kimura, *J. Bacteriol.* **153**, 163 (1983).

[17] F. M. Ausubel, R. Brent, R. E. Kingston, D. D. Moore, J. G. Seidman, A. Smith, and K. Struhl, "Current Protocols in Molecular Biology." Wiley Interscience, New York, 1991.

2. Determine the cell density by measuring the OD_{600}; the optimal density for transformation is between OD_{600} 0.5 and 1. (An OD_{600} of 1 equals about 3×10^7 cells/ml.) Overgrown cells can be diluted and allowed to proliferate for another cell division cycle (1–2 hr).

3. Harvest the cells by centrifugation at 1000*g* in a benchtop centrifuge (Sorvall, Newtown, CT) for 5 min at room temperature. Wash the cells once in 50 ml of distilled H_2O and once in 10 ml of lithium acetate solution [0.1 *M* lithium acetate, 10 m*M* Tris–Cl (pH 8.0), and 1 m*M* Na_2EDTA].

4. Resuspend the cells in 0.5 ml (1/100 initial culture volume) of lithium acetate solution. Dispense 100-μl aliquots. To each cell suspension aliquot, add 5 μl of carrier DNA (sheared salmon sperm DNA, 10 mg/ml; Sigma Chemical, St. Louis, MO) and 1 μg of plasmid DNA (pGilda-bax).

5. Add polyethylene glycol (PEG) solution [40% (w/v) PEG (MW 3300–4000; Sigma), 0.1 *M* lithium acetate, 10 m*M* Tris–HCl (pH 8.0), and 1 m*M* Na_2EDTA] to the cell–DNA mixture. Vortex vigorously for 10 sec, and incubate at 30° for 45 min with constant shaking.

6. Heat shock in a 42° water bath (VWR Scientific Products, Bridgeport, NJ) for 5–15 min. Collect the cells by centrifugation and resuspend in 200 μl of Tris–EDTA (TE) buffer.

7. Spread the transformation mixture on SD plates lacking the amino acid selected for (e.g., SD-Trp plates for pGilda-bax) and incubate in a 30° incubator (VWR Scientific Products).

8. Transformants should usually appear in 2–4 days.

Monitoring Expression of Bax by Immunoblot Analysis

The expression of Bax from the *GAL1* promoter in pGilda-bax can be induced by growing yeast in galactose-containing but glucose-free medium. The optimal induction time depends on the yeast strain used and thus should be determined empirically. Typically we induce the *GAL* promoter for 12–24 hr. The induction of Bax expression can be monitored by immunoblot analysis. Yeast lysates are prepared by the glass bead method described below.

Protocol for Preparing Yeast Extracts with Glass Beads

1. Inoculate single colonies of pGilda-Bax- or pGilda-containing yeast into 4 ml of SD-Trp liquid medium supplemented with glucose in a 15-ml Falcon tube. Grow the yeast in a 30° incubator–shaker (Lab-Line Instruments) at 250 rpm to midexponential phase (OD_{600} of 1).

2. Collect the cells by centrifugation at 1000*g* for 5 min in a benchtop centrifuge (Sorvall) and wash the cells three times with distilled H_2O. After

TABLE I

S. cerevisiae MEDIA[a]

YPD

Per liter, autoclave the following:

Bacto-Yeast extract	10 g
Bacto-Peptone	20 g
Glucose (dextrose)	20 g

For the SD and BMM media below, either 2% (w/v) glucose or 2% (w/v) galactose in combination with 1% (w/v) raffinose is included as a carbon source. These sugars can be made as 20% (w/v) stock solutions and sterilized by filtration, using bottle-top filter units (0.22 μm; Becton Dickinson and Company, Lincoln Park, NJ). To make solid media, agar (Difco Laboratories, Detroit, MI) added to a final concentration of 2% (w/v) before autoclaving and warm medium is poured into 10- or 15-cm petri dishes (Fisher Scientific, Pittsburgh, PA).

SD (synthetic dropout medium)

Per liter, autoclave 6.7 g of YNB (yeast nitrogen base without amino acids; Difco). The following amino acids and supplements are included, except those selected for when culturing yeast strains containing plasmids:

Adenine sulfate	20 mg
Uracil	20 mg
L-Tryptophan	20 mg
L-Histidine-HCl	20 mg
L-Arginine-HCl	20 mg
L-Methionine	20 mg
L-Tyrosine	30 mg
L-Leucine	30 mg
L-Isoleucine	30 mg
L-Lysine-HCl	30 mg
Phenylalanine	50 mg
Aspartic acid	100 mg
Valine	150 mg
L-Threonine	200 mg

Stock solutions of these supplements can be made at 4% [w/v; 2% (w/v) for uracil], sterilized by autoclaving; the exceptions are L-tryptophan, L-tyrosine, and L-threonine, which should be filter sterilized and added to autoclaved media after cooling to 65°.

the final wash, transfer the cells to 20 ml of SD-Trp medium supplemented with galactose in a 100-ml Erlenmeyer flask and continue incubation in a 30° incubator–shaker for the desired time (usually overnight to 24 hr).

3. Collect the cells as described above and wash once in ice-cold phosphate-buffered saline (PBS). Resuspend in 200 μl of lysis buffer [150 m*M*

TABLE I (*continued*)

BMM (Burkholder's minimum medium)

The basic components of BMM are similar to those of YNB, but the concentrations for minerals and vitamins are generally lower.

1. Per liter, autoclave before adding vitamins:

KH_2PO_4	1.5 g
$MgSO_4 \cdot 7H_2O$	500 mg
$CaCl_2 \cdot 2H_2O$	330 mg
KI	100 μg
Asparagine	2 g
Mineral stocks a–c (10,000×)	100 μl each
Vitamin block (1000×)	1 ml

2. Add required amino acids (concentrations are the same as for SD medium).

Mineral stock (10,000×)
  a. Per 50 ml:

H_3BO_3	30 mg
$MnSO_4 \cdot 7H_2O$	50 mg
$ZnSO_4 \cdot 7H_2O$	150 mg
$CaSO_4 \cdot 5H_2O$	20 mg

  b. Per 50 ml:
    $Na_2MoO_4 \cdot 2H_2O$ 100 mg
  c. Per 50 ml:
    $FeCl_3 \cdot 6H_2O$ 125 mg

Vitamin stock (1000×)
  Per 100 ml, filter sterilize the following:

Thiamine	20 mg
Pyridoxine	20 mg
Nicotinic acid	20 mg
Pantothenic acid	20 mg
Biotin	0.2 mg
Inositol	1 g

[a] Adapted from Sherman[15a] and Toh-e *et al.*[15]

NaCl, 5 mM Na$_2$EDTA, 50 mM Tris (pH 7.4), and 0.5% (v/v) Nonidet P-40 (NP-40)] containing protease inhibitors [aprotinin (0.1 TIU/ml; Sigma), pepstatin A (0.7 μg/ml in methanol; Sigma), leupeptin (50 μg/ml in PBS; Sigma), and 1 mM phenylmethylsulfanyl fluoride (PMSF, in methanol; Sigma)].

4. Transfer the samples to microcentrifuge tubes and add equal volumes of chilled acid-washed glass beads (Sigma). Vortex in a cold room for 10 min. Pause the vortexing several times, to prevent overheating of the protein sample.

5. Centrifuge the samples in a microcentrifuge in the cold room at the maximal speed for 5 min and transfer the supernatant to fresh microcentrifuge tubes. Centrifuge at maximal speed again for 20 min and save the supernatant (protein extract).

6. Subject 20 μg of cell extracts to sodium dodecyl sulfate–polyacrylamide gel electrophoresis (SDS–PAGE)/immunoblot analysis, using an antibody to the protein of interest.

Assays for Yeast Cell Death Induced by Bax

When pGilda-bax is introduced into an appropriate yeast strain, the effect of LexA-Bax fusion protein on yeast growth can be easily observed by plate tests. Yeast containing pGilda-bax are maintained on glucose-containing SD-Trp plates. Single colonies can be streaked on galactose-containing SD-Trp plates to induce LexA-Bax expression side by side with control streaks from colonies containing pGilda vector lacking *bax*. Note that after 3–4 days, little yeast growth was evident for the pGilda-bax streaks, whereas the vector control grew normally, forming a think layer of confluent cells. The drastically reduced growth rate observed on galactose medium, which induces Bax expression, indicates that the gene product is either cytotoxic to yeast or induces cell cycle arrest. Two assays can be employed to distinguish between these possibilities: (1) a vital dye exclusion assay and (2) a clonigenic assay.

Vital Dye Exclusion Assay. Analogous to mammalian cells, the viability of yeast can be detected by virtue of the ability of cells to retain vital dyes, for example, trypan blue. This assay monitors a later stage of cell death, after cells have lost membrane integrity and hence are no longer able to exclude vital dye. Other vital dyes, for example, methylene blue and erythrosin B, have also been employed.

Trypan Blue Assay Protocol

1. Inoculate a single colony of either pGilda- or pGilda-bax-containing yeast into 4 ml of glucose-containing SD-Trp medium and culture the yeast overnight in a 30° incubator–shaker to an OD_{600} of 0.3–0.4. If the cells are overgrown, dilute the culture with fresh medium and grow to this optimal density.

2. Collect the cells by centrifugation at 1000g for 5 min in a benchtop centrifuge and wash the cells three times with distilled H_2O. Resuspend the cells in 4 ml of galactose-containing SD-Trp medium and continue culturing the cells in a 30° incubator–shaker.

3. At 6-hr intervals remove 20 μl of culture, and mix it with an equal volume of 0.4% (w/v) trypan blue solution (Sigma), and use a hemocytome-

ter (Hausser Scientific, Horsham, PA) to enumerate blue and white cells, respectively, under a light microscope. Count at least 500 cells and calculate the ratio of dead (blue) cells among the total cell population.

Clonigenic Assay. To discern genuine cell death and growth arrest, a "clonigenic" (or colony formation) assay can also be performed; it is the most accurate and reliable method. As described above, the *GAL* promoter is a conditional promoter and can be efficiently shut off by glucose in the growth medium. Thus, after culturing pGilda-bax-containing yeast in galactose-containing medium for the desired period of time, yeast are transferred back to glucose-containing medium to prevent further production of the Bax protein. If Bax expression is lethal and the cells have passed the "point of no return" step in the death process, yeast cannot resume growth even in the absence of further Bax protein production. On the other hand, if Bax expression causes merely growth inhibition, cells should recover from the cell cycle arrest and resume growth when Bax expression is terminated (grown in glucose medium)

Clonigenic Assay Protocol

1–2. Steps 1 and 2 are the same as described above for the trypan blue exclusion assay.

3. After culturing the cells in galactose-containing medium, remove an aliquot of cells every 6 hr and determine the concentration of the culture by OD_{600}. Dilute the cells in distilled H_2O and spread about 300–1000 cells per glucose-containing SD-Trp plate (in the case of pGilda-bax). Spread three plates per sample.

4. Incubate the plates in a 30° incubator for 3–4 days and count the number of colonies to obtain the ratio of yeast with clonigenic ability compared with control (pGilda-containing yeast).

When performing dye exclusion and colony formation assays, an important factor is cell density before galactose induction. When the culture density is too high, i.e., $OD_{600} > 0.6$ at the beginning point of culturing in galactose medium, induction of the *GAL* promoter is inefficient. Another point to consider is that during longer culturing in liquid medium, spontaneous mutations accumulate and those mutations that render yeast resistant to Bax are selected for and yeast cell number begins to increase. To avoid this "background" problem, the cell death assays described above are best completed in 2–3 days (from the induction of Bax protein in galactose medium).

Assays for Other Apoptosis Hallmarks in Yeast. Apoptosis in mammalian cells is often accompanied by distinct morphological and physiological changes, including but not limited to membrane blebbing, externalization of phosphoserine, chromatin condensation, nuclear fragmentation, and DNA degradation first into large fragments and then into small nucleosomal fragments.[18] There are established assays to detect some of these changes in mammalian cells. Similar assays can also be performed in yeast. We have examined possible nuclear changes of *S. cerevisiae* cells expressing Bax and found that although a certain form of chromatin condensation did occur, there was no detectable nuclear fragmentation and oligonucleosomal DNA degradation as determined by 4′,6-diamidino-2-phenylindole (DAPI) stain and agarose gel electrophoresis.[8]

Screening cDNA Libraries for Bax Inhibitors in Yeast

Background

Bax-induced cell death in yeast creates opportunities for functional screening of cDNA libraries for Bax-inhibitory proteins. cDNA libraries can be introduced into yeast cells in which Bax production is conditionally regulated by the *GAL1* promoter (e.g., YEp51-Bax). Transformants are plated directly onto semisolid medium containing galactose. Colonies that form are candidates for encoding proteins capable of rescuing yeast from Bax-induced lethality. Secondary screens, however, must be performed to exclude false positives. This is most easily accomplished when the cDNA library employs the *URA3* marker for selection, because a negative selection using 5-fluoroorotic acid (5FOA) for accelerating and subsequently verifying loss of *URA3*-marked plasmids is available.[19] Thus, candidate transformants are grown in uracil-containing medium to encourage loss of the cDNA library plasmid while maintaining selection for the Bax plasmid in glucose-containing medium. After verifying loss of the *URA3*-library plasmid on 5FOA, candidate transformants are plated on galactose medium to reinduce Bax production, thus determining whether the rescue phenotype is dependent on the library cDNA. Finally, yeast DNA is prepared from the positive clones and plasmid DNA is propagated in *Escherichia coli,* followed by transformation back into yeast harboring an inducible Bax plasmid to confirm the ability of the introduced cDNAs to rescue against Bax-induced cell death. Plasmids that consistently rescue against Bax can then be subjected to DNA sequencing; however, some of these will be found to encode proteins that suppress expression of *bax* in yeast rather than suppressing function of the Bax protein in yeast. Thus, it is imperative

---

[18] A. H. Wyllie, J. F. R. Kerr, and A. R. Currie, *Int. Rev. Cytol.* **68,** 251 (1980).
[19] J. D. Boeke, J. Trueheart, G. Natsoulis, and G. R. Fink, *Methods Enzymol.* **154,** 164 (1987).

to verify by immunoblotting that Bax protein accumulates in transformants when grown in galactose-containing medium.

Transformation of cDNA Libraries

In a screen of several yeast strains, the strain BF264-15Dau (*MATa ade1 his2 leu2-3,112 trp1-1a ura3*) was found to be most susceptible to Bax-induced cell death[20] (our unpublished data, 1999). For the purposes of this protocol, our YEp51-Bax transformant of this strain is designated QX95001. The transformation procedure is similar to the one described above, with some modifications.

1. A single colony of QX95001 is inoculated into 50 ml of SD-L medium (SD medium without leucine) and grown overnight at 30° with vigorous agitation.

2. Dilute an empirically determined amount of the culture into 500 ml of SD-L medium, so that the OD_{600} is 0.1. Let the culture grow for two or three generations, which is generally 4–6 hr based on a typical doubling-time in selective medium of ~2 hr.

3. Pellet the cells by centrifugation, using a fixed-angle rotor at 4000g for 5 min. Wash the cells once with 50 ml of double-distilled H_2O (ddH_2O) and once with 50 ml of 0.1 M lithium acetate. Resuspend the cells in 1 ml of 0.1 M lithium acetate at 30° for 15 min.

4. Boil 50 μl of carrier DNA (sheared salmon sperm DNA, 10 mg/ml in ddH_2O; Sigma Chemical) for 5 to 10 min, and put on ice immediately after boiling.

5. Add the carrier DNA, 4.8 ml of 50% (w/v) PEG 3350, 0.6 ml of ddH_2O, 0.6 ml of 1 M lithium acetate, and 50–100 μg of cDNA library to the 1-ml cell suspension. Vortex vigorously for a few seconds and incubate at 30° for 30–40 min.

6. Heat the samples at 42° for 10 min.

7. Pellet the cells at 6000g for 5 min. Resuspend the cells in 1 ml of ddH_2O, and plate ~200-μl cell aliquots onto five agarose plates (150 × 15 mm; Fisher) containing appropriate selective medium. For example, if the cDNA library is cloned into a *URA3*-marked plasmid, SD-U-L containing 2% (w/v) galactose and 2% (w/v) agar should be used.

8. A 50-μl aliquot of the cells should also be serially diluted 10-fold into SD medium and plated onto SD-U-L/glucose plates to estimate the transformation efficiency. Typical transformations yield ~10^6 transformants/ml when using the methods described above, although transformation efficiencies can be affected by several factors, including the quality

---

[20] Q. Xu and J. C. Reed, *Mol. Cell* **1**, 337 (1998).

of the carrier DNA and the library DNA, the host strain, and the condition of cells.

9. Incubate the plates at 30° for 5 to 7 days (or until colonies appear).

Secondary Screening

The procedure described below assumes that the cDNA library is constructed in a *URA3*-based plasmid, and thus employs a 5FOA negative selection step. If other types of markers are used in the library plasmid, then omit the 5FOA step but confirm the loss of the library plasmid by restreaking on nonselective medium before performing the Bax resistance test. Note that it may be necessary to replate the transformants a second or even third time before the library plasmid has been entirely lost.

1. Individual transformants that grow on galactose medium are picked, patched onto SD-L medium, and allowed to grow at 30° for 2 days. This allows the cells to lose the *URA3* plasmid. (The original plates from the library screening should be retained at 4°.)

2. Patches are replica plated onto SD-L plates containing 5FOA and incubated at 30° for another 1 or 2 days. This will select against yeast cells that retain the *URA3* plasmid.[19] At this step, the patches are also replica plated on uracil-deficient, glucose-containing (SD-U/glucose) plates for later recovery of cDNA library plasmids.

3. Patched cells are replica plated onto SD-L/5FOA plates again and incubated at 30° for another 1 to 2 days.

4. Patched cells are finally replica plated onto SD-L/galactose plates and incubated at 30° for 2 to 3 days. If growth fails to occur on galactose medium, this suggests that resistance of the transformant is dependent on the library cDNA, because cells lacking the library plasmid are not immune to Bax-induced cell death.

Tertiary Screening

After determining that resistance of the yeast cells to Bax-induced cell death is dependent on the cDNA library plasmid, the library plasmids are then recovered by transformation into *E. coli* and the resulting plasmid DNA is transformed back into Bax-expressing yeast cells to determine whether the plasmid DNA will consistently give rise to transformants that can grow on galactose medium.

1. Grow a colony of the transformants whose growth on galactose medium is dependent on the library cDNA in 5 ml of SD-U medium overnight.

2. Pellet the cells at maximum speed for 20 sec in a microcentrifuge (Eppendorf, Hamburg, Germany) in a 1.5-ml polypropylene snap-cap tube (microcentrifuge tube).

3. Wash the yeast cells with 1 ml of ddH$_2$O.

4. Resuspend the yeast cells in 200 μl of lysis buffer: 2% (v/v) Triton X-100, 1% (v/v) sodium dodecyl sulfate, 100 mM NaCl, 10 mM Tris–Cl (pH 8.0), 1 mM EDTA (pH 8.0).

5. Add glass beads (Sigma) to the top of the liquid level and vortex at maximum speed for 10 min.

6. Add 200 μl of phenol–chloform–isoamyl alcohol, vortex briefly, and invert the closed tubes several times.

7. Centrifuge in a microcentrifuge at maximum speed ($\sim$12,000–16,000g) for 10 min at room temperature.

8. Recover the supernatants into a fresh 1.5-ml microcentrifuge tube and add 2.5 vol of ethanol to precipitate total yeast DNA at $-70°$.

9. Wash the DNA with 0.5 ml of 70% (v/v) ethanol, air dry, and resuspend the DNA in 20 μl of ddH$_2$O.

10. Take 5 μl of DNA for transformation of *E. coli,* [XL1-blue strain, e.g., is generally appropriate (Strategene, La Jolla)], using either electroporation or a CaCl$_2$ method.

11. Pick bacterial transformants and make plasmid DNA by any standard miniprep procedure. Perform appropriate restriction enzyme digests to distinguish between YEp51-Bax and the library cDNA plasmid.

12. Employ the recovered library plasmid to retransform QX95001 cells (which contain Yep51-Bax), and select the transformants on SD-U-L plates.

13. Streak two or three individual colonies onto an SD-U-L/Gal plate to determine whether transformants grow.

Pick those plasmids that consistently confer protection against Bax and subject them to DNA sequencing as well as immunobot analysis to verify production of Bax protein.

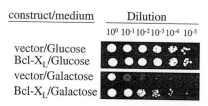

Fig. 1. Example of rescue of yeast from Bax-induced lethality by Bcl-X$_L$. Yeast strain QX95001 cells containing YEp51-Bax were transformed with either vector control plasmid (p426-GPD) or a plasmid containing Bcl-X$_L$. Cells were grown in SD-U-L medium overnight. Cultures were serially diluted 1 : 10 and 5 μl of each dilution was dropped onto plates containing either glucose or galactose medium. Plates were incubated at 30° for 3 days.

Comments

We have described methods used in our laboratory to express and study the death-inducing activity of the proapoptotic Bax protein in the budding yeast *S. cerevisiae*. The function of an antiapoptotic gene such as *bcl-2* can be assayed by its ability to inhibit Bax-induced yeast lethality, using assays similar to these described in this chapter. In this case, it is preferable to express Bcl-2 via a second expression vector carrying a different promoter to prevent promoter competition. The findings that Bcl-2 family proteins retain at least some of their death-regulatory activity in yeast afford us with the opportunity to perform structural–functional studies of Bcl-2 family proteins in yeast. Domains of a prodeath protein, e.g., Bax, that are critical for Bax to kill yeast can be initially mapped with the yeast system and then correlated with studies in mammalian cells to understand the structure–function relationships of the Bax protein. Similar structure–function analysis can be performed for an antiapoptotic member of the Bcl-2 family, for example, Bcl-2 or Bcl-X_L, by virtue of its ability to inhibit Bax-mediated lethality in yeast (Fig. 1).

The findings that expression of Bax is detrimental to yeast and that this lethal effect is most likely mediated by the intrinsic activity of Bax also create the opportunity to use yeast to identify other components of mammalian cell death machinery, either by functional cloning or by classic yeast genetics approaches. Mammalian cDNA libraries can be screened for genes that inhibit Bax-induced cell death in yeast, and the function of these gene products can be studied in both yeast and mammalian cells. Using this screen, we have identified known antiapoptotic regulators, including Bcl-2, Bcl-X_L, and Mcl-1 (Fig. 1; N. Ke and J. C. Reed, unpublished data, 1999). We were also able to identify novel human genes *BI-1* and *BI-2*, which function as Bax inhibitors in both yeast and mammalian cell.[20,21] On the other hand, yeast mutants can be isolated that render yeast resistant to Bax killing. Subsequent cloning of the wild-type yeast gene will help identify components of the Bax-mediated cell death pathway in yeast and perhaps mammalian cells as well.[22]

Acknowledgments

This work was supported by grants from the U.S. Army Breast Cancer Research Program (TR950134 to Q.X. and PR951168 to J.C.R.). We thank Tara Brown for manuscript preparation.

[21] ■. Zhang, *et al.*, submitted (2000).
[22] S. Matsuyama, Q. Xu, J. Velours, and J. C. Reed, *Mol. Cell* **1,** 327 (1998).

[28] Exploiting the Utility of Yeast in the Context of Programmed Cell Death

By CATHERINE N. TORGLER, ROBIN BROWN, and ERIC MELDRUM

Many researchers have explored the extent to which yeast can be used to dissect the mechanisms of programmed cell death in higher cells. Yeast has been used as a system to analyze protein–protein interactions and structure–function relationships, and as a cloning tool to identify novel higher eukaryote regulators of apoptosis. In addition, classic genetic strategies in yeast have been used to analyze the mechanisms of action of core pathway members. The purpose of this chapter is to describe the strategies pursued and act as a source for the technical details necessary to exploit the yeast *Saccharomyces cerevisiae* and *Schizosaccharomyces pombe* in the context of programmed cell death.

Introduction

The application of genetically tractable organisms to answer functional biology questions in mammalian cells has proved an extremely useful strategy in many fields, with success relying on the conservation of function between the genetic model systems and mammalian cells. In the context of programmed cell death (PCD), genetically tractable organisms have been exploited with notable success by the *Caenorhabditis elegans* community.[1,2] While an apoptotic signaling pathway has been genetically identified in *C. elegans,* the worm has not proved to be as useful in attempts to understand how that pathway signals information leading to caspase activation. As a consequence, many questions still remain concerning the mechanism of action of many proteins in the core apoptotic pathway originally described by Horvitz and coworkers. Therefore, in addition to experiments in higher cells a section of the cell death community has attempted to examine the extent to which other genetically tractable organisms can be used to dissect the mechanisms of action of the key players in higher cell PCD.[3] Yeast has been chosen by many researchers because of the availability of a well-

[1] R. E. Ellis, J. Yuan, and H. R. Horvitz, *Annu. Rev. Cell Biol.* **7,** 663 (1991).

[2] M. Hengartner, *in* "*C. elegans* II" (D. L. Riddle, T. Blumenthal, B. J. Meyer, and J. R. Preiss, eds.), pp. 383–415. Cold Spring Harbor Laboratory Press, Cold Spring Harbor, New York, 1997.

[3] J. C. Ameisen, *Science* **272,** 1278 (1996).

developed range of tools, its ease of analysis, and the large body of literature illustrating its utility as a genetic model system.

This review acts as a source for the technical details necessary to pursue strategies aimed at the exploitation of the yeasts *S. cerevisiae* and *S. pombe* in the context of PCD. One central point that should be made at this juncture is that yeast do not possess an apoptosis program or homologs of any of the core pathway members. How, then, can yeast be of any utility? The potential being explored by researchers is as follows:

- Yeast can be used as a system to analyze protein–protein interactions and structure–function relationships. Here the advantage lies within the fact that this experimental system is simplified by the absence of known homologs of the core pathway members (see Section II,A, Structure–Function Analysis).
- Classic genetic strategies may be used to analyze the mechanisms of action of core pathway members in yeast. Here the advantage lies within the fact that well-developed strategies can be used to dissect the mechanism of action in yeast. What remains is to determine if similar mechanisms are employed in higher cells (see Section II,C, Application of Yeast Genetics).
- Yeast can be used as a cloning tool for novel higher eukaryote regulators of apoptosis. Here the unicellular organism is used simply as a readout to enable cloning procedures to be executed in a high-throughput manner (see Section II,C,3, Suppressor Screening in Yeast).

In the following review it is hoped that sufficient technical details can be passed on to allow researchers, if they wish, to pursue any one of the above strategies with their protein(s) of interest.

I. Basic Yeast Biology and Tools

It is beyond the scope of this chapter to outline all the rudimentary technical skills necessary to work with yeast. Several excellent technical reviews exist[4] and a growing number of useful web sites are available; these can be consulted for additional information.[5,6]

A. Strains

Saccharomyces cerevisiae and *Schizosaccharomyces pombe* strains are freely available from individual researchers and also, if preferred, from the

---

[4] C. Guthrie and G. Fink (eds.), "Guide to Yeast Genetics and Molecular Biology." *Methods Enzymol.* **194** (1991).
[5] http://www.bio.uva.nl/pombe/handbook/
[6] http://www.sacs.ucsf.edu/home/HerskowitzLab/protocols/protocol.html

National Collection of Yeast Cultures (Norwich Research Park, Norwich, UK). If embarking on a genetic analysis or complicated strain construction it is worth considering the selectable markers necessary for use in the starting strain and also whether the mating type is of importance. The strains used as experimental subjects must carry nonrevertible null alleles of genes necessary for the biosynthesis of various essential components (most usually histidine, uracil, leucine, adenine, lysine, or tryptophan). All yeast expression plasmids carry a gene or "selectable marker," which complements the particular mutation in question. As a consequence, cells transformed with a particular plasmid are selected by growing on synthetic medium that does not contain the particular essential component. For *S. cerevisiae* and *S. pombe* the selectable marker genes exploited are different but the principle is the same. If an experiment is planned that requires mating and dissection of spores, then it is worth using a starting strain that can be crossed to an opposite mating type and the diploid be selected. For *S. pombe* this function is well performed by the *ade6*-M210 and *ade6*-M216 markers, which complement intragenically. As a consequence, neither haploid version will grow on plates lacking adenine but diploids will. Similar strategies can be used for *S. cerevisiae,* or haploids can be used that have different selectable markers, thus resulting in a diploid that is able to grow on plates where both components are absent.

B. Plasmids

For *S. cerevisiae* there are three different plasmid types: YEp (episomal), YCp (centromeric), and YIp (integrative). All three types are constructed from bacterial plasmids and therefore can be selected and propagated in *Escherichia coli*. In addition to the yeast selectable marker and bacterial components, YEp plasmids contain a yeast origin of replication (ARS, autonomous replication site), which promotes autonomous replication, and also a small region derived from the 2μm circle, which facilitates equal partition of plasmid molecules between mother and daughter. As a consequence of this manufacture YEp plasmids are multicopy (10–100/cell) and segregate equally on division. In addition to the selectable marker, bacterial components, and the ARS element, YCp plasmids contain a chromosomal centromere, which allows the plasmids to engage the spindle apparatus. As a consequence of this manufacture, YCp plasmids are present as a single copy per cell and segregate equally between mother and daughter. YIp plasmids do not contain an ARS, 2μm element, or centromere and contain only a yeast selectable marker and bacterial components. As a consequence, these plasmids are unable to replicate autonomously and require integration into the genome. For efficient homologous recombination at the marker locus the plasmid DNA must be linearized within the yeast selectable marker and transformed into cells. This genomic construction is usually extremely stable.

For *S. pombe* there are only multicopy and integrative plasmids. There are no single-copy centromeric plasmids, as the *S. pombe* centromere is too large for the design of a practical shuttle vector. In *S. pombe* the difference between the two types of plasmid is the presence or absence of the ARS element, but they are used in the same way as for *S. cerevisiae.*

C. Promoters

The expression of a cloned open reading frame (ORF) in a yeast expression vector can be varied in terms of the absolute quantity of protein and whether expression is regulatable or constitutive. The absolute quantity of protein expressed can be varied by choice of promoter strength or type of plasmid. In principle, genetic stability is the highest priority and, consequently, if expression is sufficient for a phenotype, an integrative plasmid is always the optimal choice. However, it may be necessary to achieve maximal possible expression and in such circumstances multicopy plasmids are an option. The reader should, however, be warned that if the expression phenotype is detrimental to yeast cells then a multicopy plasmid will recombine at a high frequency, resulting in loss of expression and reversion from the original phenotype. It is important, therefore, to apply rigorous strain maintenance principles, returning always to a confirmed glycerol stock and not using liquid/agar stocks that have been through many divisions.

Numerous transcription cassettes for the constitutive expression of a cloned ORF are available for the two yeasts. In *S. cerevisiae,* constitutive promoters commonly exploited are alcohol dehydrogenase (*ADH*), triose phosphate isomerase (*TPI*), and the 786 promoter. Inducible promoters most commonly used are the *GAL1* and *CUP1* promoters. A large selection of *S. cerevisiae* expression plasmids is commercially available (e.g., R&D Systems, Minneapolis, MN) and can also be obtained from individual researchers. In *S. pombe,* the *ADH* promoter is also a commonly used, relatively strong constitutive promoter. The cauliflower mosaic virus promoter can be used as a weak constitutive promoter or as a regulatable promoter in a specific genetic background that expresses the *tet* repressor. In this background, the addition of tetracycline to culture media will induce expression within 2–3 hr.[7] The most commonly used regulatable promoter for *S. pombe* is the *nmt1* promoter, which is repressed by thiamine.[8] Induction of expression therefore requires washing of cells in media lacking thiamine

---

[7] K. Faryar and C. Gatz, *Curr. Genet.* **21,** 345 (1992).
[8] K. Maundrell, *Gene* **123,** 127 (1993).

and, if the cells are logarithmically growing, subsequent cell divisions will decrease intracellular thiamine until, after approximately 16 hr, a critical threshold of intracellular thiamine is exceeded and expression begins. Three different strengths of the *nmt1* promoter exist.

Having chosen the type of yeast, plasmid, and promoter for the experiment, it will be necessary to transform that DNA into cells, confirm that the correct integration event has occurred (only in the case of integrative plasmids), and prove that mRNA or protein of the appropriate molecular weight is being synthesized. If regulatable promoters are being used, it is also necessary to demonstrate that expression is efficiently repressed under the appropriate conditions. Section III (Protocols) should make these necessary steps and subsequent confirmations routine and rapid.

Knowledge of the above should allow the manufacture of genetically stable strains that express the protein intended. The rest of this chapter focuses on the potential utility of such a strain.

II. Utility of Yeast in the Context of Programmed Cell Death

A. Structure–Function Analysis

Two-hybrid analysis has been used extensively by the PCD community. Many novel members of the PCD machinery have been isolated in two-hybrid screens and many interaction domains have been mapped in detail.[9] Several excellent reviews of the theoretical and technical details of two-hybrid analysis have been written and the reader should consult those for technical details.[10,11]

In addition to two-hybrid analysis, many researchers have capitalized on the fact that the interpretation of experimental observations in yeast is simplified by the absence of known homologs of the core PCD pathway members. It has therefore been demonstrated in *S. pombe* that the BH3 domain of Bak is necessary for lethality[12] and that the lethality of Bax in *S. cerevisiae* is lost if Bax is intentionally localized to an intracellular compartment other than the mitochondrion.[13] Interestingly, it has also been

[9] S. Farrow and R. Brown, *Curr. Opin. Gen. Dev.* **6**, 45 (1996).
[10] S. Fields and R. Sternglanz, *Trends Genet.* **10**, 286 (1994).
[11] C. Bai and S. J. Elledge, *Methods Enzymol* **273**, 331 (1996).
[12] B. Ink, M. Zornig, B. Baum, N. Hajibagheri, C. James, T. Chittenden, and G. Evan, *Mol. Cell Biol.* **17**, 2468 (1997).
[13] H. Zha, H. A. Fisk, M. P. Yaffe, N. Mahajar, B. Herman, and J. C. Reed, *Mol. Cell Biol.* **16**, 6494 (1996).

demonstrated that a form of Bcl-2 that is unable to interact with Bax in a two-hybrid test is able to rescue *S. cerevisiae* from Bax-mediated death.[14]

B. Morphological Consequences of Expressing Core Pathway Members

Researchers have studied the morphological consequences of expessing core pathway members in yeast in an attempt to identify phenotypic read-outs other than death and to explore whether yeast cells die on expression in a manner similar to higher cells.[12,13,15] It is our belief, however, that this particular approach has not provided convincing evidence that yeast die in a fashion similar to higher cells when proteins such as Bak, Bax, or caspases are overexpressed. It is more likely that morphological observations are consequences of yeast death rather than the activation of a primordial cell death pathway. Similarly, the danger of exploiting morphology as a selectable phenotype in heterologous expression experiments is that information gained may be pertinent only to yeast biology and not relevant to the mechanisms by which neurons or other cells reach apoptotic commitment. A more productive research effort has been to use death on induction of expression in yeast as a selectable phenotype in classic genetic strategies aimed at identifying the molecular mediators of that death in yeast. Several groups have pursued this approach and arrived at interesting models for how certain mammalian core pathway members regulate apoptotic fate in higher cells.[16–19] The next section deals with the technical requirements for this approach.

C. Application of Yeast Genetics

A classic forward genetic strategy that utilizes yeast revolves around the ability to observe a particular biology of interest and then screen for mutations and/or genes that alter that biology in interesting ways. In the context of this chapter, however, the utility of yeast lies within a reverse genetic strategy, in which one is in possession of a gene and the interest lies in the ability to quickly determine the action of the encoded protein in yeast. This is possible only if the expression of that gene creates a robust

[14] J. M. Jürgensmeier, S. Krajewski, R. C. Armstrong, G. M. Wilson, T. Oltersdorf, L. C. Fritz, J. C. Reed, and S. Ottilie, *Mol. Biol. Cell* **8**, 325 (1997).
[15] F. Madeo, E. Fröhlich, and K.-U. Fröhlich, *J. Cell Biol.* **139**, 729 (1997).
[16] C. N. Torgler, M. de Tiani, T. Raven, J.-P. Aubry, R. Brown, and E. Meldrum, *Cell Death Differ.* **4**, 263 (1997).
[17] S. Matsuyama, Q. Xu, and J. C. Reed, *Mol. Cell* **1**, 327 (1998).
[18] Q. Xu and J. C. Reed, *Mol. Cell* **1**, 337 (1998).
[19] I. Marzo, C. Brenner, N. Zamzami, J. M. Jürgensmeier, S. A. Susin, H. L. Vieira, M. C. Prevost, Z. Xie, S. Matsuyama, J. C. Reed, and G. Kroemer, *Science* **281**, 2027 (1998).

selectable phenotype that can easily be distinguished from wild type. In many of the examples discussed below the selectable phenotype is the death of yeast cells on induction of expression. However, it is worth noting that it is also possible to engineer a selectable phenotype equally useful as death in yeast. For instance, while the expression of a blue color in the context of two-hybrid interactions is perhaps the most widely exploited example of a synthetic selectable phenotype, it is possible to similarly create artificial readouts in yeast for many enzyme activities such as kinases and proteases, using, for example, fluorescent substrates.

If the selectable phenotype generated is determined to be sufficiently robust, then suppressors of that phenotype can be isolated in two different ways. Yeast genes that, when mutated, result in cells that no longer display the expression phenotype can be isolated by the procedures outlined below. In addition, yeast genes and/or higher cell cDNAs that, when coexpressed in yeast, result in cells that no longer display the expression phenotype can be isolated by the procedures outlined in Section II,C,3, Suppressor Screening in Yeast.

Two different approaches have been used to isolate genomic suppressors of lethal phenotypes in yeast. In the first, a Bax-expressing *S. cerevisiae* strain was randomly mutated to reveal that mutations in the F_0F_1-ATPase made cells resistant to lethality on Bax expression.[17] The second approach studied the lethality caused by Bak expression in the yeast *S. pombe*. In this approach Bak-interacting yeast proteins were isolated by two-hybrid screening and the requirement for that interaction in lethality was tested by specific gene replacement at the isolated locus. By this strategy it was demonstrated that an interaction between calnexin 1 (Cnx1) and Bak was necessary for Bak-mediated lethality in *S. pombe* (Fig. 1).[16] The protocols below outline the technical details involved in random mutagenesis and gene replacement strategies.

1. Demonstrating Genomic Suppression in Yeast by Gene Replacement. As mentioned above, it is possible to make a specific mutation in yeast and determine whether that mutation influences the expression phenotype. The method involved here is termed *gene replacement* and can be carried out in two different ways. In the first method, one copy of the gene of interest is disrupted in a diploid strain (see Section III,I, Gene Disruption). To isolate a haploid that expresses a mutant of interest the diploid is transformed with a mutant construct on a multicopy plasmid and the diploid dissected. The progeny of the dissection that possess the deletion marker and the selectable marker for the plasmid now express only the mutant protein. It is advisable to confirm that this is true by sequencing a number of reverse transcriptase-polymerase chain reaction (RT-PCR) products. In the second strategy a wild-type locus, in haploid cells, is replaced by a

A   Two-hybrid analysis

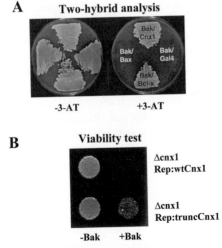

        -3-AT   +3-AT

B   Viability test

                                        Δcnx1
                                        Rep:wtCnx1

                                        Δcnx1
                                        Rep:truncCnx1

        -Bak   +Bak

FIG. 1. Bak interaction with Cnx1 is necessary for lethality in *S. pombe*. (A) To demonstrate the two-hybrid interaction between Bak and Cnx1 the *S. cerevisiae* strain Y190 was transformed with a plasmid containing the full-length *bak* cDNA fused to sequence encoding the binding domain of the yeast GAL4 protein. This strain was then transformed with a vector containing sequence encoding the yeast GAL4 DNA-binding domain fused to either the *cnx1, bcl-X*, or *bax* cDNAs. Positive two-hybrid interaction is displayed by growth selection on +3-AT. The Bak interaction domain has been mapped to the calnexin cytosolic tail. (B) The cells shown are deleted for the *S. pombe cnx1* gene. Function and viability are maintained by expression of either wild-type Cnx1 (Rep:wtCnx1) or a truncated Cnx1 that lacks the cytosolic tail (Rep:truncCnx1). These two strains are transformed with *bak* on a multicopy plasmid. As can be seen, deletion of the Bak-binding domain of Cnx1 is sufficient to make cells resistant to Bak lethality.

mutant version via a two-step homologous recombination process. In the first step the mutated gene (in an integrative vector with uracil as marker) is integrated into the genome at the position of the wild-type locus. To achieve this, cells are transformed with a plasmid that is linearized within the gene and not within the selectable marker. This homologous recombination results in two tandem copies of the gene, one of which is mutated. The second step requires excision of the wild-type gene by 5-fluoroorotic acid (5-FOA) selection (see Section III,J, Curing Plasmid DNA). In theory, half of the uracil-minus clones recovered will possess the mutation and half will be wild type at the locus of interest. Genomic PCR can be used to identify the mutated version and, in addition, Southern analysis should also be performed to confirm that large quantities of genomic DNA have not been lost in the excision.

    2. Isolating Yeast Genomic Suppressors by Random Mutagenesis. Many different mutagens exist for the mutagenesis of yeast at random positions in the genome and the best mutagens result in a high frequency of base pair substitutions without being extremely toxic. Many people use ethylmethane sulfonate (EMS) or N-methyl-N'-nitro-N-nitrosoguanidine (MNNG) because they can be added to liquid cultures. However, they are specific in their action and for this reason some people prefer ultraviolet light, as it produces a greater range of substitutions and frameshift mutations. For a more detailed description of mutagenesis techniques see *Methods in Enzymology*, Vol. 194.[20]

The following protocol for mutagenesis attempts to guide the reader through the different possibilities that may be encountered, both in terms of strain construction and mutagen chosen. It is important to note that there is no correct answer concerning how many cells should be mutated in order to completely cover all the possible rescuing mutations that could be created. The numbers provided represent what the authors routinely initiate a screen with, but it should be noted that if numerous mutations are not isolated in the same gene it is likely that full coverage has not been achieved.

1. Two different strain constructions are routinely used for this approach. If the gene expressed is in a construct integrated in the genome, two copies integrated at different sites must be employed. This eliminates the possibility that a single mutation in the integrated construct can rescue. The frequency of a simultaneous mutation in both the integrated loci is negligible. If the gene expressed is on a multicopy plasmid then it is necessary to mutate the strain and then transform the multicopy construct into the cells.

2. Before screening, the mutagen concentration and length of time of exposure should be titrated in order to achieve 50% lethality.

3. Either plate the cells directly on selective inducing agar plates or transform the cells with multicopy plasmid before plating. If mutating with UV light, it is necessary first to plate the cells and expose to UV after plating. Whatever way the cells are mutated, it is important to know how many cell equivalents have been mutated and screened. This is achieved by working with a parallel unmutated culture and following the same procedure as with the mutated cells. It is recommended that 2×10^8 cells be mutated and plated.

4. Mutated cells will not grow at a normal rate. It is a good idea to monitor the growth of colonies on the plate and mark the day when they appear. It is a matter of personal judgment how many mutated colonies to

[20] C. W. Lawrence, *Methods Enzymol.* **194,** 273 (1991).

pick. It is advisable also to monitor the appearance of unmutated colonies under expressing conditions and stop picking mutated colonies when the background from the wild-type population (if there is any) begins to grow.

5. Any mutated colony that now grows under inducing conditions could be an interesting genomic mutation or an expression defect. If screening for resistance to expression from a multicopy plasmid it is advisable to remove the plasmid (see Section III,J, Curing Plasmid DNA) and transform cells with a new plasmid, preferably with expression from a different promoter. Isolates that remain resistant to expression are genuine "hits." If screening for resistance to expression from an integrated plasmid, it is advisable to confirm wild-type quantities of protein expression by Western blotting.

6. It is necessary to determine whether the mutation is recessive or dominant. The mutated clone is mated to a wild-type opposite mating type and the resulting diploid is tested for resistance to induced expression (see Section III,M, Mating, Dissection, and Mating Type Testing). Recessive clones will now die on expression and dominant mutations will remain resistant. At this point most researchers freeze the dominant mutations for future analysis and continue to work on the recessive mutations. To clone gene mutations that cause a dominant phenotype it is necessary to create a genomic DNA library from that particular clone. This library is then transformed into the starting strain and selected for the dominant phenotype.

7. Mutagenesis procedures mutate the genome at multiple positions. For this reason it is necessary to separate the interesting mutation from the others and determine whether the resistance phenotype is caused by a single mutation. This is a process known as "outcrossing" or "backcrossing" and is performed by mating the mutated isolate with a wild-type opposite mating type. The resulting diploid is sporulated and the progeny of the cross dissected (see Section III,M, Mating, Dissection, and Mating Type Testing). The progeny are then retested for the presence of the plasmid selectable marker(s) and then for resistance. This whole procedure is performed at least twice, each time proceeding with a single progeny from the cross and storing the precross version. While performing the outcrossing, it is important to determine whether the segregation of the resistance mutation occurs with a Mendelian ratio of $2:2$ in the population that possesses the appropriate markers. If this is not apparent by the second cross it is likely that resistance is generated by the simultaneous mutation of multiple loci and such clones should be discarded.

8. Having now isolated a series of resistant clones that are recessive mutations at a single locus it is advisable to confirm wild-type expression of the protein concerned and separate mutants into complementation

groups. This identifies how many of the population of mutants isolated are mutated at different gene loci. This is done by generating an opposite mating type of each mutant, which still expresses the protein of interest under inducing conditions (see Section III,M, Mating, Dissection and Mating Type Testing). The approach is to systematically mate every mutant against the other clones isolated and test whether the diploids are rescued from the selectable phenotype. Remember that all clones are recessive mutations and therefore if the resulting diploids are still resistant to expression, the two crossed mutants must be at the same locus. In this way, the whole collection of mutants can be divided into a smaller collection of complementation groups. Each member of a complementation group may, of course, be mutated at different positions within the same gene sequence or promoter.

9. All that remains is to clone the wild-type yeast genes that, when mutated, result in loss of the expression phenotype (see below).

3. Suppressor Screening in Yeast. On isolating a recessive mutation that is no longer sensitive to expression of a protein of interest it is necessary to clone the wild-type gene by complementation. This is possible only because the mutants isolated are recessive and the wild-type gene rescues. Another possibility, which exploits the same procedure to a large extent, involves screening for suppressors of a robust selectable phenotype with multicopy or centromeric expression libraries. The advantage in this situation is that higher eukaryote cDNAs can be isolated that regulate the activity of the protein being expressed in yeast. This approach essentially uses yeast as a cloning tool and has been exploited with some success to isolate a novel mammalian apoptosis suppressor known as Bax inhibitor 1 (BI-1).[18] The procedure for both the methods described above is in essence the same.

1. The strain of interest is transformed with the library onto inducing or repressing selective plates. If the experiment is to rescue a selectable phenotype with a cDNA library, then cells can be transformed onto inducing selective plates and the library expression must be under the control of a yeast promoter. If, however, the experiment is to clone a recessive yeast mutation by complementation cloning, then care must be taken that the successfully complemented clones can be distinguished from the mutated population. For example, if the mutation phenotype is resistance to death, then a successfully rescued clone will die on selective inducing plates. In this case transformants should be plated under repressing conditions, allowed to grow, and then replica plated onto inducing plates (see Section III,L, Replica Plating). Rescued colonies will be those that grow under repressing but not expressing conditions. The library in the case of complementation

cloning is usually a yeast genomic library that uses a uracil marker. The use of the uracil marker makes it easier to cure the plasmid in later stages of the analysis (see Section III,J, Curing Plasmid DNA).

2. Depending on the precise details of the experiment and the phenotype it must be possible to distinguish between rescued colonies and the background. Positive clones should be separated and retested. Duplicated positives could either be real rescuing genes or spontaneous revertants. In order to distinguish between these two possibilities it is necessary to demonstrate that the rescuing phenotype is plasmid dependant. This is done by 5-FOA selection (see Section III,J, Curing Plasmid DNA).

3. All positives that are not rescued on inducing plates containing 5-FOA are plasmid-dependent rescuers.

4. Isolate the plasmid (see Section III,K, Plasmid Isolation from Yeast).

5. It is possible to extract, from a single positive, plasmids that appear to have different restriction maps. This is due to plasmid recombination and transformation with multiple plasmids. It is necessary to retransform the original strain with a representative plasmid from an extracted collection to demonstrate that it rescues. If it does not rescue, then it is either a false positive, or another plasmid in the extraction will be the rescuer. If the strain being extracted contains more than one episomal plasmid and there are a large number of cDNA-dependent rescuers, we recommend probing the *E. coli* colonies with the respective selectable markers on the plasmids to identify the plasmids of interest.

6. If on retest the plasmid still rescues, sequence the insert.

III. Protocols

A. *Saccharomyces cerevisiac and Schizosaccharomyces pombe Cell Culture*

To obtain a reproducible result, any experiment with yeast requires a fresh and exponentially growing culture. Strains should be frozen in 30% (v/v) glycerol and kept at −80°. Every new inoculum should be derived from the frozen stock to avoid mutational changes that may develop with time. The average doubling time of *S. cerevisiae* and *S. pombe* is 90 min and 3 hr, respectively.

B. *Saccharomyces cerevisiae Harvesting*

1. The day before the culture is required, make a fresh inoculum from a yeast glycerol stock in 5 ml of medium and shake overnight at 30°.

2. In the morning, dilute the inoculum to a concentration of 1×10^6 cells/ml in a volume of medium appropriate for the number of cells required and shake at 30°.

3. Centrifuge the cells at 1000g for 5 min when they reach approximately 5×10^6–1×10^7 cells/ml.

C. Schizasaccharomyces pombe Harvesting

1. Two days before the culture is required, make a fresh inoculum from a yeast glycerol stock in 5 ml of medium and shake overnight at 30°.

2. The next evening, dilute the inoculum to a concentration of 1–2 $\times$ 10^5 cells/ml in a volume of medium appropriate for the number of cells required and shake at 30° overnight.

3. Centrifuge the cells at 1000g for 5 min when they reach approximately 5×10^6–1×10^7 cells/ml.

D. Lithium Acetate Transformation of Saccharomyces cerevisiae

The following protocol is scaled for 20 transformations. It is recommended to always generate an excess volume of competent cells and use what is required.

1. Harvest 5×10^8–1×10^9 cells (see Section III,B, *S. cerevisiae* Harvesting).

2. Resuspend the cell pellet in 50 ml of sterile water and centrifuge the cells at 1000g for 5 min and discard the supernatant.

3. Repeat step 2.

4. Resuspend the cells in 1 ml of sterile water in an Eppendorf tube and centrifuge at 10,000g for 5 sec and discard the supernatant.

5. Resuspend the cells in 1 ml of 1$\times$ TELiAc buffer (see Section IV) and centrifuge at 10,000g for 5 sec and discard the supernatant.

6. Repeat step 5.

7. Resuspend the cells in 1 ml of 1$\times$ TELiAc buffer.

8. Distribute plasmid DNA (1–5 μg) and 2 μl of salmon sperm carrier DNA (stock, 10 mg/ml) in sterile Eppendorf tubes. Miniprep-quality DNA is sufficient.

9. Add 50 μl of cells to the plasmid DNA–carrier DNA mixture. It is also recommended to incorporate into the transformation a positive control (vector) and negative control (no plasmid DNA).

10. Vortex the mixture briefly.

11. Add 300 μl of PEG solution (see Section IV) and vortex the mixture briefly.

12. Incubate the mixture for 30–45 min at 30°.

13. Heat shock for 20 min at 42°.

14. Centrifuge the cells at 10,000g for 5 sec and wash off the PEG solution three times by resuspending the pellet in 1 ml of sterile water. The first pellet resuspension is difficult and best performed with a toothpick.

15. Resuspend the cells in 1 ml of sterile water.

16. Plate various dilutions of the transformed cells (200 μl, 50 μl, and the remainder of the cells centrifuged and resuspended in 200 μl) on appropriate selective plates.

E. Lithium Acetate Transformation of Schizosaccharomyces pombe

The following protocol is scaled for 10 transformations. It is recommended to always generate an excess volume of competent cells and use what is required.

1. Harvest 5×10^8–1×10^9 cells (see Section III,C, S. pombe Harvesting).

2. Wash the cell pellet in 50 ml of sterile water and centrifuge at 1000g for 5 min.

3. Discard the supernatant and resuspend the pellet in 50 ml of 1× TELiAc buffer.

4. Centrifuge the cells at 1000g for 5 min.

5. Repeat steps 3 and 4.

6. Discard the supernatant and resuspend the cell pellet in 1 ml of 1× TELiAc buffer.

7. Distribute the plasmid DNA (1–5 μg) and 2 μl of salmon sperm carrier DNA (10 mg/ml) in sterile Eppendorf tubes. It is also recommended to incorporate into the transformation a positive control (vector) and negative control (no plasmid DNA). Miniprep-quality DNA is sufficient.

8. Add 100 μl of cells to the plasmid DNA–salmon sperm DNA mixture.

9. Vortex the mixture and incubate at room temperature for 10 min.

10. Add 260 μl of PEG solution and vortex.

11. Incubate for 30–45 min at 30°.

12. Add 43 μl of dimethyl sulfoxide (DMSO), vortex, and incubate at 42° for 5 min.

13. Centrifuge the cells for 5 sec at 10,000g.

14. Wash the pellet three times with sterile water. The first pellet resuspension is difficult and best performed with a toothpick.

15. After three washes, resuspend in 1 ml of sterile water.

16. Plate various dilutions of the transformed cells (200 μl, 50 μl, and the remainder of the cells centrifuged and resuspended in 200 μl) on appropriate selective plates.

F. Isolation of Yeast Genomic DNA

This procedure works equally well for both S. pombe and S. cerevisiae. The yield is approximately 50–100 μg.

1. Harvest 1–5 × 10⁸ cells (see Section III,B or C, above).

2. Resuspend the pellet in 10 ml of sorbitol buffer (Section IV) supplemented with 20 μl of 14 M 2-mercaptoethanol.

3. Centrifuge at 1000g for 5 min; discard the supernatant.

4. Resuspend the pellet in 1 ml of sorbitol buffer supplemented with Zymolyase-20T (2 mg/ml, from *Arthrobacter luteus*), lysing enzymes (1 mg/ml), and 2-mercaptoethanol (14 M, 2 μl/ml).

5. Incubate the cells for at least 30 min at 37°.

6. After 30 min, confirm complete cell wall digestion by adding 10 μl of the cell suspension to 90 μl of sodium dodecyl sulfate [SDS, 1% (w/v)]. Complete cell lysis should occur and cell debris should be observed under a light microscope. If the lysis is incomplete, incubate for an additional 30 min at 37° and retest.

7. Centrifuge at 10,000g for 5 min.

8. Resuspend the pellet in 0.5 ml of 5× TE (see Section IV).

9. Add 25 μl of SDS 20%, (w/v) and invert the tube to mix (do not vortex).

10. Incubate for 10 min at 65°.

11. Add 0.5 ml of 5× TE supplemented with 335 μl of 5 M potassium acetate, mix by inversion.

12. Chill on ice for 30 min.

13. Centrifuge at 10,000g for 15 min at 4°.

14. Transfer the supernatant to a new 2-ml Eppendorf tube and add 1 ml of isopropanol.

15. Leave for 5 min at room temperature.

16. Centrifuge at 10,000g for 10 min, discard the supernatant, and drain well.

17. Resuspend the pellet in 0.5 ml of 1× TE supplemented with proteinase K (0.5 mg/ml).

18. Incubate for 2 hr at 37°.

19. Add 0.5 ml of phenol–chloroform, invert the tube several times (do not vortex), centrifuge for 5 min at 10,000g, and transfer the upper aqueous phase to a new tube (avoid interphase).

20. Repeat step 19.

21. Add 1 ml of ice-cold 100% ethanol and 30 μl of 3 M sodium acetate.

22. Chill on dry ice for 10 min (or at −80° overnight).

23. Centrifuge at 10,000g for 10 min at 4°.

24. Wash the pellet with 70% (v/v) ethanol.

25. Centrifuge at 10,000g for 5 min at 4°; discard supernatant.

26. Air dry the pellet.

27. Resuspend the pellet in 25–50 μl of 1× TE supplemented with DNase-free RNase (50 μg/ml).

28. Resuspend DNA by gently shaking the tubes overnight at 30°.

29. Measure the DNA concentration at OD_{260} of a 500- to 1000-fold dilution (OD_{260} of 1 equals 50 μg of DNA per milliliter).

G. Protein Extraction

This procedure works equally well for both *S. pombe* and *S. cerevisiae*. The yield is approximately 200 μg.

1. Harvest 5×10^7 cells (see Section III,B or C, above).

2. Resuspend the pellet in 300 μl of trichloroacetic acid [TCA, 10% (w/v)] and transfer into an Eppendorf tube.

3. Add an equal volume of acid-washed beads to the resuspended pellet.

4. Vortex for 5 min in total by alternating vortexing (1 min) and chilling on ice (30 sec).

5. Transfer the crude extract to a new tube and keep on ice.

6. Centrifuge the crude extract at 10,000g at room temperature for 20 min.

7. Remove the supernatant and resuspend the pellet completely by thoroughly pipetting up and down in 1 ml of cold acetone.

8. Centrifuge at 10,000g at room temperature for 20 min.

9. Remove the supernatant and resuspend the pellet in 100 μl of 2$\times$ SDS/sample buffer (see Section IV).

10. Boil the sample at 95° for 5 min.

11. If sample buffer changes color on addition to the extract, restore the pH by adding 1 μl of 10 M sodium hydroxide.

12. Freeze the sample at $-20°$ until analysis by SDS–polyacrylamide gel electrophoresis (PAGE).

H. RNA Extraction from Yeast

This procedure works equally well in both *S. pombe* and *S. cerevisiae* and can be performed in 1.5-ml Eppendorf tubes. The yield is approximately 100 μg. If RNA degradation occurs, use diethylpyrocarbonate (DEPC)-treated buffers, but maintaining the sample at 4° during the procedure should be sufficient.

1. Harvest 5×10^7–1×10^8 cells (see Section III,B or C, above).

2. Wash the pellet with water and discard the supernatant (this pellet can be stored at $-20°$ in 1.5-ml Eppendorf tubes).

3. Resuspend the pellet in 400 μl of Northern lysis buffer (Section IV).

4. Add 0.4 g of acid-washed glass beads to the resuspended cells.

5. Add 400 μl of phenol–chloroform.

6. Bead lyse cells at 4° by vortexing for 4 min in total by alternating vortexing (1 min) and chilling on ice (30 sec).

7. Centrifuge at 10,000g for 5 min.

8. Remove the upper aqueous phase (avoid interphase) and add to a new tube already containing 300 μl of cold phenol–chloroform.

9. Vortex and centrifuge at 10,000g for 5 min.

10. Remove the upper aqueous phase and add to a new tube already containing 800 μl of cold 100% ethanol.

11. Vortex and store overnight at −20°.

12. To measure the RNA concentration, centrifuge 50 μl of the ethanol solution at 10,000g for 10 min. Air dry the pellet and resuspend in 50 μl of water. Measure the OD_{260} of a 100- to 500-fold dilution (OD_{260} of 1 equals 40 μg of RNA per milliliter).

I. Gene Disruption

Disrupting a gene of interest requires a "deletion construct," which consists of a selectable marker gene flanked by sequences of the open reading frame that is going to be deleted. High-efficiency homologous recombination in *S. cerevisiae* allows 40 bp of flanking sequence to be enough for targeted integration. As a consequence, PCR oligonucleotides (oligos) can be designed that include 40 bp of ORF at the 5′ end and 20 bp of the selectable marker at the 3′ end. PCR amplification of the marker gene with these oligos therefore results in a marker gene flanked by 40 bp of ORF sequence. This PCR product can be transformed directly into cells. This PCR approach is most commonly used with the bacterial geneticin resistance gene *kan^r*, which, when integrated into the genome, makes *S. cerevisiae* resistant to geneticin.[21] In the case of *S. pombe* it is still more efficient to make a deletion construct by traditional restriction mapping and subcloning, where a portion of the chosen ORF has been replaced by the selectable marker (Fig. 2). The deletion construct is then excised from the plasmid and directly transformed into cells.

1. Transform cells with the "deletion construct" DNA (0.1–1 μg) by the lithium acetate method into a chosen diploid strain. As discussed, it is recommended that *S. pombe* be maintained as a diploid by *ade6*-M210/216 intragenic complementation on agar plates lacking adenine. In addition, targeted integration into the correct locus will be more efficient if the diploid is deleted for the chosen marker gene locus. The "deletion construct" will integrate into the genome by homologous recombination to replace a portion of the gene of interest with the marker gene.

[21] A. Wach, A. Brachat, R. Pohlmann, and P. Philippsen, *Yeast* **10,** 1793 (1994).

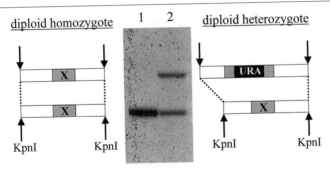

FIG. 2. Detecting gene deletion by Southern analysis. Southern analysis was performed on genomic DNA digested with an appropriate restriction enzyme (e.g., *Kpn*I) from a wild-type homozygous diploid strain (lane 1) and from deleted heterozygous diploid strain (lane 2). Shown on each side of the Southern blot (probed with gene of interest *X*), is the schematic representation of their respective genomic maps. A band shift is observed in the deleted heterozygous diploid strain as a consequence of an integration of the "deletion construct" in one of the loci.

2. Select for diploid transformants by selection for the selectable marker.

3. A Southern blot probed with the disrupted gene should be performed. If appropriate restriction sites are used, DNA from the heterozygous deleted diploid strain can be discriminated from wild type.

4. Sporulate the heterozygous deleted diploid transformant (see Section III,M,3 or III,M,4,c) and dissect the spores. If the gene is not essential for survival, four spores will germinate from an ascus and a 2:2 segregation of the marker should be observed. If the deleted gene is essential for viability, a maximum of two spores will be recovered from all dissections and none will carry the deletion marker.

J. Curing Plasmid DNA

To prove that a rescuing phenotype in yeast is plasmid dependent (and not due to a spontaneous genomic mutation), it is important to remove the plasmid from the strain and demonstrate that the original phenotype has been recovered. This is most easily performed if the plasmid selectable marker is uracil.

1. Make a fresh patch on a synthetic plate lacking the appropriate nutrients but containing uracil.

2. Replica plate the cells on a new synthetic plate containing 0.01% (w/v) 5-fluoroorotic acid (5-FOA) and supplemented with uracil. Uracil-plus cells are killed by 5-FOA, and thus only cells that spontaneously lose the plasmid are able to grow.

3. Confirmation by replica plating should be made that the resulting cells are uracil minus.

K. Plasmid Isolation from Yeast

This procedure works equally well for both *S. pombe* and *S. cerevisiae*.

1. Scrape a 1 × 1 cm patch of cells from a plate and resuspend in 100 μl of sterile water.

2. Centrifuge the cells for 4 sec at 10,000g.

3. Remove the supernatant and wash the pellet with 1 ml of water.

4. Repeat steps 2 and 3.

5. Discard the supernatant and resuspend the pellet in 100 μl of sterile water.

6. Transfer the cells to a chilled 0.2-cm cuvette for electroporation.

7. Electroporate the cells at 1800 V, 200 Ω, 50 μF.

8. Transfer the cells to an Eppendorf tube and add 10 μl of SDS (10%, w/v).

9. Add 300 μl of phenol–chloroform, invert the Eppendorf tube several times (do not vortex), and centrifuge for 5 min at 10,000g.

10. Remove the upper aqueous phase (avoid the interphase layer) and add to a new tube already containing 300 μl of phenol–chloroform. Invert the Eppendorf tube several times (do not vortex) and centrifuge for 5 min at 10,000g.

11. Repeat step 10.

12. Remove the upper aqueous phase (approximately 100 μl) to a new tube.

13. Add 10 μl of 3 M sodium acetate and 220 μl of ice-cold 100% ethanol.

14. Chill for at least 15 min at $-80°$ (sample can be kept overnight at $-80°$).

15. Centrifuge for 10 min at 10,000g.

16. Remove the supernatant and wash the pellet with 70% (v/v) ethanol.

17. Air dry the pellet.

18. Resuspend the pellet in 10 μl of sterile water and use 5 μl of transform into electrocompetant *E. coli* with a high efficiency of transformation.

L. Replica Plating

Colonies on plates can be copied multiple times by transferring them first onto a sterile velvet surface and then pressing a clean plate onto the velvet. It is important to orientate the plates relative to a fixed position on the blotter so that individual colonies within a field can be located on the different plates. In addition, it is important not to transfer too many cells onto the plates. To avoid this, cells can be lifted off again, using a clean velvet that is then discarded. Replica plating can be used to confirm which selectable markers are present in the strain, to transfer cells to promoter-inducing or noninducing conditions, and for mating type testing. After use, velvets can be washed (without detergent) and baked in foil for sterilization.

M. Mating, Dissection, and Mating Type Testing

Haploid *S. cerevisiae* and *S. pombe* can exist as two different mating types. For *S. cerevisiae* they are referred to as **a** and *α*, and for *S. pombe* they are referred to as h$^+$ and h$^-$. When opposite mating types are cocultured, pheromones secreted from the opposite mating type stimulate cells to exit the mitotic cell cycle, arrest their division, and fuse to create diploids. Diploids cannot mate but divide mitotically and, under certain growth conditions, can undergo meiosis and sporulation to generate two pairs of the haploid mating types. There are differences in the handling of *S. cerevisiae* and *S. pombe* in the context of mating and it is simpler to consider the approaches separately.

1. Saccharomyces cerevisiae Mating

1. Streak fresh patches of the strains to be crossed onto yeast extract–peptone–dextrose (YEPD) plates and grow overnight at 30°.
2. When the patches have grown, pick a pellet of cells onto the end of a toothpick. Deposit the cells onto a new YEPD plate that has been annotated with the crosses being performed.
3. Pick a pellet of cells of the opposite mating type onto the end of a fresh toothpick. Mix the two cell types with 2–5 μl of sterile water on the surface of the YEPD plate. This should create a concentrated mix of the two cell types on the surface of the plate.
4. If it is possible to isolate diploids via the complementation of nutritional markers, incubate overnight at 30° and replica plate onto the appropriate synthetic minimal plate to isolate diploids.
5. If it is not possible to isolate diploids via the complementation of nutritional markers, incubate the mating mixture of 4–6 hr at 30°. Spread a small amount of the mating mixture onto a YEPD plate and search by microscopy for dumb-bell-shaped zygotes. On locating a zygote, pick it up with the dissection needle and deposit it on a noted position on the plate. This will grow to give a diploid colony.

2. Saccharomyces cerevisiae Mating Type Testing

1. Spread patches of the strains to be tested on the surface of a YEPD plate. Innoculate two YEPD liquid cultures of mating tester strains for

which the mating type is known. Incubate the patches and liquid cultures overnight at 30°.

2. Spread an appropriate number of **a** test and α test plates with 10^5 cells/plate of the appropriate mating type.

3. Once the spread cells have soaked in thoroughly, replica plate the master plate onto the testers, remembering to change the velvet between replica plating onto the **a** and α test plates.

4. Incubate at 30° overnight.

5. Mating type is revealed by a halo of no growth on the tester lawn surrounding the patch of the opposite mating type. This occurs because the pheromone secreted by the patch causes cells from the opposite mating type to exit the cell division cycle in preparation for mating. Therefore, patches with halos around them are the opposite mating type to that of the tester lawn.

3. Dissecting Saccharomyces cerevisiae Diploids

A detailed description of the equipment and techniques necessary to dissect haploid spores can be found in *Methods in Enzymology,* Vol. 194.[22]

1. To make diploid *S. cerevisiae* sporulate, freshly growing cells are streaked out onto sporulation plates.

2. After incubation at 30° for 2 days it should be possible to detect tetrads under the microscope. The time taken for sporulation is strain specific. In the most frequently used laboratory strains, sporulation will be complete within 4 days, although in slow-sporulating strains it can take up to 2 weeks.

3. Before dissection, the wall or ascus, which contains the spores, must be digested. A pellet of sporulated cells is picked into 50 μl of zymolyase (4 mg/ml). Digestion at room temperature can be followed by microscopy and digestion is complete when asci are visible as discrete spheres arranged in diamond shapes.

4. Schizosaccharomyces pombe Mating, Dissection, and Mating Type Testing

Schizosaccharomyces pombe mating efficiency is poor in comparison with that of *S. cerevisiae* and, under the conditions in which mating is possible, *S. pombe* diploids undergo meiosis extremely rapidly, resulting in haploids. As a consequence, specific *S. pombe* genetic backgrounds have been generated in which diploids can be selected from haploids and procedures have been developed to ensure that minimal haploid contamination occurs.

---

[22] F. Sherman and J. Hicks, *Methods Enzymol.* **194,** 21 (1991).

a. Schizosaccharomyces pombe Mating. To increase the efficiency of diploid isolation from a cross it is recommended that the haploids used carry the *ade6*-M210 and *ade6*-M216 mutations. These genes complement intragenically and therefore only the diploid will be able to grow on selective plates lacking adenine.

1. Streak fresh patches of the strains to be crossed onto synthetic plates that possess the appropriate nutrients. Grow overnight at 30°. This period of growth under limiting nutrient conditions will facilitate mating.

2. When the patches have grown, pick a pellet of cells onto the end of a toothpick. Deposit the cells onto a malt extract (ME; see Section V) plate that has been annotated with the crosses being performed.

3. Pick a pellet of cells of the opposite mating type onto the end of a fresh toothpick. Mix the two cell types with 2–5 μl of sterile water on the surface of the ME plate. This should create a concentrated mix of the two cell types on the surface of the plate.

4. Incubate for 18 hr at 25°. It is important not to allow mating to proceed for too long, or diploids will undergo meiosis.

5. Scrape half the mating patch into 500 μl of sterile water and resuspend the cells by pipetting.

6. Plate various concentrations of cells (200 and 20 μl) onto synthetic plates lacking adenine and incubate at 30°.

7. After 3 days, colonies should become apparent. It is recommended that a negative control be performed, in which cells are not mixed with the opposite mating type. Comparison of experimental and control plates should make it easier to judge when colonies are beginning to grow.

8. As soon as it is possible to pick colonies, they should be removed from the selective plate to minimize entering meiosis. Colonies should be picked into sterile water and plated to give single colonies onto plates containing YES medium (Section V) supplemented with Phloxine B (5 mg/liter).

9. Diploid colonies will appear dark red on Phloxine B-containing plates (while haploids will be pale pink) afer 2–3 days. On YES plates, diploids will not undergo meiosis so readily. However, it is bad practice to culture diploid *S. pombe* for too long, as it will either undergo meiosis and sporulation or sporulation-deficient cells will accumulate. It is recommended that a glycerol suspension of the diploid be made immediately on creating it, or proceed immediately to the dissection stage.

b. Schizosaccharomyces pombe Mating Type Testing. Schizosaccharomyces pombe cells are mating type tested by mating the unknown strain to h$^+$ and h$^-$ mating type testers and then testing for the presence of a high density of tetrads, using iodine crystals. Iodine stains the starch in the spores a dark color and therefore successful matings give rise to black patches on exposure to Iodine.

1. Perform steps 1–3 of the S. pombe mating procedure. Mate, however, to both mating types in order to have a negative control.
2. Incubate for 3 days at 25°.
3. In a fume hood, shake some iodine crystals into the lid of a petri dish and place the agar plate bearing the mated patches onto the dish. Leave for 5–10 min, until a color difference becomes apparent. Patches that become black when mated with one mating type should be a light brown color when mated against the other.

c. Dissecting Schizosaccharomyces pombe Diploids

1. Diploid cells from Phloxine B plates are patched onto ME plates and incubated at 25° for 3 days. During this time cells will undergo meiosis and sporulation.
2. Under the microscope a mixture of rod-shaped S. pombe diploids can be distinguished from rod shapes that contain within them four spores arranged in a row. If sporulated cells are easily seen without searching, it is possible to dissect.
3. If sporulation has occurred, pick a pellet of cells into 500 μl of distilled water and resuspend by pipetting. Spread a line of cells onto a YES plate and confirm under a dissection microscope that cells are of the appropriate density.
4. Incubate the dissection plate at 37° for 8 hr to digest the asci. This plate can now be stored at 4° for several days until dissection can be performed.

IV. Buffer Recipes

Northern lysis buffer

NaCl	500 mM
Tris-HCl (pH 7.5)	200 mM
EDTA (pH 7.5)	10 mM
SDS	1% (w/v)

PEG solution

PEG 4000	40% (w/v)
TELiAc	1×

SDS/sample buffer (2×)

Tris-HCl	100 mM
SDS	4% (w/v)
Bromphenol blue	0.2% (w/v)
Glycerol	20% (w/v)
2-Mercaptoethanol	280 mM

2× SDS/sample buffer lacking 2-mercaptoethanol can be stored at room temperature as 1-ml aliquots. 2-Mercaptoethanol should be added just before the buffer is used from a 14 M stock (20 μl in 1 ml of buffer)

Sorbitol buffer, pH 7.5

Sorbitol	1 M
NaPO$_4$	50 mM

TE (10×)

Tris-HCl (pH 7.5)	100 mM
EDTA (pH 7.5)	10 mM

TELiAc (10×)
Lithium acetate in 10× TE   1 M

V. Culture Media

A. Saccharomyces cerevisiae Media

YEPD

Yeast extract	1% (w/v)
Peptone	2% (w/v)
Glucose	2% (w/v)
Agar	2% (w/v)

Synthetic minimal medium for S. cerevisiae (SD)

Yeast nitrogen base (no amino acids)	0.67% (w/v)
Glucose	2% (w/v)
Agar	2% (w/v)

Supplement with the appropriate essential nutrients depending on the selectable markers mutated in the particular strain

Sporulation plates for S. cerevisiae

Potassium acetate	1% (w/v)
Agar	2% (w/v)

5-FOA plates (*S. cerevisiae* and *S. pombe*): 5-FOA plates are essentially SD plates minus the appropriate amino acids but supplemented with uracil. 5-FOA is heat labile and insoluble unless heated to approximately 55°. Therefore, the following procedure is recommended for making 500 ml of medium for FOA plates:

1. Dissolve the ingredients for 500 ml of SD (plus uracil) in 300 ml of water.
2. Sterilize by autoclaving.
3. Dissolve 0.5 g of 5-FOA in 200 ml of water and dissolve by heating to approximately 55°.
4. Filter sterilize the dissolved FOA solution and, when the autoclaved medium has cooled to approximately 55°, combine the two solutions and pour the plates.

B. *Schizosaccharomyces pombe* Media

YES media

Yeast extract	0.5% (w/v)
Glucose	3% (w/v)
Agar	2% (w/v)

Supplement with adenine, histidine, leucine, uracil, and lysine hydrochloride (200 mg/liter)

Synthetic minimal medium for *S. pombe* (MM)

Potassium hydrogen phthalate	0.3% (w/v)
Na_2HPO_4	0.22% (w/v)
NH_4Cl	0.25% (w/v)
Glucose	2% (w/v)
Salts stock (50×)	2 ml
Minerals stock (10,000×)	0.01 ml
Vitamins stock (1000×)	0.1 ml
Agar	2% (w/v)

Supplement with the appropriate essential nutrients depending on the selectable markers mutated in the particular strain

Salts stock (50×) (for 1 liter)

$MgCl_2 \cdot 6H_2O$	53.3 g
$CaCl_2 \cdot 2H_2O$	0.735 g
KCl	50 g
Na_2SO_4	2 g

Minerals stock (10,000×) (for 100 ml)

H_3BO_3	0.5 g
$MnSO_4$	0.4 g
$ZnSO_4 \cdot 7H_2O$	0.4 g
$FeCl_3 \cdot 6H_2O$	0.2 g
$MoO_4 \cdot 2H_2O$	0.16 g
KI	0.1 g
$CuSO_4 \cdot 5H_2O$	0.04 g
Citric acid	1 g

Filter sterilize and store in aliquots at 4°

Vitamins stock (1000×) (for 100 ml)

Nicotinic acid	1 g
myo-Inositol	1 g
Biotin	1 mg
Pantothenic acid (Na salt)	100 mg

Filter sterilize and store as frozen aliquots

ME plates for *S. pombe* sporulation

Malt extract	3% (w/v)
Glucose	3% (w/v)
Agar	2.5% (w/v)

Supplement as for YES (except for lysine) and adjust to pH 5.5 with sodium hydroxide.

Section VII

Studying Receptors and Signal Transduction Events Implicated in Cell Survival and Cell Death

[29] Production of Recombinant TRAIL and TRAIL Receptor: Fc Chimeric Proteins

By Pascal Schneider

The tumor necrosis factor (TNF)/TNF receptor (TNFR) families of ligands and receptors are implicated in a variety of physiological and pathological processes and regulate cellular functions as diverse as proliferation, differentiation, and death. Recombinant forms of these ligands and receptors can act to agonize or antagonize these functions and are therefore useful for laboratory studies and may have clinical applications.

A protocol is presented for the expression and purification of dimeric soluble receptors fused to the Fc portion of human IgG_1 and of soluble, N-terminally Flag-tagged ligands. Soluble recombinant proteins are easier to handle than membrane-bound proteins and the use of tags greatly facilitates their detection and purification. In addition, some tags may provide enhanced biological activity to the recombinant proteins (mainly by oligomerization and stabilization effects) and facilitate their functional characterization. Expression in bacterial (for selected ligands) and eukaryotic expression systems (for ligands and receptors) was performed using M15 pREP4 bacteria and human embryonic kidney 293 cells, respectively. The yield of purified protein is about 1 mg/liter for the mammalian expression system and several milligrams per liter for the bacterial expression system. Protocols are given for a specific ligand–receptor pair, namely TRAIL (Apo-2L) and TRAIL receptor 2 (DR5), but can be applied to other ligands and receptors of the TNF family.

Introduction

TRAIL (Apo-2L) and its receptors are members of the rapidly growing TNF/TNFR families of ligands and receptors, which are involved in functions as diverse as inflammation, cell proliferation, differentiation, and death.[1] TRAIL (Apo-2L)[2,3] induces apoptosis in a variety of transformed

---

[1] C. A. Smith, T. Farrah, and R. G. Goodwin, *Cell* **76**, 959 (1994).

[2] S. R. Wiley, K. Schooley, P. J. Smolak, W. S. Din, C. P. Huang, J. K. Nicholl, G. R. Sutherland, T. D. Smith, C. Rauch, C. A. Smith, and R. G. Goodwin, *Immunity* **3**, 673 (1995).

[3] R. M. Pitt, S. A. Marsters, S. Ruppert, C. J. Donahue, A. Moore, and A. Ashkenazi, *J. Biol. Chem.* **271**, 12687 (1996).

cell lines and binds to at least five receptors.[4] TRAIL-R1 (DR4)[5] and TRAIL-R2 (DR5)[6-10] both contain an intracellular death domain and are competent to signal apoptosis. TRAIL-R3 (DcR1), a glycosylphosphatidylinositol (GPI)-anchored receptor,[6-9,11] TRAIL-R4, which contains only a partial and inactive death domain,[12] and osteoprotegerin, a dimeric soluble receptor that binds both TRAIL and RANKL (TRANCE/ODF),[13-15] are believed to act as decoy receptors for TRAIL and to modulate the activity of TRAIL-R1 and TRAIL-R2.

Several methods for the expression of recombinant proteins have been described. The strategy described here (Fig. 1) is suitable for the production of bioactive recombinant soluble ligands and receptors of the TNF/TNFR families and is illustrated with TRAIL and TRAIL-R2 as specific examples. Using this strategy, the extracellular TNF homology domain of ligands is expressed in fusion with an N-terminal Flag tag in either bacterial or mammalian expression systems, while the extracellular domain of receptors, fused to the Fc portion of a human immunoglobulin G_1,[16] is expressed in

[4] P. Golstein, *Curr. Biol.* **7,** R750 (1997).

[5] G. Pan, K. O'Rourke, A. M. Chinnayan, R. Gentz, R. Ebner, J. Ni, and V. M. Dixit, *Science* **276,** 111 (1997).

[6] G. Pan, J. Ni, Y.-F. Wei, G.-L. Yu, R. Gentz, and V. M. Dixit, *Science* **277,** 815 (1997).

[7] J. P. Sheridan, S. A. Marsters, R. M. Pitti, A. Gurney, M. Skubatch, D. Baldwin, L. Ramakrishnan, C. L. Gray, K. Baker, W. I. Wood, A. D. Goddard, P. Godowski, and A. Ashkenazi, *Science* **277,** 818 (1997).

[8] P. Schneider, J. L. Bodmer, M. Thome, K. Hofmann, N. Holler, and J. Tschopp, *FEBS Lett.* **416,** 329 (1997).

[9] M. MacFarlane, M. Ahmad, S. M. Srinivasula, T. Fernandes-Alnemri, G. M. Cohen, and E. S. Alnemri, *J. Biol. Chem.* **272,** 25417 (1997).

[10] H. Walczak, M. A. Degli-Esposti, R. S. Johnson, P. J. Smolak, J. Y. Waugh, N. Boiani, M. S. Timour, M. J. Gerhart, K. A. Schooley, C. A. Smith, R. G. Goodwin, and C. T. Rauch, *EMBO J.* **16,** 5386 (1997).

[11] M. A. Degli-Esposti, P. J. Smolak, H. Walczak, J. Waugh, C. P. Huang, R. F. DuBose, R. G. Goodwin, and C. A. Smith, *J. Exp. Med.* **186,** 1165 (1997).

[12] M. A. Degli-Esposti, W. C. Dougall, P. J. Smolak, J. Y. Waugh, C. A. Smith, and R. G. Goodwin, *Immunity* **7,** 813 (1997).

[13] W. S. Simonet, D. L. Lacey, C. R. Dunstan, M. Kelley, M. S. Chang, R. Luthy, H. Q. Nguyen, S. Wooden, L. Bennett, T. Boone, G. Shimamoto, M. Derose, R. Elliott, A. Colombero, H. L. Tan, G. Trail, J. Sullivan, E. Davy, N. Bucay, L. Renshawgegg, T. M. Hughes, D. Hill, W. Pattison, P. Campbell, S. Sander, G. Van, J. Tarpley, P. Derby, R. Lee, Amgen EST Program, and W. J. Boyle, *Cell* **89,** 309 (1997).

[14] H., Yasuda, N. Shima, N. Nakagawa, K. Yamaguchi, M. Kinosaki, S. Mochizuki, A. Tomoyasu, K. Yano, M. Goto, A. Murakami, E. Tsuda, T. Morinaga, K. Higashio, N. Udagawa, N. Takahashi, and T. Suda, *Proc. Natl. Acad. Sci. U.S.A.* **95,** 3597 (1998).

[15] J. G. Emery, P. McDonnell, M. Brigham-Burke, K. C. Deen, S. Lyn, C. Silverman, E. Dul, E. R. Appelbaum, C. Eichman. R. DiPrinzio, R. A. Dodds, I. E. James, M. Rosenberg, J. C. Lee, and P. R. Young, *J. Biol. Chem.* **273,** 14363 (1998).

[16] K. Peppel, D. Crawford, and B. Beutler, *J. Exp. Med.* **174,** 1483 (1991).

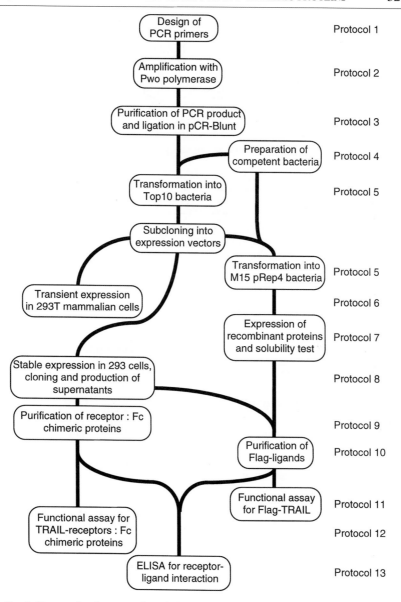

FIG. 1. Strategy for the expression of Flag–ligands and receptor: Fc chimeric proteins. This strategy is applicable to various ligands and receptors of the TNF/TNFR families. Expression in bacteria is suitable only for a limited number of ligands. Functional assays depend on the nature of the recombinant proteins and are described only for TRAIL and TRAIL-R2. Protocols for the various steps are indicated on the right.

A

pCMV

T7   HindIII   EcoRI   BamHI   SalI

```
-TA ATA CGA CTC ACT ATA GGG AGA CCC AAG CTT GAA TTC GGA TCC GTC GAC    51
  X   I   R   L   T   I   G   R   P   K   L   E   F   G   S   V   D
```

hinge, CH2 + CH3 domains of human IgG1

```
AAA ACT CAC ACA TGC CCA CCG TGC CCA GCA CCT GAA CTC CTG GGG GGA CCG   102
  K   T   H   T   C   P   P   C   P   A   P   E   L   L   G   G   P
```

```
TCA GTC TTC ···   GenBank accession n° X70421   ··· TTC TCA TGC        663
  S   V   F  ···                                   ···  F   S   C
```

```
TCC GTG ATG CAT GAG GCT CTG CAC AAC CAC TAC ACG CAG AAG AGC CTC TCC   714
  S   V   M   H   E   A   L   H   N   H   Y   T   Q   K   S   L   S
```

     NotI   XhoI   Stops   XbaI

```
CTG TCT CCG GGT AAA TGA GTG CGC GCG GCC GCT CGA GTA GAT GAC TAG TCT   765
  L   S   P   G   K   *
```

Apal   Sp6

```
AGA GGG CCC TAT TCT ATA GTG TCA CCT AAA T
```

B

pCMV

T7   HindIII   HA signal peptide

```
-TA ATA CGA CTC ACT ATA GGG AGA CCC AAG CTT AAT CAA AAC ATG GCT ATC    51
                                                             M   A   I
```

HA signal peptide   Flag

```
ATC TAC CTC ATC CTC CTG TTC ACC GCT GTG CGG GGC GAT TAC AAA GAC GAT   102
  I   Y   L   I   L   L   F   T   A   V   R   G   D   Y   K   D   D
```

Flag   PstI   SalI   XhoI   BamHI

```
GAC GAT AAA GGA CCC GGA CAG GTG CAG CTG CAG GTC GAC CTC GAG GGA TCC   153
  D   D   K   G   P   G   Q   V   Q   L   Q   V   D   L   E   G   S
```

BstXI

     EcoRI   SmaI   XbaI   ApaI

```
ACT AGT AAC GGC CGC CAG TGT GCT GGA ATT GAA TTC CCC GGG TCT AGA GGG   204
  T   S   N   G   R   Q   C   A   G   I   E   F   P   G   S   R   G
```

Sp6

```
CCC TAT TCT ATA GTG TCA CCT AAA T
  P   Y   S   I   V   S   P   K
```

FIG. 2.

C

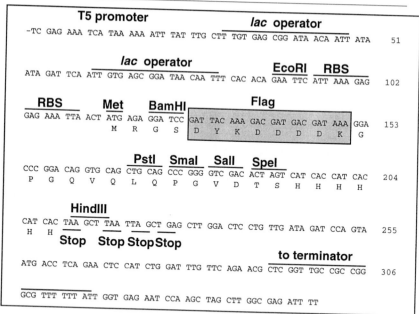

FIG. 2. Expression plasmids. (A) Mammalian expression vector for receptor : Fc. Plasmid is derived from the pCR-3 vector (InVitrogen). Only part of the human IgG₁ sequence is indicated. The extracellular portion of a receptor, including its signal peptide, must be cloned 5′ of the IgG₁ sequence, e.g., as a *Hind*III–*Sal*I fragment. (B) Mammalian expression vector for Flag–ligands. Plasmid is derived from the pCR-3 vector (InVitrogen). The predicted cleavage site of the signal peptide is indicated by an arrow. The extracellular portion of a ligand, including its stop codon, must be cloned 3′ of the Flag sequence, e.g., as a *Pst*I–*Eco*RI fragment. (C) Bacterial expression vector for Flag–ligands. Plasmid is derived from the pQE-16 vector (Qiagen). The extracellular portion of a ligand, including its stop codon, must be cloned 3′ of the Flag sequence, e.g., as a *Pst*I–*Spe*I fragment. This expression plasmid should be used in conjunction with pREP4 (Qiagen). pREP4 encodes the *lac* repressor for tight regulation of the expression. RBS, Ribosome-binding site.

mammalian cells. The resulting fusion proteins are referred to as Flag–ligands and receptor : Fc, respectively. These soluble molecules can be easily detected and purified via their tags. In addition, the biological activity of specific ligands [e.g., Fas ligand (FasL) and TRAIL] is greatly enhanced by cross-linking with anti-Flag antibodies.[17]

In this chapter, three expression vectors derived from commercially available plasmids are described (Fig. 2). The strategy for the amplification, cloning, expression, purification, and functional testing of recombinant ligands and receptors is depicted in Fig. 1. Figure 1 also refers to the 13 specific

[17] P. Schneider, N. Holler, J. L. Bodmer, M. Hahne, K. Frei, A. Fontana, and J. Tschopp, *J. Exp. Med.* **187**, 1205 (1998).

protocols described in this chapter. It is assumed that the experimenter has access to the cDNA of the ligand or receptor to be expressed and to a DNA sequencing service, and that he or she is familiar with standard molecular biology, sodium dodecyl sulfate–polyacrylamide gel electrophoresis (SDS–PAGE), and Western blotting techniques.

Expression Plasmids

Receptors of the TNFR family[18] are generally type-1 membrane proteins containing a secretion signal peptide at their N terminus. The mammalian expression vector for receptor : Fc chimeric proteins is based on the pCR-3 mammalian expression vector (InVitrogen, NV Leek, The Netherlands), modified to include the following features (Fig. 2A): (1) a multiple cloning site for insertion of the extracellular domain of the receptor of interest, including its natural signal peptide, and (2) a cassette encoding the hinge, CH2, and CH3 domains of human IgG_1, which allows easy detection and purification of the recombinant protein. The SalI–NotI cassette encoding the Fc portion of IgG_1 was kindly provided by J. Browning (Biogen, Cambridge, MA).

Ligands of the TNF family are type-2 membrane proteins. To obtain the carboxy-teminal extracellular domain of the ligand as a secreted, tagged protein, the multiple cloning site of the pCR-3 vector has been modified to provide the following elements (Fig. 2B): (1) the signal peptide of hemagglutinin (bases −9 to +45, base 1 being the A of ATG)[19] to target the recombinant protein to the secretory pathway, (2) the eight-amino acid Flag sequence[20] to facilitate detection and purification of the recombinant protein, (3) a linker sequence located between the Flag tag and the ligand, and (4) a multiple cloning site for insertion of the ligand cDNA.

For bacterial expression of ligands, a modified version of the pQE-16 vector (Qiagen, Basel, Switzerland) has been developed that presents the following features (Fig. 2C): (1) a Flag sequence following the initiating methionine, (2) a linker sequence, and (3) a multiple cloning site for insertion of the ligand cDNA. The cDNA of interest is under the control of the T5 promoter and its expression is regulated by two lac operator sequences. The expression of the recombinant protein will be tightly regulated only if the host bacteria contains the pREP4 plasmid (Qiagen), which allows

[18] J. H. Naismith and S. R. Sprang, Trends Biochem. Sci. 23, 74 (1998).
[19] M. J. Gething, J. Bye, J. Skehel, and M. Waterfield, Nature (London) 287, 301 (1980).
[20] T. P. Hopp, K. S. Prickett, V. L. Price, R. T. Liggy, C. J. March, D. P. Cerretti, D. L. Urdal, and P. J. Conlon, BioTechnology 6, 1204 (1988).

overexpression of the *lac* repressor.* Some ligands, including TRAIL, are successfully expressed in prokaryotic systems. Bacterial expression is, however, detrimental for highly N-glycosylated ligands and results in insoluble inclusion bodies. In this case, expression in mammalian cells is recommended.

Protocol 1: Design of Polymerase Chain Reaction Primers

Cloning of ligands and receptors of the TNF/TNFR families in the expression vectors described above requires the presence of suitable restriction sites flanking the cDNA of interest. These sites can be introduced by, polymerase chain reaction (PCR) amplification, using suitable primers. Primers used for amplification of TRAIL and TRAIL-R2 are shown in Fig. 3.

1. Choose the portion of the cDNA to be amplified. In principle, the sequence of ligands should start anywhere between the transmembrane domain and the first β strand of the TNF homology domain (for alignments of ligands, see Refs. 21 and 22). If the option of cloning the entire extracellular domain is chosen, one should keep in mind that processing may occur in the mammalian expression system between the Flag sequence and the TNF homology domain, preventing the detection and purification of the recombinant ligand.

For receptors, the entire extracellular domain, starting at the initiation ATG, is generally suitable.

2. Check that the restriction sites that will be used for subcloning into the expression vector are absent from the cDNA sequence to be amplified. If the site is present, either choose another or use a restriction site yielding compatible ends. As a reminder, *Pst*I is compatible with *Nsi*I, *Sal*I is compatible with *Xho*I, *Spe*I is compatible with *Xba*I and *Nhe*I, *Bam*HI is compatible with *Bgl*II, and *Sma*I is compatible with any blunt-cutting enzyme such as *Stu*I or *Eco*RV.

3. The 5' forward primer for ligands should contain the following features (5' to 3'): 3 "protecting" nucleotides** whose sequence should not

---

* Expression can be obtained in the absence of the pREP4 plasmid. However, positive bacteria are rapidly counterselected because of transcriptional leak.

[21] M. C. Peitsch and J. Tschopp, *Mol. Immunol.* **32,** 761 (1995).

[22] K. B. Tan, J. Harrop, M. Reddy, P. Young, J. Terrett, J. Emery, G. Moore, and A. Truneh, *Gene* **204,** 35 (1997).

** A few nucleotides at the 5' end of the primer that do not match the cDNA sequence are sometimes removed during the PCR. The addition of "protecting" nucleotides ensures that the restriction site will not be destroyed.

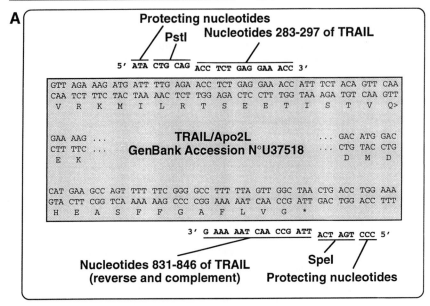

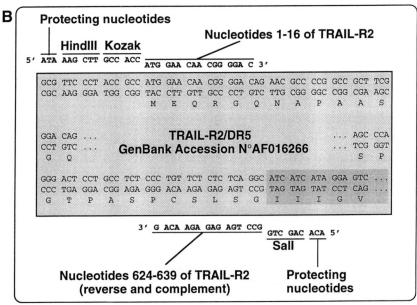

be complementary to the last 3 nucleotides of the primer, the suitable restriction site, and 14–17 nucleotides matching the sequence of the desired N terminus of the ligand. Ideally, the 3′ end of primers will be a G or a C.

The 5′ forward primer for receptors should contain the following features (5′ to 3′): 3 "protecting" nucleotides, the suitable restriction site, a Kozak consensus sequence (GCCACC),[23] and 14–17 nucleotides matching the N-terminal sequence of the receptor, including the ATG.

4. The 3′ reverse primer should contain (5′ to 3′): 3 "protecting" nucleotides, the suitable restriction site, and 14–17 nucleotides complementary to the reverse 3′ sequence of the ligand (including the stop codon) or to the 3′ portion of the extracellular domain of the receptor (do not add a stop codon).

Protocol 2: Amplification with Pwo Polymerase

Pwo polymerase (Boehringer Mannheim GmbH, Mannheim, Germany) is a high-fidelity DNA polymerase allowing minimal introduction of unwanted mutations in the amplified sequence.

1. Prepare the following PCR:

dNTP mixture (Pharmacia, Uppsala, Sweden) (2 mM each)	5 μl
5′, Forward primer 10 μM	5 μl
3′ Reverse primer 10 μM	5 μl
Pwo polymerase buffer (10×) plus MgSO₄ (provided with Pwo polymerase)	5 μl
Template (e.g., 50 ng of plasmid DNA)	1 μl
Water	28 μl
Pwo polymerase (1 U/μl)	1 μl (add last)
Mineral oil overlay if no heating cover is available	

[23] M. Kozak, *Nucleic Acids Res.* **12**, 857 (1984).

Fig. 3. Primer design for amplification of TRAIL and TRAIL-R2 extracellular domains. (A) A partial sequence of TRAIL is shown in gray. Primers used for the amplification are shown. PstI and SpeI sites present on the 5′ extensions are used for subcloning into PstI–SpeI sites of the bacterial expression vector (Fig. 2C). (B) A partial sequence of TRAIL-R2 is shown in gray. The beginning of the transmembrane domain is highlighted in dark gray (sequence IIIGV . . .). Primers used for the amplification are shown. HindIII and SalI sites present on the 5′ extensions are used for subcloning into HindIII–SalI sites of the mammalian expression vector, in fusion with the Fc domain of human IgG₁ (Fig. 2A). Note that a Kozak consensus sequence of translation initiation is introduced 5′ of the ATG.

2. Perform PCR program: 3 min at 94°; 28 cycles of 94°/55°/72°, 1 min each; and a final 10-min extension step at 72°.

Protocol 3: Purification of Polymerase Chain Reaction Product and Ligation in pCR-Blunt

Pwo polymerase generates blunt-ended PCR products that can be conveniently cloned into the pCR-Blunt vector and transformed into competent *Escherichia coli* Top 10, using a ZeroBlunt PCR cloning kit (InVitrogen). Background is low with this system because vectors religated without insert are toxic to the bacteria.

1. Analyze the totality of the PCR on an agarose gel in the presence of ethidium bromide according to standard protocols. Visualize the PCR product under long-wavelength UV (365 nm) and excise the band of interest with a razor blade.*

2. Extract the PCR product from the gel slice, using 12 μl of Qiaex II beads (Qiagen, protocol included), and elute the purified PCR product in 20 μl of water.

3. Set up the ligation reaction with

Ligation buffer 10×**	1 μl
pCR-Blunt vector (25 ng/μl)**	2 μl
Purified PCR product (replace with 6 μl of water for a control ligation)	6 μl
T4 DNA ligase (1 U/μl)**	1 μl

Incubate for 1 hr or more at 16°.

Protocol 4: Preparation of Frozen CaCl$_2$ Competent Bacteria

1. Prepare 5 ml of an overnight culture of the Top10 strain of *E. coli* (included in the ZeroBlunt PCR cloning kit) in L-broth medium. For M15

---

* If no PCR product is obtained, check the sequence of the primers and their concentration, check the identity of the template, and introduce a positive control to ensure that other reagents are suitable. Try to decrease the annealing temperature (e.g., 52°) and the concentration of MgSO$_4$ (see instructions provided with the *Pwo* polymerase). If unsuccessful, try to use *Taq* polymerase (e.g., *Taq* from Life Technologies) instead of *Pwo*, reducing the number of cycles to 20. The use of *Taq* polymerase will, however, result in an increased probability of inserting unwanted mutations during the amplification.

** These reagents are included in the ZeroBlunt PCR cloning kit (InVitrogen). Alternatively, T4 DNA ligase (Life Technologies) and the buffer provided with the enzyme (5× concentrated; use 2 μl) are also suitable. Empty, religated pCR-Blunt vector can be amplified in bacterial strains expressing the *lac* repressor (e.g., TG1) and can be prepared for ligation of PCR fragments by digestion with the restriction enzyme *Stu*I. Competent bacteria can be prepared according to protocol 4.

pREP4 bacteria (Qiagen), perform the same protocol but use L-broth medium supplemented with kanamycin (50 μg/ml).

2. Use an overnight culture to inoculate 500 ml of L-broth medium in a 2-liter flask and shake at 225–250 rpm at 37° until an OD_{600} of 0.3–0.4 is reached (takes about 2–3 hr) (use L-broth as a blank).

3. Cool the bacteria for 10 min under aeration (i.e., on a rotatory shaker in a cold room). Perform all subsequent steps at 4° or on ice.

4. Centrifuge bacteria at 3000g (4000 rpm in a 16-cm-radius rotor) for 10 min. Discard the supernatant.

5. Resuspend the pellets in a total volume of 150 ml of ice-cold, sterile 0.1 M $CaCl_2$ and leave on ice for 20 min.

6. Centrifuge the bacteria at 3000g for 10 min and resuspend the pellets in 2 ml of sterile 0.1 M $CaCl_2$ and 1 ml of sterile 50% (v/v) glycerol.

7. Freeze as 50-μl aliquots in 1.5-ml polypropylene tubes and store at −70°.*

Protocol 5: Transformation into Competent Top10 or M15 pREP4 Bacteria

1. Thaw aliquots of competent bacteria on ice.**

2. Add 2 μl of 0.5 M 2-mercaptoethanol to the competent bacteria and stir gently with a pipette tip. Do not pipette up and down.

3. Add 1–2 μl of the ligation reaction and stir gently with pipette tip. Leave on ice for 30 min.

4. Heat for 40 sec in a 42° water bath. Do not mix.

5. Place the bacteria on ice for 2 min. Add 450 μl of SOC or equivalent medium.†

6. Shake the tubes (225–250 rpm) at 37° for 1 hr.

7. Plate 100–200 μl of each culture on L-broth agar plates containing the appropriate antibiotic [kanamycin (50 μg/ml) for pCR-Blunt vectors, ampicillin (100 μg/ml) for pQE and pCR-3-derived vectors] and incubate for 10 to 18 hr at 37°.

8. Analyze the colonies. For protein expression screening in M15 pREP4, see protocol 7. Minipreparation of plasmid DNA, restriction digests, plasmid amplification, plasmid sequencing, and subcloning of the

---

* To this purpose, precool the bottoms of the tubes by immersing them in a mixture of ethanol–dry ice and add competent bacteria directly to the bottoms of the tubes. Competent bacteria can be stored for several months at −70°.

** If competent cells are being used for the first time, also perform a transformation without plasmid. No or few colonies should be obtained in this control.

† SOC medium contains 2% (w/v) tryptone, 0.5% (w/v) yeast extract, 0.05% (w/v) NaCl, 2.5 mM KCl, 10 mM $MgCl_2$, and 20 mM glucose.

insert into the expression plasmid are performed according to standard technologies. Plasmids used for sequencing and transfection purposes are prepared with Qiagen-Tip 100 or equivalent product.

Protocol 6: Transient Expression in 293T Mammalian Cells

Human embryonic kidney 293 cells (ATCC CRL 1573) stably expressing the large T antigen of simian virus 40 (SV40) (293T cells) (a gift of M. Peter, German Cancer Research Center, Heidelberg, Germany) are diluted twice weekly in the medium described below. Transient expression is useful to check the expression of a construct and to rapidly obtain small quantities of a recombinant protein. 293 cells without large T antigen may also be used. In this case, see protocol 7 for the transfection procedure.

1. The day before transfection, dilute 1 ml of a thoroughly resuspended culture of confluent 293T cells (corresponding to roughly 10^6 cells) in 7 ml of Dulbecco's modified Eagle's medium (DMEM; Life Technologies, Gaithersburg, MD) supplemented with 10% (v/v) heat-inactivated fetal calf serum (FCS) and antibiotic mixture (5 μg/ml each of penicillin and streptomycin, neomycin at 10 μg/ml; Life Technologies). Seed in a 9-cm cell culture plate (e.g., Nunc, Roskilde, Denmark).

2. Sterilize 10 μg of expression plasmid in a 1.5-ml tube (optional):

Plasmid at 1 μg/ml in H$_2$O	10 μl
H$_2$O	40 μl
Potassium acetate, pH 5.5 (3 M)	25 μl
Ethanol	190 μl

3. Mix and leave for 10 min at $-70°$.

4. Centrifuge for 3 min at 4° at 13,000g (13,000 rpm in an 8-cm-radius tabletop centrifuge). A small DNA pellet should be visible.

5. Discard the supernatant. Add 1 ml of 70% (v/v) ethanol at $-20°$. Invert the tube. Spin briefly. Carefully discard all of the supernatant while working under a sterile hood. Let the pellet dry for about 1 hr, but do not overdry.

6. Redissolve the pellet in 500 μl of sterile 250 mM CaCl$_2$.

7. Add 500 μl of 2× HeBS solution (16.4 g of NaCl, 11.9 g of HEPES acid, 0.21 g of anhydrous Na$_2$HPO$_4$, 800 ml of H$_2$O. Adjust to pH 7.05 with NaOH. Add H$_2$O to 1 liter. Sterilize by filtration)* dropwise while gently vortexing the tube.

---

* The pH of the 2× HeBS solution is critical. Whenever possible, adjust the pH of the new solution with that of a previous batch giving satisfactory results. Otherwise, prepare several solutions differing by 0.02 pH unit and test the transfection efficiency. The plasmid pEGFP-N2 encoding green fluorescent protein (Clontech, Palo Alto, CA) can be useful for this purpose if a fluorescence microscope is available.

8. Add the solution to the cells within 1 min and mix gently by swirling the plate.

9. Leave overnight. A fine precipitate should form that is, most of the time, visible under the microscope (this precipitate may have the appearance of a bacterial contamination but should not be mistaken as such).

10. Aspirate the medium, rinse once with 8 ml of phosphate-buffered saline (PBS), taking care not to detach the cells, and add 8 ml of Optimem medium lacking FCS (Life Technologies).

11. Leave the cells for 3 to 5 days. Collect the supernatant and concentrate 20-fold, down to 400 μl, using a Centricon-30 (Amicon, Easton, TX). Use 20 μl for Western blotting analysis (see protocol 8, step 11).

Protocol 7: Expression of Recombinant Proteins in M15 pREP4 Bacteria and Solubility Test

After transformation of the ligation into M15 pREP4 bacteria (protocol 5), positive colonies are identified by expression screening.

1. Inoculate 10 colonies, each into 1 ml of L-broth containing ampicillin (100 μg/ml) and kanamycin (50 μg/ml) (LB Amp/Kan). Grow overnight at 37°.

2. Inoculate 1 ml of LB Amp/Kan with 50 μl of overnight culture. At least one colony should be inoculated twice (uninduced control). Grow for 1 hr at 37° at 225–250 rpm.

3. Add 5 μl of a 100 mM aqueous solution of isopropyl-β-D-thiogalactoside (IPTG; Boehringer Mannheim) to all tubes except the uninduced control and let grow for an additional 4 hr (or overnight).

4. Harvest the bacteria. Resuspend the pellet in 300 μl of protein sample buffer containing 40 mM dithiothreitol.

5. Sonicate the samples with a sonicating probe (Sonifer 250, setting 2; Branson Ultrasonics, Danbury, CT) in order to fragment the DNA. Samples should be nonviscous and easy to pipette after this treatment. If not, sonicate again.

6. Analyze samples by SDS–PAGE on a 12% (w/v) gel, using standard techniques. Load 20 μl of sample for a 1.5-mm-wide gel with 3-mm slots. Perform the electrophoresis, stain the gel with Coomassie blue, and destain.

If necessary, the solubility properties of the recombinant protein can be checked (steps 7–12).

7. Induce a 1-ml culture of a positive clone with IPTG (see steps 2 and 3).

8. Harvest the bacteria in a 1.5-ml polypropylene tube. Resuspend the pellet in 300 μl of PBS.

9. Sonicate on ice, taking care not to heat the sample.

10. Spin the sample at 13,000g in a tabletop centrifuge. Save the supernatant (soluble protein fraction).

11. Wash the pellet with PBS and resuspend in 300 μl of PBS (insoluble protein fraction).

12. Analyze 20 μl of each sample by SDS–PAGE and Coomassie blue staining. If the protein is insoluble, try to induce bacteria at 18° or express it in the mammalian system.

Protocol 8: Stable Expression in 293 Cells, Cloning, and Production of Supernatants

Human embryonic kidney 293 cells (ATCC CRL 1573) are diluted twice weekly in DMEM-nutrient mix F12 (1:1) (Life Technologies) supplemented with 2% (v/v) FCS and antibiotics (293 medium).

1. Two days before transfection, resuspend confluent 293 cells and dilute 600 μl (corresponding to roughly 5 × 10⁵ cells) in 3.4 ml of 293 medium. Seed in a 25-cm² cell culture flask (e.g., Nunc) in 4 ml. Most cells should be attached before transfection.

2. Sterilize 10 μg of expression plasmid (see protocol 6, step 2) (optional). Redissolve the plasmid in 250 μl of sterile 250 mM CaCl₂.

3. As it is not possible to transfect in 293 medium, aspirate the medium, wash the cells once with PBS, and add 4 ml of DMEM supplemented with 10% (v/v) FCS and antibiotics.*

4. Add 250 μl of 2× HeBS to the plasmid solution while vortexing gently. Wait for 1 min and add solution to the cells. Leave overnight. A fine precipitate should form.

5. Wash the cells once with PBS and add 4 ml of 293 medium.* Leave for 1 to 3 days.

6. Resuspend the cells and dilute 1:8 in 4 ml of selection medium. Selection medium is prepared as follows:

 a. Combine 50 ml of 293 medium with 400 μl of a 100-mg/ml aqueous solution of G418 (Life Technologies).**

---

* To avoid detaching cells, hold the flask upside down and pipette liquids onto the cover of the flask, not directly onto the cells.

** G418 is provided as a powder in which the percentage of active compound varies from lot to lot. The concentration given in this protocol applies to 60% active G418 (i.e., the final concentration of active G418 in the selection medium is 480 μg/ml). G418 is toxic: wear gloves and follow the safety recommendations provided with the product.

b. Adjust the pH with approximately 20 μl of 2 M NaOH.*

c. Sterilize by filtration (0.2-μm pore size).

7. When cells have grown dense again (normally after 4 to 7 days), dilute 1 : 8 in 4 ml of selection medium.

8. When cells are confluent again (normally after 4 to 7 days), resuspend the cells and seed 10 μl in 8 ml of selection medium, in a 9-cm-diameter cell culture plate. Leave until distinct clones are visible with the unaided eye (normally after 10 to 15 days).

9. Transfer a microscope to under a sterile hood. With a 200-μl pipette, pick 12 individual clones** and transfer them to the wells of a flat-bottom 96-well plate (e.g. Costar, Cambridge, MA) containing 200 μl of selection medium. Wrap the plate in aluminum foil containing a humidified paper towel to minimize evaporation.

10. Dilute the clones on a regular basis (usually after 7–10 days, thereafter every 3–4 days) by pipetting the cells up and down and transferring 20 μl into a fresh well containing 200 μl of selection medium. Wait until the medium in the initial wells has turned completely yellow (usually 14 days).

11. Screen the conditioned supernatants (20 μl for a 1.5-mm-wide polyacrylamide gel with 3-mm slots) for the presence of the recombinant protein by Western blotting. Do not reduce samples containing receptor : Fc fusion proteins. For detection of unreduced receptor : Fc fusion proteins, use a 0.5-μg/ml solution of protein A–peroxidase (Sigma, St. Louis, MO). For detection of Flag–ligands, use a 1-μg/ml solution of anti-Flag M2 murine IgG$_1$ monoclonal antibody (Sigma), followed by goat anti-mouse IgG–peroxidase (Jackson ImmunoResearch, Milan Analitika, La Roche, Switzerland) diluted 1 : 2000. Reveal with ECL reagent (Amersham, Little Chalfont, England).

12. Amplify selected clones by transferring them successively from the 96-well plate into 25-cm^2 (4 ml, usually 7 days), 75-cm^2 (25 ml, usually 3–4 days), and 175-cm^2 (60 ml, usually 3–4 days) cell culture flasks, and a 2-liter roller bottle (e.g., Falcon, Lincoln Park, NY) (1 liter of medium, leave for 14 days).†

---

* G418 is acidic and must be titrated, otherwise cells could be killed as a result of medium acidification. Add NaOH until the medium turns orange.

** Clones are loosely attached to the plastic. They are easily detached by pushing with the pipette tip and should then be aspirated in as little medium as possible to avoid contamination by other cells. If cloning by limiting dilution is preferred, seed 10 cells per well to obtain clones in about one-third of the wells.

† Roller bottle cultures are grown at 37° in normal atmosphere (no additional CO_2) with caps completely closed and at 1 rpm. Time course studies confirmed that secretion yield correlates with the age of the culture as long as cells are alive. Most cells are dead after 14 days and longer incubations are not recommended. Caution should be taken if a possible degradation of the recombinant protein is suspected.

13. Harvest cells from the roller bottle. Remove the cells by centrifugation ($3000g$, 4000 rpm in a 16-cm-radius rotor) and filter the supernatant using a 0.45-μm pore size filter (Nalgene, Rochester, IL). Add 0.02% (w/v) NaN$_3$. Store at 4° or −20° until purification (see protocols 9 and 10).

Protocol 9: Purification of Receptor: Fc Chimeric Proteins

A one-step purification of receptor: Fc chimeric proteins is performed with HiTrap protein A columns (Pharmacia).* The purification yield depends both on the nature of the receptor and on the clone of interest. It usually ranges from 0.5 to 5 mg/liter.

1. Equilibrate the column in PBS. If necessary, perform a preelution step (see step 4).
2. Using a peristaltic pump, load the HiTrap protein A column with 1 liter of 293 cell supernatant at a flow rate of up to 4 ml/min. If the column is loaded overnight, medium may be cycled.
3. Using a syringe, wash the column with 10 ml of PBS.
4. Elute the column with 4 ml of 0.1 M citrate-NaOH, pH 2.5. Immediately reequilibrate the column with PBS [or PBS–0.02% (w/v) NaN$_3$ for long-term storage] and neutralize eluate with 1 ml of 1 M Tris-HCl, pH 8.5. Check that the pH of the neutralized eluate is between 7 and 8 by spotting 1 μl onto pH paper. [Alternatively, steps 3 and 4 can be performed with a fast protein liquid chromatography (FPLC) system.]
5. Concentrate the eluate in a Centricon-30 to less than 0.5 ml. Wash twice with 2 ml of PBS to exchange buffer.
6. Determine the protein concentration, using bicinchoninic acid (BCA) reagent (Pierce, Rockford, IL) and bovine serum albumin as a standard.
7. Dilute with PBS to a convenient concentration, sterilize by filtration at 0.2 μm, using Millex-GV low protein binding filters (Millipore, Bedford, MA), aliquot, and freeze-dry. Store at −20°. Check purity by SDS–PAGE and Coomassie blue staining (Fig. 4A).

Protocol 10: Purification of Flag–Ligands

Steps 1–8 are specific for ligands expressed in bacteria. Phase separations in Triton X-114 are designed to remove bacterial lipopolysaccharides. For 293 cell supernatants, start at step 9. The yield for ligands expressed in bacteria is about 2–3 mg/liter of culture. The expected yield for ligands produced in mammalian cells is 0.5 to 1 mg/liter of supernatant.

---

* These columns are provided with all necessary adaptors for connection to syringe, tubing, and FPLC systems.

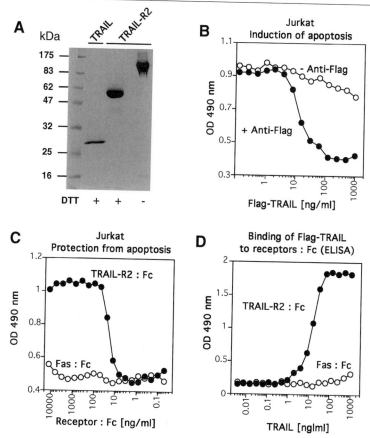

FIG. 4. Purity and functional tests for Flag–TRAIL and TRAIL-R2 : Fc. (A) Coomassie blue-stained polyacrylamide gel with 4 μg of Flag–TRAIL (reduced conditions) and 8 μg of TRAIL-R2 : Fc (reduced and unreduced conditions). The top of the gel is included in the figure. Prestained molecular mass markers are from New England BioLabs (Beverley, MA). (B) Functional test for Flag–TRAIL. Experiment was performed on Jurkat cells according to protocol 11, in the presence or absence of anti-Flag M2 antibody (2 μg/ml). (C) Functional test for TRAIL-R2 : Fc. Experiment was performed on Jurkat cells in the presence of Flag–TRAIL (100 ng/ml) and anti-Flag M2 antibody (2 μg/ml) according to protocol 12. Fas : Fc[25] was used as a negative control. (D) Binding ELISA between TRAIL-R2 : Fc and Flag–TRAIL. Fas : Fc was used as a negative control. Experiment was performed according to protocol 13.

1. Inoculate 15 ml of L-broth containing ampicillin (100 μg/ml) and kanamycin (50 μg/ml) (LB Amp/Kan) with M15 pREP4 bacteria expressing the ligand of interest. Grow overnight at 37°.

2. Dilute 10 ml of overnight culture with 500 ml of LB Amp/Kan in a 2-liter flask and grow at 37° until $OD_{600} \simeq 0.5$ (2–3 hr).

3. Cool for 1 hr at 18°. Add IPTG to a final concentration of 0.5 mM and grow overnight at 18°.

4. Harvest the bacteria by centrifugation (3000g, 10 min). Discard the medium. Freeze the pellets (corresponding to 100 ml of culture) at this stage if necessary.

5. Resuspend one pellet (corresponding to 100 ml of culture) in 40 ml of PBS and sonicate on ice (setting 1–3, intermittently for 2–3 min). Avoid heating. Alternatively, French press in a volume of 4 ml.

6. Centrifuge at 27,000g (15,000 rpm in a fixed-angle, 10.8-cm-radius rotor) for 10 min at 4°. Recover the supernatant.

7. Add 5 ml of precondensed 12% (v/v) Triton X-114[24] to the supernatant and mix gently.* Keep on ice until a homogeneous solution is obtained. Warm to 37° (this is best achieved in several 15-ml tubes rather than in a single 50-ml tube). Centrifuge for 5 min at 3000g at room temperature. Save the upper aqueous phase. Discard the Triton X-114 phase.

8. Repeat step 7.

9. Using a syringe or a peristaltic pump, load 10 ml of aqueous phase (or 500 ml of 293 cell supernatants) on a 1-ml column of M2-agarose (Sigma).** Save the flowthrough.

10. Using a syringe, wash the column with 10 ml of PBS.

11. Elute the column with 4 ml of 0.1 M citrate-NaOH, pH 2.5. Immediately reequilibrate the column with PBS [or PBS–0.02% (w/v) NaN$_3$ for long-term storage]. Neutralize the eluate with 1 ml of 1 M Tris-HCl, pH 8.5. Check that the pH is between 7 and 8 by spotting 1 μl onto pH paper. (Alternatively, steps 10 and 11 can be performed with an FPLC system.)

12. Repeat steps 9–11, until flowthrough is depleted from the recombinant protein (usually twice).

13. Concentration, buffer exchange, dosage, sterilization, and control of purity are performed as described in protocol 9, steps 5–7.

---

[24] C. Bordier, *J. Biol. Chem.* **256,** 1604 (1981).

* Precondensed Triton X-114 is obtained by subjecting a 2% (v/v) solution of commercial-grade Triton X-114 in PBS to three cycles of phase separation. One cycle consists of two steps, one at 4° and the other at 37°, the latter yielding a detergent-rich lower phase consisting of 12% (v/v) Triton X-114 and a detergent-depleted upper phase, which is discarded. The detergent-rich phase is diluted to 2% (v/v) with fresh PBS before the next cycle. For large volumes, no centrifugation is required as phase separation will occur within 16 hr at 37°. After the third cycle, the detergent phase can be stored for years at −20°.

** M2-agarose can be conveniently packed in a used HiTrap column (Pharmacia). To this purpose, remove the red top and cut 2-mm plastic at the top of the column, using a razor blade. Remove the black plug and empty the column, taking care to save the filters. Empty HiTrap columns are not available. Alternatively, use any kind of column.

Protocol 11: Functional Assay for Flag–TRAIL

This assay requires TRAIL-sensitive cell lines. Human T lymphoma Jurkat cells and BJAB Burkitt lymphoma cells are suitable for this purpose. These cells are maintained in RPMI 1640 medium (Life Technologies) supplemented with 10% (v/v) FCS and antibiotics. TRAIL is assayed in the presence of anti-Flag M2 antibodies, which greatly enhances the cytotoxic activity of Flag–TRAIL.[17] A typical result is shown in Fig. 4B.

1. Prepare 1.2 ml of a 4-μg/ml solution of anti-Flag M2 in cell culture medium, and distribute 20 × 50 μl in a 96-well cell culture plate.

2. Use the remaining medium to prepare 50 μl of a 4-μg/ml solution of Flag–TRAIL and add it to the first well. Mix by pipetting up and down.

3. Transfer 50 μl of medium from the first to the second well, mix, and continue dilutions up to the nineteenth well. Do not touch the last well (control without TRAIL). Change the pipette tip every third or fourth well.

4. Resuspend Jurkat cells at about 10^6/ml in culture medium (at least 1.2 ml; no anti-Flag) and distribute 50 μl into each well. This gives 50,000 cells per well exposed to a 2-μg/ml concentration of anti-Flag M2 (constant concentration) and 2-fold dilutions of TRAIL starting at a final concentration of 1000 ng/ml.

5. Fill empty wells with 100 μl liquid. Wrap in aluminum foil containing a wet paper towel and incubate for 16 hr at 37°.

6. Check under the microscope that cells exposed to TRAIL have undergone apoptosis.

7. Mix 25 μl of phenazine methosulfate (PMS; Sigma)* at 0.9 mg/ml in PBS with 500 μl of 3-(4,5-dimethylthiazol-2-yl)-5-(3-carboxymethoxyphenyl)-2-(4-sulfophenyl)-2H-tetrazolium, inner salt (MTS; Promega, Madison, WI) at 2 mg/ml in PBS. Add 20 μl per well and incubate at 37° for 1 to 4 hr, until sufficient color has developed.

8. Measure the absorbance at 490 nm, using an ELISA reader. Use the "shake low" function if available or take care to gently mix the plate before the measurement.

Protocol 12: Functional Assay for TRAIL Receptor: Fc
    Chimeric Proteins

This assay requires bioactive TRAIL and a TRAIL-sensitive cell line. A typical result is shown in Fig. 4C.

---

* PMS is light sensitive. Always keep protected from light.
[25] P. Schneider, J. L. Bodmer, N. Holler, C., Mattmann, P. Scuderi, A. Terskikh, M. C. Peitsch, and J. Tschopp, *J. Biol. Chem.* **272**, 18827 (1997).

1. Prepare 1.2 ml of cell culture medium containing a 4-μg/ml concentration of anti-Flag M2 and a 200-ng/ml concentration of Flag–TRAIL. Distribute 19 × 50 μl in a 96-well cell culture plate. Add medium without TRAIL to the last well.

2. Use the remaining medium to prepare 50 μl of a 50-μg/ml solution of TRAIL receptor: Fc and add it to the first well. Mix by pipetting.

3. Transfer 50 μl of medium from the first to the second well, mix, and continue dilutions up to the nineteenth well. Do not touch the last well (control without TRAIL). Change the pipette tip every third or fourth well.

4. Resuspend Jurkat cells at about 10^6/ml in culture medium without anti-Flag M2 (at least 1.2 ml) and distribute 50 μl into each well. This gives 50,000 cells per well exposed to anti-Flag M2 (2 μg/ml), Flag–TRAIL (100 ng/ml) (constant concentrations) and 2-fold dilutions of TRAIL receptor: Fc, starting at a final concentration of 12.5 μg/ml.

5–8. Perform these points according to protocol 10.

Protocol 13: Enzyme-Linked Immunosorbent Assay for
Receptor–Ligand Interaction

This protocol allows the detection of receptor–ligand binding in a totally defined system. A typical result is shown in Fig. 4D.

1. Coat an ELISA plate (96-well Maxisorp Immunoplate; Nunc) for 2 hr at 37° (or overnight at room temperature) with 100 μl of TRAIL receptor: Fc (1 μg/ml) in PBS. Prepare 12–20 wells.

2. Empty the plate. Add 300 μl of block buffer [5% (v/v) FCS in PBS] and incubate for ≥1 hr at 37°.

3. Wash three times with wash buffer [PBS plus 0.05% (v/v) Tween 20].

4. Add 100 μl of incubation buffer [0.5% (v/v) FCS in PBS] to each well. Add 100 μl of Flag–TRAIL (4 μg/ml in incubation buffer) to the first well, mix, and transfer 100 μl to the next well. Continue dilutions, but do not touch the last well (control without TRAIL). Incubate for 1 hr at 37°. This will give 2-fold dilutions of TRAIL starting at a final concentration of 2000 ng/ml.*

5. Wash three times with wash buffer.

6. Add 100 μl of anti-Flag M2 at 1 μg/ml in incubation buffer and incubate for ≥30 min at 37°.

7. Wash three times with wash buffer.

8. Add 100 μl of goat anti-mouse IgG–peroxidase (Jackson Immuno-Research) diluted 1:1000 in incubation buffer for ≥30 min at 37°.

---

* This dilution protocol is easy to perform, but the ligand may be depleted from the solution at low concentrations (binding to the receptor).

9. Wash three times with wash buffer.

10. Add 100 μl of o-phenylenediamine (OPD) reagent (Sigma; dissolve 1 tablet each of OPD and buffer–hydroxyurea–H_2O_2 in 20 ml of water).

11. Let color develop as needed.

12. Stop the reaction with 50 μl of 2 N HCl.

13. Read the OD at 490 nm with an ELISA reader.

Acknowledgments

I thank Jürg Tschopp for support, and Margot Thome and Kimberly Burns for careful reading of this manuscript. This work was supported by grants from the Swiss National Science Foundation (to J. Tschopp) and from the Swiss Federal Office of Public Health (to P. Schneider and J. Tschopp).

[30] Expression of Lymphotoxins and Their Receptor–Fc Fusion Proteins by Baculovirus

By ISABELLE ROONEY, KRISTINE BUTROVICH, and CARL F. WARE

The tumor necrosis factor (TNF) cytokine and receptor superfamily plays critical roles in immune physiology. Several members of this family, such as the lymphotoxins (LTα and LTβ), Fas ligand, and TNF, induce cell death in some normal and transformed cells, but also induce cell growth and differentiation. The receptors for these ligands, when expressed as fusion proteins with the Fc region of IgG, function as potent antagonists of biological activity. The receptor–Fc fusion protein is a highly versatile reagent that can be utilized in virtually all the formats designed for antibodies. In this chapter we describe the expression, purification, and assays for lymphotoxins and their receptors, using a recombinant baculovirus system.

Introduction

Several members of the TNF cytokine-receptor superfamily have emerged as key regulators of the innate and acquired immune response.[1–3] Initially discovered as potent cytotoxic and antitumor factors, TNF, lymphotoxin (LT), and related cytokines activate signaling pathways that also

---

[1] C. A. Smith, T. Farrah, and R. G. Goodwin, *Cell* **76,** 959 (1994).

[2] K. J. Tracey and A. Cerami, *Annu. Rev. Med.* **45,** 491 (1994).

[3] Y.-J. Liu and J. Banchereau, *J. Exp. Med.* **184,** 1207 (1996).

0076-6879/00 $30.00

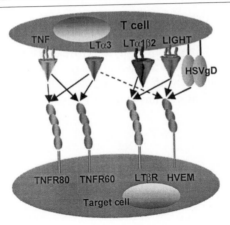

FIG. 1. The immediate members of the TNF superfamily. The diagram depicts the ligand receptor-binding interactions among these related cytokines. TNF, LTα, and LTβ genes are linked in the MHC on chromosome 6, whereas the LIGHT gene is found on chromosome 16. Herpesvirus envelope glycoprotein D (gD) is the entry factor that binds HVEM. Binding between LTα and HVEM is relatively weak. The $LTα_2β_1$ complex is a minor component with no known biological function and is not shown in this diagram.

induce resistance to apoptosis, as well as cell growth or cellular differentiation depending on the physiologic system. The TNF superfamily currently consists of more than 14 distinct ligand–receptor signaling systems. TNF, LTα, LTβ, and LIGHT form the immediate TNF family, reflected by their shared (but distinct) receptor-binding properties (Fig. 1).

The 60-kDa TNF receptor (type 1; CD120a), which binds TNF and LTα, contains in its cytosolic tail a conserved region, termed the "death domain," that is responsible for inducing apoptosis of normal and transformed cells by recruiting and activating caspases (reviewed in Ref. 4). The death domain is also found in the receptors Fas and TRAIL receptors 1 and 2. Other receptors in the immediate family, 80-kDa TNFR (TNFR80 or TNFR2), LTβR, and herpesvirus entry mediator (HVEM), exhibit complex ligand binding patterns. These receptors signal via TNF receptor-associated factors (TRAFs), a family of ring finger proteins that participate in TNFR and interleukin 1 receptor (IL-1R) activation of NF-κB (proinflammatory) and JNK (stress) signaling pathways. CD40, CD30, and CD27 also utilize TRAFs for signaling. In some settings the TRAF-binding receptors can also induce apoptosis of transformed cells. Mice genetically deficient in lymphotoxins have revealed roles in the development of peripheral

[4] C. F. Ware, S. Santee, and A. Glass, in "The Cytokine Handbook" (A. Thompson, ed.), 3rd Ed. Academic Press, San Diego, California, 1998.

lymphoid tissue and tissue organization distinct from TNF. However, the complexity of the ligand–receptor pairing and manifold signaling pathways continues to obscure a clear picture between molecular mechanisms and physiologic outcome. As immunologists the emerging challenge is to link the physiologic consequences to specific ligand–receptor systems and their signaling pathways. This is especially urgent as interest in TNF-related ligands escalates from the introduction of these ligands and receptors into the clinic.[5-7] Described in this chapter are some of the theoretical considerations and methods we have used to express lymphotoxins and their receptors suitable for research purposes.

Structural Features of Tumor Necrosis Factor Superfamily

TNF and LTα, prototypical members of the family, are compact trimers assembled from subunits that have an antiparallel β-sandwich topology, a feature that defines this protein family.[8,9] The TNF/LTα receptors are single-transmembrane glycoproteins with a conserved cysteine-rich motif in the ligand-binding ectodomain, which is shaped as an elongated structure. The crystal structure of the LTα–TNFR60 complex reveals the binding of three receptors at sites formed between adjacent LTα subunits in the trimer.[10]

TNF, like its relatives, is a type II transmembrane protein[11] that is biologically active, but is also shed from the cell surface by a metalloprotease as an active soluble mediator.[12] In contrast, LTα lacks a transmembrane domain, and as a homotrimer is exclusively secreted.[13] LTα is also retained in a biologically active form on the cell surface by assembling with LTβ

[5] D. Lienard, A. M. Eggermont, K. H. Schraffordt, B. B. Kroon, F. Rosenkaimer, P. Autier, and F. J. Lejeune, *Melanoma Res.* **4**(Suppl. 1), 21 (1994).

[6] D. Lienard, P. Ewalenko, J. J. Delmotte, N. Renard, and F. J. Lejeune, *J. Clin. Oncol.* **10**, 52 (1992).

[7] D. L. Fraker and H. R. Alexander, *Melanoma Res.* **4**(Suppl. 1), 27 (1994).

[8] E. Y. Jones, D. I. Stuart, and N. P. Walker, *Nature (London)* **338**, 225 (1989).

[9] M. J. Eck and S. R. Sprang, *J. Biol. Chem.* **264**, 17595 (1989).

[10] D. W. Banner, A. D'Arcy, W. Janes, R. Gentz, H. J. Schoenfeld, C. Broger, H. Loetscher, and W. Lesslauer, *Cell* **73**, 431 (1993).

[11] D. Pennica, G. E. Nedwin, J. S. Hayflick, P. H. Seeburg, R. Derynck, M. A. Palladino, W. J. Kohr, B. B. Aggarwal, and D. V. Goeddel, *Nature (London)* **312**, 724 (1984).

[12] R. A. Black, C. T. Rauch, C. J. Kozlosky, J. J. Peschon, J. L. Slack, M. F. Wolfson, B. J. Castner, K. L. Stocking, P. Reddy, S. Srinivasan, N. Nelson, N. Boiani, K. A. Schooley, M. Gerhart, R. Davis, J. N. Fitzner, R. S. Johnson, R. J. Paxton, C. J. March, and D. P. Cerretti, *Nature (London)* **385**, 729 (1997).

[13] P. Gray, B. Aggarwal, C. Benton, T. Bringman, W. Henzel, J. Jarrett, D. Leung, B. Moffat, P. Ng, L. Svedersky, M. Palladino, and G. Nedwin, *Nature (London)* **312**, 721 (1984).

during biosynthesis into heterotrimers.[14–16] LTα and LTβ form two distinct complexes of differing subunit ratios; LTα$_1$β$_2$, the predominant form expressed by T cells, binds with high affinity to the LTβR, whereas LTα$_2$β$_1$ binds to the TNFR and with low affinity to the LTβR.

The critical roles played by TNF and related proteins in diverse disease processes have fueled investigation to utilize these cytokines directly as drugs in disease conditions, as well as to develop reagents that specifically inhibit individual systems. Many of the ligands and receptors are membrane anchored and are not naturally released into a soluble phase. However, recombinant strategies have been developed to create proteins that are biologically active in the soluble phase and can be used as surrogates of their membrane counterpart.

Soluble Ligands

The three-dimensional profiles of TNF, LTα, and CD40 ligand serve as excellent models to design modified forms of these ligands. Three-dimensional models of the other ligands can be rendered easily by e-mail.[17] The TNF-related proteins are type II membrane proteins with a short N-terminal cytoplasmic domain (varying from 15 to 60 residues), a hydrophobic anchor that is followed by the ectodomain (Fig. 2A). The ectodomain emerges from the membrane typically with a proline-rich stalklike region of variable length. The stalk precedes the bulk of the protein, which folds into a β-sandwich structure that self-assembles into a trimer creating the receptor-binding domain. To create soluble recombinant forms of the receptor-binding domain, the cytosolic and membrane domains are truncated at a position ~10–20 residues before the start of the first β strand and replaced with a signal sequence to direct secretion. In some cases an epitope tag is inserted to identify the molecule by immunochemical methods. The reasoning for picking the truncation site is based on preserving the integrity of the first β strand, which is essential for the trimeric structure and receptor-binding properties of the ligand. This is gleaned from a survey of the natural truncation (processing) sites found for several ligands. TNF cleavage occurs at Ala–Val77, 12 residues before the start of the first β strand (Pro88). Likewise, LTα processing occurs at Gly–Val35, 27 residues before Pro63. Fas ligand is cleaved at Ser–Leu127. A good place to start engineering a

[14] M. J. Androlewicz, J. L. Browning, and C. F. Ware, *J. Biol. Chem.* **267**, 2542 (1992).
[15] J. L. Browning, A. Ngam-ek, P. Lawton, J. DeMarinis, R. Tizard, E. P. Chow, C. Hession, B. O'Brine-Greco, S. F. Foley, and C. F. Ware, *Cell* **72**, 847 (1993).
[16] J. L. Browning, K. Miatkowski, D. A. Griffiths, P. R. Bourdon, C. Hession, C. M. Ambrose, and W. Meier, *J. Biol. Chem.* **271**, 8618 (1996).
[17] N. Guex and M. C. Peitsch, *Electrophoresis* **18**, 2714 (1997).

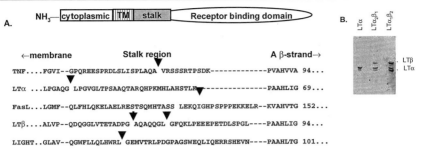

FIG. 2. (A) Cleavage and truncation sites that generate soluble forms of TNF-related ligands. *Top:* Schematic diagram of TNF-related ligands as type II membrane configuration (amino terminus is in the cytoplasm). *Bottom:* Amino acid sequence around the stalk region of several TNF-related ligands. Arrows indicate the natural cleavage sites of TNF, LTα, and Fas ligand; for LTβ and LIGHT arrows indicate where recombinant forms have been made that truncate the cytoplasmic and transmembrane regions. The A β strand is the start of the receptor-binding domain. (B) Soluble ligands purified by ion-exchange and affinity chromatography. Five micrograms of the stated ligands was subjected to SDS–PAGE on a 15% (w/v) gel and detected by Coomassie staining.

soluble ligand is ~10–20 residues before the first β-strand proline (at a convenient restriction site). LTβ, which is not naturally cleaved, was truncated at position Leu69, 18 residues before Pro88, and this form assembles with LTα into a stable trimer. LTβ expressed in the absence of LTα is not active in soluble or membrane form.[18] A stable soluble form of LIGHT has been made by truncation at Gly66; however, a slightly larger truncation (Trp80) created a protein that was relatively unstable (our unpublished observations, 1999). Truncation of type II membrane proteins at the N-terminal cytoplasmic and membrane regions requires that a leader (signal) peptide be added to the cDNA construct. The choice of signal peptide is arbitrary as the signal sequences from vascular cell adhesion molecule 1 (VCAM-1) or influenza hemagglutinin were used for soluble LTβ[16] and Fas ligand, respectively.[19]

Tumor Necrosis Factor Receptors as Soluble Fc Fusion Proteins

Soluble forms of the ectodomains of various TNF receptors, produced naturally or by genetic engineering, retain their ligand-binding activity and

[18] L. Williams-Abbott, B. N. Walter, T. Cheung, C. R. Goh, A. G. Porter, and C. F. Ware, *J. Biol. Chem.* (1997).
[19] P. Schneider, J. L. Bodmer, N. Holler, C. Mattman, P. Scuderi, A. Terskikh, M. C. Peitsch, and J. Tschopp, *J. Biol. Chem.* **272,** 18827 (1997).

can function as potent antagonists (Fig. 3). TNFRs are naturally shed from the cell surface by TNF-related processing enzyme (Adam 8 or TACE) and exist as monomers or dimers. Shope fibromavirus (poxvirus) encodes a TNFR80 homolog that is secreted as a dimer and serves the virus as an important virulence factor. Engineered dimeric forms of all the TNFR-related proteins have been made in soluble formats as fusion proteins with

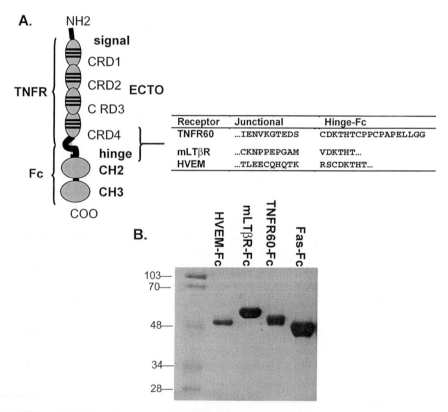

Receptor	Junctional	Hinge-Fc
TNFR60	...IENVKGTEDS	CDKTHTCPPCPAPELLGG
mLTβR	...CKNPPEPGAM	VDKTHT...
HVEM	...TLEECQHQTK	RSCDKTHT...

FIG. 3. TNF receptor–Fc fusion proteins. (A) Schematic diagram of TNFR60–Fc fusion protein, showing the cysteine-rich domains of TNFR60 and the hinge, CH2, and CH3 domains of IgG$_1$. At right is the deduced amino acid sequence of the region of the receptors where the proteins have been genetically linked to the hinge region of IgG$_1$. (B) Purified, baculovirus-produced TNF receptor–Fc fusion proteins. Proteins were analyzed on an SDS–10% (w/v) polyacrylamide gel with reducing agent. Approximately 5 μg was loaded per lane and stained with Coomassie blue. Observed molecular masses are as follows: HVEM : Fc, 54.3 kDa; mouse LTβR : Fc, 60.5 kDa; p60 : Fc, 55.5 kDa; Fas : Fc, 50.5 kDa.

TABLE I
PROPERTIES OF TNFR–Fc FUSION PROTEINS[a]

Receptor	Size (kDa)	Ligands	Half-maximal saturation ($\sim$Kd; nM)	Ref.
TNFR60–Fc	55–60	TNF	0.1500	20
		LTα	0.280; 0.650	16, 20
		LT$\alpha_1\beta_2$	1.5	16
TNFR80–Fc	75–80	TNF		
		LTα		
		LT$\alpha_2\beta_1$		
LTβR–Fc	61	LT$\alpha_1\beta_2$	0.080	16
		LT$\alpha_2\beta_1$	12	16
		LIGHT	10–20	21
Fas–Fc	43	Fas ligand	30	22
HVEM–Fc	54	LIGHT	10	21
		LTα	20	21

[a] Concentrations of ligand required for half-maximal binding to receptor–Fc proteins were determined by neutralization assay,[20,22,23] flow cytometry,[21] or by ELISA.[16]

the Fc region of IgG (Table I[16,20–23]). These latter reagents have become standard tools in research, from the discovery of novel ligands to defining the physiologic roles of these cytokines.[19,21,24–27]

The Fc region of IgG has been utilized as a fusion partner in numerous settings with many different proteins.[28,29] The Fc component of IgG$_1$ consists of the hinge, CH2, and CH3 domains. The hinge region is composed of three subdomains: upper, core, and lower hinge regions. The upper and core hinge regions are encoded by the hinge exon, whereas the lower hinge region is part of the CH2 exon. The core provides the interchain disulfide

[20] P. D. Crowe, T. L. VanArsdale, B. N. Walter, K. M. Dahms, and C. F. Ware, *J. Immunol. Methods* **168,** 79 (1994).

[21] D. N. Mauri, R. Ebner, R. I. Montgomery, K. D. Kochel, T. C. Cheung, G.-L. Yu, S. Ruben, M. Murphy, R. J. Eisenbery, G. H. Cohen, P. G. Spear, and C. F. Ware, *Immunity* **8,** 21 (1998).

[22] S. Iho, H. Shau, and S. H. Golub, *Cell Immunol.* **144,** 1 (1992).

[23] R. I. Montgomery, M. S. Warner, B. Lum, and P. G. Spear, *Cell* **87,** 427 (1996).

[24] P. D. Crowe, T. L. VanArsdale, B. N. Walter, C. F. Ware, C. Hession, B. Ehrenfels, J. L. Browning, W. S. Din, R. G. Goodwin, and C. A. Smith, *Science* **264,** 707 (1994).

[25] P. Rennert, J. L. Browning, and P. S. Hochman, *Int. Immunol.* **9,** 1627 (1997).

[26] R. Ettinger, J. L. Browning, S. A. Michie, W. van Ewijk, and H. O. McDevitt, *Proc. Natl. Acad. Sci. U.S.A.* **93,** 13102 (1996).

[27] R. Ettinger, R. Mebius, J. L. Browning, S. A. Michie, S. van Tuijl, G. Kraal, W. van Ewijk, and H. McDevitt, *Int. Immunol.* **10,** 727 (1998).

[28] A. Ashkenazi and S. M. Charnow, *Curr. Opin. Immunol.* **9,** 195 (1997).

[29] S. M. Charnow and A. Askenazi, *Trends Biochem.* **14,** 52 (1996).

linkage that stabilizes the dimer, the upper hinge region provides the tether to the receptor ectodomain, and the lower hinge region adds conformational flexibility.[30]

From X-ray diffraction data the ectodomain of the TNFR60 molecule is revealed as an elongated structure formed by extensive disulfide bonding. The TNFR60 ectodomain contains 24 cysteines divided between four cysteine-rich domains (CRDs). Three disulfide bonds in each CRD form upper and lower loops centered on the second, third, and fourth cysteines in the sequence CxxCxxC (Fig. 3). This motif is highly conserved and represents the signature motif of this receptor family. A proline-rich stretch of amino acids separates the ectodomain from the transmembrane domain. Truncation of the receptor at various residues to the C-terminal side of the last cysteine residue in the ectodomain does not seem to alter ligand-binding activity, although a systematic study has not been performed. Unpaired cysteine residues should be accounted for in the design of receptor Fc constructs. For instance, CD27 has an unpaired cysteine in the ectodomain and normally forms a dimer. In contrast to the design of soluble ligands, the endogenous signal peptide of the receptor is utilized in the construct.

The receptor–Fc fusion protein is a highly versatile reagent that can be utilized in virtually all the formats designed for antibodies. The Fc fusion protein is easily purified by affinity chromatography, using protein A or G resins and acid elution conditions. The purified protein can be used directly as an immunogen to produce polyclonal or monoclonal antibodies specific for the ectodomain of the receptor.[31] Contaminating anti-Fc antibodies are easily depleted by passage of the serum over a column containing immobilized human IgG. We have noted that antibodies specific for the fusion protein, presumably epitopes composed of residues at the receptor–Fc junction, are frequently obtained during the initial screen. The Fc region provides the utility of the molecule for direct use in flow cytometry and receptor-mediated ligand precipitation reactions (analogous to immunoprecipitation) or indirectly with antibodies in a ligand-capture enzyme-linked immunosorbent assay (ELISA).

Description of Methods

Baculovirus Expression of Proteins

Baculoviruses are DNA viruses that infect insect cells. Use of baculovirus as a vector to introduce cDNAs into insect cells results in high expression

---

[30] L. J. Harris, S. B. Larson, and A. McPherson, *Adv. Immunol.* **72,** 191 (1999).
[31] T. L. VanArsdale and C. F. Ware, *J. Immunol.* **153,** 3043 (1994).

of recombinant protein (up to 25 to 50% of total cellular protein, or 1 g of protein per 10^9 cells, although 10 to 100 mg of protein per 10^9 cells is more usual[32]). Production of mammalian proteins in this eukaryotic system allows proper folding, disulfide bond formation, glycosylation, and other posttranslational modifications.

Transfer Vector and Cloning

Sophisticated transfer vectors have been developed for introduction of cDNAs into baculovirus. Production of the proteins described here involves insertion of cDNA into the multiple cloning site of the transfer vector pVL1392 (InVitrogen, San Diego, CA), a polyhedrin gene locus-based transfer vector. This vector is used to rescue the lethality of the BaculoGold baculovirus DNA (PharMingen, La Jolla, CA) as follows.

Baculovirus Gold DNA (250 ng) is mixed with pVL1392 (1 μg) containing the inserted sequence of interest. Insect cells [0.7×10^6 TN5 B1-4 cells freshly seeded in a 25-cm^2 tissue culture flask in 5 ml of ExCell 401 medium (JRH Biosciences, Lenexa, KS)] are incubated with this mixture in the presence of Lipofectin (14 μg; Life Technology, Gaithersburg, MD) at 27°. After 4 days, the culture medium, now containing some recombinant, competent baculovirus, is collected. A shallow serial dilution in a microplate is carried out to isolate individual clones, and the recombinant baculovirus clones are tested for protein expression. These are selected for further amplification to obtain a high-titer stock solution (10^9 to 10^{10} infectious virus per milliliter). This is done by infecting freshly seeded TN5 B1-4 cells at a multiplicity of infection (MOI) of 0.5 and incubation for 4 days before harvesting the medium. An end-point titer is performed again to quantify the infectious virus.

Determination of Baculovirus End-Point Titer

Infectious virus can be measured in different ways. We employ an end-point dilution assay for cytopathic effects.[32] TN5 B1-4 cells are seeded in a microtiter plate (5000/well) and incubated with 10 μl of serial dilutions of recombinant beculovirus preparation produced as described above. After incubation at 27° for 7 days, cells are observed microscopically for cytopathic changes (cell–cell fusion and detachment from plastic).

Viral titer is calculated as the tissue culture infectious dose at 50% infection (TCID$_{50}$) by estimating the dilution yielding 50% infection. This is accomplished by using the proportionate distance (PD) estimate, PD =

---

[32] D. R. O'Reilly, L. K. Miller, and V. A. Luckow, "Baculovirus Expression Vectors: A Laboratory Manual." W. H. Freeman and Company, New York, 1992.

$(A - 50)/(A - B)$, where A is the percentage above 50% infection and B is the percent response below 50%. The log $TCID_{50}$ is the log of the dilution giving a response greater than 50% minus the PD of that response. The virus titer is the reciprocal of this value, when corrected for volume and the Poisson distribution factor (0.69), that is, PFU/ml = $TCID_{50} \times 0.69$.

Production of Recombinant Protein

Recombinant proteins are expressed in the insect cell line BT1 Tn5 B1-4 grown in ExCell 401 medium (JRH Biosciences). It has been demonstrated that ligands of the TNF family produced by insect cells, although not glycosylated to the same extent as those produced in mammalian cells, are correctly assembled and specifically bind the appopriate TNF receptors.[20] Freshly seeded cells in 175-cm² flasks are infected with recombinant baculovirus containing the desired cDNA at an MOI of 10, and incubated at 27° for 2 hr. Medium is then removed and the cells washed twice with fresh ExCell medium. Fresh medium (30 ml/flask) is added and the cells incubated for 4 days at 27°. After 1–2 days of incubation the cells show morphological change in response to baculovirus infection. The cells fuse and become loosely or nonadherent. After 4 days the medium is harvested and centrifuged for 20 min at 3000 rpm in a Sorvall (Newtown, CT) 7 tabletop centrifuge to remove cells and debris. Protease inhibitors [phenylmethylsulfonyl fluoride (PMSF, 1 mM), leupeptin (0.5 μg/ml), and EDTA (5 mM)] are added, the medium is filtered through a 0.2-μm pore size filter, and Nonidet P-40 (NP-40, 0.1%, v/v) is added. Medium can be stored frozen at $-80°$.

Purification of Ligands

Browning et al.[16] described affinity purification techniques for soluble trimeric complexes of LTα and -β, based on specific binding of these complexes to Fc constructs of the receptor proteins TNFR60 and LTβR. To preserve affinity columns we use ion-exchange chromatography to concentrate and partially purify the ligands before using receptor-affinity chromatography. A high degree of purity is obtained by these techniques (Fig. 2A).

Ion-Exchange Purification of LTα Homotrimer

Filtered baculovirus supernatant is diluted 1:2 with 20 mM Tris, adjusted to pH 7.6, and applied to a cationic ion-exchange column (SP HiTrap; Pharmacia, Uppsala, Sweden) at a flow rate of 2.5 ml/min. The column is washed with 20 mM Tris, pH 7.6 (30 ml), LTα is then eluted with a gradient of 0 to 1 M NaCl in 20 mM Tris, pH 7.6 (50-ml total volume), and 2-ml fractions are collected. LTα consistently elutes at 300 mM NaCl.

The concentration of LTα in insect cell supernatant measured by specific ELISA is routinely 18–20 mg/liter.[20] The yield of LTα by ion-exchange chromatography is 60–70% and purity (by silver-stained gel) is ~70%.

Affinity Purification of LTα

LTα partially purified by ion-exchange chromatography is further purified to homogeneity by affinity purification on a column of TNFR60–Fc coupled to Affi-Gel (Bio-Rad, Hercules, CA). Briefly, TNFR60–Fc at 2–5 mg/ml is dialyzed against 0.1 M morpholinepropanesulfonic acid (MOPS, pH 6.0) and incubated with Affi-Gel (1 ml of Affi-Gel per 2–5 mg of TNFR60–Fc) for 4 hr at 4°. Unbound sites are blocked by incubation with 50 mM Tris–HCl, pH 8.0. Partially purified LTα in phosphate-buffered saline (PBS) is applied to a 0.5-ml TNFR60–Fc column at a flow rate of 0.1 ml/min. The column is washed with 10 vol of PBS and bound protein is eluted with 3 vol of 20 mM glycine–150 mM NaCl, pH 3.0. Fractions (0.25 ml) are collected into tubes containing 25 μl of 1 M Tris, pH 8.0, to neutralize the pH. LTα-containing fractions are identified by specific ELISA and Coomassie blue-stained sodium dodecyl sulfate (SDS)–polyacrylamide gel, and pooled and then dialyzed against 1000 vol of PBS, pH 7.4, for 2 hr at 4°.

The yield of LTα by this method as measured by specific ELISA is 80–90%. Specific activity on U937 cells (units per milligram of protein, where 1 unit is the amount required to kill 50% of the cells) is 25,000 units/mg. The purity (by silver-stained gel) is close to 99%.

Expression of LT$\alpha_1\beta_2$. T cells express two heterotrimers of LTα and LTβ, with LT$\alpha_1\beta_2$ the predominant form.[33] Insect cells coinfected with baculovirus encoding LTα and sLTβ-myc produce all combinations of trimers—LTα homotrimer, LT$\alpha_2\beta_1$-myc, and LT$\alpha_1\beta_2$-myc—in proportions that depend on the ratio of LTα- and sLTβ-encoding baculoviruses (Fig. 4).[16] Browning *et al.*[16] suggest infection with LTα and SS sLTβ baculovirus at MOIs of 2 and 7, respectively. We routinely use MOIs of 1 and 3, respectively. Secretion of LT$\alpha_1\beta_2$-myc in insect cells (measured by specific ELISA) is consistently 3 to 5 mg/liter, with minimal contamination by LT$\alpha_2\beta_1$-myc.

Purification of LT$\alpha_1\beta_2$. LT$\alpha_1\beta_2$-myc is purified by a combination of ion-exchange chromatography and a modification of the affinity-purification technique described by Browning *et al.*[16] LT$\alpha_1\beta_2$ is purified by a method

[33] J. L. Browning, I. Dougas, A. Ngam-ek, P. R. Bourdon, B. N. Ehrenfels, K. Miatkowski, M. Zafari, A. M. Yampaglia, P. Lawton, and W. Meier, *J. Immunol.* **154**, 33 (1995).

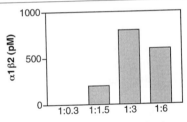

Ratio of LTα : LTβ baculovirus

Fig. 4. Effect of the ratio of LTα to LTβ viruses on production of LTα$_1$β$_2$ by Tn5 cells. Tn5 cells were infected with baculovirus encoding LTα and LTβ in the ratios shown. After 4 days medium was harvested and LTα$_1$β$_2$ was measured by specific ELISA, using LTβR–Fc as the capture molecule and monoclonal anti-LTα (9B9) for detection.

identical to that described above for LTα, except that pH 6.5 is used in buffers for loading, washing, and elution. LTα$_1$β$_2$ elutes at 300 mM NaCl. The yield is >50%. The eluate contains a mixture of LTα, LTα$_1$β$_2$-myc, and LTα$_1$β$_2$-myc, as well as other contaminating proteins.

Affinity Purification of LTα$_1$β$_2$. Ion exchange-purified LTα$_1$β$_2$-myc in PBS is cycled several times over a column of TNFR60–Fc coupled to Affi-Gel in order to remove contaminating LTα and LTα$_2$β$_1$-myc. The flowthrough (unbound material) from this procedure is then passed over a column of LTβR–Fc bound to Affi-Gel (5 mg of LTβR/ml of gel) at a flow rate of 0.1 ml/min. The column is washed with 10 vol of PBS and bound protein is then eluted with 3 vol of 20 mM glycine–150 mM NaCl, pH 3.0. Fractions of 250 μl are collected into tubes containing 25 μl of 1 M Tris-HCl, pH 8.0, to neutralize pH. LTα$_1$β$_2$-myc-containing fractions are identified by specific ELISA, pooled, dialyzed against PBS, and stored at −80°. The purity as measured by specific ELISAs and silver-stained gel is ~99% (Fig. 2B).

Expression and Purification of Receptors

The purification technique exploits the affinity of the Fc portion of the fusion protein for protein G. The pH of the filtered baculovirus (BV) supernatant (300 ml) is adjusted to 7.0 by addition of 1 M Tris-HCl, pH 7.0, to 20 mM and this supernatant is then applied to a 1-ml column of protein G–Sepharose beads (Pharmacia) at a flow rate of 2 ml/min. The column is washed with 20 mM Tris-HCl, pH 7.0 (20 ml), and bound protein is eluted with 20 mM glycine–150 mM NaCl, pH 3.0 (3 vol). Fractions of 1 ml are collected into tubes containing 50 μl of 1 M Tris-HCl, pH 8.0. Fc construct-containing fractions, identified by protein concentration (spectrophotometer at 280 nm) and confirmed by Coomassie-stained SDS–polyacrylamide gel, are pooled and then dialyzed against 1000 vol of PBS for 2 hr at 4°. The protein is stored at −80°. This method recovers 80 to 90% of protein, with purity at >98% (Fig. 3B).

Assays for Ligands

Enzyme-Linked Immunosorbent Assay

Two-site ELISAs are used to measure ligands of the TNF family. In each case, an Fc construct of the ligand's receptor is used as capture molecule. Bound ligand is then detected by serial incubations with a specific monoclonal antibody directed against either the protein itself or an epitope tag, such as the c-myc epitope, followed by goat anti-mouse IgG coupled to horseradish peroxidase (GAM IgG–HRP). The plate is developed with the colorimetric reagent 2,2′-Azino-bis(3-ethylbenzthiazoline-6-sulfonic acid) (ABTS).

Enzyme-Linked Immunosorbent Assay for Measurement of LTα

TNFR60–Fc [150 ng/well in 50 μl of binding buffer (150 mM NaCl–20 mM Tris, pH 9.6)] is immobilized in the wells of a microtiter plate (Nunc Immunoplates; Fisher Scientific, Tustin, CA) for 16 hr at 4°. After washing five times with PBS–0.1% (v/v) Tween 20, samples and standards, diluted in PBS–3% (w/v) bovine serum albumin (BSA), are applied (50 μl/well) and incubated at room temperature for 1 hr. Wells are washed five times, incubated with monoclonal mouse anti-LTα (9B9; 100 ng/ml) in PBS–BSA (50 μl/well) for 1 hr at room temperature, washed a further five times, incubated with goat anti-mouse IgG coupled to horseradish peroxidase (GAM IgG–HRP; Amersham Life Sciences) diluted 1:1500 in PBS–BSA for 1 hr at room temperature, and washed a final five times before development. H$_2$O$_2$ (1 μg/ml) is added to ABTS solution (400 μg/ml in 0.1 M sodium citrate, pH 4.2) immediately before use. Wells are incubated with ABTS solution (50 μl/well) at room temperature for 15 min to allow full color development. A solution of 2% (w/v) SDS (50 μl/well) is added to stop the reaction and color intensity is measured at 415 nm in a Spectramax 250 microtiter plate reader (Molecular Devices, Sunnyvale, CA). Note that use of TNFR60–Fc as capture reagent ensures that this method measures only active ligand.

Measurement of LTα₁β₂. The procedure is identical to that described above for LTα, except that the plate is coated with mouse LTβR–Fc (3 μg/ml), whch binds LTβ but not LTα. Detection, as described above, is with monoclonal anti-LTα 9B9. Note that the combined use of LTβR–Fc and 9B9 ensures that only complexes containing both LTα and LTβ will be detected by this method.

Measurement of $LT\alpha_2\beta_1$. Availability of an assay that specifically measures $LT\alpha_2\beta_1$ is essential to ensure that this cytokine does not significantly contaminate preparations of $LT\alpha_1\beta_2$. (The LTβR–Fc-based assay described above will detect both types of complexes.) The procedure is identical to those described above, except that TNFR60–Fc is used as capture molecule (this will bind $LT\alpha_2\beta_1$, but not $LT\alpha_1\beta_2$) and a monoclonal anti-β antibody (C37) is used as detecting agent. Thus, the assay will detect only $LT\alpha_2\beta_1$.

Cytotoxicity Assay

The HT29.14S cell line (American Tissue Culture Collection) expresses both TNFR60 and LTβR, is highly sensitive to induction of apoptosis and inhibition of proliferation by ligands of the TNF family,[34] and is commonly used to assess the cytotoxicity of the recombinant proteins described here.

HT29.14S cells are plated in microtiter plates at a density of 5000/well and incubated at 37° for 3 hr, during which cells become adherent. Cells are incubated in complete Dulbecco's modified Eagle's medium (DMEM) with varying concentrations of ligand (typically over a dose range of 1 to 0.001 nM) and interferon γ (IFN-γ, 80 U/ml) in 100 μl for 48–72 hr at 37°. Plates are wrapped with plastic wrap to prevent evaporation from the outer wells of the microplate. The degree of cytotoxicity is assessed after 48 hr (for TNF, LTα, and FasL) or 72 hr (for $LT\alpha_1\beta_2$) by MTT dye reduction assay as described below.[34]

MTT Assay

The dye reduction assay involves the use of 3-(4,5-dimethylthiazol-2-yl)-2,5-diphenyltetrazolium bromide (MTT), which is prepared as a stock solution of 5 mg/ml in PBS and filtered before use. This stock solution (20 μl) is added to each well and the plate incubated at 37° for 4 hr; live cells reduce the dye, which appears as violet intracellular crystals. The medium is then aspirated and the cells solubilized in 100 μl of acidic isopropanol [400 μl of concentrated HCl in 50 ml of 70% (v/v) isopropanol]. The reduction of dye, indicated by the purple color, is proportional to the density of live cells, as only cells with active mitochondria reduce the dye. The intensity of color is measured at 570 nm in a Spectramax microtiter plate reader (Molecular Devices).

[34] J. L. Browning, K. Miatkowski, I. Sizing, D. A. Griffiths, M. Zafari, C. D. Benjamin, W. Meler, and F. Mackay, *J. Exp. Med.* **183,** 867 (1996).

Detection of Ligands by Flow Cytometry

Detection of surface ligands by flow cytometry may provide for the identification of important cell types. For instance, surface lymphotoxin and TNF are useful markers of recently activated helper T type 1 (Th1) cells; Fas ligand can be detected on some tumors in addition to its expression by cytotoxic T cells. Baculovirus systems can be used to reconstitute the LTαβ complex on the surface of insect cells in a biologically active form (Fig. 5). Varying the ratio of recombinant viruses can produce more or less LTα₁β₂ on the cell surface, analogous to production of soluble LTα₁β₂ (Fig. 4).

After stimulation of lymphocytes or transfection of cells with ligand cDNA or baculovirus, cells for flow cytometry are harvested, washed, and incubated on ice (30 min) in Hanks' salts (pH 7.4)–20 mM HEPES buffer and 10% (v/v) bovine calf serum–0.1% (w/v) sodium azide containing TNFR–Fc fusion protein at 10–20 μg/ml. The cells are washed twice and phycoerythrin (PE)-conjugated affinity-purified goat anti-human antibody is used to detect the Fc fusion proteins. Controls for nonspecific binding to include in the assay are normal human IgG. IgG (heat aggregated, 10 μg/ml) from the appropriate species (e.g., rabbit) is included in the buffer

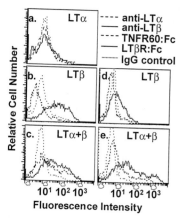

FIG. 5. Reconstitution of surface LTαβ complexes in Insect cells. Tn5 B1-4 insect cells were infected with the baculoviruses containing LTα (a), LTβ (b and d), or both LTα and LTβ (c and e). LTα and LTβ antigens were detected with anti-LTα (9B9) or anti-LTβ (B9) monoclonal antibodies and goat anti-mouse IgG-conjugated PE. Ligand formation was measured in (d) and (e) with LTβR–Fc or TNFR60–Fc and detected with goat anti-human IgG–PE. Normal mouse (a–c) or human IgG (d and e) were used as controls for nonspecific staining. The lymphotoxin-α (LTα) subunit is essential for the assembly, but not receptor specificity, of the membrane-anchored LTα₁β₂ heterotrimeric ligand. [From L. Williams-Abbott, *et al.*, *J. Biol. Chem.* **272**, 19451 (1997). Reprinted with permission.]

to block nonspecific binding to Fc receptors. Fluorescence is detected by flow cytometry (FACSCaliber, Becton Dickinson, San Jose, CA). Specific receptor–Fc binding is determined by calculating the fluorescence intensity as follows: (mean fluorescent channel) (% positive fluorescent events), where a positive event has fluorescence value >98% of the value for normal IgG. Specific fluorescence intensity represents the fluorescence intensity after subtraction of the value for control IgG. Most TNFR–Fc fusion proteins are stable at 4° for 1 week. TNFR60–Fc appears to be more labile and should be used within 1 to 2 days.

Assays for Receptor: Fc Proteins

Flow Cytometric Assay

The activity of receptor–Fc fusion protein can be conveniently measured by fluorescence-activated cell sorting (FACS) staining. The activity of Fas–Fc is measured by staining insect cells expressing Fas ligand (FasL). We use Tn5 B1-4 insect cells that have been infected with recombinant baculovirus containing FasL cDNA at an MOI of 10 and incubated at 27° for 24 hr. The FasL-expressing cells are harvested, washed in HBSS–FBS binding buffer [Hanks' balanced salt solution (HBSS, pH 7.4), 20 mM HEPES, 5% (v/v) FBS, and 0.1% (w/v) sodium azide] and resuspended at 10^7 cells/ml in HBSS–FBS. The cells (20 μl of this cell suspension) are then incubated with Fas–Fc at concentrations of 1, 5, and 25 μg/ml in HBSS–FBS (40-μl total volume) on ice for 30 min. After washing twice in HBSS–FBS, Fas–Fc bound to the cells is detected by incubation with goat anti-human IgG–PE (10 μg/ml) for 30 min at 4° in the dark, with two washes in HBSS–FBS between the incubations. As a positive control for expression of FasL we suggest staining cells with an anti-FasL monoclonal antibody, such as A11 (Alexis, San Diego, CA). After washing twice in HBSS–FBS and once in PBS, cells are suspended in PBS, and analyzed on a FACSCalibur (Beckton Dickinson) using CELLQUEST.

Neutralization Assays for Receptor: Fc Fusion Proteins

These examine the ability of TNFR–Fc to neutralize the cytolytic activity of their ligands. Assays for TNFR60–Fc and Fas–Fc have been described by Crowe et al.[20] and Brunner et al.,[35] respectively.

Procedure for TNFR60–Fc (from Crowe et al.[20]). L929 cells (2×10^4/well in 96-well flat-bottomed microtiter plates) are incubated overnight at

---

[35] T. Brunner, R. J. Mogil, D. LaFace, N. J. Yoo, A. Mahboubi, F. Echeverri, S. J. Martin, W. R. Force, D. H. Lynch, and C. F. Ware, Nature (London) 373, 441 (1995).

37° in complete medium containing mitomycin C (0.5 μg/ml). Cells are then incubated with previously determined doses of purified recombinant TNF (2, 0.2, and 0.02 nM) in the presence of graded amounts of TNFR60–Fc for 20 hr at 37°. The percentage growth inhibition is determined by the MTT dye assay.

Procedure for Fas–Fc (from Brunner et al.[35]). Functional Fas ligand (FasL) is induced in the hybridoma A1.1 cell line, which rapidly undergoes apoptosis after activation by T cell receptor ligation or by addition of phorbol ester and ionomycin.

A1.1 cells (5 × 10^5/ml) are plated on anti-CD3-coated plastic in the presence of graded concentrations of Fas–Fc for 16 hr at 37°. The percentage apoptosis is determined by staining with acridine orange/ethidium bromide fluorescence staining and microscopic examination.

Ligand Precipitation

Ligand specificity of receptors can be determined by using receptor–Fc proteins to bind and immobilize (precipitate) ligands from cell extracts. Mauri et al.[21] used this method to identify the cellular ligand for HVEM, LIGHT, expressed by activated T cells (Fig. 6).

Cells are labeled with [^{35}S]methionine and [^{35}S]cysteine as described.[20] Cells (10^7) are harvested and extracted with NP-40 (1%, v/v) nonionic detergent in buffer with 50 mM Tris, pH 7.4, containing 10 mM iodoacetamide and protease inhibitors. The detergent-soluble fraction, obtained after centrifugation, is subjected to precipitation as described.[36] Briefly, detergent extracts are precleared by addition of 10 μg of normal mouse or rabbit IgG and 20 μl of protein G–Sepharose beads for 1 hr and the beads are removed by centrifugation. The Fc fusion proteins are added to the supernatant at 10–20 μg/ml with 20 μl of protein G beads and incubated for 1–2 hr with nutation. The beads are washed four times with 1% (v/v) NP-40 buffer and once with PBS. Proteins are solubilized by adding 20 μl of 2× SDS–PAGE buffer and resolved by SDS–PAGE [12% (w/v) acrylamide] and detected by phosphoimage analysis (PhosphorImager; Molecular Dynamics, Sunnyvale, CA).

Concluding Remarks

Lymphotoxins and related ligands, and receptor–Fc fusion proteins, have been successfully expressed in most eukaryotic expression systems

---

[36] C. F. Ware, P. D. Crowe, M. H. Grayson, M. J. Androlewicz, and J. L. Browning, *J. Immunol.* **149**, 3881 (1992).

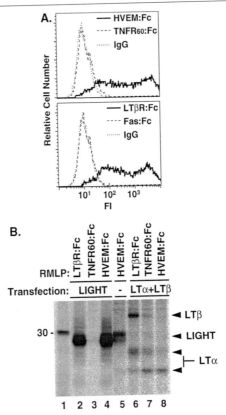

FIG. 6. Identification of the HVEM surface ligand, LIGHT. (A) HEK293 cells (10^6) were transfected with 5 μgl of LIGHT cDNA (pCDNA3 with CMV promoter), using the calcium phosphate method. After 24 hr of transfection, cells were harvested with 20 mM EDTA in PBS, and incubated with HVEM:Fc, mLTβR:Fc, TNFR:Fc, Fas:Fc, or IgG and stained with anti-huIgG–PE. Mock-transfected cells did not stain with any Fc fusion protein. (B) HEK293 cells transfected with LIGHT cDNA, or cotransfected with LTα and LTβ cDNAs (pCMD8 with CMV promoter) (as described above), were metabolically labeled and the extracts precipitated with mLTβR:Fc (lanes 2 and 6), TNFR60:Fc (lanes 3 and 7), and HVEM:Fc (lanes 4, 5, and 8) as described above. For comparison, in lane 5, the HVEM:Fc precipitation was from metabolically labeled activated II-23 cells. [From D. N. Mauri, *et al.*, *Immunity* **8**, 21 (1998). Reprinted with permission.]

including insect and mammalian systems. Each system has its own particular advantages and disadvantages. We have found that for most nonclinical investigations, expression in baculovirus offers several advantages. First, the expressed proteins are produced at reasonable levels (20–40 mg/liter for LT; 2–8 mg/liter for TNFR–Fc) with a high specific activity at minimum expense. For example, TNFR60–Fc produced in baculovirus neutralized TNF cytotoxicity at a 1:1 molar ratio[20] and when produced under serum-free conditions contamination with endotoxins is minimized. Although a systematic comparison is lacking, Fas–Fc produced by baculovirus appeared to be more stable and showed better neutralizing activity for Fas ligand in contrast to Fas–Fc produced in COS-7 cells (our unpublished observations, 1999). Another parameter to be considered in the selection of an expression system is the processing of glycoproteins, which is different for insect and mammalian cells. The importance that carbohydrate has in determining the metabolic half-life of circulating antibody may influence the usefulness of the Fc fusion proteins *in vivo*.

The introduction of TNFR–Fc fusion proteins into the clinic is in progress, yet much remains to be learned about how these fusion proteins function as drugs. Furthermore, our understanding of the effects these cytokines in different physiologic settings and diseases, including infectious disease and cancer, remains relatively minuscule. The availability of biologically active ligands and their antagonists will remain a cornerstone in the development of novel cytokine-based strategies for disease intervention.

[31] Analysis of the CD95 (APO-1/Fas) Death-Inducing Signaling Complex by High-Resolution Two-Dimensional Gel Electrophoresis

By Carsten Scaffidi, Frank C. Kischkel, Peter H. Krammer, and Marcus E. Peter

Introduction

CD95 (APO-1/Fas) belongs to a group of apoptosis-inducing receptors, also called the death receptors, that constitute a subclass of the tumor necrosis factor (TNF) receptor superfamily.[1] All death receptors are characterized by an intracellular domain, the death domain (DD), that is essential for the transduction of the death signal.

The crystal structure of TNF receptor I in complex with its ligand

---

[1] M. E. Peter, C. Scaffidi, J. P. Medema, F. C. Kischkel, and P. H. Krammer, The death receptors. *In* "Apoptosis, Problems and Diseases" (S. Kumar, ed.). Springer-Verlag, Heidelberg, Germany, p. 25 (1998).

TNF-β revealed a trimeric structure of the activated receptor.[2] Therefore, it is believed that the death receptors are triggered by trimerization. For CD95 it was demonstrated that dimerization of the receptor was not sufficient for its activation,[3,4] whereas oligomerization of the receptor, either by its natural ligand CD95L or by agonistic antibodies such as anti-APO-1,[5] induced apoptosis in sensitive cells. As the intracellular part of the death receptors does not contain any enzymatic function, the signal is transduced by proteins that interact with the receptor. Many proteins were identified that bind to the intracellular domain of the CD95 receptor by use of the yeast two-hybrid system.[6] A positive signal in this artificial system may not always reflect an *in vivo* association of a protein with the activated CD95 receptor. Therefore, we have developed a biochemical approach to identify proteins that interact with the CD95 receptor in an activation-dependent manner (Fig. 1). This method is based on coimmunoprecipitation of signaling molecules with the activated (crosslinked) CD95 receptor from metabolically labeled cells followed by high-resolution two-dimensional isoelectric focusing–sodium dodecyl sulfate–polyacrylamide gel electrophoresis (2D IEF–SDS–PAGE) analysis and autoradiography. By this method we identified a set of signaling proteins, designated cytotoxicity-dependent APO-1-associated proteins (CAP1–4), that bind only to the activated CD95 receptor and form the death-inducing signaling complex (DISC).[4] The investigation of the DISC is important because it is the first step in death receptor signaling. Here we describe the technical details of the analysis of the CD95 DISC.

Reagents

Buffer Used for Cell Lysis and Immunoprecipitation

Lysis buffer: 30 mM Tris-HCl (pH 7.5), 150 mM NaCl, 1 mM phenylmethylsulfonyl fluoride (PMSF), small peptide inhibitors,[4] 1% (v/v) Triton X-100 (Serva, Heidelberg, Germany), and 10% (v/v) glycerol

[2] D. W. Banner, A. D'Arcy, W. Janes, R. Gentz, H. J. Schoenfeld, C. Broger, H. Loetscher, and W. Lesslauer, *Cell* **73**, 431 (1993).

[3] J. Dhein, P. T. Daniel, B. C. Trauth, A. Oehm, P. Möller, and P. H. Krammer, *J. Immunol.* **149**, 3166 (1992).

[4] F. C. Kischkel, S. Hellbardt, I. Behrmann, M. Germer, M. Pawlita, P. H. Krammer, and M. E. Peter, *EMBO J.* **14**, 5579 (1995).

[5] B. C. Trauth, C. Klas, A. M. Peters, S. Matzku, P. Möller, W. Falk, K.-M. Debatin, and P. H. Krammer, *Science* **245**, 301 (1989).

[6] M. E. Peter, F. C. Kischkel, S. Hellbardt, A. M. Chinnaiyan, P. H. Krammer, and V. M. Dixit, *Cell Death Differ.* **3**, 161 (1996).

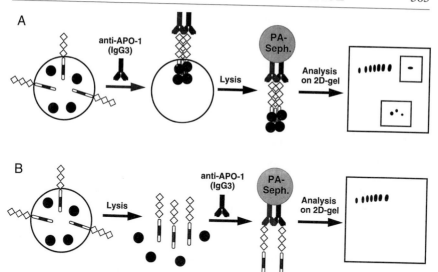

Fig. 1. Principle of CD95 DISC analysis. (A) Stimulated condition. Incubation of cells with agonistic IgG$_3$ anti-APO-1 antibody leads to receptor multimerization and binding of signaling molecules to the activated receptor. After cell lysis the receptor complex is immunoprecipitated with protein A (PA)–Sepharose and analyzed by high-resolution 2D IEF–SDS–PAGE. (B) Nonstimulated condition. Cells are first lysed and lysates are then supplemented with IgG$_3$ anti-APO-1 antibody. The monomeric receptor is immunoprecipitated and analyzed as described in (A).

Buffers Used for Isoelectric Focusing

Gel matrix: 9.5 M urea, 3.8% (w/v) acrylamide, 0.2% (w/v) bisacrylamide, 1.1% (w/v) ampholines pH 5–7 (Serva), 1.1% (w/v) ampholines pH 5–7 (Pharmacia, Uppsala, Sweden) 0.8% (w/v) ampholines pH 3.5–10 (Pharmacia), 2% (v/v) Nonidet P-40 (NP-40), 0.07% (v/v) $N,N,N,'N'$-tetramethylethylenediamine (TEMED), and 0.01% (w/v) ammonium persulfate (APS)

Upper running buffer: 20 mM NaOH

Lower running buffer: 10 mM H$_3$PO$_4$

Sample buffer: 9.8 M urea, 2% (w/v) ampholines pH 5–7 (Pharmacia), 4% (v/v) NP-40, 100 mM dithiothreitol (DTT)

Equilibration buffer: 60 mM Tris (pH 6.8), 2% (w/v) SDS, 50 mM DTT, 10% (v/v) glycerol

Buffers Used for SDS–PAGE

Agarose solution: 0.125 M Tris (pH 6.8), 0.1% (w/v) SDS, 10% (v/v) glycerol, 1% (w/v) agarose, 50 mM DTT, 0.001% (w/v) bromphenol blue

Fixing solution: 20% (v/v) methanol, 10% (v/v) acetic acid in distilled water

Methods

The first event in CD95 signaling after receptor cross-linking is the association of signaling proteins (CAP1–4) with the receptor, a process called DISC formation.[4] To analyze DISC formation we developed a biochemical approach based on the technique of coimmunoprecipitation. In this approach we made use of the agonistic anti-APO-1 antibody to cross-link CD95 on the cell surface. The receptor-bound antibody is then used for immunoprecipitation of activated CD95 after cell lysis. By doing so, only the activated receptor is immunoprecipitated, whereas the monomeric receptor, which has not been cross-linked by anti-APO-1, remains in the lysate. Therefore, this method can discriminate between activated and nonactivated CD95.

To analyze the DISC, activated CD95 receptor is immunoprecipitated from ^{35}S-labeled cells (Fig. 1A). The immune complex is then analyzed by high-resolution 2D IEF–SDS–PAGE and the immunoprecipitated receptor as well as the coimmunoprecipitated signaling molecules are visualized by autoradiography. To identify proteins that associate only with the activated CD95 receptor in a specific manner, a control immunoprecipitation is performed. Therefore, only the nonactivated (monomeric) CD95 receptor is immunoprecipitated from ^{35}S-labeled cells and analyzed in the above-described way (Fig. 1B). Specific activation-dependent signaling molecules should be present only in the immunoprecipitate of the activated receptor, whereas nonspecific proteins will be present in both immunoprecipitates.

^{35}S Metabolic Labeling of Cell Lines

The DISC analysis described here is suitable for cell lines expressing high levels of CD95 such as the B-lymphoblastoid cell line SKW 6.4, the Burkitt lymphoma Raji, the B cell lymphoma BJAB, or the leukemic T cell lines H9 and HuT78. Other cell lines may also be analyzed with some modifications of the method as described below. For each sample 3×10^7 cells are needed. To metabolically label cellular proteins with ^{35}S, cells are grown to midlog phase (about 3–5×10^5 cells/ml), washed twice with phosphate-buffered saline (PBS), and starved at 37° for 1 hr in medium lacking cysteine and methionine (GIBCO, Grand Island, NY) at a density of 2×10^6 cells/ml. After that the medium is supplemented with 0.5 mCi of [^{35}S]methionine/cysteine mix (ProMix; Amersham, Arlington Heights,

IL) per 3×10^7 cells. Labeling is carried out for 24 hr at 37° because the CD95 receptor has a relatively slow turnover rate.[7] To label the intracellular signaling molecules, labeling periods of 4 hr may be satisfactory.

Stimulation of CD95 Receptor and Cell Lysis

The CD95 receptor is activated by the agonistic IgG$_3$ anti-APO-1 monoclonal antibody (MAb), which leads to receptor cross-linking due to coaggregation of the IgG$_3$ heavy chains. For stimulation the anti-APO-1 antibody is added at a concentration of 2 μg/ml directly to the cells after metabolic labeling (a total of 30 μg of anti-APO-1 for 3×10^7 cells in 15 ml of medium). In a standard assay the stimulation time is 5 min at 37°, but the kinetics of DISC formation may also be analyzed.[8,9] As unstimulated control, 3×10^7 metabolically labeled cells are left untreated. Stimulation is terminated by cooling the cells to 4° and removing unbound anti-APO-1 antibody by washing the cells once with ice-cold PBS. For lysis the cell pellet is resuspended in 1 ml of ice-cold 1% (v/v) Triton X-100-containing lysis buffer. Other detergents may also be used; note, however, that stronger detergents may disrupt the association of the signaling molecules and the receptor, whereas mild detergents such as Brij 58 or digitonin do not efficiently solubilize the receptor complex. The cells are incubated in lysis buffer for 15 min on ice and lysates are subsequently cleared by centrifugation at 14,000g at 4°. The lysate is then directly used for immunoprecipitation.

Immunoprecipitation of CD95 Death-Inducing Signaling Complex

For the unstimulated control the lysate of untreated cells is supplemented with 3 μg of anti-APO-1. This amount is equivalent to the stimulated sample, given that approximately 10% of the antibody used for stimulation has bound to the cells. In the control lysate the anti-APO-1 antibody can bind the CD95 receptor; however, formation of the DISC is no longer possible after cell lysis. In the following immunoprecipitation procedure both the stimulated and the unstimulated lysates are treated in the same way. For removing nonspecific binding proteins lysates are incubated twice for 1 hr, while rotating at 4°, with 30 μg of an isotype-matched control

[7] M. E. Peter, S. Hellbardt, R. Schwartz-Albiez, M. O. Westendorp, G. Moldenhauer, M. Grell, and P. H. Krammer, *Cell Death Differ.* **2**, 163 (1995).
[8] J. P. Medema, C. Scaffidi, F. C. Kischkel, A. Shevchenko, M. Mann, P. H. Krammer, and M. E. Peter, *EMBO J.* **16**, 2794 (1997).
[9] C. Scaffidi, J. P. Medema, P. H. Krammer, and M. E. Peter, *J. Biol. Chem.* **272**, 26953 (1997).

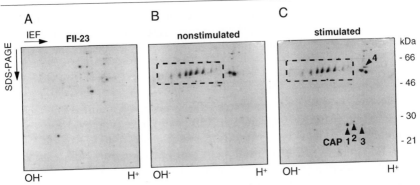

FIG. 2. Analysis of the CD95 DISC in K50 cells. K50 cells (3×10^7) were metabolically labeled with 0.5 mCi of [^{35}S]methionine/cysteine, lysed in lysis buffer, and subsequently supplemented with 3 µg of IgG$_3$ anti-APO-1 (B) or stimulated with IgG$_3$ anti-APO-1 (2 µg/ ml) for 5 min before lysis (C). Lysates were precleared with isotype-matched control antibody FII-23 covalently coupled to CNBr-activated Sepharose beads (A). Subsequently, the CD95 receptor was immunoprecipitated with protein A–Sepharose. Immunoprecipitates were analyzed by 2D IEF–SDS–PAGE and autoradiography. The dashed boxes indicate the migration position of CD95.

MAb such as the IgG$_3$ MAb FII-23[4] covalently coupled to 30 µl of CNBr-activated Sepharose 4B beads (Pharmacia). During this step no protein A– or G–Sepharose must be used, as this would already bind the anti-APO-1 antibody in the lysate. After each preclearing step the beads are separated from the lysate by centrifugation at 600g for 1 min. The beads of the last preclearing step may be saved for analysis as an additional negative control (Fig. 2A).

For precipitation of the DISC the precleared lysate is supplemented with 30 µl of protein A–Sepharose 4B beads (Sigma, St. Louis, MO) and incubated for more than 1 hr with rotating at 4°. During that time the anti-APO-1 antibody, either bound to the cross-linked CD95 receptor (stimulated condition) or the monomeric CD95 receptor (unstimulated condition), binds to the protein A–Sepharose beads, which are then separated from the lysate by centrifugation as described above. The beads are washed four times with 20 vol of ice-cold lysis buffer and centrifuged as described above. After the last wash residual lysis buffer is removed with a Hamilton syringe and dry beads may be stored at −20° until further analysis.

Two-Dimensional Gel Analysis of Immunoprecipitate

Two-dimensional gel IEF–SDS–PAGE is a standard analysis method and has been described previously in detail.[10] Briefly, first-dimension iso-

---

[10] L. A. Huber and M. E. Peter, *Electrophoresis* **15**, 283 (1994).

electric focusing is carried out in 22-cm glass tubes with a gel length of 18.75 cm and a diameter of 2.5 mm. To build the pH gradient gels are prefocused for 2 hr at 0.33 mA per gel until the voltage reaches 1200 V. Beads from the immunoprecipitation are supplemented with 150 μl of sample buffer and proteins are solubilized at 30° with shaking for 30 min. Solubilized samples are loaded together with the beads onto the prefocused tube gels and focused for 18–20 hr at 1200 V. After focusing, tube gels are ejected and equilibrated for 10 min in 5 ml of equilibration buffer. For the second dimension tube gels are placed on top of 12% SDS–12% (w/v) polyacrylamide gels (23 cm × 23 cm × 1.5 mm), using an agarose solution, and gels are run at 120 V for 15 hr. Gels are then fixed for >1 hr and subsequently amplified for 30 min with Amplify (Amersham). The dried gels are subjected to autoradiography for about 7 days at −70°, using intensifying screens.

Results

In Fig. 2 the result of a DISC analysis of CD95 receptor-transfected BL-60 cells (K50) is shown. Under nonstimulated conditions (Fig. 2B) only the CD95 receptor is immunoprecipitated; it migrates as distinct spots on a 2D gel because of different glycosylation of the molecule.[7] Additional spots likely represent background binding of nonspecific proteins, as they can also be seen in a control immunoprecipitation with an isotype control antibody (Fig. 2A). Only in the immunoprecipitate of the stimulated receptor (Fig. 2C) are the CAP proteins detectable, demonstrating that they are recruited to the CD95 receptor in a stimulation-dependent way. It is obvious from Fig. 2 that in a one-dimensional analysis of the DISC, CAP4 would not be distinguishable from the immunoprecipitated CD95 receptor and CAP1, -2, and -3 would be detected as one single band. Therefore, the 2D gel analysis is a powerful tool because it provides information about every component of the CD95 DISC.

We identified CAP1 and CAP2 as the phosphorylated DD-containing adapter molecule FADD (also called MORT-1).[11,12] At its N terminus FADD contains a death effector domain (DED), which was shown to be essential for the recruitment of CAP3 and CAP4 to the DISC.[13] CAP4 was

[11] A. M. Chinnaiyan, K. O'Rourke, M. Tewari, and V. M. Dixit, *Cell* **81,** 505 (1995).
[12] M. P. Boldin, E. E. Varfolomeev, Z. Pancer, I. L. Mett, J. H. Camonis, and D. Wallach, *J. Biol. Chem.* **270,** 7795 (1995).
[13] A. M. Chinnaiyan, C. G. Tepper, M. F. Seldin, K. O'Rourke, F. C. Kischkel, S. Hellbardt, P. H. Krammer, M. E. Peter, and V. M. Dixit, *J. Biol. Chem.* **271,** 4961 (1996).

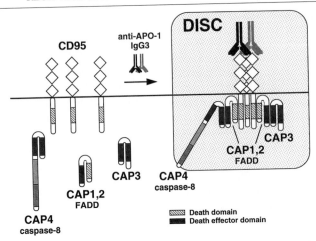

FIG. 3. Schematic representation of the CD95 DISC. See text for details.

identified to be a DED-containing protease[14] belonging to the new family of ICE-like proteases, the caspases, and was therefore named FLICE (for FADD-like ICE). It was also called MACH/Mch5[15,16] or, finally, caspase-8. The model in Fig. 3 summarizes the process of DISC formation as the first event during CD95-mediated apoptosis. Triggering of CD95 either by agonistic antibody or CD95 ligand leads to aggregation of the receptor with its intracellular death domain (DD) and the subsequent binding of the adapter molecule FADD. Through the DED of FADD, CAP3 and the effector molecule caspase-8 are recruited to the receptor complex. Recruitment of caspase-8 to the DISC leads to its proteolytic activation, which is the first step during CD95 signaling and can be detected as early as 10 sec after receptor triggering.[8,9] During activation all cytosolic caspase-8 is converted to active caspase-8 at the DISC. The active caspase-8 subunits p18 and p10 are released into the cytoplasm, where they can cleave and activate other caspases, leading to the death of the cell.

By using monoclonal antibodies against caspase-8 we demonstrated that two different caspase-8 isoforms, designated caspase-8/a and -8/b, are expressed in almost all cell lines tested.[9] With high-resolution 2D IEF–

[14] M. Muzio, A. M. Chinnaiyan, F. C. Kischkel, K. O'Rourke, A. Shevchenko, J. Ni, C. Scaffidi, J. D. Bretz, M. Zhang, R. Gentz, M. Mann, P. H. Krammer, M. E. Peter, and V. M. Dixit, *Cell* **85,** 817 (1996).

[15] M. P. Boldin, T. M. Goncharov, Y. V. Goltsev, and D. Wallach, *Cell* **85,** 803 (1996).

[16] T. Fernandes-Alnemri, R. C. Armstrong, J. Krebs, S. M. Srinivasula, L. Wang, F. Bullrich, L. C. Fritz, J. A. Trapani, K. J. Tomaselli, G. Litwack, and E. S. Alnemri, *Proc. Natl. Acad. Sci. U.S.A.* **93,** 7464 (1996).

SDS–PAGE we demonstrated that both isoforms are recruited to the CD95 DISC, although caspase-8/b almost completely comigrated with an unspecific spot that was also detected in the unstimulated control (Fig. 4A). Prolonging the time of CD95 receptor stimulation before DISC immunoprecipitation resulted in the detection of two new DISC components, which we named CAP5 and CAP6 (Fig. 4B). We identified CAP5 and CAP6 as being the DED-containing prodomain of caspases-8/a and -8/b, respectively.[8,9] The prodomain remains partly bound to the CD95 receptor after proteolytic activation of caspase-8. Therefore the method described here not only detects binding of signaling molecules to the activated CD95 receptor but is also suited to detect proteolytic activation of caspase-8.

Discussion

DISC formation is the first event after CD95 cross-linking, resulting in activation of caspase-8 and initiation of a signaling cascade leading to apoptosis.[8] Cells that are defective in DISC formation, e.g., because of ectopic expression of a truncated CD95 receptor[4] or a dominant negative mutant of FADD,[13] are resistant to CD95-mediated apoptosis. Therefore, DISC formation is an essential step of CD95 signaling. The method de-

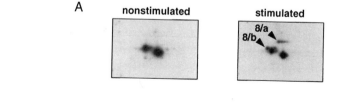

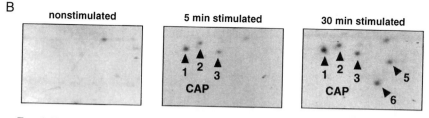

FIG. 4. Two isoforms of caspase-8 are recruited and activated by the CD95 DISC. (A) K50 cells (3×10^7) were treated as described in Fig. 2. Arrowheads indicate migration positions of caspases 8/a and 8/b. (B) BJAB cells (3×10^7) were metabolically labeled and left either unstimulated or stimulated with IgG$_3$ anti-APO-1 (2 μg/ml) for the indicated periods of time. The CD95 DISC was analyzed as described in Fig. 2. Arrowheads indicate the migration position of CAP 1, 2, 3, 5, and 6. Only the relevant parts of the 2D gels are shown.

scribed here is able to investigate this signaling step and to visualize the components of the CD95 DISC in a single analysis.

In the following some modifications of the above-described technique are discussed to make it suitable for other applications. Cell lines expressing high levels of CD95 are most often studied for DISC analysis; however, low-expressing cell lines can also be analyzed by prolonging the exposure time of the 2D gel or by increasing the number of cells per sample.[8] In the latter case more preclearing steps are recommended during immunoprecipitation to reduce nonspecific signals. We described cell lines that form only a few DISCs, although they express high levels of CD95 on their surface.[17] In these cells only a little caspase-8 is activated at the receptor level whereas strong caspase-8 activation can be observed downstream of a signaling step involving mitochondria. This type of caspase activation can be blocked by Bcl-2 overexpression. Only in these cell lines, designated type II cells, is Bcl-2 effective in inhibiting CD95-mediated apoptosis. The reason why type II cells form only a few DISCs is unknown at present. So far, only the T cell lines Jurkat and CEM have been identified to be type II cells. Therefore, DISC analysis in these cell lines is more difficult and can be performed only by increasing the number of cells used per analysis. When cell numbers are limiting, the unstimulated control may be replaced by subjecting the stimulated lysate to an additional immunoprecipitation after the protein A precipitation step, using anti-APO-1 antibody covalently coupled to Sepharose-4B beads. By doing so only the monomeric CD95 receptor will be immunoprecipitated and can therefore serve as a negative control.[4]

Specific monoclonal antibodies have become available for the DISC components FADD and caspase 8.[9] These reagents make DISC analysis much easier, as the detection of the receptor-associated molecules can now also be done by Western blotting with anti-FADD (Transduction Laboratories) and anti-caspase-8 monoclonal antibodies.[18] CD95 is, so far, the only death receptor for which DISC analysis has been established as a standard method for investigating early apoptotic signaling events. Several reports have demonstrated an involvement of the CD95 system in apoptosis induction by cytotoxic drugs.[19–21] In addition, it has been shown that resis-

[17] C. Scaffidi, S. Fulda, F. Li, C. Friesen, A. Srinivasan, K. J. Tomasseli, K.-M. Debatin, P. H. Krammer, and M. E. Peter, *EMBO J.* **17,** 1675 (1998).
[18] C. Scaffidi, P. H. Krammer, and M. E. Peter, "Methods: A Companion to *Methods in Enzymology*," **17,** 287 (1999).
[19] M. Müller, S. Strand, H. Hug, E. M. Heinemann, H. Walczak, W. J. Hofmann, W. Stremmel, P. H. Krammer, and P. R. Galle, *J. Clin. Invest.* **99,** 403 (1997).
[20] S. Fulda, H. Sieverts, C. Friesen, I. Herr, and K. M. Debatin, *Cancer Res.* **57,** 3823 (1997).
[21] C. Friesen, I. Herr, P. H. Krammer, and K. M. Debatin, *Nature Med.* **2,** 574 (1996).

tance of tumor cells toward such drugs can be caused by a defect in CD95 signal transduction.[21] Therefore, DISC analysis of CD95 may be useful in determining the block in, e.g., drug-resistant cancer cells.

Acknowledgments

This work was supported by grants from the Deutsche Forschungsgemeinschaft, the Bundesministerium für Forschung und Technologie, and the Tumor Center Heidelberg/Mannheim.

[32] Measurement of Ceramide Levels by the Diacylglycerol Kinase Reaction and by High-Performance Liquid Chromatography–Fluorescence Spectrometry

By Ron Bose and Richard Kolesnick

Introduction

The sphingomyelin (SM) pathway is a ubiquitous, evolutionarily conserved signaling system analogous to conventional systems such as the cAMP and phosphoinositide pathway.[1-3] Ceramide, which serves as second messenger in this pathway, is generated from SM by the action of a neutral or acidic sphingomyelinase (SMase), or by *de novo* synthesis through the enzyme ceramide synthase. Direct targets for ceramide action have been identified, including ceramide-activated protein kinase, ceramide-activated protein phosphatase, and protein kinase Cζ, which couple the SM pathway to well-defined intracellular signaling cascades. The SM pathway induces differentiation, proliferation, or growth arrest depending on the cell type. Often, however, the outcome of ceramide signaling is apoptosis. Mammalian systems respond to diverse stresses with ceramide generation, and studies show that yeast manifest a form of this response.[4,5] Thus, ceramide signaling is an older stress response system than the caspase/apoptotic pathway, and hence these two pathways must have been linked later in evolution. Signaling of the stress response through ceramide appears to

[1] L. A. Peña, Z. Fuks, and R. Kolesnick, *Biochem. Pharmacol.* **53**, 615 (1997).
[2] Y. A. Hannun, *Science* **274**, 1855 (1996).
[3] S. Spiegel, D. Foster, and R. Kolesnick, *Curr. Opin. Cell Biol.* **8**, 159 (1996).
[4] G. M. Jenkins, A. Richards, T. Wahl, C. Mao, L. Obeid, and Y. Hannun, *J. Biol. Chem.* **272**, 32566 (1997).
[5] G. B. Wells, R. C. Dickson, and R. L. Lester, *J. Biol. Chem.* **273**, 7235 (1998).

play a role in development of human diseases including ischemia–reperfusion injury, insulin resistance and diabetes, atherogenesis, septic shock, and ovarian failure.[1-3] Further, ceramide signaling mediates the therapeutic effect of chemotherapy and radiation in some cells.[1-3] An understanding of mechanisms by which ceramide regulates physiological and pathological events in specific cells may provide new targets for pharmacologic intervention.

Cellular ceramide content has generally been measured by one of four assays: the diacylglycerol (DAG) kinase assay,[6] a method involving lipid charring,[7] methods involving derivatization of ceramide followed by high-performance liquid chromatography (HPLC),[8] or a variety of radiolabeling techniques.[9] Direct comparisons between these techniques have reproducibly yielded the same result.[10] In this chapter, we detail the methodology for measuring ceramide with diacylglycerol kinase, an enzymatic assay, and by a nonenzymatic biochemical technique involving deacylation of ceramide and its derivitization with o-phthalaldehyde (OPA).

Methods

Diacylglycerol Kinase

Principle of the Diacylglycerol Kinase Reaction. The *Escherichia coli* enzyme diacylglycerol kinase phosphorylates ceramide with $[\gamma\text{-}^{32}P]ATP$ as the phosphate donor to generate $[^{32}P]$ceramide 1-phosphate. This reaction occurs in a mixed micelle and is performed under conditions in which the reaction goes to completion. $[^{32}P]$Ceramide 1-phosphate is then extracted, resolved from other reaction products by thin-layer chromatography, and quantitated by liquid scintillation counting, in comparison with a standard curve.

Materials. Ceramide type III and diethylenetriaminepentaacetic acid (DTPA) are obtained from Sigma (St. Louis, MO). Cardiolipin is from Avanti Polar Lipids (Alabaster, AL). $[\gamma\text{-}^{32}P]ATP$ (3000 Ci/mmol) is from NEN Life Sciences (Boston, MA). *Escherichia coli* diacylglycerol kinase is

---

[6] P. P. Van Veldhoven, W. R. Bishop, and R. M. Bell, *Anal. Biochem.* **183,** 177 (1989).

[7] J. B. Marsh and D. B. Weinstein, *J. Lipid Res.* **7,** 574 (1966).

[8] A. H. Merrill, Jr., E. Wang, R. E. Mullins, W. C. L. Jamison, S. Nimkar, and D. C. Liotta, *Anal. Biochem.* **171,** 373 (1988).

[9] R. N. Kolesnick, *Prog. Lipid Res.* **30,** 1 (1991).

[10] A. Tepper, G. R. Boesen-de Cock, E. de Vries, J. Borst, and W. van Blitterswijk, *J. Biol. Chem.* **272,** 24308 (1997).

from Biomol (Plymouth Meeting, PA) or from Calbiochem (San Diego, CA). Octyl-β-D-glucopyranoside is from Calbiochem. Silica gel 60 thin-layer chromatography plates are from Whatman (Clifton, NJ). The bath sonicator (Branson 1200) is from Fisher Scientific (Danbury, CT).

Stock Solutions. Methanolic KOH (0.1 *N*) is prepared by dissolving KOH in 100% methanol. Buffered saline solution (BSS) consists of 135 m*M* NaCl, 1.5 m*M* CaCl$_2$, 0.5 m*M* MgCl$_2$, 5.6 m*M* glucose, and 10 m*M* HEPES, pH 7.2. Reaction buffer (2×) consists of 100 m*M* NaCl, 100 m*M* imidazole, 2 m*M* EDTA, and 25 m*M* MgCl$_2$ and is pH 6.5.

Assay Procedure. Lipids from tissue culture cells are extracted with 1 ml of chloroform–methanol–1 *N* HCl (100 : 100 : 1, v/v/v) in the presence of a total volume of 0.3 ml of aqueous solution (serum-free medium or saline). Serum contains significant amounts of ceramide and should be removed before extraction of cells. The resulting organic phase is separated and dried under N$_2$. This lipid extract is subjected to mild alkaline hydroly-sis, by adding 0.5 ml of 0.1 *N* methanolic KOH, and incubating for 1 hr at 37°, to digest glycerophospholipids. Samples are reextracted with 0.5 ml of chloroform, 0.27 ml of BSS, and 30 μl of 100 m*M* EDTA and the organic phase is dried under N$_2$.

The following reaction mixture should be prepared immediately before the running of the assay. One hundred microliters of reaction mixture is required per sample. It can be prepared in bulk and then aliquoted to each sample. Cardiolipin (150 μg) is dried under N$_2$, 20 μl of 1 m*M* DTPA is added, and this mixture is sonicated by a bath sonicator for 2 min. The mixture should now appear milky and homogeneous. Octyl-β-D-glucopy-ranoside (825 m*M*, 6.2 μl) is added and the solution should now appear clear or mildly cloudy. To this mixture is added 50 μl of 2× reaction buffer, 7.2 μl of 10 m*M* imidazole, 0.8 μl of 1 m*M* DTPA, 2 μl of 100 m*M* dithiothreitol (DTT), 1 μl of 100 m*M* ATP, 3.5 μg of diacylglycerol kinase, 10 μCi of [γ-^{32}P]ATP (3000 Ci/mmol), and sufficient distilled water to bring the volume to 100 μl.

Reaction mixture (100 μl) is added to each dried lipid sample. Samples are vortexed gently and incubated for 30 min at room temperature. The reaction is terminated by extraction of lipids with 1 ml of chloroform–methanol–1 *N* HCl (100 : 100 : 1), 170 μl of buffered saline solution, and 30 μl of 100 m*M* EDTA. The lower organic phase is dried under N$_2$. Ceramide 1-phosphate is resolved by thin-layer chromatography on silica gel 60 plates using a solvent system of chloroform–methanol–acetic acid (65 : 15 : 5, v/v/v), detected by autoradiography, and incorporated ^{32}P is quantified by liquid scintillation counting. The level of ceramide is deter-mined by comparison with a standard curve of 50–1000 pmol of ceramide type III, which is generated concomitantly. Under the assay conditions described above, the reaction goes to completion, and hence is not subject to potential activators contained within a biologic sample.[6]

High-Performance Liquid Chromatography–Fluorescence Spectrometry Assay

Principle of High-Performance Liquid Chromatography–Fluorescence Spectrometry Assay. Ceramide is deacylated by strong alkaline digestion to sphingosine, which contains a primary amine. *o*-Phthalaldehyde (OPA) reacts with the primary amine of sphingosine to form a fluorescent compound that is separated by HPLC and quantitated by fluorescence spectrometry. This method represents a modification of the procedure described by Merrill and co-workers for quantitation of sphingosine and other sphingoid bases.[8]

Materials and Equipment. Ceramide type III, sphingosine, and OPA are obtained from Sigma. A 14-carbon analog of sphingosine is from Matreya (Pleasant Gap, PA). HPLC equipment is from Waters (Millipore, Milford, MA) and includes a Nova Pak C_{18} column (60 Å, 4 μm, 3.9 × 150 mm), model 420/420AC/fluorescence detector, and model 746 data module integrator.

Assay Procedure. Lipids from tissue culture cells are extracted with 1 ml of chloroform–methanol–1 N HCl (100 : 100 : 1, v/v/v) in the presence of a total volume of 0.3 ml of aqueous solution (serum-free medium or saline). Serum contains significant amounts of ceramide and should be removed before extraction of cells. The resulting organic phase is separated and dried under N_2. Standards of 50–2000 pmol of ceramide type III are prepared fresh. An internal standard of 500 pmol of a 14-carbon analog of sphingosine is added to both samples and standards to estimate recovery. Samples and standards are dried under N_2 and 0.5 ml of 1 M KOH in 90% (v/v) CH_3OH is added. Tubes are sealed with screw caps and heated at 90° for 1 hr to quantitatively convert ceramide into sphingosine. This digestion procedure does not convert complex sphingolipids, such as sphingomyelin, galactosylceramide, or glucosylceramide, into sphingosine. After digestion, samples and standards are extracted with 0.5 ml of 1 M HCl in CH_3OH, 1.0 ml of $CHCl_3$, and 0.75 ml of 1 M aqueous NaCl. The organic phase is dried under N_2 and redissolved in 0.1 ml of CH_3OH. OPA solution should be prepared fresh daily by dissolving 10 mg of OPA in 0.2 ml of ethanol and then adding 10 μl of 2-mercaptoethanol followed by 19.8 ml of 3% (v/v) aqueous boric acid (adjusted to pH 10.5 with KOH). OPA solution is added to the tubes (0.1 ml/tube) and they are incubated at room temperature for 5 min. A total of 0.5 ml of Solvent A [CH_3OH–5 mM aqueous potassium phosphate (pH 7.0) (90:10, v/v)] is added and samples and standards are resolved by reverse-phase HPLC, using a Nova Pak C_{18} column and isocratic elution with solvent A. The following settings are employed: injection volume, 20 μl; solvent flow rate, 0.6 ml/min; cycle time, 30 min; fluorescence excitation wavelength, 340 nm; and emission wavelength, 455 nm. Free sphingosine is measured by use of milder digestion conditions, 0.1 M KOH in 100% CH_3OH at 37° for 1 hr, with subsequent steps identical to those above. Retention times of various sphingosines depend on alkyl chain length: the 14-, 16-, 18-, and 20-carbon analogs of sphingosine elute in 3.6, 5.3, 8.1, and 13.7 min, respectively. The 18-carbon sphingosine is the predominant form contained within mammalian ceramides. Quantitation is performed by comparison of fluorescent peaks of the samples with the standard curve coupled with correction for recovery by use of the internal standard. Use of an internal standard is necessary because with large sample numbers, the total HPLC run time can be long,

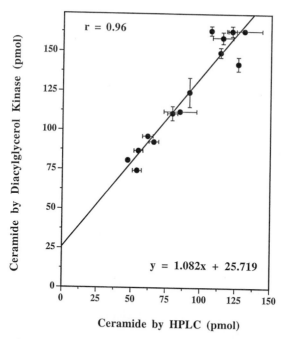

FIG. 1. Comparison of the diacylglycerol kinase assay with the HPLC–fluorescence spectrometry assay. Stimulated or unstimulated SL2 cells were extracted and samples were divided for ceramide determination by both diacylglycerol kinase and HPLC–fluorescence spectrometry assays. Data (mean ± SEM) represent a compilation of multiple determinations from two experiments.

creating the potential for small degrees of fluorescent product degradation. Formally, the fluorescence signal represents the sum of ceramide and free sphingosine, but the contribution of free sphingosine is minor, as ceramide levels are 10- to 20-fold greater than free sphingosine in mammalian cells.

Conclusions

A comparison of ceramide levels measured by both the diacylglycerol kinase and HPLC methodologies is shown in Fig. 1. In our laboratory, the level of detection of ceramide by either method is about 50 pmol. Figure 1 shows that, with minor differences, these two techniques yield similar results. One report showed that these two techniques measure similar ceramide levels up to 750 pmol.[11]

Although neither technique is preferred, both have advantages. The diacylglycerol kinase reaction can be performed with up to 50 samples at a time with relative ease, and requires no sophisticated apparatus. Alternately, derivatization with OPA does not use radioactivity or a biologic assay.

[11] M. Garzotto, M. White-Jones, Y. Jiang, D. Ehleiter, W. C. Liao, A. Haimovitz-Friedman, Z. Fuks, and R. Kolesnick, *Cancer Res.* **58**, 2260 (1998).

[33] Measurement of Ceramide Synthase Activity

By RON BOSE and RICHARD KOLESNICK

Introduction

Ceramides can be synthesized *de novo* by acylation of the sphingoid bases sphingosine, dihydrosphingosine (sphinganine), or phytosphingosine via the action of ceramide synthase (sphingosine *N*-acyltransferase; EC 2.3.1.24) yielding ceramide, dihydroceramide, and phytoceramide, respectively.[1] Dihydroceramide can be oxidized to ceramide by introduction of a *trans*-4,5 double bond by the enzyme dihydroceramide desaturase.[2] While the quantity of phytoceramide in mammalian cells is limited, phytoceramide is the major ceramide in yeast.[3]

[1] R. N. Kolesnick, *Prog. Lipid Res.* **30**, 1 (1991).
[2] C. Michel, G. van Echten-Deckert, J. Rother, K. Sandhoff, E. Wang, and A. H. Merrill, *J. Biol. Chem.* **272**, 22432 (1997).
[3] R. L. Lester and R. C. Dickson, *Adv. Lipid Res.* **26**, 253 (1993).

The ceramide synthase pathway can be stimulated by drugs and ionizing radiation, typically over a period of hours.[4,5] Treatment of cells with the chemotherapeutic drug daunorubicin resulted in prolonged ceramide generation in P388, HL-60, and U937 cells.[4,6] Further, the fungal toxin fumonisin B1 (FB1), which Merrill and co-workers have documented as a specific inhibitor of ceramide synthase activity,[7] blocked daunorubicin-induced synthase activation, ceramide generation, and cell death. The mechanism of activation of the enzyme by daunorubicin has not yet been determined although a signal from damaged DNA may be involved. Consistent with this hypothesis, studies show that metabolic incorporation of ^{125}I-labeled 5-iodo-2′-deoxyuridine ([^{125}I]dURd), which produces DNA double-strand breaks, signaled *de novo* ceramide synthesis by posttranslational activation of ceramide synthase.[5]

Ceramide synthase is a membrane-bound enzyme and has not yet been molecularly identified. Ceramide synthase activity has been detected in only two intracellular compartments, the endoplasmic reticulum (ER)[8] and the mitochondria.[9] Its role at the ER is almost certainly related to the requirement for ceramide synthesis for generation of higher sphingolipids. The function of ceramide synthase at the mitochondria is presently unknown. However, the central role of mitochondria during the commitment phase of the apoptotic process[10] and the wealth of information linking ceramide signaling of apoptosis to Bcl-2,[11] suggest the function of ceramide synthase is to signal stress in this compartment. Consistent with this observation, in a screen for genes involved in apoptosis, Paumen *et al.*[12] isolated carnitine palmitoyltransferase I (CPT I). CPT I is located in the outer mitochondrial membrane and catalyzes the transfer of long-chain fatty acids into the mitochondria for β oxidation.[12] In the presence of a CPT I inhibitor, palmitate was rerouted for *de novo* ceramide synthesis, apparently via

[4] R. Bose, M. Verheij, A. Haimovitz-Friedman, K. Scotto, Z. Fuks, and R. N. Kolesnick, *Cell* **82**, 405 (1995).

[5] W.-C. Liao, A. Haimovitz-Friedman, R. Persaud, M. McLaughlin, D. Ehleiter, N. Zhang, M. Gatei, M. Lavin, R. Kolesnick, and Z. Fuks, *J. Biol. Chem.* **274**, 17908 (1999).

[6] M. Boland, S. Foster, and L. O'Neill, *J. Biol. Chem.* **272**, 12952 (1997).

[7] A. J. Merrill, Jr., E. Wang, T. Vales, E. Smith, J. Schroeder, D. Menaldino, C. Alexander, H. Crane, J. Xia, D. Liotta, F. Meredith, and R. Riley, *Adv. Exp. Med. Biol.* **392**, 297 (1996).

[8] E. C. Mandon, I. Ehses, J. Rother, G. van Echten, and K Sandhoff, *J. Biol. Chem.* **267**, 1144 (1992).

[9] H. Shimeno, S. Soeda, M. Yasukouchi, N. Okamura, and A. Nagamatsu, *Biol. Pharm. Bull.* **18**, 1335 (1995).

[10] N. Zamzami, C. Brenner, I. Marzo, S. A. Susin, and G. Kroemer, *Oncogene* **16**, 2265 (1998).

[11] R. N. Kolesnick and M. Kronke, *Annu. Rev. Physiol.* **60**, 643 (1998).

[12] M. Paumen, Y. Ishida, M. Muramatsu, M. Yamamoto, and T. Honjo, *J. Biol. Chem.* **272**, 3324 (1997).

ceramide synthase, resulting in FB1-inhibitable apoptosis. A better understanding of the mechanisms regulating ceramide synthase action should help elucidate the role of ceramide in some forms of stress-induced apoptosis.

Methods

Principle of the Ceramide Synthase Assay

Microsomal membrane preparations are incubated with sphingoid base, usually sphinganine, and radiolabeled palmitoyl-coenzyme A, and the resulting radiolabeled dihydroceramide is quantified.

Materials and Equipment

Ceramide type III, palmitoyl-coenzyme A, and defatted bovine serum albumin are obtained from Sigma Chemical (St. Louis, MO). Sphinganine is from Biomol (Plymouth Meeting, PA). [1-^{14}C]Palmitoyl-coenzyme A (55 mCi/mmol) is from American Radiolabeled Chemicals (St. Louis, MO). Diacylglycerol is from Serdary/Doosan Research Laboratory (Englewood Cliffs, NJ). Silica gel 60 thin-layer chromatography plates are from Whatman (Clifton, NJ).

Stock Solutions

Homogenization buffer consists of 25 mM HEPES (pH 7.4), 5 mM EGTA, 50 mM NaF, leupeptin (10 μg/ml), and soybean trypsin inhibitor (10 μg/ml). A stock solution of 2 mM sphinganine in 100% ethanol is prepared fresh every 3–7 days.

Assay Procedure

Microsomal membranes are prepared by collecting 75 × 10^6 cells via centrifugation, washing once with ice-cold phosphate-buffered saline (PBS), followed by resuspension of cells into 300 μl of homogenization buffer. Cells are disrupted using an ice-cold, motor-driven Tenbroeck tissue homogenizer and lysates are centrifuged at 800g for 5 min. The resulting supernatant is centrifuged at 250,000g for 30 min.[13] The microsomal membrane pellet is resuspended in 1 ml of cold homogenization buffer and the protein concentration measured by the method of Bradford.[14] Membranes should be prepared fresh on the day of experiment.

[13] J. Liu, S. Mathias, Z. Yang, and R. N. Kolesnick, *J. Biol. Chem.* **269**, 3047 (1994).
[14] M. M. Bradford, *Anal. Biochem.* **72**, 248 (1976).

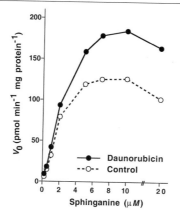

	V_{max} (pmol/min/mg)	K_m (μM)
Daunorubicin	240	4.0
Control	140	2.7

Fig. 1. Activation of ceramide synthase by daunorubicin. P388 murine leukemia cells at an initial density of 1.0×10^6/ml were treated with 10 μM daunorubicin for 6 hr. Microsomal membranes were prepared and ceramide synthase activity assayed as described. Dihydroceramide was resolved by thin-layer chromatography, detected by comigration with ceramide standards, and quantified by liquid scintillation counting. Values represent the mean of determinations from four separate experiments. K_m and V_{max} were calculated by Eadie–Hofstee analysis.

Assays of ceramide synthase activity are based on the procedure of Harel and Futerman.[15] Microsomal membrane protein (75 μg) is incubated in a 1-ml reaction mixture containing 2 mM MgCl$_2$, 20 mM HEPES (pH 7.4), 20 μM defatted bovine serum albumin, varying concentrations (0.2–20 μM) of sphinganine, 70 μM unlabeled palmitoyl-coenzyme A, and 3.6 μM (0.2 μCi) [1-^{14}C]palmitoyl-CoA. Sphinganine is dried under N$_2$ from a stock solution in 100% ethanol and dissolved by sonication in the reaction mixture prior to addition of microsomal membranes and palmitoyl-CoA. Important controls include microsomes that have been boiled for 20 min and samples lacking sphinganine. The reaction is started by addition of palmitoyl-CoA and stopped after 1 hr at 37° by extraction of lipids with 2 ml of chloroform–methanol (1:2, v/v). A total of 500 μl of lower phase is removed and dried under N$_2$. The resulting lipid film is resuspended into 50 μl of chloroform–methanol (1:1, v/v) containing ceramide (type III from bovine brain,

[15] R. Harel and A. H. Futerman, *J. Biol. Chem.* **268,** 14476 (1993).

1 mg/ml) and diacylglycerol (1 mg/ml). Forty microliters of this solution is loaded onto a silica gel thin-layer chromatography plate. Dihydroceramide is resolved from free radiolabeled fatty acid, using a solvent system of chloroform–methanol–3.5 N aqueous ammonium hydroxide (85:15:1, v/v/v), identified by iodine vapor staining on the basis of comigration with ceramide type III standards, and quantified by liquid scintillation counting. Relevant R_f values in this system include palmitoyl-CoA, 0.0; sphingosine, 0.17; palmitic acid, 0.1–0.2; ceramide, 0.66; and diacylglycerol, 0.78. The velocity of the reaction is linear for at least 2 hr and the amount of palmitoyl-coenzyme A consumed does not exceed 5% of total under these experimental conditions.

Results

Activation of ceramide synthase by daunorubicin treatment of P388 cells is shown in Fig. 1. Cells were treated with 10 μM daunorubicin for 6 hr and then microsomal membranes were prepared. Eadie–Hofstee transformation of the data revealed a 70% increase in the apparent V_{max} of the enzyme with little or no change in the K_m for sphinganine.

[34] Measurement of Sphingomyelinase Activity

By ERICH GULBINS and RICHARD KOLESNICK

Introduction

Three forms of sphingomyelinase (SMase) have been described that manifest acid, neutral, or basic pH optima for maximal enzyme activity.[1] Acid SMase mediates diverse biologic functions in transmembrane signaling, membrane homeostasis, intracellular trafficking, and even in the regulation of lipoprotein deposition in vessel walls.[1–4] A rapid activation of acid SMase with concomitant release of ceramide has been shown upon cellular stimulation via CD95, the 55-kDa tumor necrosis factor (TNF) receptor or environmental stress, e.g., ionizing radiation, to list a few examples.[5–10]

[1] Y. A. Hannun, *Science* **274,** 1855 (1996).
[2] R. N. Kolesnick and M. Krönke, *Annu. Rev. Physiol.* **60,** 643 (1988).
[3] J. P. Slotte and E. L. Biermann, *Biochem. J.* **250,** 653 (1988).
[4] X. Xu and I. Tabas, *J. Biol. Chem.* **250,** 24849 (1991).
[5] S. Schütze, K. Potthoff, T. Machleidt, D. Berkovic, K. Wiegmann, and M. Krönke, *Cell* **71,** 765 (1992).

Acid SMase may be critical for CD95-triggered apoptosis in some cell types, since B lymphocytes from Niemann–Pick disease type A patients, who suffer from an inborn defect of acid SMase activity, display attenuated apoptosis after CD95 cross-linking.[11] Likewise, acid SMase knockout mice showed defects in endothelial cell apoptosis on irradiation or treatment with bacterial lipopolysaccharide.[10,12]

A second enzyme involved in the degradation of cellular SM is neutral SMase.[1] This enzyme is less well characterized than acid SMase. The activity of neutral SMase is increased on ligation of the 55-kDa TNF receptor, CD95, L-selectin, or after oxidative stress.[13-15] Neutral SMase may be suppressed by glutathione, a reduction in the level of which may signal neutral SMase activation during stress-induced apoptosis.[16] Adam-Klages and coworkers showed that activation of neutral SMase via TNF depends on a short motif in the cytoplasmic tail of the TNF receptor, which binds an adaptor protein termed FAN.[13] FAN seems to transmit the activation signal to the neutral SMase, although the exact mechanism is unknown. Definitive evidence of the role of neutral SMase in signaling will have to wait for characterization of this enzyme(s) at the molecular level. In this regard, Tomiuk et al.[17] described the isolation of a protein, that, when overexpressed, increased cellular neutral SMase activity. Whether this enzyme is

[6] M. G. Cifone, R. DeMaria, P. Roncali, M. R. Rippo, M. Azuma, L. L. Lanier, A. Santoni, and R. Testi, J. Exp. Med. 180, 1547 (1994).

[7] C. G. Tepper, S. Jayadev, B. Liu, A. Bielawska, R. A. Wolff, S. Yonehara, Y. A. Hannun, and M. F. Seldin, Proc. Natl. Acad. Sci. U.S.A. 92, 8443 (1995).

[8] E. Gulbins, R. Bissonette, A. Mahboubi, W. Nishioka, T. Brunner, G. Baier, G. Baier-Bitterlich, C. Byrd, F. Lang, R. Kolesnick, A. Altman, and D. Green, Immunity 2, 341 (1995).

[9] B. Brenner, K. Ferlinz, M. Weller, H. Grassmé, U. Koppenhoefer, J. Dichgans, K. Sandhoff, F. Lang, and E. Gulbins, Cell Death Differ. 5, 29 (1998).

[10] P. Santana, L. A. Pena, A. Haimovitz-Friedman, S. Martin, D. Green, M. McLoughlin, C. Cordon-Cardo, E. Schuchman, Z. Fuks, and R. Kolesnick, Cell 86, 189 (1996).

[11] R. DeMaria, M. R. Rippo, E. Schuchman, and R. Testi, J. Exp. Med. 187, 897 (1998).

[12] A. Haimovitz-Friedman, C. Cordon-Cardo, S. Bayoumy, M. Garzotto, M. McLoughlin, R. Gallily, C. K. Edwards, E. H. Schuchman, Z. Fuks, and R. Kolesnick, J. Exp. Med. 186, 1831 (1997).

[13] S. Adam-Klages, D. Adam, K. Wiegman, S. Struve, W. Kolanus, J. Schneider-Mergener, and M. Krönke, Cell 86, 937 (1996).

[14] M. G. Cifone, P. Roncaioli, R. DeMaria, G. Carnarda, A. Santoni, G. Ruberti, and R. Testi, EMBO J. 14, 5859 (1997).

[15] B. Brenner, H. Grassmé, C. Müller, F. Lang, C. P. Speer, and E. Gulbins, Exp. Cell. Res. 243, 123 (1998).

[16] B. Liu and Y. A. Hannun, J. Biol. Chem. 272, 16281 (1997).

[17] S. Tomiuk, K. Hofman, M. Nix, M. Zumbansen, and W. Stoffel, Proc. Natl. Acad. Sci. U.S.A. 95, 3638 (1998).

actually neutral SMase will require additional experimentation. Basic SMase, which is found in the gastrointestinal tract, is involved in digestion but not signaling, and is not discussed further in this chapter.

Methods

Acid Sphingomyelinase

Principle. The activity of the acid SMase is measured as the degradation of radioactive SM to ceramide and phosphorylcholine. Since the choline group is soluble in water and contains the radioactive label, released radioactivity can be easily separated from the radioactive SM substrate, which partitions into the organic phase during modified Folch extraction. Acid SMase is discriminated from the neutral or basic activity by performing the assay at pH 5.0. The activity of the acid SMase can be detected in immunoprecipitates of the enzyme as well as in whole cell lysates.[9,18]

Materials. [^{14}C]Sphingomyelin (54.5 mCi/mmol) is purchased from NEN Life Science Products (Frankfurt, Germany) and protein A- or G-coupled agarose is from Santa Cruz (Heidelberg, Germany). All other reagents are obtained from Sigma (Deisenhofen, Germany). For bath sonication of cells, a high-intensity ultrasonic processor (high-intensity ultrasonic processor; Sigma) with a 2-mm microtip (Sigma) is employed. SM micelles are prepared in a bath sonicator (1200; Branson, Danbury, CT). Silica gel 60 thin-layer chromatography plates are from Merck (Darmstadt, Germany).

Assay Procedure. Immunoprecipitates of acid SMase are obtained after partial lysis of cells (1–10 × 10^6 cells/sample, depending on the level of expression of acid SMase) in ice-cold buffer containing 50 mM Tris (pH 7.4), 10 mM bacitracin, 1 mM benzamidine, 1 mM Na$_3$VO$_4$ aprotinin and leupeptin (10 μg/ml each), soybean trypsin inhibitor (0.1 mg/ml), and 0.2% (v/v) Triton X-100 (buffer A). The following steps should be performed at 4° to avoid inactivation or proteolytic degradation of the enzyme. After addition of lysis buffer, samples are immediately sonicated three times for 10 sec each, avoiding heating of the sample. Cell debris should be pelleted by a 5-min centrifugation at 600g, and the supernatant transferred to a new tube containing an equal amount of 50 mM Tris (pH 7.4), 3% (v/v) Nonidet P-40 (NP-40), 1% (v/v) Triton X-100, 1 mM Na$_3$VO$_4$, and aprotinin and leupeptin (10 μg/ml each) (buffer B). Acid SMase is subsequently immunoprecipitated by incubation for 4 hr at 4° with antisera, and immune complexes are immobilized by addition of 20 μl of protein

[18] H. Grassmé, E. Gulbins, B. Brenner, K. Ferlinz, K. Sandhoff, K. Harzer, F. Lang, and T. F. Meyer, *Cell* **91**, 605 (1997).

A- and G-coupled agarose for 60 min. Since some agarose preparations contain an acid SMase activity as a contaminant, the agarose must be tested for purity prior to use. Immunoprecipitates are washed three times in buffer B followed by three washes in 50 mM sodium acetate (pH 5.0), 0.2% (v/v) Triton X-100, 1 mM Na$_3$VO$_4$, and aprotinin and leupeptin (10 μg/ml each). Washed immunoprecipitates are resuspended into 30 μl of [^{14}C]sphingomyelin (100 nCi/sample, 60 μM final concentration) in 250 mM sodium acetate (pH 5.0), 1.3 mM EDTA, 0.05% (v/v) NP-40 (assay buffer) at 37°. Since the substrate [^{14}C]sphingomyelin is insoluble in water, it is first dried by SpeedVac (Savant, Hicksville, NY) and solubilized into micelles in assay buffer, using a Branson bath sonicator for 10 min. Samples are incubated for 30 min and extracted with 800 μl of CHCl$_3$–CH$_3$OH (2:1, v/v) and 250 μl of H$_2$O. An aliquot of the upper phase is collected and radioactivity reflecting the degradation of [^{14}C]sphingomyelin is quantified by liquid scintillation counting. During the collection of the upper phase care should be taken to avoid contamination with lower phase. Figure 1 shows that immunoprecipitates of acid SMase from Jurkat cells display a K_m for sphingomyelin of 2 μM, and a V_{max} of 4.3 μmol/mg protein/hr. Typically, triggering CD95 results in a threefold increase in the V_{max} of acid SMase to 13.5 μmol/mg protein/hr with no change in the K_m.

To measure acid SMase activity in whole-cell lysates, cells are collected into a buffer consisting of 0.1% (v/v) Triton X-100, 50 mM sodium acetate, 0.1 mM Na$_3$VO$_4$ and aprotinin and leupeptin (10 μg/ml each). Adherent cells are directly scraped into this buffer, while suspension cells should be centrifuged at 500g for 3 min and the supernatant discarded prior to addition of buffer. Extracts are then sonicated as described above, centrifuged at 600g for 5 min, and the supernatants are added to an equivalent volume (50–100 μl) of [^{14}C]SM (100 nCi/sample) in 250 mM sodium acetate (pH 5.0), 1.3 mM EDTA, and 0.05% (v/v) NP-40. [^{14}C]SM is prepared, samples are handled, and activity is quantified as described above.

An alternative protocol has been described by Krönke and colleagues.[5,19] Cells are stimulated, washed, and resuspended in 0.2% (v/v) Triton X-100. The samples are incubated for 15 min at 4° to permit lysis. Samples are then centrifuged at 16,000g and an aliquot of the supernatant is incubated for 2 hr at 37° with [^{14}C]SM in 250 mM sodium acetate (pH 5.0) and 1 mM EDTA. Radioactive phosphorylcholine is then extracted with CHCl$_3$–CH$_3$OH (2:1, v/v) and 0.3 vol of H$_2$O. The labeled phosphorylcholine is determined in the aqueous phase as described above. The K_m and V_{max} measured in lysates of HEK 293 cells is 11 μM and 21 nmol/mg

[19] K. Wiegmann, S. Schütze, T. Machleidt, D. Witte, and M. Krönke, *Cell* **78**, 1005 (1994).

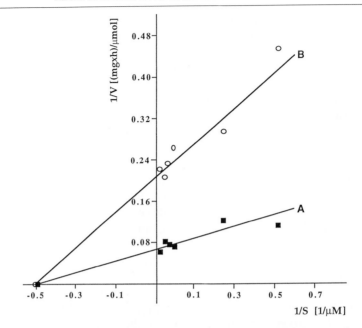

Fig. 1. Kinetic parameters of the acid SMase. The activity of acid SMase was determined after immunoprecipitation of the enzyme from Jurkat cells stimulated for 2 min with a monoclonal anti-CD95 antibody (CH-11) (A) or left unstimulated (B). Kinetic parameters for the acid SMase are presented as a double-reciprocal Lineweaver–Burk plot. Protein was estimated, after resolution of immunoprecipitated acid SMase by SDS–10% (w/v) PAGE, by comparison of silver-stained bands with a standard curve generated from known quantities of glutathione S-transferase.

protein/hr, respectively. The V_{max} increases to 33 nmol/mg protein/hr after 3 min of TNF stimulation (100 ng/ml) while the K_m remains unchanged.[20]

Results from Schissel et al.[21] show that a portion of the acid SMase is secreted by many cells. The activity of this form of the acid SMase depends on the addition of Zn^{2+} (0.1 mM) to the assay. Otherwise, assay conditions are similar to those described above.

Neutral Sphingomyelinase

Principle. The assay employs principles similar to that for acid SMase. Activity of neutral SMase is also determined by degradation of the substrate

[20] R. Schwandner, K. Wiegmann, K. Bernardo, D. Kreder, and M. Krönke, *J. Biol. Chem.* **273,** 5916 (1998).

[21] S. L. Schissel, E. H. Schuchman, K. J. Williams, and I. Tabas, *J. Biol. Chem.* **271,** 18431 (1996).

[^{14}C]SM followed by organic extraction and measurement of [^{14}C]phosphorylcholine in the aqueous phase. Since the neutral SMase is inactive at low pH, altering the pH permits discrimination of this activity from acid SMase.

Materials. All chemical reagents are obtained as described above.

Assay Procedure. Cells are resuspended into 1.5 vol of an ice-cold buffer 1 [20 mM HEPES (pH 7.4), 2 mM EDTA, 10 mM MgCl$_2$, 5 mM dithiothreitol (DTT), 0.1 mM Na$_3$VO$_4$, 10 mM freshly prepared glycerophosphate, 7.5 mM ATP, leupeptin and aprotinin (10 μg/ml each), and 0.2% (v/v) Triton X-100] and sonicated three times for 10 sec each as described for acid SMase. The enzyme assay is initiated by addition of 0.5 vol of buffer 2 [20 mM HEPES (pH 7.4), 10 mM MgCl$_2$, 0.2% (v/v) Triton X-100, and [^{14}C]SM (100 nCi/sample) (final concentration, 60 μM)]. The commercial substrate [^{14}C]SM is dried in a SpeedVac prior to the assay to remove organic solvent, resuspended in buffer 2, and bath sonicated for 10 min. Samples are incubated for 90 min, extracted with 1 vol of CH$_3$Cl$_3$–CH$_3$OH–CH$_3$COOH (4:2:1, v/v/v) and 1 vol of H$_2$O, vortexed, and centrifuged, and an aliquot of the supernatant is quantified by liquid scintillation counting. The appearance of radioactive material in the aqueous phase reflects release of the ^{14}C-labeled choline headgroup from the [^{14}C]SM substrate.

A second protocol used to measure neutral SMase activity has been described by Wiegmann and co-workers[19]: Cells, terminated by immersion in a methanol–dry ice bath, are washed in phosphate-buffered saline (PBS), and pellets are resuspended in 20 mM HEPES (pH 7.4), 2 mM EDTA, 10 mM MgCl$_2$, 5 mM DTT, 0.1 mM Na$_3$VO$_4$, 0.1 mM Na$_2$MoO$_4$, 30 mM p-nitrophenyl phosphate, 10 mM freshly prepared glycerophosphate, 750 μM ATP, 10 μM leupeptin, 1 mM phenylmethylsulfonyl fluoride (PMSF), 10 μM pepstatin, and 0.2% (v/v) Triton X-100 for 5 min. Thereafter, cells are homogenized with an 18-gauge needle, centrifuged at 500g, and supernatants incubated for 2 hr with [^{14}C]SM (200 nCi/sample) in 20 mM HEPES (pH 7.4) and 10 mM MgCl$_2$. Samples are extracted with CH$_3$Cl$_3$–CH$_3$OH (2:1, v/v) and the release of [^{14}C]phosphorylcholine is determined as above.

Discussion and Conclusions

A major problem with the acid SMase assay is the initial centrifugation step after lysis and sonication of the cells. Unpublished data suggest that the acid SMase may be located in a detergent-resistant/insoluble fraction, and thus the protein is pelleted during centrifugation at high gravities (E. Gulbins, unpublished observation, 1999). Thus, for the determination of the activity in whole-cell lysates the centrifugation step should be avoided or performed at low speed (e.g., 800g). Since the cellular location of the neutral SMase is unknown, centrifugation is also avoided during determination of neutral SMase activity. However, the centrifugation step is necessary in the immunoprecipitation protocol to remove cell debris, including nuclei, and should be performed at low speed. Complete solubilization of the protein has then to be achieved for immunoprecipitation by addition of stronger detergents [3% (v/v) NP-40 and 1% (v/v) Triton X-100]. The advantage of the immunoprecipitation protocol is greater purification and thus higher specific activity.

Stimulation of cell surface receptors often results in a two- to threefold increase in the V_{max} of the acid SMase activity with little or no change in the K_m. Greater activation (up to fourfold) is achieved after infection of cells with pathogenic bacteria, e.g., *Neisseria gonorrhoeae*.[18] Elevation of the neutral SMase activity is usually less and rarely exceeds twofold.

[35] Assays for JNK and p38 Mitogen-Activated Protein Kinases

By TATSUHIKO SUDO and MICHAEL KARIN

Introduction

A large body of knowledge has accumulated regarding the role of mitogen-activated protein kinases (MAPKs) in converting extracellular stimuli into cellular responses.[1-4] There are three main groups of MAPKs, and each group consists of at least several isoforms.[5-7] Among them the JNK and p38 subgroups were suggested to play an important role in apoptosis.[8,9] Although it also has many other functions, including the control of cell

[1] C. S. Hill and R. Triesman, *Cell* **80,** 199 (1995).

[2] M. Cobb and E. J. Goldsmith, *J. Biol. Chem.* **270,** 14843 (1995).

[3] M. Karin, *J. Biol. Chem.* **270,** 16483 (1995).

[4] C. J. Marshall, *Cell* **80,** 179 (1995).

[5] T. G. Boulton, S. Nye, D. Robbins, N. Ip, E. Radz, M. Cobb, and G. Yancopoulos, *Cell* **65,** 663 (1991).

[6] S. Gupta, T. Barrett, A. J. Whitmarsh, J. Cavanagh, H. K. Sluss, B. Dérijard, and R. J. Davis, *EMBO. J.* **15,** 2760 (1996).

[7] Y. Jiang, H. Gram, M. Zhao, L. New, J. Gu, L. Feng, F. DiPadova, R. J. Ulevitch, and J. Han, *J. Biol. Chem.* **272,** 30122 (1997).

[8] Z. Xia, M. Dickens, J. Raingeaud, R. J. Davis, and M. E. Greenberg, *Science* **270,** 1326 (1995).

[9] D. Yang, C.-Y. Kuan, A. J. Whitmarsh, M. Rincon, T. S. Zheng, R. J. Davis, P. Rakic, and R. A. Flavell, *Nature (London)* **389,** 865 (1997).

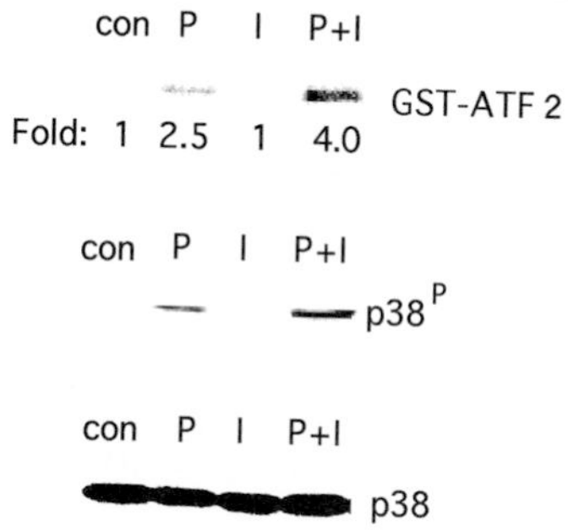

con P   I   P+I

Fold: 1 2.5 1 4.0   GST-ATF 2

con P   I   P+I

— —p38^P

con P   I   P+I

p38

FIG. 1. p38 activation by ionomycin and phorbol ester (phorbol myristate acetate, PMA) treatment of Jurkat cells. Jurkat cells were treated with 2 μM ionomycin (I) and/or 10 nM PMA (P) for 15 min, lysed, and an immune complex kinase assay was performed as described in text with GST–ATF2 as a substrate. *Top:* p38 kinase activities (i.e., phosphorylation of GST–ATF2). Fold stimulation of kinase activity in comparison with control (given an arbitrary value of 1.0) is indicated. *Middle:* Phosphorylation status of p38 detected by Western blotting of the same cell extracts (10 μg of protein per lane) with an antibody specific to the phosphory-lated form of p38 (New England BioLabs). *Bottom:* Total p38 content as detected by Western blotting of the same gel used above, with p38α-specific polyclonal antibody.

proliferation and differentiation,[10] JNK was originally identified as an up-stream kinase for c-Jun,[11,12] a major component of AP-1 transcription fac-tors, and also suggested to activate ATF2/CRE-BP1 by phosphorylation of its activation domain.[13] The p38 subgroup was simultaneously identified as either an antiinflammatory drug (CSAID)-binding protein,[14] a lipopoly-saccharide (LPS)-activated kinase,[15] or a stress-responsive kinase.[16] Al-though each pathway has unique regulatory features, various extracellular stimuli, especially proinflammatory cytokines such as tumor necrosis factor (TNF) or interleukin 1 (IL-1), often activate these kinases simultaneously. By using a class of pyridinyl imidazole compounds, such as SB 202190[14] or

[10] A. Minden and M. Karin, *Biochim. Biophys. Acta* **1333**, F85 (1997).
[11] M. Hibi, A. Lin, T. Smeal, A. Minden, and M. Karin, *Genes Dev.* **7**, 2135 (1993).
[12] B. Dérijard, M. Hibi, I.-H. Wu, T. Barrett, B. Su, T. Deng, M. Karin, and R. J. Davis, *Cell* **76**, 1025 (1994).
[13] S. Gupta, D. Campbell, B. Dérijard, and R. J. Davis, *Science* **267**, 389 (1995).
[14] J. C. Lee, J. T. Laydon, P. C. McDonnell, T. F. Gallagher, S. Kumar, D. Green, D. McNulty, M. J. Blumenthal, J. R. Heys, S. W. Landvatter, J. E. Strickler, M. M. McLaughlin, I. R. Siemens, S. M. Fisher, G. P. Livi, J. R. White, J. L. Adams, and P. R. Young, *Nature* (*London*) **372**, 739 (1994).
[15] J. Han, J. D. Lee, L. Bibbs, and R. J. Ulevitch, *Science* **265**, 808 (1994).
[16] J. Rouse, P. Cohen, S. Trigon, M. Morange, A. Alonso-Llamazares, D. Zamanillo, T. Hunt, and A. R. Nebreda, *Cell* **78**, 1027 (1994).

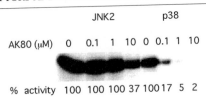

FIG. 2. Inhibition of p38α activity by SK202190 (AK80). Bacterially expressed GST–p38 or JNK2 expressed in baculovirus was preincubated with the indicated concentrations of SK202190 (AK80) and assayed for their ability to phosphorylate full-length c-Jun as described in text. The phosphorylated proteins were separated by SDS–PAGE and visualized by autoradiography.

SB 203580,[17] it has become possible to attribute certain cellular functions to the drug-sensitive forms of p38. Unfortunately, however, there are still no potent and specific inhibitors for p38δ and for the different JNK isoforms. It should also be noted that at levels exceeding 20 μM the p38 inhibitors also inhibit JNK activity.[18]

Kinase Assays

Immune Complex Assay

Cultured cells are lysed in lysis buffer containing 20 mM Tris (pH 7.6), 0.5% (v/v) Nonidet P-40 (NP-40), 250 mM NaCl, 3 mM EDTA, 1 mM dithiothreitol (DTT), 0.5 mM phenylmethylsulfonyl fluoride (PMSF), 20 mM β-glycerophosphate, 1 mM sodium orthovanadate, and leupeptin (1 μg/ml). The lysates are cleared by centrifugation at 15,000g for 15 min at 4°. This can usually be done with a microcentrifuge. Epitope-tagged kinases produced by transient transfection or endogenous kinases are immunoprecipitated from the supernatants (100 μg of protein) by incubation for 1 to 12 hr with primary antibody prebound to protein A beads. The pellets are washed twice with lysis buffer and twice with a kinase buffer composed of 25 mM HEPES (pH 7.5), 20 mM β-glycerophosphate, 20 mM PNPP, 20 mM MgCl$_2$, 2 mM DTT, and 0.1 mM sodium orthovanadate. The kinase reactions are initiated by addition of 5 μCi of [γ-^{32}P]ATP, 20 μM "cold" ATP, and 1 μg of substrate protein, such as glutathione S-transferase (GST)–c-Jun(1–79) or GST–ATF2(1–254) for JNK and p38, respectively. The reaction mixture in a final volume of 25 μl is incubated at 30° for 30 min and the reaction is terminated by addition of sodium dodecyl sulfate

[17] A. Cuenda, J. Rouse, Y. N. Doza, R. Meier, P. Cohen, T. F. Gallagher, P. R. Young, and J. C. Lee, *FEBS Lett.* **364**, 229 (1995).
[18] E. Jacinto, G. Werlen, and M. Karin, *Immunity* **8**, 31 (1998).

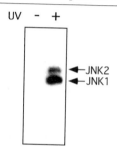

FIG. 3. In-gel kinase assay for JNK activation. HeLa cells were either not treated (–) or exposed to UV-C radiation (40 J · m⁻²). After 15 min the cells were lysed and 20-μg samples of cleared lysate protein were separated by electrophoresis on an SDS–polyacrylamide gel containing GST–c-Jun(1–79) substrate (1 mg/ml). After renaturation the in-gel kinase assay was conducted as described. The migration positions of the 46-kDa (JNK1) and 55-kDa (JNK2) forms of JNK are indicated.

(SDS) sample buffer and boiled for 5 min in tightly capped tubes. The reaction products are separated by electrophoresis on SDS–12% (w/v) polyacrylamide gels and the ^{32}P-labeled proteins are visualized by autoradiography (Fig. 1, top). A good correlation is observed between kinase activities and the phosphorylation status of p38, which is detected by Western blotting with an anti-phospho-p38 antibody (Fig. 1, middle). Similar correlation is usually seen between JNK activity measured by this assay and its phosphorylation status determined by Western blotting with anti-phospho-JNK antibodies (data not shown). Like other MAPKs, the p38s and the JNKs are activated by dual phosphorylation at conserved threonine and tyrosine residues by specialized MAPK kinases (MAPKKs).[19,20]

Comment. The specific inhibitory effect of SB202190 was examined in an *in vitro* kinase assay using recombinant proteins, such as JNK2 expressed in a baculovirus system and bacterially expressed GST–p38. Full-length c-Jun expressed in *Escherichia coli*[21] was employed as substrate for both kinases (Fig. 2). The sites phosphorylated by p38 are located within the DNA-binding domain of c-Jun and are thus different from the JNK phosphoacceptor sites: Ser-63 and -73, which are located within the *trans*-activation domain of c-Jun.

[19] A. Lin, A. Minden, H. Martinetto, F. X. Claret, C. Lange-Carter, F. Mercurio, G. L. Johnson, and M. Karin, *Science* **268**, 286 (1995).
[20] Z. Wu, E. Jacinto, and M. Karin, *Mol. Cell. Biol.* **17**, 7407 (1997).
[21] D. B. Smith and K. S. Johnson, *Gene* **67**, 31 (1988).

In-Gel Kinase Assay

Cell lysates are prepared as described above and 10–50 μg of lysate protein or immuneprecipitates prepared from an equivalent amount of a protein lysate prepared from nontreated and treated cells are separated by electrophoresis on SDS–polyacrylamide gels that were polymerized in the presence of substrate protein (0.5–1 mg/ml). The substrates for detection of JNK and p38 activities are GST–c-Jun(1–79) and GST–ATF2, respectively. The separated proteins are denatured and renatured within the gel as follows. After electrophoresis, the gel is washed twice with 100 ml of 20% (v/v) 2-propanol, 50 mM HEPES (pH 7.6) for 30 min. The gel is then washed twice with buffer A [50 mM HEPES (pH 7.6), 5 mM 2-mercaptoethanol] for 30 min followed by incubation in 200 ml of buffer A supplemented with 6 M urea for 1 hr at room temperature. Next, the gel is sequentially soaked for 15 min each time in buffer A containing 3, 1.5, and 0.75 M urea. After washing several times with 100 ml of buffer A containing 0.05% (v/v) Tween 20 at 4°, the kinase reaction is initiated by immersing the gel in 10 ml of kinase buffer supplemented with 0.1 mCi of [γ-^{32}P]ATP for 2 hr at 30° with gentle shaking. The reaction is terminated by washing with a fixing solution containing 1% (w/v) sodium pyrophosphate and 5% (w/v) trichloroacetic acid until the background level of radioactivity is less than 1000 cpm. The phosphorylated substrates are visualized by autoradiography (Fig. 3).

Preparation of Substrate Proteins

GST-c-Jun(1–79) and GST-ATF2(1–254) were expressed and purified as described,[21] with one slight modification necessary for optimization of GST-ATF2 expression. Bacteria harboring pGEX-ATF2(1–254) were grown in Luria Broth until $A_{600} = 0.8$, then IPTG was added to 0.1 mM and the bacteria were cultured for 10–12 additional hr at 25°. GST-fusion proteins were purified by affinity chromatography on GSH-agarose.[21]

[36] Assaying for IκB Kinase Activity

By Joseph A. DiDonato

Introduction

The NF-κB/Rel transcription factors serve as key regulators of a wide variety of target genes that includes immune regulatory genes and growth control, cell survival, and cell death genes.[1–3] NF-κB/Rel exists in various combinations of homo- and heterodimeric DNA-binding subunits derived from the expression of a family of five related genes.[4] NF-κB is held in the cytoplasm in an inactive state by a group of inhibitor molecules collectively termed IκBs.[5] IκBα, IκBβ, and IκBε are the three most widely studied of the IκBs. Cellular stimulation by a variety of inducing agents including proinflammatory cytokines, ultraviolet light (UV), phorbol esters, mitogens, and infection by some viruses and bacteria results in the proteolysis of the IκBs by the 26S proteasome and release of NF-κB.[1,6–8] Other extracellular stimuli such as reoxygenation of cells or treatment of cells with pervanadate (a tyrosine phosphatase inhibitor) result in tyrosine phosphorylation of IκBα and cause its dissociation from NF-κB, thereby allowing activation of the transcription factor.[9] The newly liberated NF-κB is rapidly translocated to the nucleus, where it activates its target genes.

The key event in NF-κB activation is its release from the IκBs, which in most instances occurs by proteolysis. Signal-induced IκB proteolysis is triggered only after the inducible phosphorylation of the IκBs at conserved amino-terminal serine residues.[4,10] It has been unclear whether phosphorylation at these conserved serines resulted from the action of an induced IκB kinase(s) or the inactivation of a phosphatase. We now know that for

[1] M. Grilli, J. J. Chiu, and M. J. Lenardo, *Int. Rev. Cytol.* **143,** 1 (1993).
[2] A. A. Beg and D. Baltimore, *Science* **274,** 782 (1996).
[3] M. Wu, H. Lee, R. E. Bellas, S. L. Schauer, M. Arsura, D. Katz, M. J. FitzGerald, T. L. Rothstein, D. H. Sherr, and G. E. Sonenshein, *EMBO J.* **15,** 4682 (1996).
[4] I. M. Verma, J. K. Stevenson, E. M. Schwarz, D. Van Antwerp, and S. Miyamoto, *Genes Dev.* **15,** 2723 (1995).
[5] S. T. Whiteside and A. Israel, *Semin. Cancer Biol.* **8,** 75 (1997).
[6] P. A. Baeuerle and D. Baltimore, *Cell* **87,** 13 (1996).
[7] L.Ghoda, X. Lin, and W. C. Greene, *J. Biol. Chem.* **272,** 21281 (1997).
[8] G. J. Schouten, A. C. Vertegaal, S. T. Whiteside, A. Israel, M. Toebes, J. C. Dorsman, A. J. van der Eb, and A. Zantema, *EMBO J.* **16,** 3133 (1997).
[9] V. Imbert, R. A. Rupec, A. Livolsi, H. L. Pahl, E. B. Traenckner, C. Mueller-Dieckmann, D. Farahifar, B. Rossi, P. Auberger, P. A. Baeuerle, and J. F. Peyron, *Cell* **86,** 787 (1996).
[10] T. S. Finco and A. S. Baldwin, *Immunity* **3,** 263 (1995).

a number of extracellular stimuli, especially tumor necrosis factor α (TNF-α) and interleukin 1 (IL-1), there is an inducible IκB kinase activity that exists as a high molecular weight ($\sim$900,000) complex (IKK). The IKK complex was found to contain the two related kinases IKKα and IKKβ (also known as IKK1 and IKK2, respectively), and these subunits are believed to be the catalytic polypeptides.[11–14] Both IKKα and IKKβ are strongly activated by the proinflammatory cytokines TNF-α and IL-1, with kinetics that are consistent with the physiologically relevant phosphorylation of IκBα prior to its degradation.[11–13] IKK complexes containing IKKα and IKKβ are activated to a lesser degree by the phorbol ester tetradecanoyl phorbol acetate (TPA) or phorbol myristate acetate (PMA) and UV.[11–13] Biochemically, the identity of additional IKK components and the exact stoichiometry of IKKα and IKKβ in the IKK complex remain to be determined. In addition, it will be important to determine if homodimeric IKKα or IKKβ–IKK complexes exist and, if so, what their function(s) are.

Rationale

To ascertain if there is IκB kinase activity induced by a specific stimulus in a particular cell type under examination, the first step should be to examine the cellular extract for NF-κB DNA-binding activity. Assuming that there is NF-κB DNA-binding activity in response to the stimulus, IκB abundance levels should then be examined by immunoblot analysis. Immunoblot analysis using specific IκB antibodies will reveal whether the IκBs are degraded. If IκB is being degraded in response to stimulus, it is a strong indication that the IKK activity described above is being induced. On the other hand, if IκB is not being degraded, it would indicate that perhaps the alternative mechanism of tyrosine phosphorylation and dissociation of IκBα is responsible for the activation of NF-κB. In this chapter, we focus only on determining IKK activity that is induced and leads to IκB degradation (but see Potential Advantages/Disadvantages and Problems).

Previous studies showed that the IκB kinase activity responsible for TNF-α- and IL-1-induced degradation of IκB resulted in the phosphoryla-

[11] J. A. DiDonato, M. Hayakawa, D. M. Rothwarf, E. Zandi, and M. Karin, *Nature (London)* **388,** 548 (1997).

[12] E. Zandi, D. M. Rothwarf, M. Delhase, M. Hayakawa, and M. Karin, *Cell* **91,** 243 (1997).

[13] F. Mercurio, H. Zhu, B. W. Murray, A. Shevchenko, B. L. Bennett, J. Li, D. B. Young, M. Barbosa, M. Mann, A. Manning, and A. Rao, *Science* **278,** 860 (1997).

[14] C. H. Regnier, H. Y. Song, X. Gao, D. V. Goeddel, Z. Cao, and M. Rothe, *Cell* **90,** 373 (1997).

[15] J. D. Woronicz, X. Gao, Z. Cao, M. Rothe, and D. V. Goeddel, *Science* **78,** 866 (1997).

tion of both conserved serines.[16] Furthermore, substitution of threonines for these conserved serines was shown to block phosphorylation and NF-κB activation.[16] Therefore, the bona fide IKK activity cannot utilize threonine as a phosphoacceptor. This observation was utilized to make a series of glutathione S-transferase (GST) fusion protein substrates to be used in determining authentic IKK activity. The GST fusion proteins contain either the wild-type IκBα amino acid sequence (aa 1–54) (WT), or a 32A/36A alanine substitution, which exchanges alanines in place of serines at positions 32 and 36 (32A/36A), or threonine substitutions at positions 32 and 36 (32T/36T). Authentic IKK activity will phosphorylate only the wild-type IκBα substrate. Phosphorylation of either the 32A/36A or 32T/36T substrate indicates a bogus IκB kinase activity.

To examine extracts of stimulated cells for IKK activity, this chapter describes two approaches. In the first approach, either IKKα or IKKβ is immunoprecipitated from cellular extracts and is used as a source of kinase in a subsequent kinase reaction utilizing the various GST–IκBα fusion protein substrates. The second approach describes how to chromatographically analyze cellular extracts for IKK activity.

Solutions and Materials

Solution I
  HEPES-NaOH (pH 7.6), 20 mM
  β-Glycerophsphate, 40 mM
  NaF, 20 mM
  Na$_3$VO$_4$, 1 mM
  p-Nitrophenyl phosphate (PNPP), 20 mM
  Dithiothreitol (DTT), 1 mM
  Nonidet P-40 (NP-40), 0.1% (v/v)
  Bestatin, leupeptin, aprotinin, pepstatin (10 μg/ml)
  Phenylmethylsulfonyl fluoride (PMSF), 1 mM
Kinase buffer
  HEPES-NaOH (pH 7.6), 20 mM
  β-Glycerophosphate, 20 mM
  MgCl$_2$, 10 mM
  Na$_3$VO$_4$, 0.1 mM
  PNPP, 10 mM
  NaCl, 50 mM
  DTT, 2 mM
  Bestain, leupeptin, aprotinin, pepstatin, 10 μg/ml

---

[16] J. A. DiDonato, F. Mercurio, C. Rosette, J. Wu-Li, H. Suyang, S. F. Ghosh, and M. Karin, *Mol. Cell Biol.* **16,** 1295 (1996).

Buffer A: 20 mM Tris-HCl, 20 mM NaF, 20 mM β-glycerophosphate, 20 mM PNPP, 500 mM Na$_3$VO$_4$, 2.5 mM sodium metabisulfite, 5 mM benzamidine, 1 mM EDTA, 0.5 mM EGTA, 1 mM PMSF, 10% (v/v) glycerol, pH 7.6, supplemented with aprotinin (20 μg/ml), leupeptin (2.5 μg/ml), bestatin (8.3 μg/ml), pepstatin (1.7 μg/ml), and 0.05% (v/v) NP-40

Buffer B: Buffer A plus 0.1% (v/v) Brij 35

ATP column buffer: 50 mM HEPES-NaOH (pH 7.3), 50 mM β-glycerophosphate, 60 mM MgCl$_2$, 1 mM Na$_3$VO$_4$, 1.5 mM EGTA, 1 mM DTT

IKKα and IKKβ antibodies: Commercially available from Santa Cruz Biotechnology (Santa Cruz, CA) and from PharMingen (San Diego, CA)

Normal rabbit serum and γ-linked ATP-Sepharose affinity resin: Available from Sigma Chemicals (St. Louis, MO)

Protein A–Sepharose, Mono Q and Superose-6 columns: From Pharmacia (Uppsala, Sweden)

Micro-concentrators: From Pall/Filtron (Ann Arbor, MI)

All other chemicals and protease inhibitors: Reagent grade, from Calbiochem (San Diego, CA)

Immunoprecipitations

Cells are left untreated or stimulated for 10 min with TNF-α (20 ng/ml) and then harvested by decanting the tissue culture medium and quickly washing the cells three times (5 ml each) with ice-cold phosphate-buffered saline (PBS). The cells are then scraped into 1 ml of ice-cold PBS and transferred to a 1.5-ml microcentrifuge tube and pelleted for 30 sec at 2000g, and the PBS is completely removed by vacuum. The cell pellet is then resuspended in solution I supplemented to 0.42 M NaCl and 10% (v/v) glycerol ($\sim$100 μl/10^6 cells) and placed on a rotator in the cold for 20 min, then centrifuged at 13,000g for 15 min in the cold. The supernatant is transferred to a fresh tube and the protein concentration is determined. The lysates can also be used in electromobility shift assays (EMSAs) at this point to initially screen for NF-κB DNA-binding activity. *Note:* It is much easier and faster to perform the EMSA first and then determine whether to proceed further. All steps listed below are performed in the cold (4°) or on ice unless otherwise indicated.

1. Equal amounts (typically 200–300 μg) of whole cell extract (WCE) from nonstimulated and TNF-treated cells are added to prechilled microcentrifuge tubes containing enough solution I to bring the NaCl concentra-

tion to ~100 mM. Normal rabbit serum (2 μl) and 30 μl of a 50:50 (v/v) slurry of protein A–Sepharose beads (equilibrated in solution I) are then added to the diluted lysates. The lysates are then rotated in the cold for 1 hr, and spinning the tubes at 2000g for 15 sec then pellets the protein A beads. The supernatants are removed and placed into new chilled microcentrifuge tubes.

2. Anti-IKKα or anti-IKKβ or an irrelevant control antibody (between 1 and 2 μg) is added to the lysates and they are placed back on the rotator for 2–3 hr. Protein A–Sepharose [20 μl of a 50:50 (v/v) slurry] is then added to the lysates and they are placed back on the rotator for an additional 1 hr.

3. The Protein A–Sepharose immunopellet is sedimented by centrifugation at 2000g for 15 sec and the supernatant is removed.

4. The immunopellet is washed twice with 1 ml of solution I containing 0.4 M NaCl and pelleted as described above.

5. The immunopellets are then washed with solution I containing 0.4 M NaCl and 2.0 M urea, the beads are pelleted as described above, and the supernatant is removed.

6. The pellet is then washed twice with 1 ml each of kinase buffer lacking MgCl$_2$ but supplemented with 1 mM PMSF and 0.5 mM DTT. After the last wash, the supernatant is removed.

7. The immunopellet is washed and equilibrated for 5 min with kinase buffer (1 ml) and the beads are then pelleted as described above. The supernatant is then carefully decanted and the last traces of supernatant are then taken off the immunopellet (without removing any of the beads!). This last step leaves only the immunopellet (~10 μl of protein A beads).

8. A kinase buffer master mix supplemented with unlabeled ("cold") ATP and with [γ-^{32}P]ATP (6000 Ci/mmol) and containing either GST–IκBα(1–54) WT, 32A/36/A, or 32T/36T as substrate is added to the immunopellets, bringing the final reactions to 20 μM ATP, 5 μCi of [γ-^{32}P]ATP, and 3 μg of GST–IκBα substrate. Reactions are placed at 30° for 30 min and then on ice until sodium dodecyl sulfate–polyacrylamide gel electrophoresis (SDS–PAGE) sample-loading buffer can be added. The samples are then heated to 100° for 5 min.

9. Samples are then fractionated on SDS–10% (w/v) polyacrylamide gels until the free isotope is almost run off the gel (free isotope runs with the yellow of PNPP). The free isotope region is excised from the gel and discarded, and the rest of the gel is stained with Coomassie blue and then destained to visualize the proteins and demonstrate equal loading of substrate. Also, the IgG bands can be observed and this is a further indication that equal amounts of antibody (hence, immunocomplex) are contained in each lane. The gels are soaked in destain solution supplemented with

5% (v/v) glycerol for 30 min and then dried. Dried gels are then autoradiographed. Alternatively, the radioactivity in the substrate bands can be quantitated on a phosphoimager.

Chromatographic Purification of Inducible IκB Kinase

IKK activity can be purified approximately 25,000-fold by using the three-step purification scheme described below. For most applications this amount of enrichment is not necessary and a simple anion-exchange (Mono Q) chromatographic step coupled with a gel-filtration (Superose-6) step will yield ~500-fold enrichment of the IKK activity. Tissue culture cells (3–5×10^7 cells), e.g., HeLa cells, are left either nontreated or stimulated with TNF-α (20 ng/ml) for 10 min and then washed with ice-cold PBS and harvested as described above for immunoprecipitation. All steps described below are done in the cold (4°) or on ice. Cell pellets are resuspended in ice-cold buffer A (1.0 ml) and allowed to swell on ice for 10 min before being lysed with a Dounce homogenizer. The lysate is then centrifuged at 8000g to pellet the nuclei. The supernatant is removed and then centrifuged at 100,000g in a Beckman (Fullerton, CA) Ti 50.1 rotor (38,000 rpm) for 60 min. The supernatant is then removed (this is the S100 fraction) and either flash frozen in liquid nitrogen or stored on ice for subsequent fractionation in a fast protein liquid chromatography (FPLC) system (Pharmacia).

1. The first step in the chromatographic fractionation of the S100 supernatant IKK activity is chromatography on the anion-exchange resin Mono Q, previously equilibrated in buffer B. The lysate (no more than 10 mg of total protein) is filtered through a 0.45-μm pore size cellulose acetate syringe tip filter to remove any particulates and is then loaded onto the Mono Q column at 0.5 ml min⁻¹. The column is washed with 5 ml of buffer B at a flow rate of 1 ml min⁻¹ and developed with a linear 100–300 mM NaCl gradient in buffer B, also at a flow rate of 1 ml min⁻¹. The fraction size is 0.3 ml.

2. Input and each fraction are analyzed in kinase assays with the WT and 32A/36A GST–IκBα substrates as described above. Typically, 5–10 μl of each fraction is assayed. Kinase assays are fractionated on SDS–10% (w/v) polyacrylamide gels, stained, destained, and then dried as described above. Fractions that contain IKK activity that phosphorylates the wild-type IκB substrate but not of the 32A/36A mutant are identified, pooled, and the volume determined.

3. The pooled active fractions are diluted 1 : 1 with ATP column buffer and loaded by gravity onto a 0.5-ml γ-linked ATP–Sepharose affinity col-

umn.[17] The flowthrough is passed back over the affinity column four times. The ATP affinity column is then washed sequentially with 2 ml of ATP column buffer, 0.05% (v/v) Brij 35; 2 ml of ATP column buffer, 0.05% (v/v) Brij 35, 250 mM NaCl; and eluted with 5 ml of ATP column buffer, 0.05% (v/v) Brij 35, 250 mM NaCl, 10 mM ATP. The fraction size is 0.5 ml. Fractions eluted with ATP are buffer exchanged with 1.5 ml of ATP column buffer on Pall/Filtron microconcentrators with a 10-kDa cutoff in order to reduce the cold ATP concentration (to ~100 μM ATP) prior to assaying for IKK activity. Typically, 3–5 μl of each fraction is assayed. IKK activity is detected as described above and active fractions are pooled.

4. Pooled active fractions are diluted 1:3 in buffer B and loaded back onto the Mono Q column as in step 1. The Mono Q column is washed with 1 ml of buffer B and then with 3 ml of buffer B containing 175 mM NaCl at a flow rate of 1 ml min⁻¹. The IKK activity is then eluted from the column with 3 ml of buffer B, 0.3 M NaCl, also at a flow rate of 1 ml min⁻¹, and the fraction size is 0.3 ml. Fractions are assayed for IKK activity as described in step 2. Active fractions are pooled and concentrated on Pall/Filtron microconcentrators as described above until the volume is approximately 0.25 ml.

5. Concentrated pooled fractions (0.20 ml) from step 4 are chromatographed on a Superose-6 gel-filtration column (1.0 × 30 cm; Pharmacia) equilibrated in buffer B with 300 mM NaCl, run at a flow rate of 0.3 ml min⁻¹, and 0.6-ml fractions are collected. Samples (3–5 μl) of the fractions are assayed for IKK activity as described above, pooled, and then concentrated, also as described above. The concentrated IKK can be stored for 3 months at 4° without losing activity. Alternatively, the IKK preparation can be aliquoted, flash frozen, and stored at −80°. Typically, silver stain analysis of purified IKK from the gel-filtration column shows that approximately 10% of the total protein are the 84- and 87-kDa polypeptides, which correspond to IKKα and IKKβ, respectively.[11] To prepare larger quantities of IKK from 100 mg or more of starting S100 supernatant, use the chromatographic protocol detailed in Ref. 11.

Potential Advantages/Disadvantages and Problems

Potential problems that may arise in using the immunoprecipitation assay for IKK activity are relatively few. In fact, immunoprecipitation of IKK activity containing IKKα and IKKβ is the most rapid and efficient way to determine if a specific stimulus can activate IKK activity in the cell

[17] S. P. Davies, S. A. Hawley, A. Woods, D. Carling, T. A. Haystead, and D. G. Hardie, *Eur. J. Biochem* **223**, 351 (1994).

type of interest. The main problem that one may encounter in detecting IKK activity could arise from the stringency of the wash buffer containing urea. Some antibodies to IKKα, IKKβ, or other IKK components may not be able to interact with the IKKs efficiently at the higher (2–3 M) urea concentrations and therefore a lower concentration of urea may be required in the wash buffer. Although this problem has not yet been observed, there are relatively few IKKα and IKKβ antibodies available at this time, and this problem may arise once a wider variety of antibodies is available. In addition, antibodies directed against the extreme amino-terminal end of the IKKs could interfere with subsequent IKK activity unless the kinase is first eluted from the antibody with the specific antigenic peptide.[18]

Chromatographic purification of IKK activity by the three-step purification scheme described above is laborious and requires a minimum of 3–5 mg of starting cellular extract. In comparison, immunoprecipitation of IKK activity can be accomplished with as little as 50 μg of whole cell exract in about one-twentieth the time. The advantage of purifying IKK activity chromatographically would be when a particular stimulus activates NF-κB DNA-binding activity yet does not result in activation of IKK activity containing IKKα or IKKβ. Utilization of the various GST–IκBα substrates along with the additional GST–IκBα mutant 42F, where Tyr-42 is mutated to a phenylalanine, should allow one to chromatographically delineate the nature of the particular IKK activity. To date, the only physiologically relevant stimulus that results in tyrosine phosphorylation of IκBα is when cells are reperfused with oxygen after an ischemic episode. The site of IκBα phosphorylation appears to be at Tyr-42.[9] The kinase responsible for this activity is thought to be a Lyn-related kinase.[9]

[18] F. Mercurio, personal communication (1999).

[37] Assays for Akt

By THOMAS F. FRANKE

An increasing number of publications have underscored the importance of the serine/threonine kinase Akt in the regulation of cell survival, proliferation, and insulin-dependent metabolic cell responses. Critical to the understanding of Akt signaling in cells are experimental methods that assess its activation and phosphorylation state. In this chapter, we evaluate the most commonly used techniques to examine Akt activity. Immunocomplex kinase assays that utilize Akt-specific substrates are described, as is the use of phosphospecific antibodies directed against Akt phosphorylation sites. Furthermore, we introduce coupled enzyme assays that indirectly measure the activity of Akt by examining the activity of Akt substrates.

Introduction

The serine/threonine kinase Akt was originally identified as the product of the cellular homolog of the v-*akt* oncogene.[1,2] Akt has also been cloned on the basis of its similarity to protein kinases A and C. Thus, Akt has been called RAC protein kinase (RAC-PK; related to protein kinase A and C) or protein kinase B (PKB) by other authors.[3,4] To date, Akt or PKB is most commonly used to refer to the enzyme. A variety of homolog kinases have been identified and form the Akt kinase family. It includes Akt (which is in fact Akt1), Akt2 (Akt2$_a$ and Akt2$_b$), and Akt3, or PKBα, PKBβ (PKBβ_1 or PKBβ_2), and PKBγ, respectively.[5] Akt enzymes are found in all higher eukaryotic organisms and Akt-related kinases have been isolated from *Caenorhabditis elegans, Drosophila melanogaster,* and other invertebrate species.[6,7] Akt enzymes exist in all mammalian species tested and also in other vertebrates including *G. gallus.*[2,3,8–11]

Studies have indicated that Akt is critical in transducing signals that originate from extracellular stimuli. Akt is involved in intracellular responses following antiapoptotic signals, and metabolic and proliferative cell activation.[5,12,13] Extracellular stimuli that trigger Akt activation in cells include growth factors, such as platelet-derived growth factor (PDGF), epidermal growth factor, insulin-like growth factor I/II, insulin, and others.[12] Several cytokines including interleukin 2 (IL-2) and IL-3 lead to Akt

[1] S. P. Staal, *Proc. Natl. Acad. Sci. U.S.A.* **84,** 5034 (1987).

[2] A. Bellacosa, T. F. Franke, M. E. Gonzales-Portal, K. Datta, T. Taguchi, J. Gardner, J. Q. Cheng, J. R. Testa, and P. N. Tsichlis, *Oncogene* **8,** 745 (1993).

[3] P. F. Jones, T. Jakubowicz, F. J. Pitossi, F. Maurer, and B. A. Hemmings, *Proc. Natl. Acad. Sci. U.S.A.* **88,** 4171 (1991).

[4] P. J. Coffer and J. R. Woodgett, *Eur. J. Biochem.* **201,** 475 (1991).

[5] J. Downward, *Curr. Opin. Cell Biol.* **10,** 262 (1998).

[6] S. Paradis and G. Ruvkun, *Genes Dev.* **12,** 2488 (1998).

[7] T. F. Franke, K. D. Tartof, and P. N. Tsichlis, *Oncogene* **9,** 141 (1994).

[8] P. F. Jones, T. Jakubowicz, and B. A. Hemmings, *Cell Regul.* **2,** 1001 (1991).

[9] J. Q. Cheng, A. K. Godwin, A. Bellacosa, T. Taguchi, T. F. Franke, T. C. Hamilton, P. N. Tsichlis, and J. R. Testa, *Proc. Natl. Acad. Sci. U.S.A.* **89,** 9267 (1992).

[10] H. Konishi, T. Shinomura, S.-I. Kuroda, Y. Ono, and U. Kikkawa, *Biochem. Biophys. Res. Commun.* **205,** 817 (1994).

[11] H. Konishi, S.-I. Kuroda, M. Tanaka, H. Matsuzaki, Y. Ono, K. Kameyama, T. Haga, and U. Kikkawa, *Biochem. Biophys. Res. Commun.* **216,** 526 (1995).

[12] T. F. Franke, D. R. Kaplan, and L. C. Cantley, *Cell* **88,** 435 (1997).

[13] T. F. Franke and L. C. Cantley, *Nature (London)* **390,** 116 (1997).

activation.[14,15] Akt is also activated subsequent to stimulation of cells with neurotrophins, by G protein-coupled receptor systems that transduce signals via $G\alpha_q$, $G\alpha_i$, and $G\beta\gamma$ subunits and by integrin-dependent mechanims.[16-19] Common to all these different signal transduction systems is their ability to induce the activation of phosphatidylinositol 3-kinase (PI3K), which in turn generates second-messenger molecules. These second messengers then activate Akt and phosphatidylinositol-dependent kinases (PDKs) that phosphorylate Akt, resulting in activated Akt.[5,12] PI3K-independent mechanisms of Akt activation exist. Although the activation of Akt by different forms of cellular stress is controversial, other studies have suggested a role for Ca^{2+}-dependent kinases in the activation of Akt that is independent of PI3K activity.[20,21]

The exploration of Akt signal transduction mechanisms depends on the ability to assay its activation state, either by determining its phosphorylation state or by determining the ability of Akt to phosphorylate substrates. We describe those methods that we have developed to analyze Akt in cells.

Akt Immunocomplex Assays

Activity Assays for Endogenous Akt

Akt activity can be assayed by purifying Akt from cell lysates and performing immunocomplex kinase assays that measure Akt-dependent phosphotransfer onto exogenous substrates as an indication of Akt enzymatic activity.[22] This method requires the immunoprecipitation of Akt by using Akt-specific antibodies that recognize the native protein. The activity

---

[14] N. N. Ahmed, H. L. Grimes, A. Bellacosa, T. O. Chan, and P. N. Tsichlis, *Proc. Natl. Acad. Sci. U.S.A.* **94**, 3627 (1997).

[15] Z. Songyang, D. Baltimore, L. C. Cantley, D. R. Kaplan, and T. F. Franke, *Proc. Natl. Acad. Sci. U.S.A.* **94**, 11345 (1997).

[16] M. Andjelkovic, H. S. Suidan, R. Meier, M. Frech, D. R. Alessi, and B. A. Hemmings, *Eur. J. Biochem.* **251**, 195 (1998).

[17] C. Murga, L. Laguinge, R. Wetzker, A. Cuadrado, and J. S. Gutkind, *J. Biol. Chem.* **273**, 19080 (1998).

[18] M. Delcommenne, C. Tan, V. Gray, L. Rue, J. Woodgett, and S. Dedhar, *Proc. Natl. Acad. Sci. U.S.A.* **95**, 11211 (1998).

[19] W. G. King, M. D. Mattaliano, T. O. Chan, P. N. Tsichlis, and J. S. Brugge, *Mol. Cell. Biol.* **17**, 4406 (1997).

[20] S. Yano, H. Tokumitsu, and T. R. Soderling, *Nature* (*London*) **396**, 584 (1998).

[21] H. Konishi, H. Matsuzaki, M. Tanaka, Y. Ono, C. Tokunaga, S.-I. Kuroda, and U. Kikkawa, *Proc. Natl. Acad. Sci. U.S.A.* **93**, 7639 (1996).

[22] T. F. Franke, S.-I. Yang, T. O. Chan, K. Datta, A. Kazlauskas, D. K. Morrison, D. R. Kaplan, and P. N. Tsichlis, *Cell* **81**, 727 (1995).

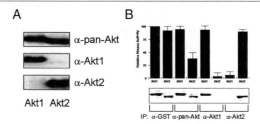

FIG. 1. (A) The specificity of Akt antibodies to recognize different Akt isoforms was determined by using cell lysates from Sf9 cells infected with baculovirus directing the expression of GST–Akt1 and GST–Akt2. Cell lysates were incubated with C-terminal pan-Akt antibody, and Akt1- and Akt2-specific antibodies. Western blots were performed according to standard procedures and developed by ECL. (B) The activity of GST–Akt1 and GST–Akt2 was determined by immunocomplex kinase assays with C-terminal pan-Akt, and Akt1- and Akt2-specific antibodies, using synthetic substrate peptide (UBI). Whereas the pan-Akt antibody recognizes both Akt isoforms in Western blot procedures, its efficiency to precipitate isoform-specific activities varies significantly between Akt1 and Akt2 when compared with precipitation using GST antibody (UBI).

of Akt in immunoprecipitates is then measured by its ability to phosphory-late exogenous substrates including histone H2B. Our laboratory uses a polyclonal Akt antibody that has been generated against a polypeptide resembling C-terminal Akt sequences.[22] Similar antibodies are also available from a variety of commercial sources including Upstate Biotechnology (UBI, Lake Placid, NY) and Santa Cruz Biotechnology (Santa Cruz, CA).

A concern could be raised when using the C-terminal peptide sequences since significant sequence overlap exists with one of the major Akt phos-phorylation sites on Ser-473. However, we have not detected interference between phosphorylation of Akt and the ability of our C-terminal antibod-ies to immunoprecipitate Akt activity. Additional antibodies have been raised against different parts of the molecule, and monoclonal antibodies are also commercially available (Transduction Laboratories, Lexington, KY). Another concern regards the specificity of Akt antibodies and their cross-reactivity with other Akt isoforms.[22] To increase further the specificity of the Akt1 and Akt2 antibodies, antibodies have been generated against those peptide sequences that are less conserved between different Akt isoforms.[23] The ability of these antibodies to distinguish Akt isoforms is demonstrated in Fig. 1A and B. Thus, our C-terminal antibody against Akt cross-reacts with the Akt2 isoform in Western blots (Fig. 1A), but it immunoprecipitates Akt2 with much less efficiency than Akt1 (Fig. 1B). Similar antibodies are commercially available from Santa Cruz and UBI

[23] T. F. Franke, D. R. Kaplan, L. C. Cantley, and A. Toker, *Science* **275,** 665 (1997).

and they are used for Western blotting and immunocomplex assays. An antibody for immunocomplex kinase assays of Akt3 activity is currently available from UBI.

Histone H2B Immunocomplex Kinase Assays for Akt

Depending on the cell system being examined, Akt1 and Akt2 activities are increased rapidly (after 2–5 min) when cells are stimulated with growth factors that increase PI3K activity.[22] Other mechanisms of Akt regulation follow other time courses and it is necessary to determine the optimal concentration range necessary for Akt activation and its time course in each signaling system. Following PDGF stimulation of NIH 3T3 cells, Akt activity is induced rapidly and transiently. Typically, NIH 3T3 cells are grown in 100-mm dishes until greater than 90% confluency is reached. Cells are then starved overnight in the absence of serum before being stimulated with PDGF-BB[22] (50–400 ng/ml). After 5 min of treatment, cells are rinsed twice briefly with ice-cold phosphate-buffered saline (PBS).

All the following steps are carried out in the cold. Cells are lysed for 10 min in 1 ml of Nonidet P-40 (NP-40) lysis buffer composed of 137 mM NaCl, 20 mM Tris-HCl (pH 7.4), 10% (v/v) glycerol, and 1% (v/v) NP-40. The buffer also contains protease and phosphatase inhibitors [20 mM NaF, 1 mM Na$_3$VO$_4$, 1 mM NaPP$_i$, 1 mM phenylmethylsulfonyl fluoride, aprotinin (2 μg/ml), and leupeptin (2 μg/ml)]. Lysates are precleared by centrifugation and preabsorbed with protein A–agarose slurry. A fraction of the lysate is removed for subsequent analysis by Western blotting. Immunoprecipitation with C-terminal Akt antibody (diluted 1 : 400) is carried out for 3 hr at 4°. Protein A–agarose is added for an additional 1 hr. Immunoprecipitates are washed three times with NP-40 lysis buffer, once with ice-cold water, and once with Akt kinase buffer (20 mM HEPES-NaOH, 10 mM MgCl$_2$, 10 mM MnCl$_2$, pH 7.4). Kinase assays are performed in 30 μl of 1× Akt kinase buffer containing 10 μCi of [γ-^{32}P]ATP (3000 Ci/mmol), 1 μM cold ATP, and 1 mM dithiothreitol (DTT). After 15 min at 30°, kinase assays are stopped by the addition of 10 μl of 4× sodium dodecyl sulfate–polyacrylamide gel electrophoresis (SDS–PAGE) loading buffer and separated in an SDS–12.5% (w/v) polyacrylamide gel. The gels are dried or transferred onto Immobilon-P membranes (Millipore, Bedford, MA). After transferring the gel onto Immobilon-P membranes and autoradiography to determine ^{32}P incorporation, Akt protein quantities are determined by Western blotting with Akt-specific antibodies to ensure equal amounts of protein in different immunoprecipitates.

Peptide Immunocomplex Kinase Assays for Akt

To determine Akt transphosphorylation activity, exogenous substrates are used. We use histone H2B, having compared it with other substrates.[22]

Histone H2B is commercially available from Boehringer Mannheim (Indianapolis, IN) and it is added as an exogenous substrate to a final amount of 0.5–1 μg/assay. Kinase reactions are performed as described above. In an alternative assay system, Akt-specific substrate peptides are used at a final concentration of 10 μM. Some Akt substrate peptides resemble the site in GSK3 that is phosphorylated by Akt.[24] A synthetic peptide resembling the optimized Akt phosphorylation motif is also available and is a more specific substrate for Akt.[25]

After kinasing substrate peptides for 10 min, the reaction is stopped by addition of a 0.5 vol of 40% (w/v) trichloroacetic acid (TCA) and spotted onto P81 phosphocellulose paper. Phosphocellulose paper strips are washed four or five times in 0.75% (v/v) phosphoric acid and the rate of radioactive incorporation measured with a liquid scintillation counter (Beckman, Fullerton, CA). Whereas the latter assay technique has the advantage of high-throughput capability and less time requirement, the analysis of phosphorylation of exogenous substrate is useful in evaluating the quality of the immunoprecipitates. As an additional modification of the use of exogenous substrates, recombinant protein substrates resembling physiological substrate of Akt have been generated and they include Bad, GSK3 (Fig. 2), and human caspase 9.[5] Although histone H2B phosphorylation is likely to be unphysiological, the site that is phosphorylated in histone H2B resembles an optimal Akt phosphorylation motif.[25]

Kinase Activity Determination of Exogenously Transfected Akt

As a modification of the protocol described above, the activity of transfected Akt in cells is determined with antibodies that recognize specific epitope tags fused to Akt and mutants of Akt.[22] We use hemagglutinin (HA) epitope-tagged forms of Akt that are precipitated with a monoclonal antibody (diluted 1:1000) against HA (available from Boehringer Mannheim). To facilitate precipitation of HA–Akt immunocomplexes, protein G–Sepharose is added to protein A–Sepharose in a 1:1 mixture. The amount of cold ATP in the kinase reaction is increased to 5 μM. By expressing kinase-deficient mutants of Akt including HA–Akt(K179M), coprecipitating kinase activities can be identified.[22]

The rigorous starvation regimen necessary to obtain a low background of Akt activity is not always compatible with cell viability. Thus, modifications have been made to the original Akt kinase protocol by increasing the stringency and/or number of washing steps, by changing the concentration

[24] D. A. E. Cross, D. R. Alessi, P. Cohen, M. Andjelkovich, and B. A. Hemmings, *Nature* (*London*) **378**, 785 (1995).

[25] D. R. Alessi, F. B. Caudwell, M. Andjelkovisc, B. A. Hemmings, and P. Cohen, *FEBS Lett.* **399**, 333 (1996).

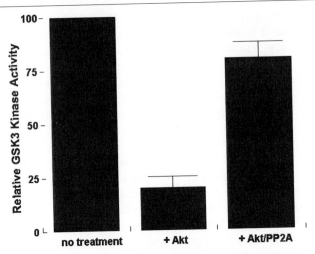

FIG. 2. GSK3 activity assays are performed in a coupled enzyme reaction after phosphorylation of recombinant GSK3 with activated GST–Akt1. The phosphorylation of prephosphorylated CREBtide peptide by GSK3 was assessed. Where indicated, Akt-phosphorylated GSK3 was dephosphorylated with PP2A to release Akt-dependent inhibition.

of cold ATP in the kinase reaction, or by including inhibitors of other kinases such as the protein kinase inhibitor (PKI).[26]

Immunoblotting

To assess the level of protein expression, cell lysates are subjected to SDS–PAGE and transferred onto Immobilon-P membranes (Millipore), using a wet-transfer blotting apparatus (Pharmacia, Uppsala, Sweden).[22] Membranes are blocked in a solution of 3% (w/v) bovine serum albumin (BSA) and 1% (w/v) ovalbumin in Tris-buffered saline (TBS) for 1 hr at room temperature. Incubations with the primary antibody are carried out at 4° in TBS containing 0.2% (v/v) Tween 20 (TBST). Akt and HA antibodies are used at a dilution of 1:1000. After washing and incubation with horseradish peroxidase-conjugated goat anti-rabbit or anti-mouse secondary antibodies purchased from Boehringer Mannheim (1:20,000 dilution), the blots are processed by enhanced chemiluminescence (ECL; Amersham, Arlington Heights, IL).

To simplify the analysis of Akt activity in cells, activation-specific anti-

[26] H. Dudek, S. R. Datta, T. F. Franke, M. J. Birnbaum, R. Yao, G. M. Cooper, R. A. Segal, D. R. Kaplan, and M. E. Greenberg, *Science* **275**, 661 (1997).

bodies have been generated that recognize phosphorylated residues Thr-308 and Ser-473, which are critical for Akt activation.[27] The Western blot protocol for these antibodies includes a blocking step with 5% (w/v) nonfat dry milk, and the antibody incubation steps are performed in TBST containing 3% (w/v) bovine serum albumin to reduce the background. Anti-phospho-Thr-308 antibody staining concurs well with PDK1 activation of Akt, but remarkable phosphorylation is detected on residue Ser-473 even in the absence of activating signals (Fig. 3B). The protocol for using these phosphorylation-specific antibodies, which is included by the manufacturer, recommends an extended starvation period to detect Akt activity changes in cell lysates. It remains to be determined if the required starvation conditions are agreeable with cell viability in systems that require Akt signal transduction to regulate downstream biologic responses.

Assays Using Recombinant Akt

A significant improvement in the understanding of Akt interaction with upstream activating signals and downstream substrates has been gained through the use of purified, recombinant Akt enzyme generated from baculovirus-infected *Spodoptera frugiperda* (Sf9) cells.[23] A six-histidine (His_6) or glutathione *S*-transferase (GST) tag is added at the N terminus of the Akt enzyme to simplify enzyme purification by using affinity chromatography. An example of purified GST–Akt1 and GST–Akt2 enzyme is shown in Fig. 3A. After elution and dialysis, biologically active protein is quantified and the relative enzyme activity is determined by peptide kinase assays using an optimal Akt substrate peptide.[25] We have defined 1 U of Akt as the activity that incorporates 1 mol of phosphate into the optimized Akt substrate peptide at 30° in 1 min when using the kinase assay conditions described above (at a linear reaction range). Purified enzyme can be used to phosphorylate Akt substrates by largely eliminating those contaminating kinase activities that could be present in Akt antibody immunoprecipitates.

The activity of wild-type Akt expressed in Sf9 cells is low, but it can be greatly enhanced by coexpression with activating Akt kinases such as PDK1 (Fig. 3A) or by coexpression with PI3K. Alternatively, mutated forms of Akt have been generated in our laboratory that contain mutated Akt phosphorylation sites or activating mutations in the Akt protein, or that carry a myristoylation signal at the N terminus.[22,23] Sf9 cells are homogenized by douncing in B-TER protein extraction buffer (Pierce, Rockford, IL), and the insoluble fraction is separated by centrifugation. The Akt

[27] D. R. Alessi, M. Andjelkovic, B. Caudwell, P. Cron, N. Morrice, P. Cohen, and B. A. Hemmings, *EMBO J.* **15,** 6541 (1996).

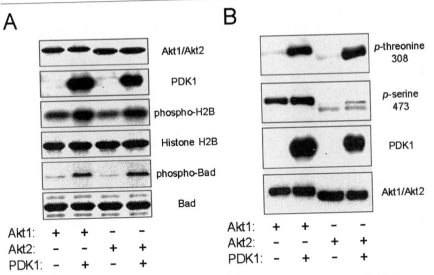

FIG. 3. (A) Baculoviruses directing the expression of GST–Akt1 and GST–Akt2 were coexpressed in Sf9 cells with a baculovirus directing the expression of recombinant human PDK1 and cell lysates were processed to perform Akt immunocomplex kinase assays. The amounts of GST–Akt1 and GST–Akt2 were determined by Coomassie blue staining of SDS–polyacrylamide gels. The expression of PDK1 was determined by immunoblot analysis, using a specific PDK1 antibody. The phosphorylation of histone H2B and recombinant GST–Bad was determined by immunocomplex kinase assays. (B) The expression and phosphorylation of GST–Akt1 and GST–Akt2 in the cell lysates from (A) was determined by the use of pan-Akt antibody and antibodies against phospho-Thr-308 and phospho-Ser-473 of Akt1. These antibodies cross-react with phosphorylated GST–Akt2, but such is not unexpected given the sequence homology between the two Akt isoforms.

protein is purified from the supernatant by affinity chromatography. After dialysis against PBS, Akt enzyme generated by this procedure efficiently phosphorylates various Akt substrates.

As a modification, activity of recombinant Akt is induced *in vitro* by adding synthetic derivatives of products of PI3K to the enzyme directly.[23,28–30] Under these conditions, Akt activation is greatly enhanced by the presence of recombinant PDK1 enzyme in the reaction mixture.[31]

[28] A. Klippel, W. M. Kavanaugh, D. Pot, and L. T. Williams, *Mol. Cell. Biol.* **17**, 338 (1997).
[29] D. R. Alessi, S. R. James, C. P. Downes, A. B. Holmes, P. R. J. Gaffney, C. B. Reese, and P. Cohen, *Curr. Biol.* **7**, 261 (1997).
[30] D. Stokoe, L. B. Stephens, T. Copeland, P. R. J. Gaffney, C. B. Reese, G. F. Painter, A. B. Holmes, F. McCormick, and P. T. Hawkins, *Science* **277**, 567 (1997).
[31] L. Stephens, K. Anderson, D. Stokoe, H. Erdjument-Bromage, G. F. Painter, A. B. Holmes, P. R. J. Gaffney, C. B. Reese, F. McCormick, P. Tempst, J. Coadwell, and P. T. Hawkins, *Science* **279**, 710 (1998).

Typically, synthetic phosphoinositides (10 μM) are sonicated using a cup-horn in the presence of phosphatidylserine and phosphatidylcholine in 10 mM HEPES (pH 7.0) containing 1 mM EDTA. The resulting micelles are added directly to the Akt kinase reaction buffer that contains recombinant PDK1. Akt activation is then assayed by phosphorylation of synthetic Akt substrate peptides or exogenous substrates such as histone H2B that are not substrates for PDK1.

Coupled Akt Enzyme Assays

It remains unclear if the Akt activity measured by Akt phosphorylation assays using exogenous substrates such as histone H2B always correlates with the ability of Akt to regulate downstream physiological substrates in cells. Additional factors are likely to influence the regulatory effects of Akt on downstream factors and they are possibly independent of Akt activity. These factors could include cofactors, localization effects of Akt and its substrates, etc. We have begun to examine Akt interactions with physiological substrates such as GSK3 and Bad. In some incidences, Akt phosphorylation inhibits downstream effectors (in the case of GSK3, Bad, and human caspase 9), but phosphorylation by Akt also activates a number of physiological substrates.[24,32–34] After Bad phosphorylation, binding of phosphorylated Bad protein to 14-3-3 molecules changes its intracellular localization and inhibits its proapoptotic properties.[32,33] In contrast, Akt phosphorylation of GSK3 and human caspase 9 changes their enzymatic activities.

Akt-Dependent GSK3 Activity Assay

GSK3 is produced as a His$_6$-tagged recombinant enzyme in Sf9 cells and purified by affinity chromatography. Purified GSK3 enzyme is also commercially available from New England BioLabs (Beverly, MA). If GSK3 is used as exogenous substrate in assays that use immobilized Akt enzyme, MnCl$_2$ is omitted from the kinase buffer. After phosphorylation, GSK3 in the supernatant is separated from the Akt enzyme. Subsequently, GSK3 kinase reactions are performed in GSK3 kinase buffer [20 mM Tris-HCl (pH 7.5), 10 mM MgCl$_2$, 5 mM dithiothreitol]. The phosphorylation of synthetic substrate peptides (200 μM) by GSK3 is examined in the presence of 200 μM ATP and 10 μCi of [γ-^{32}P]ATP (3000 Ci/mmol)

[32] S. R. Datta, H. Dudek, X. Tao, S. Masters, H. Fu, Y. Gotoh, and M. E. Greenberg, Cell **91**, 231 (1997).

[33] L. del Paso, M. Gonzales-Garcia, C. Page, R. Herrera, and G. Nuñez, Science **278**, 687 (1997).

[34] J. Deprez, D. Vertommen, D. R. Alessi, L. Hue, and M. H. Rider, J. Biol. Chem. **272**, 17269 (1997).

at 30° as an end point. The GSK3 substrate peptides used in these assays resemble GSK3 phosphorylation motifs in phosphatase inhibitor 2 and cAMP-responsive element binding protein (CREB).[35,36] Peptide phosphorylation is assayed as described using the P81 phosphocellulose paper-binding method. As a control, GSK3 is treated with protein phosphatase 2A (PP2A) prior to performing the GSK3 kinase assay. Typically, GSK3 activity is reduced to <20% after Akt phosphorylation, but GSK3 activity is restored to >80% after treatment with PP2A when using these experimental conditions.

Acknowledgments

This work was supported in part by an American Cancer Society's Institutional Research Grant and by the Herbert Irving Comprehensive Cancer Center at Columbia University.

---

[35] Q. M. Wang, I. K. Park, C. J. Fiol, P. J. Roach, and A. A. DePaoli-Roach, *Biochemistry* **33**, 143 (1994).

[36] W. J. Ryves, L. Fryer, T. Dale, and A. J. Harwood, *Anal. Biochem.* **264**, 124 (1998).

Section VIII

Other Methods

[38] Measurement of Cellular Oxidation, Reactive Oxygen Species, and Antioxidant Enzymes during Apoptosis

By LISA M. ELLERBY and DALE E. BREDESEN

Introduction

Programmed cell death is a form of cellular suicide that involves a series of characteristic morphological changes,[1] and is deliberately invoked by the cell to ensure that unnecessary or harmful cells are eliminated during development or cellular assault. These morphological changes include chromatin condensation, oligonucleosomal DNA fragmentation, plasma membrane blebbing, and cell shrinkage. Along with morphological changes during apoptosis, a number of biochemical changes occur such as alterations in reactive oxygen species (ROS),[2–4] a decrease in oxygen consumption,[5] loss of normal cytochrome c function,[5,6] and changes in the cellular oxidation–reduction status.[7–9] The exact role ROS, antioxidant enzymes, and cellular redox changes play in signaling events in apoptosis is complex, depending on both cell type and apoptotic stimuli. Some experimental work indicates ROS are important regulators of apoptosis in upstream signaling pathways, while other studies clearly indicate reactive oxygen species are generated in downstream events after the activation of caspases and the release of cytochrome c. In Fig. 1, we present a model that can account for three phases of apoptosis in which ROS and cellular redox changes may occur and we note possible feedback loops through the mitochondrial permeability transition, which is known to be modulated by reactive oxygen species as well as numerous antioxidant enzymes.

ROS and cellular redox changes can have profound effects on numerous

[1] J. F. R. Kerr, A. H. Wyllie, and A. R. Currie, *Br. J. Cancer* **26,** 239 (1972).

[2] H. Albrecht, J. Tschopp, and C. V. Jongeneel, *FEBS Lett.* **351,** 45 (1994).

[3] D. J. Kane, T. A. Sarafian, R. Anton, H. Hahn, E. B. Gralla, J. S. Valentine, T. Ord, and D. E. Bredesen, *Science* **262,** 1274 (1993).

[4] S. Tan, Y. Sagara, Y. Liu, P. Maher, and D. Schubert, *J. Cell Biol.* **141,** 1423 (1998).

[5] A. Krippner, A. Matsuno-Yagi, R. A. Gottlieb, and B. M. Babior, *J. Biol. Chem.* **271,** 21629 (1996).

[6] J. Yang, X. Liu, K. Bhalla, C. N. Kim, A. M. Ibrado, J. Cai, T.-I. Peng, D. P. Jones, and X. Wang, *Science* **275,** 1129 (1997).

[7] G. B. Pierce, R. E. Parchment, and A. L. Lewellyn, *Differentiation* **46,** 181 (1991).

[8] L. M. Ellerby, H. M. Ellerby, S. M. Park, A. L. Holleran, A. N. Murphy, G. Fiskum, D. J. Kane, M. P. Testa, C. Kayalar, and D. E. Bredesen, *J. Neurochem.* **67,** 1259 (1996).

[9] J. Cai and D. P. Jones, *J. Biol. Chem.* **273,** 11401 (1998).

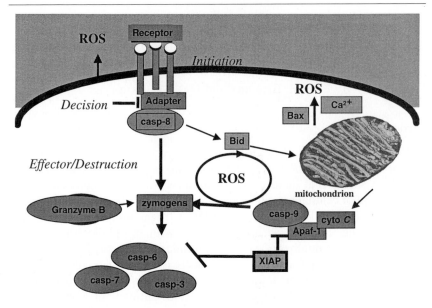

Fig. 1. Reactive oxygen species (ROS) are generated at various points in the cell death process. Amplification of ROS can occur at various stages through cell–cell mediated ROS generation, mitochondrial generation, and destruction of critical organelles in the cell.

signaling pathways, and evidence indicates that these alterations are in-volved in all three phases or steps of apoptosis—initiation, decision phase, and effector/destruction phase. For example, activation of redox sensitive kinases, transcription factors, depletion of glutathione (GSH), and redox-sensitive enzymes such as sphingomyelinase (Smase) all take place during the initiation phase or effector/decision phases of apoptosis.[10-13] Evidence for this kind of regulation in apoptosis is as follows: (1) Apoptosis signal-regulating kinase 1 (ASK1) is activated (dimerized) by hydrogen peroxide and inhibited by antioxidants; (2) redox-regulated transcription factor p53 induces apoptosis with the expression of a variety of proteins involved in the generation of ROS[11]; (3) generation of reactive oxygen species and changes in cellular redox state have been proposed to be critical events in

[10] Y. Gotoh and J. A. Cooper, *J. Biol. Chem.* **273,** 17477 (1998).
[11] K. Polyak, Y. Xla, J. L. Zweler, K. W. Kinzler, and B. Vogelstein, *Nature (London)* **389,** 300 (1997).
[12] L. Ghibelli, C. Fanelli, G. Rotilio, E. Lafavia, S. Coppola, C. Colussi, P. Civitareale, and M. R. Ciriolo, *FASEB J.* **12,** 479 (1998).
[13] B. Liu, N. Andrieu-Abadie, T. Levade, P. Zhang, L. M. Obeid, and Y. A. Hannun, *J. Biol. Chem.* **273,** 11313 (1998).

tumor necrosis factor α (TNF-α)-induced death[14]; and (4) depletion of GSH has been shown to precede the onset of apoptotic cell death in a number of apoptotic paradigms.[12] Interestingly, GSH has been shown to inhibit Smase activity directly. Smase catalyzes the hydrolysis of sphingomyelin to ceramide.[15] Other stimuli such as growth factor withdrawal,[16] human immunodeficiency virus infection,[17] and ceramide[18] result in the production of reactive oxygen species from the mitochondria, and it is likely that these events will correlate with the release of mitochondrial apoptotic factors such as cytochrome c and apoptosis-inducing factor (AIF).[6,19]

Release of cytochrome c during apoptosis results in a change in normal mitochondrial transport, which features a low level (1–5%) of superoxide production (approximately 1–5% of dioxygen consumption), to a state in which electrons are increasingly transferred to dioxygen, with a resulting increase in production of superoxide. Apoptosis during the destruction phase can also result in global changes in both glutathione and pyridine redox status. Oxidation of the cellular glutathione pool and the depletion of NADH and release of NAD^+ may be common redox alterations necessary for the completion of the apoptotic pathway. Under normal conditions, pyridine nucleotides inside the mitochondria are more reduced than in the cytosol. Only a few percent of the total cellular NADH concentration is located in the cytosol of a cell. The $NADH/NAD^+$ ratio in the heart cytosol has a standard redox potential of -226 mV. Substantial effects on cellular activity would result from the release of NAD^+ into the cytosol because of the particularly high levels of pyridine nucleotides in the mitochondria when compared with the cytosol. Superoxide production resulting from the release of cytochrome c from the mitochondria, conditions that provoke the mitochondrial permeability transition (MPT), or damage to mitochondria may contribute to release of NAD^+. Because of its uncoupling effect on the respiratory chain, MPT causes an immediate depletion of NADH and NADPH. This change, along with oxidation of the cellular GSH/GSSG pool, affects many different enzymatic reactions, as well as gene expression.

[14] K. Schulze-Osthoff, A. C. Bakker, B. Vanhaesebroeck, R. Beyaert, W. A. Jacob, and W. Fiers, *J. Biol. Chem.* **267,** 5317 (1992).
[15] I. Singh, K. Pahan, M. Khan, and A. K. Singh, *J. Biol. Chem.* **273,** 20354 (1998).
[16] L. J. S. Greenlund, T. L. Deckwerth, and E. M. Johnson, *Neuron* **14,** 303 (1995).
[17] G. H. W. Wong, T. McHugh, R. Weber, and D. V. Goeddel, *Proc. Natl. Acad. Sci. U.S.A.* **88,** 4372 (1991).
[18] C. Garcia-Ruiz, A. Colell, M. Mari, A. Morales, and J. C. Fernandez-Checa, *J. Biol. Chem.* **272,** 11369 (1997).
[19] S. A. Susin, N. Zamzami, M. Castedo, T. Hirsch, P. Marchetti, A. Macho, E. Daugas, M. Geuskens, and G. Kroemer, *J. Exp. Med.* **184,** 1331 (1996).

Either of these events could contribute to the destruction phase in which the dying cells are broken down.

In this chapter, we describe methods to measure reactive oxygen species and cellular redox changes during apoptosis. These parameters are clearly affected during apoptosis, and in many cases appear to play central roles in the apoptotic process.

Assessment of Levels of Free Radicals

Free radicals generated in cultured cells can be quantified by several approaches. We have measured the level of free radicals in apoptotic cells by monitoring the oxidatively sensitive compound 2,7-dichlorofluorescin diacetate (DCFH-DA; Molecular Probes, Eugene, OR).[3] DCFH-DA enters cells, where it is cleaved by esterases to produce DCFH, which is converted to the fluorescent product 2,7-dichlorofluorescein (DCF) derivatives by peroxides generated in the cultured cells. Cells are plated at a density of 10^5/well in 96-well tissue culture plates that have been coated with poly-L-lysine. Cells are washed three times with Hanks' balanced saline solution (HBSS) and then with HBSS buffer containing DCFH-DA (1 μg/ml). Each well contains a 100-μl total volume. Plates are read on a fluorescence plate reader (i.e., plate format fluorimeter), with an excitation wavelength of 485 nm and an emission wavelength of 530 nm, at 15-min intervals for 90 min at room temperature.

General Considerations for Processing Cultured Cells
Undergoing Apoptosis

There are a number of important factors that should be considered when collecting mammalian cells undergoing apoptotic cell death for biochemical analysis. For example, it is critical when determining various biochemical parameters that the conditions for apoptotic induction result in a substantial population of apoptotic cells and that the process of collecting these cells result in minimal losses of the apoptotic cells. Another important consideration is that the initial conditions for growth of the mammalian cells to be utilized in biochemical assays during cell death should start out under conditions in which >98% of the cells are not apoptotic. This can be determined by staining the cells with trypan blue, acridine orange, annexin V, or other stains. A third consideration is the confluence of the cells. For typical experiments with adherent cells, cells should be plated on 100-mm culture plates at 15–20% confluence (with equal numbers of cells), and then allowed to grow to 50–70% confluence. Experiments should be carried out under similar conditions, and growth of cells to high confluency (>95%)

for biochemical measurements is not recommended because of alterations in serum levels and other important biochemical parameters. To induce cell death, conditions should be utilized that maximize the percentage of apoptotic cells relative to nonapoptotic cells. Collection of cells is described below.

Protein Determination

Protein concentrations are determined by the Coomassie assay (Pierce, Rockville, MD) or another similar assay. Bovine serum albumin (BSA) may be used as a protein standard.

Enzymatic Assays

For determination of enzymatic activities, monolayers of cells on 100-mm cell culture plates are carefully scraped into medium with a sterile disposable cell lifter, and spun at $1200g$ in Eppendorf tubes for 10 min at 4°. Cells washed once with phosphate-buffered saline (PBS) are resuspended in 400 μl of phosphate buffer [50 mM sodium phosphate, 0.5% (v/v) Triton X-100, pH 7.5] and sonicated for two 15-s bursts. Sonicates are spun for 10 min at $15,000g$ and protein concentration determination of the resulting supernatants should be carried out immediately. Enzymatic assays can be performed immediately or after storage at $-70°$.

Catalase ($H_2O_2 : H_2O_2$ oxidoreductase; EC 1.11.1.6) activity of the extracts (20 μl, 5–10 mg/ml) can be measured by monitoring the disappearance of hydrogen peroxide at 240 nm.[20] A stock hydrogen peroxide solution should be prepared: 25 ml of 0.05 M potassium phosphate buffer, pH 7.0, with 85 μl of 30% (v/v) H_2O_2. Assay solution should be prepared in a cuvette by mixing 0.6 ml of 0.05 M phosphate, pH 7.0, with 0.3 ml of stock hydrogen peroxide solution. The absorbance of the hydrogen peroxide assay solution should be ~ 0.29 at 240 nm. Crude extract is added to assay solution and the change in absorbance as a function of time is measured for 1 min. The specific activity can be calculated from the following equation: $k = 2.3/60$ sec (log $A_{\text{initial}}/A_{\text{final}}$); units $= k/6.9 \times 10^{-3}$/mg of protein in crude extract.

Superoxide dismutase (superoxide : superoxide oxidoreductase; EC 1.15.1.1) activity (10 μl, crude extract diluted to 0.1 to 0.2 mg/ml) can be measured by monitoring the autooxidation of 6-hydroxydopamine ac-

[20] H. Luck, "Methods of Enzymatic Analysis" (H. V. Bergmeyer, ed.), pp. 885–892. Verlag Chemie Press, Weinheim, 1965.

cording to Heikkila[21] and Heikkila and Felicitas.[22] Superoxide dismutase (SOD) will inhibit the autooxidation of 6-hydroxydopamine by consuming superoxide generated during this process. This enzymatic assay is particularly useful in measuring SOD activity in crude extracts (SOD assays based on cytochrome c are not specific for SOD and, therefore, although useful for assaying SOD activity in relatively pure preparations of SOD, are less useful for measuring SOD activity in crude lysates). 6-Hydroxydopamine (hydrobromide, MW 250 g/mol; Sigma, St. Louis, MO) solution should be prepared and used immediately. To prepare the solution, argon purge 10 ml of double-distilled H_2O (ddH_2O) with 50 μl of concentrated perchloric acid (HClO$_4$) for 15 min. Add 25 mg of 6-hydroxydopamine (0.01 M) to the argon-purged acid solution and store the stock on ice in a foil-wrapped tube. Assays of cellular extracts should be carried out at 37° in 0.05 M sodium phosphate, 0.1 mM diethylenetriaminepentaacetic acid (DETA-PAC; Sigma), pH 7.4. The kinetics of the autooxidation of 6-hydroxydopamine should be monitored at 490 nm (0–0.5 min) under conditions that result in linear kinetics. Typically, 15 μl of stock 6-hydroxydopamine is added to 1 ml of 0.05 M sodium phosphate, 0.01 mM DETAPAC (pH 7.4) with the absorbance range set at 0–0.2 absorbance units. Record the autooxidation of 6-hydroxydopamine at 490 nm in triplicate, and calculate the slopes. Assay crude extract (~10 μl, crude extract diluted to 0.1 to 0.2 mg/ml) under conditions that result in 50% inhibition of the autooxidation of 6-hydroxydopamine (repeat three to five times). Data can be presented as micrograms of protein resulting in 50% inhibition or a standard curve can be generated with purified superoxide dismutase.

The glutathione reductase (NAD[P]H : oxidized-glutathione oxidoreductase; EC 1.6.4.2) activity of crude extracts (20 μl, 5–10 mg/ml) may be assayed spectrophotometrically by following NADPH oxidation at 340 nm at 25°.[23] The assay buffer contains the following: 1 ml of 0.05 M Tris, 1 mM EDTA (pH 8.0), 77 μl of GSSG (50-mg/ml stock GSSG in ddH_2O), 19 μl of NADPH (NADPH at 10 mg/ml in 0.5% NaHCO$_3$). Report data as glutathione reductase units per milligram of protein.

Glutathione peroxidase (glutathione : hydrogen-peroxide oxidoreductase; EC 1.11.1.9) activity of crude extracts (20 μl, 5–10 mg/ml) may be determined with a coupled assay in which the rate of *tert*-butyl hydroperoxide-dependent NADPH oxidation at 340 nm is monitored according to

[21] R. E. Heikkila, Autoxidation of 6-hydroxydopamine. *In* "CRC Handbook of Methods for Oxygen Radical Research" (R. A. Greenwald, ed.), pp. 233–235. CRC Press, Boca Raton, Florida, 1985.

[22] R. E. Heikkila and C. Felicitas, *Anal. Biochem.* **75**, 356 (1976).

[23] J. W. Anderson, C. H. Foyer, and D. A. Walker, *Planta* **158**, 442 (1983).

Gunzler and Flohe.[24] Assays are carried out as follows: 0.96 ml of 0.05 M potassium phosphate, 1 mM EDTA (pH 7.0), 10 μl of GSH (100 mM), 10 μl of yeast glutathione reductase (0.01 U/ml), and 10 μl of NADPH (15 mM in 0.1% NaHCO$_3$) are mixed together in a cuvette. The hydroperoxide-independent NADPH consumption rate at 340 nm is recorded for 3 min at 37°. Then, 10 μl of tert-butyl hydroperoxide is added to the reaction, mixed, and the overall rate at 340 nm is recorded. The same procedure is repeated with crude extracts added to the reaction mixture. This process assesses the nonenzymatic rate of oxidation of GSH, i.e., by factors that consume either hydroperoxide or NADPH other than glutathione peroxidase, and thus can be subtracted from the total rate.

Reduced Glutathione and Protein Thiols

For determination of glutathione, monolayers of mammalian cells on 100-mm plates are collected as described above. After removal of residual liquid, 300 μl of 5.5% (w/v) sulfosalicylic acid is added directly onto the cells. The precipitated protein and acid extracts are transferred to an Eppendorf tube, briefly vortexed, and centrifuged at 15,000g for 2–5 min. The acid-soluble fraction is used to measure glutathione, while the protein pellet is used for protein thiol analysis and determination of protein concentration. Total glutathione is measured according to the method of Tietze.[25] First, a standard curve is generated by measuring various concentrations of GSSG in solution (1 mM in 0.5% NaHCO$_3$, standard curve 0.05 to 10 nmol), utilizing the following recycling assay. In a disposable cuvette the following are mixed: 1.0 ml of 0.1 M potassium phosphate (pH 7.0), 50 μl of NADPH (6 mg of NADPH in 1.5 ml of 0.5% NaHCO$_3$), 20 μl of 5,5'-dithiobis-(2-nitrobenzoic and dithiobis-(2-nitrobenzoic acid) (DTNB) (1.5 mg/ml of 0.5% NaHCO$_3$), and sample or standard. Then, 20 μl of glutathione reductase (7 units) is mixed into the reaction mixture and the absorbance change at 412 nm is monitored for 2 min at 30°. After generation of a standard curve (slope versus GSH concentration), the concentration of total GSH can be determined. Oxidized glutathione (GSSG) is measured by the method of Griffith,[26] in which 2-vinylpyridine is used to derivatize the reduced form of glutathione. To derivatize the GSH, 100 μl of the sulfosalicylic-extracted sample is incubated with triethanolamine (6 μl; mix well) and then with vinylpyridine (2 μl). Vortex for 1 min. The pH of the solution should be

[24] W. A. Gunzler and L. Flohe, in "Handbook of Methods for Oxygen Radical Research" (R. A. Greenwald, ed.). CRC Press, Boca Raton, Florida, 1985.
[25] F. Tietze, Anal. Biochem. **27**, 502 (1969).
[26] O. W. Griffith, Anal. Biochem. **106**, 207 (1980).

verified (pH ~6.0 to 7.0) and the reaction carried out at room temperature for 1 hr. Assay the sample as described with 30–50 μl of sample, and prepare a standard curve with vinylpyridine-treated standards.

Protein thiol samples are prepared by resuspending acid-precipitated protein pellets into 0.1 M Tris, 0.5% (w/v) sodium dodecyl sulfate (SDS), pH 7.0. Protein thiols are measured with DTNB as described.[27] DTNB solution is prepared as follows: DTNB (1.7 mg/ml) in 0.2 M Tris, 0.02 M EDTA, pH 8.0. A thiol standard curve is generated against known amounts of GSH solution (1 mg/ml) by measuring the DTNB reaction at 412 nm. The protein concentration of samples is measured with Coomassie reagent from Pierce.

Measurement of Oxidized and Reduced Pyridine Nucleotide Levels and ATP Levels

Pyridine nucleotides and ATP levels can be measured simultaneously by reversed-phase high-performance liquid chromatography (HPLC) as described by Stocchi *et al.*[28] on an HPLC Supelcosil LC-18 column. Samples for HPLC analysis are prepared by rapid alkaline extraction. Prior to lysis, monolayers of cells on 100-mm plates are washed with ice-cold Hanks' buffer. The cells are then rapidly scraped and suspended in 300 μl of Hanks' buffer and 100 μl of 1 M KOH. The sample is immediately deproteinized by vortexing. After 3 min on ice, 400 μl of ice-cold water is added to the sample, and the solution is spun in a CF 50A Amicon (Danvers, MA) membrane and centrifuged at 1400g for 10 min. The pH of the solution is immediately adjusted to pH 6.5 with 1 M KH$_2$PO$_4$. The CF 50A cones should be prepared before collecting samples as follows: Soak the cones in 0.1 N NaOH for 1 hr, then rinse thoroughly with distilled water, and set up in centrifuge tubes. The samples may be analyzed immediately by HPLC or stored at $-70°$ for later analysis. To verify proper sample handling, known amounts of ATP or other nucleotides should be included in the processing procedure to verify good recovery and stability of nucleotides. HPLC is carried out under two different conditions. To measure ATP, ADP, NAD$^+$, NADH, NADPH, and NADP$^+$, the mobile phase should be prepared with two eluants: 0.1 M KH$_2$PO$_4$, pH 6.0 (buffer A) and 0.1 M KH$_2$PO$_4$, pH 6.0 with 15% (v/v) methanol (buffer B). The conditions for resolution of the nucleotides and nucleosides are the following: 9 min at 100% buffer A, 6-min ramp to 25% buffer B, 2.5-min ramp to 90% buffer

[27] J. Sedlak and R. H. Lindsay, *Anal. Biochem.* **25,** 192 (1968).

[28] V. Stocchi, L. Cucchiarini, M. Magnani, L. Chiarantini, P. Palma, and G. Crescentini, *Anal. Biochem.* **146,** 118 (1985).

B, 2.0-min ramp to 100% buffer B, then hold for 6 min. The flow rate is 1.3 ml/min and detection is carried out at 254 nm. Quantification should be carried out with injection of standard solutions. To measure NADPH and NADH, 50 μl of sample is injected and separated with mobile phase [0.1 M KH$_2$PO$_4$ (pH 6.0), 10% (v/v) methanol] and monitored at 340 nm. Quantification can be carried out by injection of known amounts of standard solutions.

Conclusion

We have utilized a number of the methods described here to demonstrate that the cellular oxidation–reduction potential is shifted toward more reducing in association with the expression of *bcl-2* in the neural cell lines GT1-7 and PC12.[8] This reductive shift includes an increase in the GSH/GSSG ratio, which is the major determinant of the cellular thiol redox status; an enhancement of protein free thiol groups, which is a reflection of the increased GSH/GSSG ratio; and an increase in the ratio of reduced to oxidized pyridine nucleotides. These results now are likely to be related to the ability of Bcl-2 to block the release of cytochrome *c* from the mitochondria and thus the subsequent redox-related events in apoptosis.[6] Further studies utilizing the methods described here should be useful in ordering and elucidating the role cellular redox and reactive oxygen species play in apoptosis.

Acknowledgment

This research was supported by NIH AG12282 to D.E.B. and NIH NS40251 to L.M.E.

[39] Volume Regulation and Ion Transport during Apoptosis

By Carl D. Bortner and John A. Cidlowski

The loss of cell volume is a defining characteristic of programmed cell death, which separates apoptosis from necrosis. Assessment of cell volume during apoptosis can easily be accomplished by flow cytometry, by examining changes in the light-scattering properties of living cells that are directly proportional to cell size and density. Flow cytometry quantitates only relative changes in cell size; however, when used in conjunction with Coulter

(Hialeah, FL) counter analysis, more precise cell volume data can be obtained. Flow cytometric analysis of multiple characteristics during apoptosis, such as changes in intracellular ion concentrations, membrane integrity, and DNA content, along with changes in cell size, permits sorting of various subpopulations of apoptotic cells. Below, we describe methods for rapid examination of cell volume and ion changes during apoptosis.

Introduction

The loss of cell volume, or cell shrinkage, is recognized as a unique and universal characteristic of apoptosis, which is in stark contrast to the cellular swelling that occurs during the accidental cell death process known as necrosis. Many of the biochemical features of apoptosis have been extensively studied, but only more recently has the loss of cell volume become an area of active research interest.[1-4] One of the more critical aspects of cell shrinkage during apoptosis may not be the actual reduction in cell volume, but the underlying movement of ions, which may facilitate cell shrinkage and effect other apoptotic events.

Cell volume is normally controlled within narrow limits. Any change in cell volume can prompt a cellular compensation response, involving a movement of either intracellular or extracellular ions.[5,6] When cells are placed in a hypertonic environment, shrinkage occurs because of the loss of osmotically obligated water. However, over a period of time diverse cell types compensate for the volume loss by activating a regulatory volume increase (RVI) response. This response allows for an influx of ions, with the concomitant movement of water into the cells to achieve a near-normal cell size. In contrast to the response to hypertonic conditions, exposure to a hypotonic environment results in cellular swelling. Many cell types compensate for this increase in intracellular water by activating a regulatory volume decrease (RVD) response. During the RVD response, the loss of intracellular ions permits the movement of osmotically obligated water from the cell to achieve a near-normal cell size. During apoptosis, it appears that activation of these volume regulatory responses is either inhibited or overridden during apoptosis, since compensation for the volume loss associated with programmed cell death does not occur.[2]

[1] R. S. P. Benson, S. Heer, C. Dive, and A. J. M. Watson, *Am. J. Physiol.* **270**, C1190 (1996).
[2] C. D. Bortner and J. A. Cidlowski, *Am. J. Physiol.* **271**, C950 (1996).
[3] F. M. Hughes, Jr., C. D. Bortner, G. D. Purdy, and J. A. Cidlowski, *J. Biol. Chem.* **272**, 30567 (1997).
[4] C. D. Bortner, F. M. Hughes, Jr., and J. A. Cidlowski, *J. Biol. Chem.* **272**, 32436 (1997).
[5] E. K. Hoffmann, *Curr. Top. Membr. Transp.* **30**, 125 (1987).
[6] M. Al-Habori, *Int. J. Biochem.* **26**, 319 (1994).

The ability to rapidly assess changes in cell volume is integral to the study of apoptosis, because of the dynamic nature of this mode of cell death. Flow cytometry is a powerful technique that can measure both physical and fluorescent properties of single cells in real time, and has proved invaluable in the study of apoptosis. In particular, physical attributes of unstained, living cells, such as cell size and cytoplasmic granularity, can be easily measured by monitoring their light-scattering properties at the single-cell level.[5] Light scattered in the forward direction (forward-scattered light) is roughly proportional to cell size. As a cell shrinks or loses cell volume, a decrease in the amount of forward-scattered light is observed. Likewise, cellular swelling results in an increase in the amount of forward-scattered light. Light scattered at a 90° angle from the axis of the laser beam (side-scattered light) is proportional to the amount of granularity of the cells. For example, granular cells such as monocytes, have a greater tendency to scatter light at a 90° angle than lymphocytes. In addition to the physical characteristics of the cell, the incorporation of a variety of fluorescent dyes allows for additional apoptotic characteristics, such as intracellular ion concentrations, membrane integrity, and DNA content, to be assessed in relation to the light-scattering properties.

Flow cytometry is of great value to the study of apoptosis, since both changes in cell volume and the underlying movement of ions can be rapidly measured in real time. Additional use of a Coulter counter coupled to a multisizer permits quantitative measures of cell volume. Finally, the ability to sort cells of defined physical states of apoptosis should add to the understanding of the complex programmed cell death process. Below, we describe various protocols for assessing changes in cell volume during apoptosis, using flow cytometry and Coulter counter techniques.

Flow Cytometric Assay for Cell Volume Analysis

Flow cytometric examination of light-scattering properties of a cell is a rapid means of qualitatively assessing cell size. Forward-scatter (cell size) versus side-scatter (cell granularity) dot plots, along with forward-scatter histograms, are used to determine changes in cell size, which is comparable to changes in cell volume. This technique requires minimal sample preparation and enables a large number of cells to be analyzed in a relatively short period of time. This is an extremely useful and efficient process, particularly when large numbers of experimental samples are to be examined, or in determining the relative degree of cell death in a population of cells. Below, we describe a data acquisition protocol we use to study lymphocytes undergoing cell death in response to a variety of apoptotic stimuli; it involves the use of a Becton Dickinson (San Jose, CA) FACSort equipped with

CellQuest software. However, this protocol can be employed to study apoptosis in many different cell types, using a variety of flow cytometry equipment.

Data Acquisition

1. Place approximately 5×10^5 control cells (from a stock of 5×10^5 to 1×10^6 cells/ml) into a 6-ml round-bottom, polystyrene tube (Falcon 2058; Becton Dickinson Labware, Lincoln Park, NJ).

2. Examine the control sample of cells by excitation at 488 nm, and generate a forward-scatter (FSC) versus side-scatter (SSC) dot plot and an FSC histogram.

3. For the instrument settings, set the FSC signal amplification mode to linear and adjust the detector voltage and amplification gain to position the control population of cells on the dot plot. SSC signal amplification mode can be set to either linear or log scales and adjusted, using the detector voltage and amplification gain for this parameter. Use the forward-scatter histogram to ensure the entire cell population is within the confines of the forward scatter-versus-side scatter dot plot. Once these instrument settings are set, they can be saved and used for all subsequent experimental samples.

4. Remove 1 ml of each experimental sample from the culture flasks and place each into a 6-ml round-bottom, polystyrene tube (Falcon 2058).

5. Examine each experimental sample by exciting the cells at 488 nm and observing their position on a forward-scatter versus side-scatter dot plot and a forward-scatter histogram.

Data Analysis

Once the data for the light-scattering properties of the cells have been collected, an FSC histogram can be used to analyze the changes in cell volume under a variety of experimental conditions. Individual FSC histograms can be shown side by side for the various samples. In addition, a margin can be placed on the FSC histograms denoting the lower cutoff point of the control population, such that variations from one experimental sample to another can be easily observed on different histograms. Many flow cytometry computer programs also allow for overlaying several FSC histograms onto a single plot, thus permitting the direct examination of various experimental conditions. By examining the FSC versus SSC dot plot, changes in cellular granularity can also be determined in relation to cell size. Furthermore, the data can be transferred to a graphic program that would allow for the examination of samples in a three-dimensional format.

Figure 1 shows an example of the changes in cell size of apoptotic cells that are detected by flow cytometry. S49 Neo cells (a murine immature T cell line) were either untreated (control) or treated for 48 hr with $10^{-7} M$ dexamethasone (Dex) prior to flow cytometric analysis. Both the FSC versus SSC dot plot and the FSC histogram for the Dex-treated cells clearly show the presence of two individual populations of cells: one of a control cell size and one of a smaller or shrunken cell size. The major advantage of this technique is the speed and simplicity in which many samples can be analyzed in a relatively short period of time. No sample preparation is necessary and the cells can be examined in the medium in which they are cultured. Flow cytometric measurements of cell size are best suited for cells with a nearly spherical shape, such as lymphocytes. However, relative changes in the size of other cell types can also be determined by flow

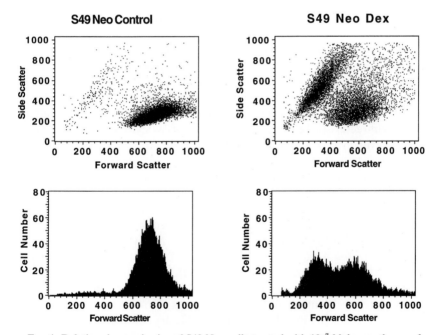

FIG. 1. Relative changes in size of S49 Neo cells treated with $10^{-7} M$ dexamethasone for 48 hr, as determined by flow cytometry. The forward-scatter versus side-scatter dot plot, and the corresponding forward-scatter histogram for the control cells, show a single population of cells. When cells were treated with dexamethasone, two distinct populations of cells are observed on the forward-scatter versus side-scatter dot plot. One population is similar to the control cells and a second population of decreased forward scatter and increased side scatter represents the shrunken apoptotic cells. The corresponding forward-scatter histogram also reflects these two populations of cells.

cytometry if control cells are carefully examined to ensure proper setting and understanding of the light-scatter parameters of these cells.

A reduction in forward-scattered light as determined by flow cytometry is not necessarily a specific marker of apoptosis, and should not be used as a sole criterion for programmed cell death. Assessment of apoptosis through the use of additional methods, such as light microscopy, electron microscopy, DNA ladders, or the incorporation of various fluorescent markers for cell death, should always accompany light-scatter analysis of apoptotic cells. Although the above procedure employs the use of live cells, we have found also that changes in cell size can be determined in fixed apoptotic cells. For example, DNA content and light-scattering properties of apoptotic cells can be simultaneously determined by flow cytometry. As cells undergo apoptosis, DNA degradation occurs, which can be detected by fixing the cells in 70% (v/v) ethanol and staining them with propidium iodide (PI). PI enters the fixed cells, where it binds to the DNA. Thus, cells that have more or less DNA will contain more or less of the PI stain, respectively, and can be analyzed as a cell cycle histogram. The occurrence of degraded DNA results in a subdiploid DNA peak on these histograms. Figure 2 (see color insert) shows the relationship between cell volume and DNA degradation in S49 Neo cells treated with a variety of apoptotic agents. Under each apoptotic condition, the shrunken population of cells directly correlates with the cells that have degraded DNA. Thus, multiple cellular characteristics can be determined by flow cytometry at the single-cell level even in fixed apoptotic cells. Nevertheless, once a model system has been established and fully characterized, a single flow cytometric parameter, specifically a decrease in forward-scattered light, can be an important indicator of the extent of apoptosis.

As a note of caution, the light-scatter detectors on individual cytometers may be positioned slightly differently by different manufacturers. Specifically, the angle of the forward-scatter detector may vary slightly from one instrument to the next such that the degree of scattered light may vary in the same experimental model system with different instruments. Therefore, it is appropriate to compare only light-scattering studies done on the same instrument with identical instrument settings. Flow cytometry measures only relative changes in cell size on an arbitrary scale of forward-scatter units, and therefore quantitative measurements of cell size and cell volume can be obtained by Coulter counter analysis.

Coulter Counter Assay for Cell Volume Analysis

The use of a Coulter counter coupled to a Coulter multisizer has been a traditional means of determining changes in cell volume. A Coulter

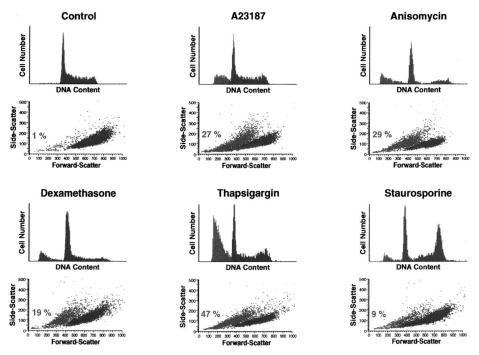

FIG. 2. Relationship between cell volume and DNA degradation. S49 Neo cells were treated with various apoptotic agents, fixed, and then examined by flow cytometry. Analysis of the DNA content and light-scattering properties of 7500 control cells showed a normal DNA content profile and a single population of cells on a forward scatter-versus-side scatter dot plot. Gates set on a PI-based area-versus-width dot plot were used to separate and analyze the DNA content in this sample to determine the position of the cells with various DNA integrity on the light-scatter dot plot by multicolor analysis. An initial gate was set on the single population of control cells observed on the area-versus-width dot plot. A second gate was set for cells that have a reduced area and width. These gates were then also used to examine cells treated with the various apoptotic agents. Analysis of the DNA content in the control sample showed that cells with a normal DNA profile had a large cell size (shown in green), while the small proportion of cells in the subdiploid region of the DNA content histogram, representing degraded DNA, had a small or shrunken cell size (shown in red). The number in red shows the percentage of shrunken cells with degraded DNA. On treatment with various apoptotic stimuli, the number of cells comprising the subdiploid region of the DNA content histogram increased and directly correlated to the position of the smaller size cells on the forward scatter-versus-side scatter dot plots. Therefore, only the shrunken population of cells contained DNA that had been degraded. [Reprinted with permission from C. D. Bortner, *et al., J. Biol. Chem.* **272,** 32436 (1997).]

counter can been used to determine the number of cells in a given sample per unit volume. However, the coupling of a Coulter counter to a Coulter multisizer permits the determination of the volume per cell for the sample. Coulter volume is an electronic measurement of cell volume based on electrolyte displacement by a cell as it passes through an orifice. This electrolyte displacement is directly proportional to the volume of the cell and allows for a precise, quantitative measure of cell volume. Below, we describe the protocol we use to assess changes in cell volume during apoptosis by Coulter analysis.

Data Acquisition

1. Prior to the acquisition of data from the experimental samples, a background analysis of the carrier solution (Isoton II; Coulter) should be performed to determine the background count. This should be no more than 50–100 counts per 0.5 ml of solution.

2. Remove 5 ml of sample from the culture flask (approximately 5×10^5 to 1×10^6 cells/ml) and place into a 20-ml blood dilution vial (Simport Plastics, Quebec, Canada). A minimum of 1000 counts per sample is typically required for an accurate assessment.

3. Add 15 ml of Isoton II (Coulter) to the blood dilution vial for a final volume of 20 ml.

4. Analyze on a Coulter multisizer II set to count 0.5 ml of sample, using an orifice of 100, μm in diameter and 75, μm in length. Set the lower and upper size limits to encompass the entire cell population, but allow for changes in cell size associated with apoptosis.

5. Record the total number of cells counted and total cumulative volume of the sample.

Data Analysis

Once the data have been recorded for all samples, the volume per cell can be calculated by dividing the total cumulative volume by the total number of cells. This number represents the average volume per cell for the entire population of cells, expressed as cubic meters, and gives a quantitative value for changes in cell volume.

Although Coulter counter analysis of apoptotic cells can give a quantitative value for changes in cell volume, one loses the flexibility and speed that is achieved with flow cytometry. The inability to discriminate other cellular characteristics, particularly the various subpopulations of cells, at any given time during apoptosis limits the use of Coulter analysis in the study of programmed cell death. However, for precise values of cell volume, the Coulter counter assay is the method of choice.

Flow Cytometric Assay of Intracellular Ions Involved in Apoptotic Cell Shrinkage

Underlying the characteristic change in cell volume during apoptosis is a movement of ions, specifically potassium and sodium, that can be examined by flow cytometry.[3,4] We have used two fluorescent indicators for analyzing the intracellular sodium and potassium concentration of cells undergoing apoptosis. The sodium indicator, sodium-binding benzofuran isophthalate (SBFI-AM), and the potassium indicator, potassium-binding benzofuran isophthalate (PBFI-AM), were originally designed by Minta and Tsien,[8] and are commercially available from Molecular Probes (Eugene, OR). These dyes are acetoxymethyl (AM) ester derivatives capable of permeating the cell membrane. Once inside the cell, the lipophilic blocking AM ester groups are cleaved by nonspecific esterases, restoring the charged, free acid form of the dye, which is retained in the cell. Since these dyes must be prepared fresh each time they are used, we find it advantageous to purchase them in individually packaged 50-μg aliquots. Both SBFI and PBFI are excited in the 340- to 360-nm range, thus requiring a flow cytometer equipped with a UV laser. We use a FACSVantage (Becton Dickinson) flow cytometer equipped with a Coherent INNOVA Enterprise laser for simultaneous 488 nm and UV excitation, with emission of the ionic dyes detected at 425 nm. We also incorporate PI (which in contrast to the fixed cells does not enter living or viable cells) into the samples to determine cells that have loss their membrane integrity. These techniques should also be readily adaptable to other ionic probes having similar properties. We have used the following protocol to examine the intracellular sodium and potassium concentrations in apoptotic lymphoid cells.

Dye Preparation and Loading

1. To 50 μg of either PBFI-AM or SBFI-AM, add an equal volume of dimethyl sulfoxide (DMSO) and 20% (w/v) Pluronic F-127 (Molecular Probes) for a final dye concentration of 2.5 mM. The addition of Pluronic F-127 is used to facilitate cell loading to the cytoplasm.

2. One hour prior to the time of examination, split each group of control and experimental cells into two equal volumes.

3. To one group of cells, add PBFI-AM to a final dye concentration of 5 μM. Repeat this for the second group of cells using SBFI-AM, again to a final dye concentration of 5 μM. Once the dye has been added, keep the cells out of direct light as much as possible.

[7] C. L. Willman and C. C. Stewart, *Semin. Diagn. Pathol.* **6**, 3 (1989).
[8] A. Minta and R. Y. Tsien, *J. Biol. Chem.* **264**, 19449 (1989).

4. Load the cells for 1 hr by incubating the samples at 37° under the appropriate CO_2 condition. Both time and temperature, along with the initial concentration of the dyes and the presence or absence of Pluronic F-127 for loading the dyes, are all variables for efficient loading of the cells. Individual examination of each of these parameters will need to be accomplished for different cell types.

5. After the dyes have been loaded, add PI to each sample for a final concentration of 10 μg/ml.

Data Acquisition

1. After the cells have been loaded with the various fluorescent indicators, remove 1 ml of the control and experimental cells from their culture flasks and place into individual 6-ml round bottom, polystyrene tubes (Falcon 2058).

2. Examine the control sample first by exciting the cells at 488 nm and observing their position on a forward-scatter versus side-scatter dot plot and a forward-scatter histogram. Set the instrument settings for these cells as described above under the section Flow Cytometric Assay for Cell Volume Analysis.

3. Once the instrument settings for the light-scattering properties have been set, examine these control cells for their potassium concentration by UV excitation (laser line 340–360 nm). We have used a 425-nm filter in front of the FL-4 detector, which is positioned solely for the UV laser to detect the changes in Na^+/K^+ dye fluorescence. Examine the cells on a forward-scatter versus side-scatter FL-4 (425 nm) dot plot and an FL-4 (425 nm) histogram. Set the FL-4 (425 nm) signal amplification mode to log and adjust the detector voltage and fine gain to position the control population of cells on the dot plot. Use the FL-4 (425 nm) histogram to make sure the entire cell population is within the confines of the forward-scatter versus side-scatter FL-4 (425 nm) dot plot. Once these instrument settings are determined, they can be saved and used for all subsequent experimental samples.

4. Examine the control cells for their membrane integrity with 488-nm excitation of PI and detection with a 575-nm filter positioned in front of the FL-3 detector. Examine the cells on an FL-4 (425 nm) versus FL-3 (575 nm) dot plot and an FL-3 (575 nm) histogram. Set the FL-3 (575 nm) signal amplification mode to log and adjust the detector voltage and fine gain to position the control population of cells on the dot plot. Use the FL-3 (575 nm) histogram to make sure the entire cell population is within the confines of the FL-4 (425 nm) versus FL-3 (575 nm) dot plot. Once these instrument settings are in place, they can again be saved and used for all subsequent experimental samples.

5. Examine each experimental sample by flow cytometry, using the above-established instrument settings, and observe the positions on the various dot plots and histograms.

Data Analysis

Once the data have been collected, the various plots and histograms can be analyzed as described above under the section Flow Cytometric Assay for Cell Volume Analysis. With multiparameter acquisition, light-scattering properties, as well as intracellular ion and membrane integrity characteristics, can be examined on a multitude of plots by multicolor analysis to detect the various subpopulations of cells. For example, a population of cells on a PBFI (K^+) fluorescence versus PI fluorescence dot plot can be gated and their position on an FSC versus SSC or FSC versus PBFI (K^+) fluorescence dot plot can be determined.

Figure 3 shows an example of the relationship between cell size and intracellular potassium in control and dexamethasone (glucocorticoid)-treated lymphoid (S49 Neo) cells. The top panels for each sample show a forward-scatter versus PBFI (K^+) fluorescence dot plot. Two individual populations of cells are observed: one population with a high forward scatter and a high K^+ concentration, and one with a low forward scatter and a low K^+ concentration. The appearance of two populations of cells in this plot is similar to what is observed in the forward-scatter versus side-scatter analysis. However, the population of cells with a reduced forward-scatter and low K^+ concentration represents both viable and nonviable shrunken apoptotic cells. The addition of PI to these samples allows for removal of cells that have lost their membrane integrity from the analysis, as shown on the PBFI (K^+) fluorescence versus PI fluorescence dot plots (Fig. 3, bottom). Here three populations of cells are observed: (1) a population with a high PBFI (K^+) fluorescence and low PI fluorescence, which represents the normal cell population; (2) a population with a decrease in PBFI (K^+) fluorescence and low PI fluorescence, representing cells that have lost potassium, but retain their membrane integrity (the viable apoptotic cells); and (3) a population of low PBFI (K^+) fluorescence and high PI fluorescence, representing apoptotic cells that have lost their membrane integrity. This latter population of cells can thus be eliminated from further ion analysis, since ionic measurements are inaccurate in the absence of an intact cell membrane.

The use of flow cytometry has several advantages in the examination of intracellular ions involved in the regulation of cell volume during apoptosis. The ability to resolve various subpopulations of cells on the basis of their light-scattering or fluorescent properties permits a detailed

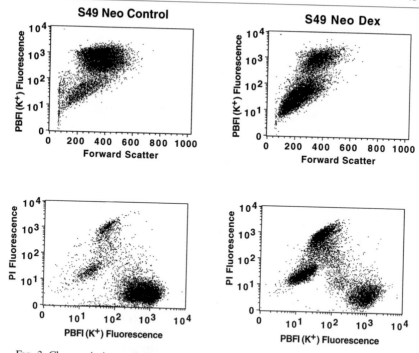

FIG. 3. Changes in intracellular potassium and membrane integrity during apoptosis. S49 Neo cells were treated with $10^{-7} M$ dexamethasone for 48 hr and examined by flow cytometry. The forward-scatter versus PBFI (K^+) fluorescence dot plots show two populations of cells: one population of high forward scatter and high PBFI (K^+) fluorescence representing the normal cells, and one population of low forward scatter and low PBFI (K^+) fluorescence representing the shrunken, apoptotic cells. The presence of dexamethasone enhances the number of shrunken apoptotic cells. The PBFI (K^+) fluorescence versus PI fluorescence dot plots show the presence of three populations of cells: one population of high PBFI (K^+) fluorescence and low PI fluorescence representing the normal cells, a second population of low PBFI (K^+) fluorescence and low PI fluorescence representing apoptotic cells that have lost potassium but maintain their membrane integrity, and a third population of low PBFI (K^+) fluorescence and high PI fluorescence representing apoptotic cells that have lost both potassium and membrane integrity. Again, the presence of dexamethasone enhances the apoptotic cell populations.

examination of the cell death process. Since the membrane integrity of cells can also be observed, ion analysis during a dynamic process such as apoptosis can be readily accomplished with the viable populations of cells. A disadvantage of this techniques is the inability to calibrate the dyes to determine a quantitative intracellular ion concentration by flow cytometry. However, the ability to examine qualitative changes of intracellular ions in cells at different stages of apoptosis makes this assay a useful tool in the study of cell death.

Sorting of Apoptotic Cells by Flow Cytometry

Apoptosis is a dynamic and asynchronous process, with only a limited number of cells undergoing cell death at any given time. As described above, flow cytometry allows for the separation and examination of various subpopulations of cells at different stages of apoptosis. Depending on the parameter of interest, individual populations of apoptotic cells can be distinguished according to their cell size, cell density, or various other characteristics such as membrane integrity, intracellular ions, and DNA content, by the incorporation of numerous fluorescent dyes. A powerful way to further analyze these subpopulations of cells is to physically separate or sort these cells. Cell sorting by flow cytometry can extend the types of assays that are currently done on the entire population of cells to confirm or initiate biochemical analysis of subpopulations of apoptotic cells at various stages of the cell death process.

One way in which we have examined changes in cell volume during apoptosis is to combine flow cytometry with Coulter analysis to exploit the advantages of each assay method. Since the flow cytometric light-scattering properties of individual cells allow for the discrimination of normal versus shrunken apoptotic cells, sorting of these populations permits individual examination of cell volume by Coulter analysis. As described above, Coulter analysis gives an average volume per cell for the entire cell population, which during apoptosis includes both normal and shrunken cells. Therefore, the actual decrease in cell volume of the shrunken apoptotic cells can be grossly underestimated. Flow cytometry can be used to sort the shrunken cells from the nonshrunken cells, whereupon a precise measure of cell volume can be obtained for each individual population of cells by Coulter analysis. In addition, we have extended the use of sorting to further analyze individual populations of apoptotic cells for their DNA and RNA content by agarose gel electrophoresis, their intracellular ion concentration by atomic absorption or mass spectrometry, and their caspase activity by fluorometric assays.

Because of the dynamic nature of programmed cell death, several considerations must be made when sorting subpopulations of apoptotic cells. For instance, the percentage of cells that make up the population of interest, the time required to sort this population, and the number of cells needed for the pending assay need to be carefully considered prior to the sort. We have determined several strategies to employ during cell sorting to allow recovery of a workable population of apoptotic cells. First, the incorporation of PI (10 μg/ml) as one of the fluorescent markers during the sorting of apoptotic cells allows for the elimination of cells that have lost membrane integrity. This particularly increases the ability to recover DNA and RNA from the subpopulation of apoptotic cells. Second, we have determined that cooling the collection device and the tubes into which the cells are being sorted dramatically increases the stability of the apoptotic cells. Since the cells of interest may be a substantially small percentage of the entire population, several hours of sort time may be required to recover the number of cells needed for further analysis. Holding the collected cells at approximately 4°, both on and off the machine, has enhanced our ability to perform biochemical assays on these rare apoptotic populations. Finally, processing the cells as soon as possible after the sort facilitates the ability to obtain accurate and measurable biochemical traits of the apoptotic cells.

Conclusion

The loss of cell volume during apoptosis is a fundamental characteristic of programmed cell death that has only recently become an area of active research. The ability to quickly assess the relationship of cell shrinkage to the various other characteristics of apoptosis can be easily achieved by flow cytometry. Flow cytometric separation of various subpopulations of apoptotic cells offers flexibility that cannot be readily matched in other assay systems. In addition, the ability to couple flow cytometry, particularly the physical sorting of cells, to other biochemical assays allows for a more precise analysis of the apoptotic features in a given population of cells.

[40] Assays for Transglutaminases in Cell Death

By Gerry Melino, Eleonora Candi, and Peter M. Steinert

Several *in vivo* and *in vitro* experimental model systems demonstrate a direct relationship between the expression and activity of tissue transglutaminase [tTG; also called transglutaminase type 2 (TGase 2)] and programmed cell death or apoptosis. This is based on mRNA and protein studies, sense and antisense transfections, identification of N^ε-(γ-glutamyl)-lysine cross-links in extracted apoptotic bodies, and in blue mouse experiments. In the epidermis, apoptosis occurs under particular conditions in the proliferative basal layer with the involvement of the tTG enzyme. However, in epidermal keratinocytes other TGases (TGase 1, TGase 3, and perhaps TGase X) are normally activated in a terminal differentiation

METHODS IN ENZYMOLOGY, VOL. 322

program, called *cornification*, that leads to cell death. These cells perform their functions after death, providing an elastic physical and permeability barrier to the skin. In fact, TGase 1 mutations cause the skin disease lamellar ichthyosis. Because all TGases share strong similarities in structure and function, being involved in mechanisms of cell death, this chapter describes the current assays for TGases at the mRNA, protein, and enzymatic levels. We also describe procedures to produce, purify, and characterize recombinant TGases, to identify mutation in disease, to isolate cross-linked bodies, and to analyze the N^ε-(γ-glutamyl)-lysine isopeptide cross-links. Finally, we discuss general rules for the interpretation and comparison of these events in cell death.

Introduction

Transglutaminases (TGase; EC 2.3.2.13) catalyze a Ca^{2+}-dependent acyl transfer reaction in which the γ-carboxamide group of a peptide-bound glutamine residue serves as an acyl donor. Primary amino groups of a variety of compounds such as polyamines, or commonly the ε-NH_2 side chain amino group of a protein-bound lysine residue, may function as acyl acceptors with the subsequent formation of substituted γ-amides.[1-7] This results in the formation of mono- or bis-N,N-(γ-glutamyl)-polyamine or N^ε-(γ-glutamyl)-lysine isopeptide bonds (Fig. 1). This reaction thus forms a usually irreversible cross-link to produce large, stable, permanent macromolecular assemblies of proteins. However, it has been shown that some TGases (at least TGase 2, also called tissue transglutaminase or tTG, and coagulation factor XIIIa) are able to catalyze the hydrolysis of the N^ε-(γ-glutamyl)-lysine linkage between short peptides only.[8] In the presence of less than saturating levels of a suitable primary amine or in the absence of an amine, water can act as the acyl acceptor with formation of peptide-bound glutamic acid instead.[1,2] Relatively few proteins contain glutamine residues that form acyl–enzyme intermediates. This capability is influenced by the amino acid sequence, charge, and local secondary structure sur-

[1] J. E. Folk and S.-I. Chung, *Adv. Enzymol. Relat. Areas Mol. Biol.* **38,** 109 (1973).
[2] J. E. Folk and S.-I. Chung, *Adv. Enzymol.* **38,** 109 (1973).
[3] J. E. Folk, *J. Biol. Chem.* **244,** 3707 (1969).
[4] J. E. Folk and S.-I. Chung, *Methods Enzymol.* **113,** 358 (1985).
[5] S. Beninati and J. E. Folk, *Adv. Exp. Med. Biol.* **231,** 79 (1988).
[6] D. Aeschlimann and M. Paulsson, *J. Biol. Chem.* **266,** 15308 (1991).
[7] J. E. Folk, *Annu. Rev. Biochem.* **49,** 517 (1980).
[8] K. N. Parameswaran, X.-F. Cheng, P. T. Velasco, J. H. Wilson, and L. Lorand, *J. Biol. Chem.* **272,** 10311 (1997).

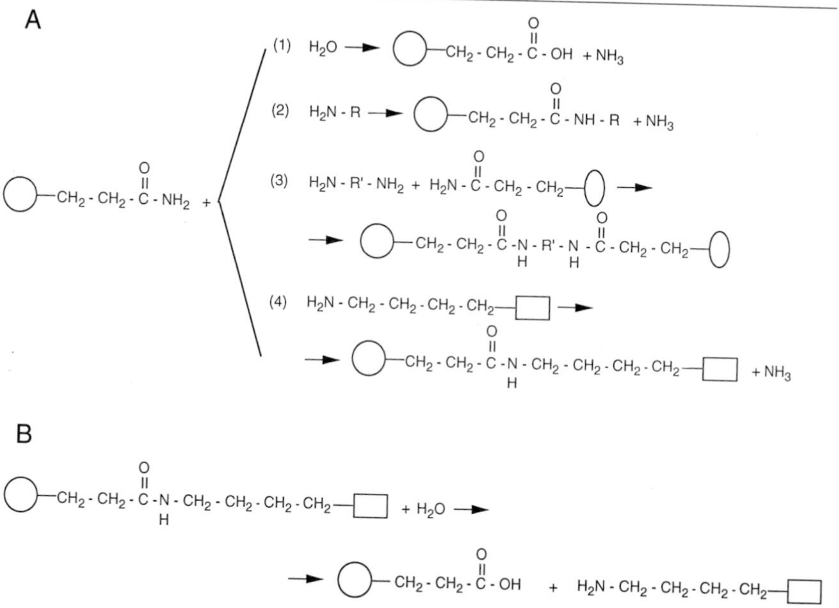

FIG. 1. Reactions catalyzed by TGases. The substrates carrying the TGase-reactive side chains are illustrated as circles (glutamines) or squares (lysines). (A) Four transamidation possibilities are shown. (1) shows the hydrolytic nature of this group of enzymes, or when the acyl derivative is transferred to water rather than an amine; (2) shows reaction involving a simple primary amine to form an amide; (3) shows reaction with a polyamine to form a secondary substituted amide; (4) shows reaction with a lysine, to form the isodipeptide bond. (B) Hydrolysis of an isodipeptide bond that is possible in short peptides only.[8]

rounding the susceptible glutamine residue, conditions that are not yet well defined.[4] However, denaturation or proteolysis of a nonreacting protein can convert it to a TGase substrate.[1,2,9]

This Ca^{2+}-dependent enzymatic activity was discovered in the livers of a number of mammals.[10,11] Studies leading to this discovery were stimulated by earlier reports of a Ca^{2+}-dependent system of guinea pig liver that promoted covalent attachment of L-lysine, through its ε-NH$_2$ group, to protein.[12] The name *transglutaminase* was assigned by Waelsh and collabo-

[9] E. Tarcsa, L. N. Marekov, J. Andreoli, W. W. Idler, E. Candi, S.-I. Chung, and P. M. Steinert, *J. Biol. Chem.* **272**, 27893 (1997).

[10] D. D. Clarke, M. J. Mycek, A. Neidle, and H. Waelsh, *Arch. Biochem. Biophys.* **71**, 277 (1957).

[11] H. Waelsh, *Arch. Biochem. Biophys.* **77**, 227 (1958).

[12] H. Boorsok, C. L. Deasy, A. J. Haagen-Smit, G. Keighley, and P. H. Powy, *J. Biol. Chem.* **179**, 689 (1949).

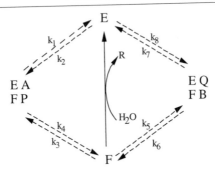

FIG. 2. Proposed double-displacement catalytic mechanism for TGases. The letter A represents the first substrate that reacts with the enzyme, and P represents the first product released, ammonia. The acyl-enzyme intermediate is shown as F; it may react with an amine (B) to form peptide-bond glutamic acid (R). If B is a peptide-bound lysine, Q becomes an N^ε-(γ-glutamyl)-lysine cross-link.[3]

rators[13] in order to distinguish this enzymatic activity involving peptide-bound glutamines from that of other enzymes that affect the carboxamide of free glutamines (glutaminases, γ-glutamyltransferase, glutamine synthetases).

Mechanism of Action

A modified double-displacement mechanism (Fig. 2) for the TGases has been proposed on the basis of chemical and kinetic findings.[3] In this mechanism, peptide-bound glutamine is designed as A, the first substrate to add to enzyme, and ammonia is designed P, the first product released. The acyl–enzyme intermediate, F, may react with an amine (B) to form peptide-bound glutamic acid, R (hydrolysis). If B is a peptide-bound lysine, Q becomes an N^ε-(γ-glutamyl)-lysine cross-link. Thus three macromolecules (two protein substrates and the enzyme) must come into contact in a highly oriented fashion at some stage in order to form the TGase-catalyzed intermolecular cross-link. This consideration becomes more intriguing when one examines, for example, the patterns of cross-linking of certain substrates by different TGases. Distinctly different products involving different glutamines and lysines, and involving intrachain or interchain cross-linking or both, have been found in cross-linking experiments with fibrin,[14–16] lori-

---

[13] M. J. Mycek, D. D. Clark, A. Neidle, and H. Waelsch, *Arch. Biochem. Biophys.* **84,** 528 (1959).

[14] S.-I. Chung and J. E. Folk, *J. Biol. Chem.* **247,** 2798 (1972).

[15] S.-I. Chung, *Ann. N.Y. Acad. Sci.* **202,** 240 (1972).

[16] S.-I. Chung and J. E. Folk, *Proc. Natl. Acad. Sci. U.S.A.* **69,** 303 (1972).

crin,[17] and certain small, proline-rich proteins[18,19] when different TGases were used. These data seem to indicate that both the conformation of the protein substrate as well as the conformation within the active site of the enzyme are important determinants of the disposition of the cross-links.

Classification

To date, seven distinct TGase gene products have been characterized in the human genome:

1. TGase 1, TG-K, or "keratinocyte" TG, about 92 kDa
2. TGase 2, tTG, or "tissue" TG, about 80 kDa
3. TGase 3, TG-E, or "epidermal" TG, about 77 kDa or 50 and 27 kDa
4. Coagulation factor XIIIa, about 80 kDa
5. Band 4.2, inactive, found in the erythrocyte membrane
6. TGase 4, TG-P, or "prostate" TG, about 80 kDa
7. TGase *X*, about 81 kDa

In Fig. 3 is shown a comparison of the amino acid sequences of the seven known members of the human TGase family.

Involvement of Transglutaminases in Apoptosis

Further discussion in this chapter is confined to those TGases involved in cell death processes, that is, programmed cell death or apoptosis (tTG or TGase 2), and terminal differentiation/cornification in stratified squamous epithelia (TGases 1, 3, and *X*).

Several groups have shown in different *in vivo* and *in vitro* experimental model systems that there is a tight association between the expression of tTG (we use this abbreviation rather than TGase 2, in keeping with the literature) and apoptosis.[20] Table I shows the comparison of events during cell death. In fact, the assay for tTG has a different interpretation, depending on its procedure and correlation with other events during cell death. While at the mRNA and inactive protein levels tTG might be present in normal cells, similar to caspases, the enzymatic activation occurs during the actual killing. The isopeptide cross-link can be detected in the apoptotic cell (similar to caspases, poly(ADP-ribose)polymerase and DNA degradation) and also after the disappearance of the apoptotic body itself, a phe-

---

[17] E. Candi, G. Melino, G. Mei, E. Tarcsa, S.-I. Chung, L. N. Marekov, and P. M. Steinert, *J. Biol. Chem.* **270**, 26382 (1995).

[18] T. Kartasova, B. J. C. Cornelissen, and P. van der Putte, *Nucleic Acids Res.* **15**, 5945 (1997).

[19] E. Candi, E. Tarcsa, W. W. Idler, T. Kartasova, L. N. Marekov, and P. M. Steinert, *J. Biol. Chem.* **276**, 7226 (1999).

[20] G. Melino and M. Piacentini, *FEBS Lett.* **430**, 59 (1998).

```
TGase1   MMDGPRSDVG-RWGGNPLQPPTTPSPEPEPEPDGRSRRGGGRSFWARCCGCCSCRNAADDDWGPEPSDSRGRGSSSGTRRPGSRGSDSRRPVSRGSGVNA   99
fXIIIa   MSETSRTAFGGRRAVPPNNSNAAEDVVPRGVNLQE   35

TGase2   MAEELVLERCDLELET---NGRDHHTADLCREKLVVRRGQPFWLTLHFEG--RNYQASVDSLTFSVVTGPAPSQEAGTKARFPLRDAVEEGDWT   89
TGase1   AGDGTIRGGMLVVNGVDLLSSRSDQNR-RHHTDEYEYDELIVRRGQPFHMLLLL-S--RTYESS-DRITLELLIGNNPEVGKGTHVIIPVGKG-GSGGWK   193
TGase3   MA-ALGVQSINWQKAF---NRQAHHTDKFSSQELILRRGQNFQV-LMIMN--KGLGSN-ERLEFIDTTGPYPSESAMTKAVFPLS-NGSSGGWS   85
TGaseX   MAQGLEVALTDLQSSR---NNVRHHTEEITVDHLLVRRGQAFNLTLYFRN--RSFQPGLDNIIFVVETGPLSDLALGTRAVFSLARHHSPSPWI   89
TGase4   MMDASKELQVLHIDFLN---QDNAVSHHTWEFQTSSPVFRRGQVFHLRLVL-N--QPLQS-YHQLKLEFSTGPNPSIAKHTLVVLDPRTPSDHYNWQ   90
fXIIIa   DLPTVELQGFLNVTSVHLFKERWDTNKVDHHTDKYENNKLIVRRGQSFYVQIDF-S--RPYDPRRDLFRVEYVIGRYPQENKGTYIPVPIVSELQSGKWG   132
Band-4.2 MGQALGIKSCDFQAARNNEE---HHTKALSSRRLFVRRGQPFTIILYFRAPVRAFLPALKKVALTAQTGEQPSKINRTQATFPISSLGDRKWWS   90

TGase2   ATVVDQQDCTLSLQLTTPANAPIGLYRLSLEASTGYQGS--SFVL-GHFILLFNAWCPADAVYLDSEEERQEYVLTQQGFIYQGSAKFIKNIPWNFGQFQ   186
TGase1   AQVVKASGQNLNLRVHTSPNAIIGKFQFTVRTQSDAGEFQLPFDPRNEIYILFNPWCPGDIVYVDHEDWRQEYVLNESGRIYYGTEAQIGERTWNYGQVD   293
TGase3   AVLQASNGNTLTISISSPASAPIGRYTMALQI--FSQGGISSVKL-GTFILLFNPWLNVDSVFMGNHAEREEYVQEDAGIIFVGSTNRIGHIGWNFGQFE   182
TGaseX   AWLETNGATSTEVSLCAPPTAAVGRYLLKIHIDSF-QGSVTAYQL-GEFILLFNPWCPEDAVVLSDSEPQRQEYVMNDYGFIYQGSKNWIRPCPWNYGQFE   187
TGase4   ATLQNESGKEVTVAVTSSPNAILGKYQLNVKT-----GNHILKSEENILYLLFNPWCKEDMVFMPDEDERKEYILNDTGCHYVGAARSIKCKPWNFGQFE   185
fXIIIa   AKIVMREDRSVRLSIQSSPKCIVGKFRMYVAVWTPYGVLRTSRNPETDTYILFNPWCEDDAVYLDNEKEREEYVLNDIGVIFYGEVNDIKTRSWSYGQFE   232
Band-4.2 AVVEERDAQSWTISVTTPADAVIGHYSLLLQVSGRKQLLL------GQFTLLFNPWNREDAVFLKNEAQRMEYLLNQNGLIYLGTADCIQAESWDFGQFE   184

TGase2   DGILDICLILLDVNPKFLKNAGRDCSRRSSPVYVGRVGSGMVNCNDDQGVLLGRWDNNYGDGVSPMSWIGSVDILRRWKNHGCQRVKYGQCWVFAAVACT   286
TGase1   HGVLDACLYILDRR------GMPYGGRGDPVNVSRVISAMVNSLDDNGVLIGNWSGDYSRGTNPSAWVGSVEILLSYLRTGY-SVPYGQCWVFAGVTTT   385
TGase3   EDILSICLSILDRSLNFRRDAATDVASRNDPKYVGRVLSAMINSNDDNGVLNGNWSENYTDGANPAEWTGSVAILKQWNATGCQPVRYGQCWVFAAVMCT   287
TGaseX   DKIIDICLKLLDKSLHFQTDPATDCALRGSPVYVSRVVCAMINSNDDNGVLIGNWTGDYEGGTAPYKWTGSAPILQQYYNTK-QAVCFGQCWVFAGILTT   278
TGase4   KNVLDCCISLL-------TESSLKPTDRRDPVLVCRAMCAMMSFEKGQGVLIGNWTGDYEGGTAPYKWTGSAPILQQYYNTK-QAVCFGQCWVFAGILTT   278
fXIIIa   GDILDTCLYVMDRA------QMDLSGRGNPIKVSRVGSAMVNAKDDEGVLVGSWDNIYAYGVPPSAWTGSDILLEYRSSEN-PVRYGQCWVFAGVFNT   325
Band-4.2 GDVIDLSLRLLSKD------KQVEKWSQPVHVARVLGALLHFLKEQRVLPTPQTQATEGEALLNKRRGSVPILRQWLTGRGRPVYDGQAWVLAAVACT   276

TGase2   VLRCLGIPTRVVTNYNSAHDQNSNLLIEYFRNEFGEIQGDKS-EMIWNFHCWVESWMTRPDLQPGYEGWQALDPTPQEKSEGTYCCGPVPVRAIKEGDLS   385
TGase1   GLRCLGLATRTVTNFNSAHDTDTSLTMDIYFDENMKPLEHLNHDSVWNFHVWNDCWMKRPDLPSGFDGWQVVDATPQETSSGIFCCGPASVSIKNGLVY   485
TGase3   ALRSLGIPSRVITNFNSAHDTDRNLSVDVYYDPMGNPL-DKGSDSVWNFHVWNEGWFVRSDLGPPYGGWQVLDATPQEMSNGVYCCGPASVRAIKEGEVD   387
TGaseX   VMRCLGIPTRVITNFDSGHDTDGNLIIDEYYDNTGRILGNKKKDTIWNFHVWNECWMARKDLPPAYGGWQVLDATPQEMSNGVFCCGPSPLTAIRKGDIF   373
TGase4   VLRALGIPARSVTGFDSAHDTERNLTVDTYVNENGEKITSMTHDSVWNFHVWDAWMKRP-----YDGWQAVDATPQESQGVFCCGPASVQAIKHGHVC   425
fXIIIa   FLRCLGIPARIVTNYFSAHDNDANLQMDIFLEEDGNVNSKLTKDSVWNYHCWNEAWMTRPDLPVGFGGWQAVDSTPQENSDGMYRCGPASVQAIKHGHVC   425
Band-4.2 VLRCLGIPARVVTTFASAQGTGGRLLIDEYYNEEGLQNEGQRGRIWIFQTSTECWMTRPALPQGYDGWQILDPSAPNGGGVLGSCDLVPVRAVKEGTVG   376

TGase2   TKYDAPFVFAEVNADVVDWI-QQDDGSV--HKSINRSLIVGLKISTKSVGRDEREDITHTYKYPEGSSEEREAFTRANHLNKLAEKEETG----------   472
TGase1   MKYDTPFIFAEVNSDKVYWQ-RQDDGSF--KIVVVEEKAIGTLIVTKAISSNMREDITYLYKHPEGSDAERKAVE---------------------   557
TGase3   LNFDMPFIFAEVNADRITWLYDNTTGKQ--WKNSVNSHTIGRYISTKAVGSNARMDVTDKYKYPEGSDQERQVFQKA--LGKLKPN---------   463
TGaseX   LNYDTPFVFSMVNADCMSWLVQG-GKE--QKLHQDTSSVGNFISTKSIQSDERDDITENYKYEEGSLQERQVFLKA--LQKLKARSFHGSQRGAELQPS   481
TGase4   IVYDTPFVFSEVNGDRLIWLVKMVNGQEELHVISMETTSIGKNISTKAVGQDRRRDITYEYKYPEGSSEERQVMDHAFLL-----------------   453
fXIIIa   FQFDAPFVFAEVNSDLIYIT-AKKDGTH--VVENVDATHIGKLIVTKQIGGDGMMDITDTYKFQEGQEEERLALE--------------------   497
Band-4.2 LTPAVSDLFAAINASCVVWKCCEDGT---LELTDSNTKYVGNNISTKAVGSDSRDCEDITQNYKYPEGSLQEKEVLERV-------EKEMEREKDNGIRPP   466

TGase2   -----------------------MAMRIRVG-QSMNMGSDFDVFAHITNNTAEEYVCRLLLCARTVSYNGILGPECGTKYLLNLTLEPFSEKSVPL-C   545
TGase1   --TAAAHGSKPNVYAN--RGSA-EDVAMQVEAQ-DAV-MQGDLMVSVMLINHSSSRRTVKLHHLYLSVTFYTGVSGTIFK-ETKKEVELAPGAWDRVTM-P   648
TGase3   --TPFAATSSMGLETEEQEPSIIG----KLKVA-GMLAVGKEVNLVLLLKNLSRDTKTVTVNMTAWTIIYNGTLVHEV-WKDSATMSLDPEEEAEHPI-K   554
TGaseX   RPTSLSQDSPRSLHTPSLRPSDVVQSLKFKLL-DPPNMGQDICFVLLALNMSSQFKDLKVNLSAQSLLHDGSPLSPF-WQDTAFITLSPKEAKTYPC-K   578
TGase4   -----------LSSEREHRQPVKENFLHMSVQSDDVLLGNSVNFTVILKRKTAALQNVNILGSFELQLYTGKKMAKLCDLNKTSQIQGQVSEVTLTLDS   541
fXIIIa   --TALMYGAKKPLNTEGVMKSR-SNVDMDFEVE-NAV-LGKDFKLSITFRNNSHNRYTITAYLSANITFYTGVPKAEFK-KETFDVTLEPLSFKKEAV-L   590
Band-4.2 -----------SLETAS-------PLYLLLK-APSSLPLRGDAQISVTLVNHSEQEKAVQLAIGVQAVHYNGVLAAKL-WRKKLHLTLSANLEKIITIG-   555

TGase2   ILYEKYRDCLTESNLIKVRALLVEPVINSYLLAAERDLYLENPEIKIRILGEPKQKRKLVAEVSLQNPLPVALEGCTFTVEGAGLTEEQKTVEIPDPVEAG   645
TGase1   VAYKEYRPHLVDQGAMLLNVSGHVKESGQVLAKQHTFRLRTPDLSLTVLGAAVVGQECEVQIVFKNPLPVTLTNVVFRLEGSGL-QRPKILNVGD-IGGN   746
TGase3   ISYAQYERYLKSDNMIRITAVCKVEDPKTQTEPDLTPLTLEVLNEARVRKPVNVQMLFSNPLDEPVRDCVLMVEGSGLLLGNLKIDVP-TLGPK   652
TGaseX   ISYSQYSQYLSTDKLIRISALGEEKSSPEKILVNKIITLSYPSITINVLGAAVVNQPLSIQVISFNPLSEQVEDCVLTVEGSGLFKKQQKVFLG-VLKPQ   677
TGase4   KTYINSLAILDDEPVIRGFIIAEIVESKEIMASEVFTSNQYPEFSIELPNTGRIGQLLVCNCIFKNTLAIPLTDVKFSLESLGI--SSLQTSDHGTVQPG   639
fXIIIa   IQAGEYMGQLLEQASLHFFVTARINETRDVLAKQKSTVLTIPEIIIKVRGTQVVGSDMTVTVQFTNPLKETLRNVWVHLDGPGV-TRPMKKMFRE-IRPN   688
Band-4.2 LFFSNFERNPPENTFLRLTAMATHSESNLSCFAQEDIAICRPHLAIKMPEKAEQYQPLTASVSLQNSLDAPMEDCVISILGRGLIHRERSYRFRS-VWPE   654

TGase2   EEVKVRMDLVPLHMGLHKLVVNFESDKLKAVKGFRNVIIGPA   -687
TGase1   ETVTLRQSFVPVRPGPRQLIASLDSPQLSQVHGVIQVDVAPAPGDGGFFSDAGGDSHLGETIPMASRGGA   -816
TGase3   ERSRVRFDILPSRSGTKQLLADFSCNKFPAIKAM-LSIDVAE   -693
TGaseX   HQASIILETVPFKSGQRQIQANMRSNKFKDIKGYRNVYVDFAL   -720
TGase4   ETIQSQIKCTPIKTGPKKFIVKLSSKQVKEINAQKIVLITK   -685
fXIIIa   STVQWEEVCRPWVSGHRKLIASMSSDSLRHVYGELDVQIQRRPSM   -733
Band-4.2 NTMCAKFQFTPTHVGLQRLTVEVDCNMFQNLTNYKSVTVVAPELSA   -700
```

FIG. 3. Amo acid alignments of seven known human TGases. This alignment scheme attempts to maximize homologies.

nomenon unique to TGases. In a broad sense these interpretations of tTG in apoptosis can translate for other TGases in cornification. The tTG-dependent formation of stable bis-N,N-(γ-glutamyl)-polyamine or N^ε-(γ-glutamyl)-lysine cross-links leads to protein polymerization conferring resistance to breakage and chemical attack, including insolubility in sodium dodecyl sulfate and chaotropic agents.[21] The tTG-catalyzed cross-links

[21] L. Fesus, V. Thomazy, F. Autuori, M. C. Ceru, E. Tarcsa, and M. Piacentini, *FEBS Lett.* **245**, 150 (1989).

TABLE I
COMPARISON OF EVENTS DURING CELL DEATH

	Normal	Commitment	Preapoptotic[a]	Apoptotic	Postapoptotic
TGase					
mRNA	Possible	Possible	Present	Present	No
Protein	Possible	Possible	Present	Present	No
Activity	No	No	No	Present	No
N^ε-(γ-Q)-K dipeptide	No	No	No	Present	Present[b]
Other marker					
Morphology	No	No	No	Present	No
Mitochondrial PT	No	No	No	Present	No
Cytochrome c release	No	No	No	Present	No
Caspase (mRNA, protein)	Present	Present	Present	Present	Present[c,d]
Caspase (activity)	No	No	No	Present	No
PARP degradation	No	No	No	Present	No
DNA fragmentation	No	No	No	Present	Present[d]

[a] For TGases, the events occurring in the apoptotic cell largely correspond to the cornification in terminally differentiated keratinocytes. Other markers involved in apoptosis are not shared or are not well established in the epidermal differentiation process.
[b] Released in the urine or shed by the skin.
[c] As cleaved active enzyme and cleaved substrates.
[d] Only in the apoptotic bodies.
Definitions: PT, permeability transition; PARP, poly(ADP-ribose)polymerase.

found in the apoptotic bodies are irreversible. Endoproteases able to hydrolyze these cross-links have not been identified in vertebrates.[20] Thus, after degradation of apoptotic bodies by phagolysosomes, the isodipeptide cross-link itself is detectable in the cell culture medium and the plasma. The expression of tTG indeed correlates with apoptosis in blue mice.[22]

However, a direct relationship between tTG and apoptosis has been established.[23-26] The overexpression of a tTG cDNA (3.3 kbp) in human neuroblastoma cells induces the characteristic features of cells undergoing apoptosis.[27] tTG transfectants show a large reduction in their growth capac-

[22] L. Nagy, V. A. Thomazy, M. M. Saydak, J. S. Stein, and P. J. A. Davies, Cell Death Differ. 4, 534 (1997).
[23] M. Piacentini, L. Fesus, M. G. Farrace, L. Ghibelli, L. Piredda, and G. Melino, Eur. J. Cell Biol. 54, 246 (1991).
[24] M. Piacentini, M. Annicchiarico-Petruzzelli, S. Oliverio, L. Piredda, J. L. Biedler, and G. Melino, Int. J. Cancer 52, 271 (1992).
[25] G. Melino, F. Bernassola, R. A. Knight, M. T. Corasaniti, G. Nistico, and A. Finazzi-Agro, Nature (London) 388, 432 (1997).
[26] L. Piredda, A. Amendola, V. Colizzi, P. J. A. Davies, M. G. Farrace, M. Fraziano, V. Gentile, I. Uray, M. Piacentini, and L. Fesus, Cell Death Differ. 4, 463 (1997).
[27] G. Melino, M. Annicchiarico-Petruzzelli, L. Piredda, E. Candi, V. Gentile, P. Davies, and M. Piacentini, Mol. Cell. Biol. 14, 6584 (1994).

ity both *in vitro* and *in vivo* when xenografted into severe combined immunodeficient (SCID) mice.[28] Conversely, transfection with an expression vector containing the human tTG cDNA in the antisense orientation results in a decrease in spontaneous as well as retinoic acid-induced apoptosis.[27] Thus tTG plays a crucial role in the killing process, both by promoting the cross-linking of vital cytoskeletal structures resulting in condensation of the cytoplasm,[28] as well as stabilizing the apoptotic body by formation of an insoluble protein scaffold preventing the release of harmful intracellular components during the later stages of the cell death processes.

In stratified squamous epithelia such as the epidermis, apoptosis as most commonly known, including involvement of the tTG enzyme, occurs in the inner proliferative basal layer in lichenoid disease and in psoriasis, and in epidermis after injury by ultraviolet irradiation. However, three different enzymes (TGases 1, 3, and perhaps X) are utilized in a terminal differentiation program leading to cell death in stratified squamous epithelia including the epidermis and its derivatives. It has been argued that this process is in fact a specialized form of apoptosis elaborated more recently during evolution in order to adjust to the rigors of terrestrial life.[29] In this case, the dead cells are retained as squames, which in time are sloughed from the surface to be replaced by newly terminally differentiated, dead keratinocytes. TGases 1, 3, and X are responsible for the cross-linking together by N^{ε}-(γ-glutamyl)-lysine bonds of a series of novel structural proteins such as loricrin, small proline-rich proteins, involucrin, and filaggrin, to form a specialized structure termed the *cornified cell envelope* (CE), which is essential for barrier function.[30,31] Also, cytoskeletal proteins (mostly keratin intermediate filaments) become cross-linked to the CE to form a coordinated structure.[32] These processes occur concomitantly with events that are remarkably similar to apoptosis. For example, dramatic increases in the intracellular calcium concentration not only activate these TGases but also endonucleases, resulting in digestion of chromatin into nucleosome-size fragments, disintegration of the nucleus, and cell death. In conclusion, we can say that in stratified squamous epithelia there are two types of cell death, which may involve similar and/or other unrelated mechanisms: (1) cornification or terminal differentiation followed by death, where the dead cells are retained and perform their function after death (resistance, barrier function); and (2) apoptosis, mainly in the basal layers, which,

[28] Z. Nemes, R. Adany, M. Balazs, P. Boross, and L. Fesus, *J. Biol. Chem.* **272**, 20597 (1997).
[29] R. R. Polakowska and A. R. Haake, *Cell Death Differ.* **1**, 19 (1994).
[30] P. M. Steinert and L. N. Marekov, *J. Biol. Chem.* **270**, 17702 (1995).
[31] P. M. Steinert, T. Kartasova, and L. Marekov, *J. Biol. Chem.* **273**, 11758 (1998).
[32] E. Candi, E. Tarcsa, J. J. DiGiovanna, J. G. Compton, P. M. Elias, L. M. Marekov, and P. M. Steinert, *Proc. Natl. Acad. Sci. U.S.A.* **95**, 2607 (1998).

similar to other tissues, is a defensive reaction to damage or regulates developing and renewing cells.

In this context most of the methods described for determination of CE assembly and structure are also applicable to the study of other cell death processes.

Detection of Transglutaminases

Reverse Transcription-Polymerase Chain Reaction Amplification of Transglutaminases

Using the reverse transcriptase-polymerase chain reaction (RT-PCR) technique, TGases can be detected from different tissues. Figure 3 shows the high homology and identity that these enzymes have within the family at the amino acid level; similar homologies exist at the nucleic acid level.

Detection of Transglutaminase 1

The *TGM1* gene is important because of its involvement in the congenital skin disease lamellar ichthyosis (LI). Table II[33] lists the specific primers used for amplification by PCR of all exons on 10 fragments for this gene.

Polymerase Chain Reaction Conditions. The protocol utilizes a hot start technique. DNA (200 ng of total genomic DNA, correspondingly less if cDNA is used) and primers (100 nM), each in 20 μl, are mixed, denatured at 100°, and held at 84° for 5 min. Ten microliters of a cocktail is added to yield a 50-μl reaction containing a final concentration of 200 μM dGTP, dATP, dCTP, and dTTP, 10 mM Tris-HCl (pH 7.0), 50 mM KCl, 1.5 mM MgCl$_2$, and 1.5 U of *Taq* polymerase. After denaturation at 95° for 60 sec, reactions are amplified at 94° for 60 sec, reannealed at 60° for 30 sec, and elongated at 72° for 90 sec with an extension of 3 sec for each of 30 cycles. In some cases, a second PCR may be necessary. The preceding reaction is diluted 1:100 and 5 μl is reamplified using the same conditions but in a 100-μl volume. In this case, another set of primers may be used, designed to be located or "nested" within the first primer set, in order to improve the specificity of the amplification reaction.

Detection of Other TGM Genes or Their cDNAs

Owing to different nucleotide sequences, it is possible to design PCR primer pairs that are specific for each of the other family members (Table

[33] L. J. Russell, J. J. DiGiovanna, N. Hashem, J. G. Compton, and S. J. Bale, *Am. J. Hum. Genet.* **55,** 1146 (1994).

TABLE II

SEQUENCES OF RT-PCR PRIMERS FOR ALL EXONS OF THE *TGM1* GENE

Exon number	Sequence pairs[a]	Nucleotide location in *TGM1* gene
1	5'-ACTTGGGCTGCAACAGAACTCGG	Unpublished
	5'-GCACGGCCTCTGATAGTGTGG	262–242
2, 3	5'-ACTGGCTGGACTACCTGGTTA	872–892
	5'-GCCTCTCCCCACCAAACATAG	1627–1607
4	5'-GTCCCAGGCTCCATCCCCTCTCCT	2501–2524
	5'-TCCCTGTCTTTCCCTCCCATCTAC	2861–2838
5, 6	5'-GGTGAGGCCCAGGGAGAGAAC	3145–3165
	5'-CAGGAGGAGGGTGGGGGTGTG	3630–3610
7	5'-TTTAGGGGTAAGGGGTGGTTG	3898–3918
	5'-CTGTAGGGCCCGGGCCACTCCT	4136–4115
8, 9	5'-ACTTGCCCCAACCCATGCCTTG	4538–4559
	5'-TGTTAATCAGGTGGGGGAGATAAG	5036–5013
10, 11	5'-TCGCATCCCTCTCCGCCTTCTCAG	7231–7254
	5'-CACTTGGCAGGAACACTTGTTGTG	8066–8043
12	5'-GAAACTGCACCTCTACCTCTCAGT	8303–8326
	5'-GATGGGGCCGGTATTCCTTGTA	8595–8574
13, 14	5'-CCAAGCAGCACACCTTCCGTCT	8662–8663
	5'-GATAGTGAGGCAGGGCAAGAACAT	9441–9418
15	5'-TCCTGACTGCCCTGGATCAGACTT	13751–13774
	5'-TAGCATCTGTTCCCCCAAGTGAAG	14212–14189

[a] In all pairs, the first is the forward primer. Data from GenBank accession number M86360, and Ref. 33.

III[34–36]). Also listed are oligonucleotides specific for the recently cloned TGase *X*. Likewise, alignment and comparison of the seven known *TGM* genes at the nucleotide level have revealed several conserved regions, including in particular the highly homologous active site sequences. This has enabled use of a consensus pair of PCR primers to amplify a 230-bp fragment that should be of general use for the identification of any known or suspected new TGase family member (Table IV).

Recommended Polymerase Chain Reaction Conditions. The protocol should utilize a hot start technique. Component conditions are identical to those listed above. After denaturation at 94° for 45 sec, the reaction is amplified after annealing at 55° for 2 min, and elongation at 72° for 3 min.

[34] V. Gentile, M. Thomazy, M. Piacentini, and H. R. Rice, *J. Cell Biol.* **119,** 463 (1992).

[35] I.-G. Kim, S.-C. Lee, J.-M. Yang, S.-I. Chung, O. W. McBride, and P. M. Steinert, *J. Invest. Dermatol.* **103,** 137 (1994).

[36] D. Aeschlimann, M. K. Koeller, B. L. Allen-Hoffmann, and D. F. Mosher, *J. Biol. Chem.* **273,** 3452 (1998).

TABLE III
Sequences of Oligonucleotides Specific for Different *TGM* Genes

Human gene	Sequence pairs[a]	Nucleotide numbers	Size of product (bp)	Ref.
TGM2	5'-CCAGCCTGCTGAGAGCCC	2213–2230	852	34
	5'-CAGTGGACTCAGCGTCAG	3047–3065		
TGM3	5'-TCCTCTCTGAAACTTGGCTTT	424–444	1481	35
	5'-GAAAATCATCTGCACGTTCAC	1885–1905		
	or			
	5'-CAAGCGGATGATGTCTTTATG	470–491	1450	35
	5'-GTCCAGGGGGTTGGAGAAAAT	1899–1920		
TGMX	5'-TAGATGAGTATTATGACAACACAGGCAGG	703–731	531	36
	5'-GGGCTGTCCTGGCTCAGTGATGTGGGC			

[a] In all pairs, the first is the forward primer. Note that, in these three cases, the human product will cross-hybridize with mouse.

Up to 37 cycles are recommended, with the last cycle containing an extended elongation period of 10 min. Reamplification as described above may be desirable.

Likewise, specific restriction enzyme sites that are conserved between species and that are diagnostic for six different *TGM* gene family members in humans have been identified: *Bsp*1286I and *Nco*I for TGase 1; *Sca*I for tTG; *Bcl*I and *Nco*I (*Ava*) for TGase 3; *Bst*EII for band 4.2 protein; *Eco*RI

TABLE IV
Degenerate PCR Primers for Amplification of Any or Potential New Members of *TGM* Gene Family[a]

Forward primer
  Consensus sequence: 5'-TATGGCCAGTGCTGGGT_TTTGCTGG_GT
  Suggested primer: 5'-TACGGCCAATGCTGGGTTTTCGCIGCAGT
                             T   G   T   GG
                                                       C
                                                       T

Reverse primer
  Consensus sequence: 5'-TGGATGA_GAGGCC_GACCTGC_C___GG
  Suggested primer: 5'-CCAGGGIGAAGATCAGICCTCGCCATCCA
                             G   C G   T TTT
                             C   T
                             T

[a] Data are from Ref. 36.

for factor XIIIa; and *Tth*111I for TGase 4. This set of specific restriction enzymes will also work with known rat or mouse *TGM* sequences.

Fluorescence Assays to Detect Transglutaminase Activity in Living Cells or in Sections

Fluorescein cadaverine or monodansylcadaverine are primary amine donors that are useful for the detection of active TGases in living cells and in cryostat sections. Fluorescein cadaverine, for instance, when added at 0.5 mM to the culture medium, diffuses throughout and labels cells in culture without cytotoxicity.[37] After appropriate fixation with methanol, fluorescein cadaverine-labeled cells can be observed by direct fluorescence microscopy, allowing visualization of the substrates for active TGases. Simultaneous detection of TGases with specific antibodies and of the fluorescein cadaverine label provides information on activity as well as the localization of the active TGase(s) and their substrates. The sensitivity of the technique can be improved by using an anti-monodansylcadaverine antibody.[6,38–41]

Immunohistochemical Detection of Transglutaminase Activity in Cryostat Sections

Procedure. Cryostat sections (5 μm) of tissue (e.g., skin biopsy[39–41]) are incubated with 1% (w/v) bovine serum albumin in 0.1 M Tris-HCl (pH 8.2) for 30 min at room temperature. TGase 1, 3, or X (other TGases are comparatively rare in the skin) activity is detected by subsequent incubation in Tris buffer containing 12 μM monodansylcadaverine and 5 mM CaCl$_2$. Negative control sections are incubated with 2 mM putrescine or monodansylcadaverine and 10 mM EDTA instead of CaCl$_2$. Parallel sections are incubated in the same buffer containing tTG (from guinea pig) (15 μg/ml) as a positive control. Both endogenous and exogenous enzyme reactions are allowed to proceed for 1 hr at room temperature and then stopped by washing the slides for 5 min in phosphate-buffered saline (PBS) containing 10 mM EDTA, and two further washes in PBS only. Incorporated dansyl label is detected by incubation for 1 hr at room temperature with polyclonal

[37] M. Lajemi, S. Demignot, L. Borge, S. Thenet-Gauci, and M. Adolphe, *Histochem. J.* **29,** 593 (1997).

[38] D. Aeschlimann, A. Wetterwald, H. Fleisch, and M. Paulsson, *J. Cell Biol.* **120,** 1261 (1993).

[39] M. Raghunath, B. Hopfner, D. Aeschlimann, U. Luthi, M. Meuli, S. Altermatt, R. Gobet, L. Bruckner-Tuderman, and B. Steinmann, *J. Clin. Invest.* **98,** 1171 (1996).

[40] D. Aeschlimann and M. Paulsson, *J. Biol. Chem.* **266,** 15308 (1991).

[41] D. Hohl, D. Aeschlimann, and M. Huber, *J. Invest. Dermatol.* **110,** 268 (1998).

anti-dansyl antibody.[38] After three washes in PBS, bound antibody is detected with FITC-coupled pig anti-rabbit IgG diluted 1:30 in PBS for 30 min. Mounting of the slides and fluorescent visualization are done by standard procedures.

Reagents. Tris-HCl buffer, 0.1 M, pH 8.2; bovine serum albumin (BSA), 1 g/100 ml in the Tris buffer; monodansylcadaverine; ethylenediaminetetraacetic acid (EDTA), 10 mM; CaCl$_2$, 5 mM; putrescine, 2 mM; tTG, purified from guinea pig liver (see below), commercially available from Sigma (St. Louis, MO); PBS; fluorescein isothiocyanate (FITC)-coupled anti-rabbit IgG second antibody (e.g., from Dako, Carpinteria, CA).

Immunohistochemical and Immunocytological Detection of Transglutaminases

Immunocytochemistry and Immunohistochemistry with Tissue Transglutaminase Antibodies

Several tTG antibodies have been described that can be used for the procedure described below, and for Western blotting methods.[42–45]

Procedure. Adherent cells are grown on chamber slides and then fixed in cold fixative [4% (w/v) paraformaldehyde in PBS] for 5 min. Slides are washed three times in PBS, and then in distilled water, and at this point they can be stored at −20°. To block endogenous peroxidases, the slides are incubated with 3% (w/v) H$_2$O$_2$ for 5 min and then rinsed in PBS. Slides are then covered with normal goat serum for 5 min to block nonspecific binding sites, followed by the anti-tTG antibody diluted 1:100,[44] and incubated in a humid chamber overnight at 4°. Slides are washed three times in PBS and incubated with secondary antibody, biotinylated goat anti-rabbit IgG (diluted 1:200), for 20 min at room temperature. After washing, the slides are incubated with streptavidin–horseradish peroxidase complex for 20 min at room temperature, washed in PBS, and treated with 2.5 ml of freshly prepared substrate solution consisting of 3% (w/v) amino-9-ethylcarbazole, 3% (w/v) H$_2$O$_2$, and 0.1 M sodium acetate (pH 5.5) buffer at room temperature, until the color is developed (3–40 min). Slides are rinsed well in distilled H$_2$O$_2$, counterstained in Mayer's hemalum, and mounted in aqueous mounting medium. The same protocol can be used for cells grown in suspension after cytocentrifugation onto a slide coated

[42] P. A. J. Davies, M. P. Murtaugh, W. Y. Moore, G. S. Johnson, Jr., and D. Lucas, *J. Biol. Chem.* **260,** 5166 (1985).

[43] K. N. Lee, P. J. Birckbichler, and L. Fesus, *Prep. Biochem.* **16,** 321 (1986).

[44] L. Fesus and G. Arato, *J. Immunol. Methods* **94,** 131 (1986).

[45] T. Iwaki, M. Miyazono, T. Hitotsumatsu, and J. Tateishi, *Am. J. Pathol.* **145,** 776 (1994).

with gelatin. The same procedure may also be applied for tissue fragments fixed in paraformaldehyde and included in low melting point paraffin.[46]

Reagents. PBS; paraformaldehyde (4%, w/v), prepared in PBS; H_2O_2 (3%, w/v); biotinylated goat anti-rabbit IgG (diluted 1 : 200); sodium acetate buffer (pH 5.5), 0.1 M; 3-amino-9-ethylcarbazole (3%, w/v); Mayer's hemalum stain; aqueous mounting medium.

Preparation and Use of Transglutaminase 1 and 3 Antibodies

A short construct of 467 amino acid residues, which retained TGase 1 activity,[47] was purified as described below and used as an immunogen for the production of a polyclonal antibody in goats.[48] This antibody has been well characterized in indirect immunofluorescence, immunoprecipitation, and Western blot analyses.[47–49] A mouse monoclonal antibody against TGase 1, B.C1, is also commercially available from Biomedical Technologies (Stoughton, MA).[50] A polyclonal antibody prepared in rabbits against the guinea pig skin TGase 3 has been described.[51] The polyclonal antibodies should be affinity purified before use.

Affinity Purification Procedure. The TGase antigen is first chemically coupled onto a commercially available support (e.g., Reacti-Gel; Pierce, Rockford, IL), using the manufacturer protocol, and equilibrated into PBS. The crude serum is then mixed with the immobilized antigen, gently shaken overnight at 4°, and packed into a column (5 × 1 cm). After washing the column with several volumes of 0.5 M NaCl in PBS, the bound specific antibody is eluted with 1 vol of 0.1 M glycine-HCl buffer (pH 3.0). The eluate is immediately neutralized by addition of 0.1 vol of 2 M Tris. Bovine serum albumin is then added to make a 1% (w/v) solution, dialyzed against PBS, and concentrated by about 10-fold. Typically, the resulting antibodies can be used at titers of 1 : 100 or so. They are stored frozen in aliquots.

Procedure for Immunoprecipitation. Tissues or cultured cells are extracted into a buffer of 0.1 M Tris–acetate (pH 7.5), 0.15 M NaCl, 1 mM EDTA containing the protease inhibitors leupeptin (1 mM), phenylmethylsulfonyl fluoride (1 mM), calpain inhibitor (10 μM), and aprotinin (0.1 unit/ml), and separate cytosolic and membrane-anchored TGase forms are recovered as desired (see below). Aliquots of 200 μl are then mixed

[46] L. Fesus, Z. Nemes, L. Piredda, A. Madi, M. di Rao, and M. Piacentini, *in* "Techniques in Apoptosis" (T. G. Cotter and S. J. Martin, eds.), p. 21. Seattle, WA, 1996.

[47] S.-Y. Kim, I.-G. Kim, S.-I. Chung, and P. M. Steinert, *J. Biol. Chem.* **269,** 27979 (1994).

[48] S.-Y. Kim, S.-I. Chun, K. Yoneda, and P. M. Steinert, *J. Invest. Dermatol.* **103,** 211 (1995).

[49] S.-Y. Kim, S.-I. Chung, and P. M. Steinert, *J. Biol. Chem.* **270,** 18026 (1995).

[50] S. M. Thacher and R. H. Rice, *Cell* **40,** 685 (1985).

[51] H.-C. Kim, M. S. Lewis, J. J. Gorman, S.-C. Park, J. E. Girard, J. E. Folk, and S.-I. Chung, *J. Biol. Chem.* **265,** 21971 (1990).

with 5–10 μl of affinity-purified antibody and mixed for 1–16 hr. Protein A-conjugated agarose beads (ImmunoPure Plus; Pierce) are preequilibrated in the preceding buffer and resuspended as a 1:1 slurry, of which 20 μl is then added to the primary antibody mixtures. After incubation by shaking for 1 hr, the conjugated beads are collected by centrifugation at 10,000g for 1 min, and washed in buffer to remove unabsorbed proteins. An alternative method is to immobilize by covalent cross-linking the anti-TGase antibody onto protein G beads with dimethyl pimelidate-HCl (Pierce), using the manufacturer recommended procedures. After washing as described above, the bound TGase may then be released by use of 0.1 M glycine-HCl buffer (pH 3.0) and immediately neutralized as described above. If a Western blotting experiment is planned, in both procedures the washed beads may be simply boiled in sodium dodecyl sulfate–polyacrylamide gel electrophoresis (SDS–PAGE) loading buffer.

Procedure for Immunofluorescence. Normal adult human skin is frozen immediately and cut into 4- to 6-μm sections. Sections are fixed with acetone for 10 min at −20° or in 4% (w/v) paraformaldehyde for 10 min at 23°. After fixation, the slides are rinsed with PBS and treated with 2% (w/v) gelatin for 30 min. To reduce nonspecific binding, sections are then incubated at 23° with TGase antibody (diluted with PBS to 1:200) for 12–16 hr at 4°, rinsed again three times, and incubated with the secondary antibody (fluorescein-conjugated rabbit anti-goat IgG, diluted 1:300) for 30 min at 23°. Slides are mounted with 90% (v/v) glycerol and analyzed in a photomicroscope equipped with epifluorescence illumination and phase-contrast devices.

Reagents. PBS; gelatin (2%, w/v); fluorescein-conjugated rabbit anti-goat IgG (diluted 1:300); glycerol (90%, v/v); paraformaldehyde (4%, w/v).

Expression, Isolation, and Purification of Transglutaminases

Recombinant Transglutaminases

Expression of TGases in eukaryotic or prokaryotic systems has been carried out using mainly TGase 1 or deletion fragments thereof in bacterial[47] and baculovirus[52] systems. There is no apparent reason why full-length or partial fragments of other members of the TGase family cannot be expressed in these systems by similar methods.

[52] E. Candi, G. Melino, A. Lahm, R. Ceci, A. Rossi, I.-G. Kim, B. Ciani, and P. M. Steinert, *J. Biol. Chem.* **273**, 13693 (1998).

Expression of Transglutaminases in Bacteria

Procedure. A cDNA clone encoding a desired TGase or fragment thereof should be generated by PCR amplification. After confirmation of its sequence, it is then ligated into the pET-11a bacterial expression vector (Novagen, Madison, WI). For this vector system, PCR products should be designed with an extension of 16 nucleotides containing an *Nhe*I restriction enzyme site at the 5' end (which provides the in-frame ATG initiation codon) and a termination codon TGA followed by a *Bam*HI site at the 3' end flanking the desired coding sequences. The clone is then transformed into host *Escherichia coli* B strain BL21/DE3. In this system the bacterial host does not posttranslationally modify the expressed recombinant protein. Moreover, the host cells do not produce proteases, allowing recovery of intact undamaged expression products. This system has been used successfully for the expression of other intact proteins that have proved to be exquisitely sensitive to proteolysis. The bacteria are grown in LB medium supplemented with ampicillin or carbinicillin (50 μg/ml) to an $OD_{600\ nm}$ of 0.6, and protein expression is induced with 1 mM isopropyl-1-thio-β-D-galactopyranoside. After 3 hr the bacteria are harvested by centrifugation at 5000g for 10 min. In this system, >90% of the TGase 1 activity partitions into inclusion bodies, but significant activity, perhaps suitable for some biochemical assays,[47] is soluble.

Reagents. Escherichia coli strain BL21/DE3; expression vector pET-11a; isopropyl-1-thio-β-D-galactopyranoside (IPTG); Luria–Bertani (LB) medium; Tris–acetate buffer, 0.1 M (pH 8.0); ampicillin or carbinicillin; NaCl, 0.1 M; protease inhibitors leupeptin, phenylmethylsulfonyl fluoride, calpain inhibitor, and aprotinin.

Expression of Transglutaminases in Baculovirus in Insect Cells

This is a eukaryotic system, and so the recombinant proteins may be posttranslationally modified.

Procedure. A cDNA clone encoding a desired TGase is cloned into the baculovirus vector pVL1392 (PharMingen, San Diego, CA) by insertion at the *Eco*RI sites. The TGase cDNAs in the pVL1392 vector are under the transcriptional control of the strong baculovirus polyhedrin promoter. The TGase recombinant virus is obtained by cotransfection of each vector with the modified AcNPV virus DNA (Baculogold DNA; PharMingen). Baculogold DNA carries a lethal deletion and does not encode viable virus particles by itself. Cotransfection of this DNA with a complementing plasmid construct rescues the lethal deletion of this virus DNA and reconstitutes, by homologous recombination, viable virus particles inside the transfected insect cells. Insect cells (Sf9) are grown in Grace's insect medium supplemented with 10% (v/v) fetal calf serum. In the cotransfection experiments 2×10^6 cells are plated in 60-mm dishes. Two micrograms of each vector is mixed with 0.5 μg of linearized viral Baculogold DNA and incubated for 10 min at 23°. After incubation, 1 ml of transfection buffer containing 25 mM HEPES (pH 7.1), 125 mM CaCl$_2$, and 140 mM NaCl is added to each tube and mixed. The DNA/transfection buffer mixtures are added to plates that contain 1 ml of regular medium, and incubated at 27° for 3–4 hr. At the end of the transfection, the medium is replaced with fresh Grace's insect medium and the cells are grown for 5 days at 27°. In this way, the medium containing the recombinant virus clones is amplified to produce a high-titer stock solution, and the exact titers ($1–8 \times 10^8$ PFU/ml) are evaluated by end-point dilution. For protein production, cells are maintained and infected, either as monolayer or in suspension, in a rotary shaker, with $1–4 \times 10^8$ PFU/ml of culture at a density of $1.5–2 \times 10^6$ cells (>95% viability) per milliliter of culture. In the case of the TGase 1 system, most of the recombinant enzyme (about 90%) is retained in the membrane fractions of the Sf9 cells. The expressed recombinant human TGase 1 is modified by both palmitoylation and myristoylation, as for the native enzyme isolated from human keratinocytes.[52]

Reagents. Baculovirus vector pVL1392; AcNPV virus DNA; Baculogold DNA (PharMingen); Sf9 insect cells; Grace's insect medium supplemented with fetal calf serum (10%, v/v); transfection buffer [HEPES (25 mM, pH 7.1), CaCl$_2$ (125 mM), NaCl (140 mM)].

Purification of Transglutaminases

Purification of Recombinant Transglutaminase 1

Procedure. Harvested bacterial pellets are suspended in a 10-fold volume of lysis buffer of 0.1 M Tris–acetate (pH 8.0), 0.1 M NaCl, 1 mM EDTA containing protease inhibitors [1 mM leupeptin, 1 mM phenylmethylsulfonyl fluoride, 10 μM calpain inhibitor, and aprotinin (0.1 unit/ml)], and incubated for 1 hr with lysozyme (20 mg/ml).[47] After sonication for 30 sec, the lysate is centrifuged for 20 min at 30,000g. The clarified extracts are applied to a 10 × 4 cm column of Q-Sepharose (for a bacterial pellet from 1 liter of culture) that has been equilibrated with buffer A containing 0.05 M Tris–acetate (pH 7.5), 1 mM EDTA, 1 mM phenylmethylsulfonyl fluoride, and 1 mM benzamidine. After washes with four column volumes of buffer A, a linear 0–0.5 M NaCl gradient for 1 liter is applied. The recombinant TGase 1 elutes between 0.17 and 0.3 M salt concentration. Column fractions containing the bulk of enzyme activity are pooled and concentrated with Amicon (Danvers, MA) membranes (YM-10), and applied to a 20 × 2 cm Superose 6 column (Pharmacia, Uppsala, Sweden) that has been equilibrated with buffer B containing 0.05 M Tris–acetate (pH 6.8), 20 mM NaCl, 1 mM EDTA, and 1 mM phenylmethylsulfonyl fluoride. Column fractions containing enzyme activity are collected and further purified on a 5 × 0.5 cm Mono Q column with a linear gradient of 0–0.4 M NaCl in buffer A. The purified recombinant TGase 1 enzyme is used as soon as possible after purification because it begins to lose activity within 16 hr during storage at 4°.[47]

Reagents. Tris–acetate buffer (0.1 M, pH 8.0); NaCl, different concentrations; protease inhibitors leupeptin, phenylmethylsulfonyl fluoride, calpain inhibitor, and aprotinin; benzamidine; ethylenediaminetetraacetic acid; lysozyme; Tris–acetate buffer (0.05 M, pH 7.5).

Purification of Transglutaminase 1 from Normal Human Epidermal Keratinocytes

Procedure. Normal human epidermal keratinocytes may be grown in any one of several well-established systems. After several days at confluency, the terminally differentiating cells are harvested, washed, and suspended (0.5 ml/60-mm dish) in buffer A containing 0.25 M sucrose, 0.1 M Tris–acetate (pH 7.5), and the protease inhibitors leupeptin (1 mM), phenylmethylsulfonyl fluoride (1 mM), calpain inhibitor (10 μM), and aprotinin (0.1 unit/μl). The cells are lysed by sonication at 4° for 30 sec, and centrifuged at 10,000g for 20 min at 4°. The supernatant contains the cytosolic forms of the TGase 1 enzyme, while the pellet contains the membrane-anchored enzyme.[53] The membrane-anchored enzyme may be released from the membranes in two ways.[53,54] The first method dissolves the membranes by resuspension of the above pellet in 0.5 ml/60-mm dish of cells of buffer A containing 0.1% (v/v) Triton X-100, incubated at 4° for 15 min, and clarified again by centrifugation at 10,000g for 20 min. The second method hydrolyzes the thio-ester linkages by which the TGase 1 enzyme is anchored to the membranes, by treatment with 1 M NH$_2$OH-HCl for 1 hr at 4° in buffer A (0.5 ml/60-mm dish), followed by centrifugation as described above. The cytosolic or membrane-anchored TGase 1 enzymes from these respective supernatants are then diluted with 2 vol (Triton method) or 5 vol (NH$_2$OH-HCl method) of a binding buffer containing 0.1 M Tris-HCl (pH 8.6), 0.15 M NaCl, 0.02% (w/v) NaN$_3$. To immunoprecipitate the TGase 1 antigens, this diluted solution is incubated for 4 hr at 4° with immobilized anti-TGase

[53] P. M. Steinert, S.-I. Chung, and S.-Y. Kim, Biochem. Biophys. Res. Commun. 221, 101 (1996).
[54] P. M. Steinert, S.-Y. Kim, S.-I. Chung, and L. N. Marekov, J. Biol. Chem. 271, 26242 (1996).

1 beads. After the beads are washed with 5 vol of the binding buffer, the antigens are released with 1 vol of buffer (0.2 M glycine-HCl, pH 2.3) and 0.15 M NaCl, immediately neutralized with 0.1 vol of 2 M Tris base, and desalted by gel filtration on a Sephadex G-10 column.[47] Alternatively, the immunoprecipitation method directly employing the affinity-purified TGase antibody followed by protein A beads may be used here: for the Triton method, the diluted solution may be passaged over a column to remove the detergent prior to adding antibody.[55] The various enzyme forms may then be resolved by chromatography on the Mono Q column as described above.

Reagents. Normal human epidermal keratinocytes; sucrose, 0.25 M; Tris–acetate buffer (0.1 M, pH 7.5); protease inhibitors as described above; Tris-HCl buffer (0.1 M, pH 8.6); NaCl, 0.15 M; and NaN$_3$.

Purification of Transglutaminase 3 from Guinea Pig Skin

Procedure. Purification of the enzyme has been achieved in five steps.[51]

Step 1. Frozen guinea pig skins (Pel-Freez Biologicals, Rogers, AR) are freed of any remaining subcutaneous fat and ground to a fine powder in a mill in liquid N$_2$. Skin powder (200 g) is homogenized in a Polytron for 3 min in 1 liter of buffer A containing 0.1 M Tris–acetate buffer (pH 7.5), 10 mM EDTA, 5 mM benzamidine hydrochloride, 2 mM leupeptin, 1 mM phenylmethylsulfonyl fluoride, and 200 trypsin inhibitor units of aprotinin per liter. The homogenate is centrifuged for 30 min at 30,000g at 4° and the pellet is extracted two additional times with 1-liter portions of buffer A as described, and the three extracts are combined. All further purification steps are performed at room temperature.

Step 2. DEAE-cellulose treatment: The combined extracts are passed onto a 5 × 7 cm column of DE52 that has been equilibrated with buffer B containing 0.05 M Tris–acetate buffer (pH 6.0) and 1 mM EDTA. The column effluent is collected, together with 500 ml of buffer B column wash. Note that the TGase 3 enzyme is neutral in charge and does not bind. The pooled effluent wash is brought to 75% saturation (516 g/liter) with (NH$_4$)$_2$SO$_4$ and stored for 16 hr. The resulting precipitate is collected by centrifugation at 20,000g for 10 min, dissolved in 150 ml of buffer C containing 5 mM Tris–acetate buffer (pH 6.0) and 1 mM EDTA, and dialyzed against the same buffer.

Step 3. Heparin–Sepharose chromatography: The dialyzed solution is applied to a 5 × 14 cm column of heparin–Sepharose CL-4B that has been

[55] R. Schmidt, S. Michel, B. Shroot, and U. Reichert, *J. Invest. Dermatol.* **90**, 467 (1988).

equilibrated with buffer C. The column is washed with buffer C until the effluent shows an absorbance of less than 0.02 at $OD_{280 \text{ nm}}$, and a 1-liter linear salt gradient, 0–0.25 M NaCl in buffer C, is passed through the column. The fractions eluting at ~0.1 M NaCl, which display TGase activity after incubation with dispase (see below), are pooled and the solution is concentrated to 2 ml by ultrafiltration with a YM-10 membrane.

Step 4. Exclusion chromatography: This concentrated solution is applied to a 2.6 × 98 cm column of Bio-Gel A-0.5m that has been equilibrated with buffer C containing 0.15 M NaCl. The fractions containing TGase activity are pooled and dialyzed against buffer C.

Step 5. S-Sepharose chromatography: The resultant solution is applied to a 2.6 × 17 cm column of S-Sepharose that has been equilibrated with buffer C. The column is washed with 200 ml of a buffer containing 5 mM Tris–acetate (pH 7.5) and 1 mM EDTA. Finally, the TGase 3 is eluted with a buffer containing 50 mM Tris–acetate buffer (pH 8.0) and 1 mM EDTA. Examination of the fractions by SDS–PAGE after staining with Coomassie blue will identify those containing a single band of TGase 3 protein of about 77 kDa. These fractions are pooled and concentrated to approximately 1 mg/ml. The protein is stable when stored under sterile conditions at room temperature for up to 1 month.

Reagents. Frozen guinea pig skins; Tris–acetate buffer (0.1 M, pH 7.5); Tris–acetate buffer (0.05 M, p6.0); Tris–acetate buffer (5 mM, pH 6.0); Tris–acetate buffer (5 mM, pH 7.5); Tris–acetate buffer (50 mM, pH 8.0); ethylenediaminetetraacetic acid (EDTA), different concentrations; protease inhibitors benzamidine hydrochloride, leupeptin, phenylmethylsulfonyl fluoride, and aprotinin; $(NH_4)_2SO_4$; NaCl, different concentrations.

Activation of Transglutaminase 3

This enzyme (sometimes termed pro-TGase 3) is usually recovered as the full-length form, which has little or no activity. It requires activation by proteolysis to become active.

Procedure. Activation of pro-TGase 3 is usually carried out with the protease dispase. The proenzyme (0.24 mg/ml) is incubated with dispase (1 unit/ml) for 10 min at 25° in 0.05 M Tris–acetate buffer (pH 6.0).[51,52] Under these conditions 80–90% of the proenzyme is cleaved and activated. The efficiency of activation can also be monitored by Coomassie staining: the activated TGase 3 will show two protein bands with mobilities of 50 and 27 kDa compared with the 77-kDa band of the proenzyme.

Pro-TGase 3 could also be activated with other proteases (proteinase K, trypsin, thrombin) but with less efficency.[51]

Purification of Tissue Transglutaminase from Liver

Procedure. The following procedure for the purification of tTG was developed for unfrozen guinea pig livers.[56] All steps are carried out at 4°.

Step 1. Freshly excised or previously frozen liver (200 g) is resuspended in 500 ml of 0.25 M sucrose and homogenized with a Polytron homogenizer at medium speed. The homogenate is centrifuged at 100,000g for 1 hr.

Step 2. Chromatography on DEAE-cellulose: The supernatant is immediately pumped into a 3.5 × 10 cm column of DEAE-cellulose equilibrated with a buffer containing 5 mM Tris-HCl (pH 7.5) and 2 mM EDTA. The column is then washed with 200 ml of equilibrating buffer, and the protein is eluted by the use of a 1.5-liter linear gradient of 0–1 M NaCl in the same buffer. The enzyme elutes from the column between 0.25 and 0.4 M NaCl. The fractions containing TGase activity are combined.

Step 3. Protamine precipitation and extraction: The pooled fractions are made to 1% (w/v) protamine sulfate and allowed to stand for 16 hr. The precipitate, which contains the enzyme, is collected by centrifugation for 15 min at 15,000g. The pellet is washed by resuspension in 10 ml of 0.2 M Tris–acetate buffer (pH 6.0), and clarified by centrifugation. This pellet is extracted with 20 ml of buffer containing 5 mM Tris-HCl (pH 7.5), 0.05 M (NH$_4$)$_2$SO$_4$, and 2 mM EDTA, homogenized with a Polytron homogenizer at low speed, and repelleted. This step is repeated two more times. The combined supernatants are made to 20 mM EDTA and 3 M (NH$_4$)$_2$SO$_4$.

Step 4. Exclusion chromatography: The precipitate is collected after 30 min by centrifugation and is dissolved in 1 to 2 ml of buffer containing 10 mM Tris–acetate (pH 6.0), 1 mM EDTA, and 1.6 M KCl. The clear supernatant is loaded on a 2.5 × 100 cm column of 10% (w/v) agarose equilibrated with the same buffer. The enzyme, which elutes after about one-half of the column volume, is collected. The fractions with highest activity are collected and combined. The enzyme is stored at a final concentration of 10–20 mg/ml at −20°.

Reagents. Guinea pig livers (unfrozen); sucrose, 0.25 M; Tris-HCl (5 mM, pH 7.5); ethylenediaminetetraacetic acid (EDTA), different concentrations; NaCl, different concentrations; protamine sulfate (1%, w/v); Tris–acetate buffer (0.2 M, pH 6.0); Tris–acetate buffer (10 mM, pH 6.0); (NH$_4$)$_2$SO$_4$; KCl, 1.6 M.

---

[56] J. E. Folk and S.-I. Chung, *Methods Enzymol.* **113**, 358 (1985).

Assays for Transglutaminase Activity

Transglutaminase Assays

TGase activity can be assayed with various primary amine donors. Here we describe essentially three types: first, using radiolabeled putrescine; second, using fluorescent monodanylcadaverine; and third, using 5-(biotin-amido)-pentylamine.

Transglutaminase Assay Using Radiolabeled Putrescine

This procedure has been modified many times since its introduction[47,56] and is based on measurement of incorporation of radiolabeled putrescine into a casein by action of TGases. Variations include the use of modified caseins, and different radiolabeled primary amines (methylamine, glycine ethyl esters) (Table V).

Procedure. Reactions are performed at 37° in 0.5 ml of buffer containing 0.1 M Tris–acetate (pH 7.5), 1% (w/v) succinylated casein, 1 mM EDTA, 10 mM CaCl$_2$, 5 mM dithiothreitol, 0.15 mM NaCl, 0.5 μCi of [^{14}C]putrescine (118 Ci/mol), and the protease inhibitors benzamidine hydrochloride (5 mM), leupeptin (2 mM), phenylmethylsulfonyl fluoride (1 mM), and aprotinin (0.2 unit/ml), and 50-μl portions of enzyme solution. After

TABLE V
AMINES TESTED AS SUBSTRATES FOR tTG^a

Amine substrate	Incorporation (μmol/100 μg protein)
Putrescine	8.9
Phenylethylamine	6.3
Glycinamide	4.9
Histamine	4.2
Methylamine	2.3
Ethanolamine	2.3
Ammonia	1.5
L- or D-Lysine	0.5
Tyrosinamide	0
Glycine	0
Glycyl-L-leucine	0
γ-Aminobutyric acid	0

^{a} Reactions are carried out under standard conditions for TGase assay.[56] The amines were all ^{14}C-labeled except for glycinamide (both ^{14}C- and ^{15}N-labeled glycinamide tested) and ammonia (^{15}N-labeled). From Ref. 1.

1 hr, the reaction is terminated by addition of 4.5 ml of cold 7.5% (w/v) trichloroacetic acid. The trichloroacetic acid-insoluble precipitates are collected onto GF/A glass fiber filters by filtration, and washed free of unincorporated radioactive amine with cold 5% (w/v) trichloroacetic acid, air dried, and counted by liquid scintillation. The specific activity is therefore expressed as picomoles of putrescine per minute per milligram of enzyme.

Reagents. Tris–acetate buffer (0.1 M, pH 7.5); succinylated casein; ethylenediaminetetraacetic acid; CaCl$_2$, 10 mM; dithiothreitol, 5 mM; NaCl; [^{14}C]putrescine (118 Ci/mol); trichloroacetic acid, 7.5 and 5% (w/v); protease inhibitors benzamidine hydrochloride, leupeptin, phenylmethylsulfonyl fluoride, and aprotinin.

Transglutaminase Assay Using Monodansylcadaverine

TGase activities can be measured by monitoring the rate of increase in fluorescence during TGase-catalyzed incorporation of monodansylcadaverine into N,N-dimethylcasein.[57] The excitation filter is 360 nm, while the emission filter is 490 nm.

Procedure. Incubations are carried out at 37° in a Millipore (Bedford, MA) 96-well low-fluorescence CytoPlate in 125-μl reaction mixtures containing 50 mM Tris-HCl (pH 7.5), 0.1 mM monodansylcadaverine, N,N-dimethylcasein (2 mg/ml), 1 mM dithiothreitol, 1 mM CaCl$_2$, and 0–0.124 μM enzyme.

Reagents. N,N-Dimethylcasein; Tris-HCl (50 mM, pH 7.5); monodansylcadaverine; dithiothreitol; CaCl$_2$.

Transglutaminase Assay Using 5-(Biotinamido)-pentylamine

This is a solid-phase microtiter plate assay (ELISA).[58]

Procedure. Microtiter plates are coated with 200 μl of N,N-dimethylcasein (10–20 mg/ml), sealed with plastic, and stored overnight at 4°. Plates may also be coated for 1 hr at 37°. After the unbound casein is discarded, the well surface is blocked with nonfat dry milk (0.5%, w/v) in 0.1 M Tris-HCl (pH 8.5) for 30 min and washed two times in the same buffer. Reagents are added to the well in the following order: 5 mM CaCl$_2$, 10 mM dithiothreitol (DTT), 0.5 mM 5-(biotinamido)-pentylamine, enzyme, and 0.1 M Tris-HCl (pH 8.5), to obtain a final volume of 200 μl/well. After incubation for 30 min at 37°, the liquid is discarded and the reaction is stopped by washing twice with 350 μl of 200 mM EDTA, followed by two washes with 350 μl

[57] L. Lorand, O. Lockridge, L. K. Campbell, R. Myhrman, and J. Bruner-Lorand, *Anal. Biochem.* **44**, 221 (1971).

[58] T. F. Slaughter, K. E. Achyuthan, T.-S. Lai, and C. S. Greenberg, *Anal. Biochem.* **205**, 166 (1992).

of 0.1 mM Tris-HCl (pH 8.5). A freshly prepared streptavidin–alkaline phosphatase solution (0.25 mg/ml) is diluted 1:150 with nonfat dry milk (0.5%, w/v) in 0.1 M Tris-HCl (pH 8.5), of which 250 μl is added per well and incubated for 1 hr at room temperature. Each well of the plate is then washed once with 350 μl of 0.01% (v/v) Triton X-100, followed by four 350-μl washes with 0.1 M Tris-HCl (pH 8.5). Then 200 μl of 0.1 M Tris-HCl (pH 8.5) and 50 μl of phosphatase substrate are added to each well. A kinetic measurement of absorbance at 405 nm is determined at 30-sec intervals for a period of 30 min, using a Micro Plate reader. TGase activity is expressed as units of optical density (OD/min).

Reagents. N,N-Dimethylcasein, 10–20 mg/ml; nonfat dry milk (0.5%, w/v); Tris-HCl (0.1 M, pH 8.5); monodansylcadaverine; dithiothreitol; CaCl$_2$; ethylenediaminetetraacetic acid, 200 mM; Triton X-100, 0.01% (v/v).

Transglutaminase Active Site Titration

To estimate accurately the concentration of TGases present in cell extracts or partially purified fractions, and/or to determine their specific activities, the active site of the TGase proteins can be alkylated with [14C]iodoacetamide.[16] The alkylation of the single active site cysteine residue results in the complete loss of enzyme activity.[1,2] Moreover, this cysteine is labeled about 1000-fold more rapidly than other cysteines. Thus, measurement of the amount of incorporated label provides a valid estimate of the molar amount of TGase enzyme. The reaction uses 5 μM iodoacetamide, which represents a 50- to 500-fold molar excess of reagent over enzyme, and is sufficient to cause a 40–60% loss of activity within 15 min.

Procedure. Reaction mixtures (200 μl) containing 0.1 M Tris–acetate (pH 8.0), 1 mM CaCl$_2$, 5 μM [14C]iodoacetamide, and aliquots of the enzymes, are incubated for 15 min at 25°. Reactions are terminated at the addition of dithiothreitol (10 mM) and EDTA (10 mM). The alkylated enzyme is separated from the unreacted [14C]iodoacetamide by exclusion chromatography on a Superose-6 column, and finally counted to calculate the amount of TGase protein present. As an experimental control to establish that only TGase protein is alkylated, the [14C]carboxymethyl enzyme may be reacted with immobilized anti-TGase antibody beads. The amount of label recovered from this immunoprecipitation reaction should be identical to that obtained by exclusion chromatography.[47]

Reagents. [14C]Iodoacetamide, 5 μM; Tris–acetate buffer (0.1 M, pH 8.0); CaCl$_2$; dithiothreitol, 10 mM; ethylenediaminetetraacetic acid, 10 mM.

Transglutaminase Inhibitors

On the basis of kinetic studies, it was established that the conserved active site cysteine residue is critically involved in the cross-linking reaction.

Like many enzymes with an active site thiol group, TGases are inactivated by alkylating agents such as halomethyl carbonyl derivatives.[59,60] The intrinsically high reactivity of these agents toward thiol groups, however, limits their utility. Also, alkyl isocyanates are covalent inhibitors of TGases but have the disadvantage of lacking specificity.[61] Primary amines such as cystamine, methylamine, and dansylcadaverine, which act as competitive inhibitors, have been used to inhibit TGase activity.[62] However, they are active only in the millimolar range and thus have limited utility because of their high reactivity or cytotoxicity. More recently, a more specific generic TGase inhibitor (LTB-2; Syntex Research, Palo Alto, CA) has been developed that is useful for cell culture systems at a concentration of 0.1 mM. Given the growing understanding of natural substrates, it may be possible to develop inhibitors more specific to each TGase enzyme. For example, by screening peptides designed from natural substrates, two highly specific peptides that function as competitive inhibitors of the TGase 1 enzyme have been described.[63] In this case, the inhibitors prevented the formation of insoluble cell envelope structures in *in vitro* keratinocyte cultures.

In Vivo Regulation of Tissue Transglutaminase by Ca^{2+}, GTP, Polyamines, S-Nitrosylation

The tTG enzyme binds guanine nucleotides and hydrolyzes GTP.[64] The intracellular activity of tTG is also regulated by the steady state level of putrescine.[26] More recently, we demonstrated that nitric oxide (NO) may regulate cell death occurring through either apoptosis or necrosis, depending on its concentration and length of exposure.[25] This balance might be achieved also through the S-nitrosylation of the tTG protein. In conclusion, the activity of tTG is controlled by the intracellular levels of calcium, GTP, NO, and polyamines.

Tissue Transglutaminase Binds GTP

A more complex role for tTG has been described.[65] tTG functions as the 74-kDa subunit ($G_{\alpha h}$) associated with the 50-kDa β subunit ($G_{\beta h}$) of the GTP-binding protein G_h. The G_h dimer acts in association with the rat liver α_1-adrenergic receptor in a ternary complex that contains the α_1-

[59] J. J. Holbrook, R. D. Cooke, and I. B. Kingston, *Biochem. J.* **135**, 901 (1973).
[60] S.-I. Chung, M. S. Lewis, and J. E. Folk, *J. Biol. Chem.* **249**, 940 (1974).
[61] M. Gross, N. K. Whetzel, and J. E. Folk, *J. Biol. Chem.* **250**, 7693 (1975).
[62] D. M. Chuang, *J. Biol. Chem.* **256**, 8291 (1981).
[63] S.-Y. Kim, W.-M. Park, S.-W. Jung, and J. Lee, *Biochem. Biophys. Res. Commun.* **233**, 39 (1997).
[64] K. E. Achyuthan and C. S. Greenberg, *J. Biol. Chem.* **262**, 1901 (1987).
[65] H. Nakaoka, D. M. Perez, K. J. Baek, T. Das, A. Husain, K. Misono, M. J. Im, and R. M. Graham, *Science* **264**, 1593 (1994).

agonist. Thus, the $G_{\alpha h}$ (hence tTG; hereafter also referred to as tTG/$G_{\alpha h}$) is a multifunctional ubiquitous "housekeeping" protein that, by binding GTP in a $G_{\alpha h}$–GTP complex, can modulate both receptor-stimulated phospholipase C (PLC) activation and transglutamination. $G_{\alpha h}$ represents a novel class of GTP-binding proteins that participate in the receptor-mediated signaling pathway. The novelty of the action of the $G_{\alpha h}$ component is that, through this complex regulatory mechanism, receptor-stimulated GTP binding may prevent the activation of harmful components of the genetic death program. Indeed, the Ca^{2+}-dependent cross-linking activity of tTG is finely tuned by GTP levels, which in turn regulates secondary messengers such as the production of inositol 1,4,5-trisphosphate (IP_3) and sn-1,2-diacylglycerol (DAG) from phosphatidylinositol 4,5-biphosphate (PIP_2). The GTP-binding activity of tTG/$G_{\alpha h}$ actively prevents activation of the lethal cross-linking activity of tTG; moreover, it provides a parallel integrated pathway to regulate the free intracellular Ca^{2+} concentration that is required for the activation of tTG. Note that intracellular Ca^{2+} levels of normal cells are typically in the micromolar range; tTG becomes functional as a cross-linking enzyme only when Ca^{2+} concentrations are raised by about two orders of magnitude, as commonly occurs in a dying cell. Thus, the fine modulation of the tTG/$G_{\alpha h}$ protein by GTP and Ca^{2+} (and possibly additional molecules such as free putrescine and other polyamines) may explain how cells are able to survive with high tTG/$G_{\alpha h}$ protein levels in their cytoplasm. Cell death prevention could also be achieved by the DAG-dependent activation of protein kinase C (PKC). Therefore, the fine regulation of the effector elements of death shows an extreme plasticity.

[γ-^{32}P]GTP Hydrolysis Assays

Procedure. The following is modified slightly from a previously published procedure.[66] Hydrolysis of [γ-^{32}P]GTP is measured in a reaction mixture (50 μl) containing 50 mM Tris-HCl (pH 7.5), 1 mM MgCl$_2$, 1 mM DTT, 2 μCi of [α-^{32}P]GTP (30 Ci/mmol), and 4 μM unlabeled GTP. The reaction is started by addition of tTG and is allowed to proceed at 37° for 30 min. Preliminary experiments should establish conditions under which the ^{32}P$_i$ release is linear over 50 min of incubation. The reaction is terminated by the addition of 750 μl of ice-cold monobasic sodium phosphate (50 mM) containing 5% (w/v) activated charcoal. The reaction tubes are then centrifuged for 2 min at 10,000g in a microcentrifuge and 400 μl of the supernatant is used for determination of ^{32}P$_i$ release by scintillation counting.

---

[66] A. Kikuchi, T. Yamashita, M. Kawata, K. Yamamoto, K. Ikeda, T. Tanimoto, and Y. Takai, *J. Biol. Chem.* **263**, 2897 (1988).

Reagents. $[\gamma\text{-}^{32}P]GTP$ (30 Ci/mmol); Tris-HCl (50 mM, pH 7.5); $MgCl_2$; dithiothreitol; GTP, 4 μM; monobasic sodium phosphate, 50 mM; activated charcoal.

Measurement of Phospholipase C Activity

Another way to monitor $tTG/G_{\alpha h}$ action is to estimate PLC activation.[65]

Procedure. The 69-kDa PLC (100 ng) and tTG (5 μg) proteins are combined in a phospholipid mixture (phosphatidylcholine–phosphatidylethanolamine–phosphatidylserine, 3 : 1 : 1) in an ice bath to reconstitute lipid vesicles.[67] The vesicles are incubated with 1 mM $MgCl_2$ in the absence or presence of 5 μM GTP-γ-S at 30° for 30 min before evaluation of PLC activity. The samples are then incubated at 30° for 10 min in the presence of $[^3H]PIP_2$ (500 cpm/nmol) and various concentrations of $CaCl_2$ (0–16 μM). The IP_3 formation is then measured.[67,68]

Reagents. Phospholipase (PLC), 100 ng; tTG, 25 μg; $MgCl_2$; GTP-γ-S; $[^3H]PIP_2$ (500 cpm/nmol); $CaCl_2$, different concentrations.

Regulation of Tissue Transglutaminase by Sense and Antisense Transfections

To study the role of tTG in physiological cell death, the cDNA for tTG has been transfected in sense and antisense orientations in neuroblastoma cells.[27]

Procedure. A full-length clone consisting of 135 bp of the 5′-noncoding region, 2061 bp of coding region, and 1058 bp of the 3′ untranslated region of the human tTG cDNA is inserted into the pSG5 DNA eukaryotic expression plasmid in the sense orientation. This plasmid, which contains the simian virus 40 early gene promoter and polyadenylation signal sequences, is linearized with *Eco*RI. For cotransfection experiments, plasmid pSV2-Neo, containing the neomycin resistance gene under the control of the same promoter, is used as a selectable marker. The transfected cells are selected by growing in medium containing G418 (400 μg/ml), and cell clones are obtained from individual G418-resistant colonies. For antisense transfections 1.0 kbp of the human tTG cDNA is inserted into pSG5. This fragment contains 135 bp of 5′ noncoding sequence and 865 bp of coding sequence, and may be generated by linearizing the full-length clone by digestion with *Eco*RI, and then the 1.0-kb insert is obtained by *Bal*I digestion. Also, antisense transfections are stabilized by cotransfecting the antisense vector with pSV2-Neo. DNA transfections in neuroblastoma cells are performed

---

[67] M. J. Im, C. Gray, and A. J. Rim, *J. Biol. Chem.* **267,** 8887 (1992).
[68] T. Das, K. Baek, C. Gray, and M. J. Im, *J. Biol. Chem.* **268,** 27398 (1993).

by calcium phosphate precipitation by standard procedures.[69] Briefly, about 10^6 cells/dish are exposed to 5 μg of plasmid DNA for 6 hr. After 2 days, cells are replaced in medium containing G418, and the transfected clones are isolated 4 weeks later.

Reagents. tTG cDNA (human endothelial cell, clone hTG-1); pSG5 DNA eukaryotic expression plasmid; pSV2-Neo plasmid; gentamicin (G418); calcium phosphate buffer.

Isolation of Cross-Linked Structures

Isolation of Liver Apoptotic Bodies

Apoptotic bodies were detected first in rat hepatocytes.[21]

Procedure. Hepatocytes may be isolated by standard procedures[70] with yields of about 1×10^8 cells/g liver. After sedimenting by centrifugation (600g for 10 min), they are suspended at 4° in 25 ml of lysis buffer containing 10 mM Tris-HCl (pH 7.5), 10 mM KCl, 2 mM MgCl$_2$, Triton X-100 (0.5%, v/v), 0.2 mM phenylmethylsulfonyl fluoride to inhibit proteases, and 0.4 mM iodoacetamide to inhibit TGase activity. After centrifugation (600g for 5 min), the pellet is washed in cold lysis buffer three times, and then resuspended in 6 M guanidine hydrochloride to dissolve nuclei and proteins. A subsequent centrifugation (600g for 10 min) results in a pellet consisting of insoluble structures that are formed by the population of apoptotic hepatocytes.

By phase-contrast microscopy, they appear as spherical or wrinkled structures, similar to the cornified envelopes formed by keratinocytes[71] (see below). By scanning electron microscopy, they appear as irregular, globular entities of relatively smooth surface. By flow cytometry, about 70–75% are 12–13 μm in diameter, 20–25% are 15–16 μm in diameter, and the remaining 5–10% may vary more widely, from 8 to 30 μm. The yield is 3.8–5.2 $\times$ 10^4/g of liver,[21] corresponding to 2–3/10^4 liver cells, which correlates well with the number of cells undergoing apoptosis at any one time in normal liver. These apoptotic bodies are insoluble in solvents such as 8 M urea with or without 5% (v/v) 2-mercaptoethanol, boiling in 2% (w/v) SDS, as well as various nonionic detergents. They may be dissolved only after extensive proteolysis[21] (see below).

Reagents. Tris-HCl (10 mM, pH 7.5); KCl; MgCl$_2$; Triton X-100; phenylmethylsulfonyl fluoride; iodoacetamide; guanidine hydrochloride, 6 M.

[69] J. Sambrook, E. F. Fritsch, and T. Maniatis, "Molecular Cloning: A Laboratory Manual," 2nd Ed. Cold Spring Harbor Laboratory Press, Cold Spring Harbor, New York, 1989.
[70] P. Moldeus, L. Hodberg, and S. Orrenius, *Methods Enzymol.* **52,** 60 (1978).
[71] H. Green, *Harvey Lect.* **74,** 101 (1980).

Isolation of Epithelial Cornified Cell Envelopes

Virtually all stratified squamous epithelia form a structure generically termed the (cornified) cell envelope (CE), which is essential for barrier function.

Procedure. The following procedure[30,31] for human epidermal stratum corneum also works well with other internal epithelial tissues including buccal mucosa, gingiva, esophagus, rodent forestomach, etc. The epidermis (or epithelium) is separated from the underlying dermis or connective tissue by heating at 60° for 30 sec, laid flat on a petri dish surface, and peeled away with fine forceps. This tissue may be stored at −70° indefinitely. In the case of the epidermis, the tissue is extracted in a buffer containing 8 M urea, 50 mM Tris-HCl (pH 8.0), and 1 mM EDTA (25 ml/g tissue). This dissolves the bulk of the keratin. The extract is filtered through nylon gauze (mesh size, 0.1 mm) to collect the stratum corneum sheets. In the case of other epithelial tissues, this filtration step may not be necessary; the urea extract should be centrifuged at 5000g for 5 min to pellet the insoluble material. CEs are then prepared from this pellet or the stratum corneum retentate by exhaustive boiling and sonication in a buffer containing 0.1 M Tris-HCl (pH 8.0), 2% (w/v) SDS, 20 mM dithiothreitol, and 5 mM EDTA (25 ml/g original tissue),[72] and pelleted at 5000g for 10 min to recover the crude CEs. To remove as much of the adherent keratin protein material as possible, this pellet is resuspended in 2 ml and centrifuged at 5000g through 20 ml of 20% (v/v) Ficoll solution in the same buffer. This step may be repeated at least once more for epidermal tissue. Finally, the CE fragments are washed three times in phosphate-buffered saline to remove the bulk of the SDS and reducing agent. Amino acid analysis of acid-hydrolyzed samples may be performed to confirm the purity of the CEs (Table VI).

Reagents: Tris-HCl (50 mM and 0.1 M, pH 8.0); ethylenediaminetetraacetic acid (EDTA); sodium dodecyl sulfate (SDS); dithiothreitol; Ficoll; urea; phosphate-buffered saline.

Detection and Quantitation of N^ε-(γ-Glutamyl)-lysine Isodipeptide Cross-Links

Recovery of Isodipeptide from Cross-Linked Structures, Cells, or Plasma

The recovery of the isodipeptide from tissues is based on its resistance to cleavage by proteases, simply because it does not possess a Cα atom

---

[72] R. H. Rice and H. Green, *Cell* **11**, 417 (1977).

TABLE VI

COMPARISONS OF AMINO ACID COMPOSITIONS OF LIVER APOPTOTIC BODIES AND EPIDERMAL CORNIFIED ENVELOPES[a]

Amino acid	Cornified envelopes		Apoptotic bodies	
	Human	Rat	Rat	Dog
Asx	0.7	0.3	1.4	1.6
Glx	7.1	4.8	9.3	9.6
Ser	9.7	8.9	8.3	8.2
Gly	23.2	11.4	23.7	23.4
His	2.6	4.4	3.5	3.9
Thr	1.6	1.2	2.1	2.3
Ala	4.1	7.4	5.9	13.7
Arg	3.1	8.6	4.3	6.9
Pro	14.6	8.5	10.6	10.2
Tyr	7.1	4.7	12.5	3.0
Val	2.5	7.5	3.7	3.0
Cys	1.4	1.6	0.1	0.3
Met	ND	ND	ND	ND
Ile	4.6	8.0	2.0	2.0
Leu	4.1	14.3	5.6	4.4
Phe	2.3	2.3	3.3	2.3
Lys	11.2	6.1	3.7	3.3

[a] ND, Not determined. From Ref. 74.

(Fig. 1). Thus cross-linked structures are extensively digested with a battery of proteases to reduce the proteins to single amino acids and the free isodipeptide itself. The proteases commonly used do not contain it. The following is a procedure for epidermal CEs that has been used with modifications for many tissues.[21,72–74]

Procedure. CEs are resuspended in a buffer containing 0.1 M N-ethylmorpholine acetate (pH 8.0) (1 mg/ml), and made to 3% (w/w) proteinase K and/or V8 protease. Digestion is continued at 37° with further additions of 3% (w/w) proteases every 12 hr for 3 days. The reaction is then made to 1% (w/w) carboxypeptidase Y, digested for 6 hr, and the process is repeated at least once more. In highly cross-linked tissues such as epidermal CEs or hair follicle inner root sheaths, which contain up to 3% isodipeptide, it may be necessary to deproteinize the digest by passage through a short desalting G10 Sephadex column.

[73] E. Tarcsa, N. Kedei, V. Thomazy, and L. Fesus, *J. Biol. Chem.* **267**, 25648 (1992).
[74] D. Hohl, T. Mehrel, U. Lichti, M. L. Turner, D. R. Roop, and P. M. Steinert, *J. Biol. Chem.* **266**, 6626 (1991).

Precipitation of proteins from plasma, cell extracts, or proteolytic digestion products may be done with 10% (w/v) trichloroacetic acid or 1% (saturated) picric acid.

Reagents. N-Ethylmorpholine acetate (0.1 *M*, pH 8.0); proteinase K; V8 protease; carboxypeptidase Y.

Detection and Quantitation of Isodipeptide

Any of the following methods may be used, depending on available equipment.

Amino Acid Analysis

Procedure. The deproteinized samples (plasma, culture fluid, proteolytic digests) are resolved on an amino acid analyzer as for any other protein hydrolysate. In most systems [e.g., Beckman (Fullerton, CA) 6300 amino acid analyzer], the isodipeptide elutes near methionine, and is well separated from other amino acids. In this case, quantitation may be done by also analyzing (after acid hydrolysis) a sample of the original cross-linked material. The synthetic N^ε-(γ-glutamyl)-lysine isodipeptide may be purchased from Accurate Chemical (Westbury, NY) as a standard.

A variation of this method, adapted for cultured cells or when the amount of cross-link is low[73,74] can involve labeling with [³H]lysine. After total enzymatic digestion, the amount of [³H]lysine in the isodipeptide peak and in free lysine itself is measured to estimate the degree of cross-linking.

High-Performance Liquid Chromatography

Procedure. To remove free lysine, 1 ml of filtered sample (plasma, culture fluid, proteolytic digests) is loaded onto a 2.4 × 0.4 cm column of AG 5OW-X8 cation-exchange resin (ammonia form), and eluted with 1 ml of distilled water. The 2-ml total is dried and may be stored at −20°. The sample is then desalted on a silica high-performance liquid chromatography (HPLC) column, using water as eluent.[21,74,75] The fractions collected at the position of isodipeptide are derivatized by phenylisothiocyanate and then analyzed by HPLC on a 3.9 mm × 30 cm μBondapak C_{18} column, using the following elution system: eluent A is 0.14 *M* sodium acetate with triethanolamine (pH 6.35, 0.5 ml/liter); eluent B is acetonitrile; isocratic separation at 10% B is carried out for 10 min followed by a linear gradient up to 25% within 2 min and subsequent isocratic separation for 8 min.

[75] E. Tarcsa and L. Fesus, *Anal. Biochem.* **188,** 135 (1990).

Fluorimetric Assay

The following technique has been developed for quantification of the isodipeptide.[76] The purified isodipeptide is treated with γ-glutamylcyclotransferase, an enzyme that catalyzes its conversion into free lysine and 5-oxo-L-proline.[77] The L-lysine is then decarboxylated to form cadaverine, which forms a stable derivative with fluorescamine that is measured fluorimetrically.[78]

Procedure. Samples (plasma, culture fluid, proteolytic digests) are deproteinized by a centrifugal microseparator (Centrifree, MW cutoff 5000; Amicon), and free lysine is removed on a column of AG 5OW-X8 as described above. Samples are then dissolved in a buffer of 0.2 M N-ethylmorpholine acetate (pH 7.4) and 0.1 ml is tested in duplicate. An experimental control is 2.5 nmol of synthetic N^ε-(γ-glutamyl)-lysine isodipeptide. Then γ-glutamylcyclotransferase is added to samples of up to 0.1 ml (0.1-U/ml final concentration) and the mixtures are incubated overnight at 37°. At this point the lysine content in each test tube is measured[78]: 0.3 ml of pyroxidal phosphate (33 mg/liter in 0.1 M malate buffer, pH 6.0) and 0.1 ml of lysine decarboxylase (type VII, 1.25 U/ml in 0.1 M malate buffer, pH 6.0) are added to each tube and incubated at 37° for 60 min. Reactions are stopped by the addition of 1 ml of 5 M NaOH. The cadaverine formed is extracted by adding 2 ml of pentan-1-ol and shaking the samples for 5 min. Samples are centrifuged (1000g, 10 min) and 1 ml of the upper phase is added to 1 ml of sodium borate buffer (25 mM, pH 8.0) and 1 ml of fluorescamine solution (0.1 g/liter in acetone) while vortexing. Fluorescence intensity is measured at 390-nm excitation and 471-nm emission within 1 hr. The isodipeptide concentration can be calculated by comparing the fluorescence intensity of the sample with the standard.

Reagents. N-Ethylmorpholine acetate (0.2 M, pH 7.4); synthetic N^ε-(γ-glutamyl)-lysine; γ-glutamylcyclotransferase; pyroxidal phosphate (33 mg/liter); malate buffer (0.1 M, pH 6.0); lysine decarboxylase (type VII); NaOH, 5 M; pentan-1-ol; sodium borate buffer (25 mM, pH 8.0); fluorescamine (0.1 g/liter in acetone).

Identification and Characterization of Residues Used in *in Vitro* or *in Vivo* Cross-Linking Reactions

The identification of cross-linked residues utilized by TGases *in vitro* has been carried out mostly with the TGase 1, 3, and tTG enzymes in a

[76] J. Harsfalvi, E. Tarcsa, M. Udvard, and L. Fesus, *Fibrinolysis* **8**, 378 (1994).

[77] M. L. Fink and J. E. Folk, *Methods Enzymol.* **94**, 347 (1983).

[78] J. M. Tiller and D. L. Bloxam, *Anal. Biochem.* **131**, 125 (1983).

wide variety of experimental systems using natural as well as artificial substrates. In this section we describe methods for analyzing the cross-linking sites of substrates involved in epidermal differentiation by the TGase 1 and TGase 3 enzymes, utilizing a combined *in vivo*[30] and *in vitro*[9,17] approach. Studies have reported that in aged cultures of fibroblasts or keratinocytes, which presumably contain significant numbers of apoptotic cells, cytoskeletal proteins such as actin,[28] vimentin,[79] or keratins[80] become cross-linked. Cross-linked residues involved in apoptotic body formation have not been identified yet, although an involucrin-like protein is thought to be a major component of liver apoptotic bodies.[74] These emerging data suggest that there are overlapping molecular mechanisms in the cell death processes of cornification and apoptosis.

In Vivo Cross-Links of Epithelial Cornified Cell Envelopes

The purpose of these experiments is to identify not only the proteins that are cross-linked by the endogenous TGases in the living tissue or cells, but also the location of the glutamine and lysine residue(s) on the proteins. Thus proteolysis conditions should be determined empirically for each type of CE of interest to optimize recovery of peptides of size suitable for informative protein sequencing. The following offers some general guidelines.

Procedure. CEs are resuspended (1 mg/ml) in a buffer of 50 mM Tris-HCl (pH 8.0) and 5 mM CaCl$_2$ and are first digested with stirring with trypsin (sequencing grade, 1% by weight, at 37°; Sigma) for 6 hr. After removal of the solubilized material by centrifugation at 14,000g for 10 min, and washing in buffer, the CE remnant is pelleted, washed, and redigested with proteinase K (3% by weight, at 37°; Promega, Madison, WI) for 3 hr. The CE remnant is pelleted, washed, and may be redigested for as many as five additional 3- to 6-hr time intervals. The amounts of soluble peptide material released into the supernatant at each proteolysis interval can be quantitated by amino acid analysis after acid hydrolysis. Aliquots are resolved by HPLC, using a reversed-phase ultrasphere ODS column (4.5 × 50 mm) with a gradient of 0–100% acetonitrile containing 0.08% (w/v) trifluoroacetic acid. The extent of proteolytic digestion correlates with the size of the peptides recovered, and thus should be optimized empirically for each type of CE used for the experiment. In general, peptides that elute with <25% (v/v) acetonitrile are <10 residues long and are likely to contain

[79] A. V. Trejo-Skalli, P. T. Velasco, S. N. Murthy, L. Lorand, and R. D. Goldman, *Proc. Natl. Acad. Sci. U.S.A.* **92,** 8940 (1995).
[80] S. Clement, A. V. Trejo-Skalli, L. Gu, N. O. Ku, L. Lorand, and R. D. Goldman, *J. Invest. Dermatol.* **109,** 778 (1997).

few cross-links, while those that are eluted by >30% (v/v) acetonitrile contain >10 residues. Ideally, the peptides should contain cross-links that adjoin two or more "peptide branches." Each peptide is collected from the HPLC chromatogram, neutralized, and concentrated to about 10 pmol/μl in preparation for protein sequencing by Edman degradation methods.

Reagents. Tris-HCl (50 mM, pH 8.0); CaCl$_2$; trypsin; proteinase K; acetonitrile (0–100%, v/v); trifluoroacetic acid.

In Vitro Cross-Linking of Substrates by Transglutaminases

Some protein substrates have only one or a few reactive glutamine and/ or lysine residues. In this case, such reactive residues may be identified *in vitro* by enzymatic attachment of a labeled marker, e.g., a polyamine such as [^{3}H]putrescine (to identify glutamines), or a labeled peptide containing a glutamine donor (to identify lysines), followed by peptide fractionation and sequence analyses. Other substrates may possess several glutamines and lysines that may be used to cross-link to each other; that is, they are "complete" substrates. The following may be adapted for virtually any known or potential TGase substrate.

Cross-Linking Reaction of Complete Substrates

Procedure. Full-length or deletion constructs of human TGase 1 are expressed in bacteria or baculovirus and purified as described above. Guinea pig liver tTG may be purchased commercially (Sigma) or purified as described above. Guinea pig epidermal TGase 3 may be isolated and activated as described above, or a cloned version may be expressed in bacteria or baculovirus. For *in vitro* cross-linking studies using loricrin as a model of a complete TGase substrate,[17] the purified unlabeled or ^{35}S-labeled loricrin (up to 10 nmol/ml) is equilibrated by dialysis into a buffer of 50 mM Tris-HCl (pH 7.5), 50 mM NaCl, 1 mM dithiothreitol, and 1 mM EDTA. The reactions are made to 5 mM CaCl$_2$ to initiate the reactions at 37°. In analytical cross-linking experiments, 25 μg (1 nmol) of ^{35}S-labeled loricrin is utilized in a 100-μl reaction volume. To standardize reactions for comparisons of different TGase enzymes, the same amount of enzymatic activity must be used for each enzyme. These activities are measured by [^{3}H]putrescine (26 Ci/mmol) incorporation into succinylated casein: a useful amount of activity is that which incorporates 1000 dpm/min (0.45 pmol/min) into the casein. Aliquots are withdrawn at selected times and the reactions are stopped by the addition of 7 mM EDTA. The amount of enzyme to be used and the reaction times required to drive the reaction to completion should be determined empirically with each substrate, as the kinetic efficiencies of the reactions vary widely between different TGases and substrates. The cross-linked products are separated on 4–12% (w/v) polyacrylamide gels and analyzed by autoradiography. In preparative experiments (for isolation of peptides for sequencing) with TGases, 4 to 10-nmol amounts of loricrin may be reacted.

Reagents. Purified TGases 1, 3, and tTG; Tris-HCl buffer (50 mM, pH 7.5); NaCl; dithiothreitol; ethylenediaminetetraacetic acid (EDTA), different concentrations; $CaCl_2$; [³H]putrescine, 26 Ci/mmol; succinylated casein.

Characterization by Peptide Mapping and Amino Acid Sequencing of in Vitro Cross-Linking Reactions

Procedure. The above cross-linking reactions of loricrin[17] with TGases 1 and 3 are stopped by addition of EDTA (7 mM), made to 10 mM $CaCl_2$, and digested with proteinase K (3% enzyme to loricrin protein by weight) (Promega) for 3 hr at 37°. Peptides may be resolved by HPLC, using a reversed-phase ultrasphere ODS column (4.5 × 250 mm) with a gradient of 0–100% (v/v) acetonitrile containing 0.08% (w/v) trifluoroacetic acid. A control sample of un-cross-linked substrate should be digested and resolved similarly. New peaks that appear in the former reactions with respect to the latter, which presumably contain cross-links, are collected and concentrated to 10 μl. Depending on the type of sequencing machine available, samples of 0.5–100 pmol of peptides may be covalently attached to a polyvinylidene difluoride (PVDF) solid support (Sequelon-AA; Millipore), and then sequenced to completion (generally for 5–15 Edman degradation cycles). Where possible, it is useful to estimate the amount of the cross-linked peptide by amino acid analysis before sequencing. In our experience, a good estimate of the peptide amount is possible on the basis of the size of its $A_{220\,nm}$ trace from the HPLC column.

Reagents. Ethylenediaminetetraacetic acid (EDTA); $CaCl_2$; acetonitrile; trifluoroacetic acid.

Cross-Linking Reaction to Identify Reactive Lysines

Procedure. The following was developed to identify the reactive lysine residue(s) of keratin intermediate filament proteins.[32] A peptide of sequence Val-Ser-Ser-Gln-Gln-Val-Thr-Gln-Ser-Cys-Ala has been identified as a powerful donor for the TGase 1, 2, and 3 enzymes.[30] Similar peptides have been reported for other TGase enzymes. The peptide of choice can be substoichiometrically labeled by N-acetylation of its amino terminus by [³H]acetic anhydride. The reaction should contain, in a 100-μl volume, 0.4 nmol of keratin substrate protein, a 0.5- to 5-fold molar excess of labeled peptide, in a buffer of 0.1 M Tris–acetate (pH 8.3), 1 mM dithiothreitol,

5 mM CaCl$_2$, and sufficient TGase enzyme to incorporate about 4 pmol of putrescine into succinylated casein per minute (typically 50–100 pmol of protein). The amount of substrate and enzyme should be adjusted empirically to allow complete reaction in 1–2 hr. The reactions are stopped by addition of EDTA (to 10 mM). To identify the lysines labeled by the peptide, the reaction products are then digested with a protease, e.g., trypsin (1:30, w/w), for 6 hr at 37°, and the products resolved by HPLC, using methods similar to those described above. Labeled or shifted peaks in comparison with a control of the substrate protein will allow identification of cross-linked peptides, which may then be recovered for sequencing.

Cross-Linking Reaction to Identify Reactive Glutamines

Procedure. The method to be used involves incorporation of [^{3}H]- or [^{14}C]putrescine into the substrate of interest instead of succinylated casein, and may be adapted from the general TGase assay procedure described above. For example, in the case of reaction of TGases 1, 3, and tTG with the proteins loricrin[17] and trichohyalin,[9] reactions contained about 100 μg of substrate per milliliter. After termination of the reaction with EDTA, the protein is digested with protease (Asp-N), fractionated by HPLC, and shifted peaks of peptides that are candidates for labeling by putrescine are recovered for protein sequencing.

Postsynthetic Processing of Transglutaminases 1 and 3

Both the TGase 1 and TGase 3 enzymes undergo processing after synthesis. In the case of TGase 3, the only known modification is proteolytic activation (see above). In the case of the TGase 1 enzyme, several modifications occur that affect its function during epidermal differentiation. These include proteolytic activation[49] and lipid acylation[53,54] required for its attachment to membranes.

Detection of Transglutaminase 1 Processed Forms

TGase 1 processed forms have been detected mainly by two procedures: by immunoprecipitation with antibodies (see above), because they can also react with cleaved forms of the enzyme; and/or by fast protein liquid chromatography (FPLC).[49,53,54] These procedures may be augmented by metabolically labeling with [^{35}S]cysteine/methionine, or by addition of a histidine tag to the C-terminal end of the enzyme, which does not interfere with TGase 1 function.[52]

Separation of Transglutaminase 1 Forms by Fast Protein Liquid Chromatography

Procedure. After isolation on TGase 1 forms, immunoprecipitated samples are chromatographed by FPLC on a 0.5 × 5 cm Mono Q FPLC column equilibrated in a buffer of 50 mM Tris–acetate (pH 7.5), using 60 ml of a 0–0.5 M NaCl linear gradient. The fractions are collected into 0.5-ml aliquots. Fractions were tested for TGase activity and after concentration were resolved on 10–20% gradient SDS–polyacrylamide gels, dried, and visualized after 3–15 days of autoradiography. Visualized bands reveal the full-length enzyme of an apparent 106-kDa form, and the 67- and 33-kDa processed forms. A 10-kDa fragment, which contains the amino-terminal portion of the TGase 1, may be difficult to detect because of sensitivity to degradation.[53]

Attachment of a Polyhistidine Region onto the Carboxy-Terminal End of Transglutaminase 1 cDNA and Immunoprecipitation Using an Anti-His Tag Antibody

Procedure. A tag of six histidine residues can be attached to the carboxy-terminal end of the full-length wild-type human TGase 1 cDNA by appropriate engineering by PCR of the baculovirus expression vector system (see above). The PCR primers are designed with an extension of 16 nucleotides containing a *Not*I site at the 5′ end (plus primer, 5′-ATA AGA ATG CGG CCG CAT GGA TGG GCC ACG TTC CGA TGT GGG CCG TTG-3′), and an extension of 18 nucleotides that encode six histidine residues followed by a termination codon TGA and *Bam*HI site at the 3′ end (minus primer, 5′-TGC TCT AGA CTA ATG ATG ATG ATG ATG ATG AGC TCC ACC TCG AGA TGC CAT AGG GA-3′). The clone is also assembled into the mammalian expression vector pCMV for transfection into epidermal keratinocytes (see above). Because the keratinocytes also express the histidine-rich proteins profilaggrin and filaggrin, which are coprecipitated by the polyhistidine monoclonal antibody (Sigma), the cells should be cultured with [^{35}S]cysteine/methionine, as these proteins do not contain cysteine or methionine.[81] After transfections and labeling, the keratinocytes are harvested and TGase 1 forms are extracted as described. Cell lysate fractions are incubated with the anti-His tag antibody (1:50 dilution in phosphate-buffered saline) for 12 hr at 4° by gentle shaking. Protein A-conjugated beads may then be used for immunoprecipitation as described above. Finally, the adsorbed antigens and primary antibodies are harvested either by boiling with SDS sample buffer containing 10% (v/v) 2-mercapto-

[81] S.-Q. Gan, W. W. Idler, O. W. McBride, N. G. Markova, and P. M. Steinert, *Biochemistry* **29**, 8432 (1990).

ethanol and resolved by electrophoresis, or the adsorbed active TGase proteins are recovered from the washed beads by elution buffer (Pierce).

Reagents. Oligonucleotides; protein A-conjugated beads; polyhistidine monoclonal antibody (Sigma); Tris-buffered solution (TBS); 2-mercapto-ethanol; elution buffer (Pierce).

Lipid Attachment

The TGase 1 enzyme exists in epithelial cells in multiple cytosolic and membrane-anchored forms, although most remains attached to membranes. This is because the enzyme is acylated by N-myristoylation of its amino-terminal glycine residue, and through S-myristoylation and/or S-palmitoyla-tion of a net of one or two of a cluster of five cysteine residues on the amino-terminal 10-kDa portion. The enzyme partitions between the mem-branes and cytosol during differentiation by variable S-acylation.[53]

Identification and Visualization of Lipid Modifications

Procedure. Normal human epidermal keratinocytes may be grown in a variety of methods in high-Ca^{2+} medium, which induces terminal differenti-ation. Cells may be metabolically labeled with [1-^{14}C]palmitate (50 μCi/ml) or [1-^{14}C]myristate (650 Ci/mol), added 4 hr before harvesting as described above. The lipid modifications can be ascertained by autoradiography of immunoprecipitation products of cytosolic and Triton X-100-extracted membrane-anchored pellets of cells. The bands corresponding to lipid modi-fication may be quantitated in a computing densitometer.

Reagents. Normal human epidermal keratinocytes; $CaCl_2$; [1-^{14}C]palmi-tate (650 Ci/mol); [1-^{14}C]myristate (650 Ci/mol); Triton X-100, 0.1% (v/v).

Analyses of Mutations in *TGM* Genes in Genetic Diseases

To date, mutations in the genes encoding the blood clotting factor XIIIa and TGase 1 enzymes have been found to cause genetic diseases in humans. No diseases have been described for the *TGM2*. Likewise, no mutations have been found in the *TGM3* or *TGMX* genes. Mutations in the *TGM1* gene that restrict or eliminate TGase 1 activity cause the congenital reces-sive disorder lamellar ichthyosis.[82,83] The following are some general proce-dures that are useful for the identification of mutations in the *TGM1* gene

---

[82] M. Huber, I. Rettler, N. Bernasconi, E. Frenk, S. M. P. Lavrijsen, D. Shorderet, and D. Hohl, *Science* **276,** 525 (1995).
[83] L. J. Russell, J. J. DiGiovanna, N. Hashem, J. G. Compton, and S. J. Bale, *Nature Genet.* **9,** 279 (1995).

from DNA samples of individuals suspected or at risk of acquiring this disease.

DNA Isolation and Polymerase Chain Reaction Analyses

Procedure. DNA may be obtained in a variety of ways, such as directly from fetal tissues, patient blood cells, and processed by standard methods. In some cases DNA may be obtained by gently swabbing the buccal mucosa for 30 sec with a cytobrush.[84] The cells obtained are placed directly in 50 mM NaOH, heated for 10 min to 95°, neutralized with 1.0 M Tris-HCl (pH 8.0), and stored at 4°.

Typically, about 200 ng of total genomic DNA (or 5 μl of the above solution) is required for subsequent PCR analyses. PCRs are performed with a 100 nM concentration of each primer; dGTP, dATP, dTTP (200 μM each), and 25 μM dCTP; 0.1 μl of [^{32}P]dCTP (3000 Ci/mmol); 1.5 U of AmpliTaq DNA polymerase; and 1× buffer containing 10 mM Tris-HCl (pH 8.3), 50 mM KCl, 1.5 mM MgCl$_2$, and gelatin (1 mg/ml). Typical PCR conditions are as follows: denaturation at 95° for 5 min; amplification at 94° for 1 min, 55° for 1 min, and 72° for 1.5 min, for 35 cycles; and a final extension at 72° for 10 min.

The microsatellite markers utilized for linkage of lamellar ichthyosis to chromosome 14 are D14S50; D14S283; MyH7, TGM1, D14S64; D14S264; D14S275; and D14S80.[33] Amplified products are diluted 1:1 with 100% formamide and analyzed by electrophoresis of 4 μl on a 6% (w/v) denaturing polyacrylamide gels. The gels are dried and autoradiograms are exposed overnight at −70°.

The specific primer pairs required to amplify all 15 exons of the *TGM1* gene are listed in Table II. In general, primary PCRs using these primers should be reamplified with the same primer pairs. These secondary PCR products are separated in 1.5% (w/v) agarose gels and purified from the gel. The PCR products may then be directly sequenced with the fmol DNA sequencing system (Promega), following the manufacturer instructions.

Reagents. NaOH, 50 mM; Tris-HCl (1.0 M, pH 8.0); primers, dGTP, dATP, dTTP, and dCTP; [^{32}P]dCTP (3000 Ci/mmol); AmpliTaq DNA polymerase; Tris-HCl (10 mM, pH 8.3); KCl; MgCl$_2$; gelatin; formamide, 100%.

Single-Strand Conformation Polymorphism

SSCP is a gel electrophoresis method that allows the detection, but not the characterization, of mutations and polymorphisms. It is based on the

---

[84] B. Richards, J. Skoletsky, A. P. Shuber, R. Balfour, R. C. Stern, and H. L. Dorkin, *Hum. Mol. Genet.* **2,** 159 (1993).

principle that the electrophoretic mobility of a molecule within a gel matrix is sensitive to the size, charge, and shape of the molecule. A single nucleotide difference between two similar sequences is sufficient to alter the folded structure of one relative to the other. This conformational change is detected as a mobility difference on gel electrophoresis. Absolute characterization of the mutations can be performed by eluting appropriate bands from SSCP gels, reamplifying the eluted DNA by PCR, followed by DNA sequencing. Successful SSCP analysis generally requires that the PCR product be <400 bp long.

[41] Anoikis

By STEVEN M. FRISCH

Introduction

We have found that integrins control apoptosis in epithelial cells, and have named this phenomenon *anoikis,* the ancient Greek word meaning "homelessness."[1] Epithelial cells that have lost interaction with their extracellular matrix—or that interact with matrix through inappropriate integrin types[2] or other molecules—cannot colonize elsewhere, because of anoikis. This process is developmentally critical, and its loss can contribute significantly to tumor malignancy (reviewed in Ref. 3).

Indeed, several oncogenes abrogate anoikis, and a tumor suppressor gene that programs tumor cells to become epithelial cells appears to work by conferring anoikis sensitivity.[1] Thus, it is important to have reliable assays for the effects of compounds or genes on anoikis. Several types of assays are described here. The most time-consuming and conventional is stable expression of new genes in epithelial cells and assaying the resulting cell lines for anoikis relative to the parental cells. We used this assay to show that anoikis is suppressed by focal adhesion kinase (FAK) and is promoted by the kinase MEKK-1,[4,5] perhaps through the Jun-N-terminal kinase pathway. This chapter specifies the gene transfer and apoptosis assay methods that work especially well for epithelial cells. In addition, we have developed methods for assaying the effects of genes on anoikis in transient

[1] S. M. Frisch and H. Francis, *J. Cell Biol.* **124,** 619 (1994).
[2] N. Boudreau, C. Sympson, Z. Werb, and M. Bissell, *Science* **267,** 891 (1995).
[3] S. Frisch and E. Ruoslahti, *Curr. Opin. Cell Biol.* **9,** 701 (1997).
[4] S. Frisch, K. Vuori, E. Ruoslahti, and P. Y. Chan, *J. Cell Biol.* **134,** 793 (1996).
[5] M. Cardone, G. Salvesen, C. Widmann, G. Johnson, and S. Frisch, *Cell* **90,** 315 (1997).

assays that, while fairly new, appear to be highly reliable and are certainly faster.

Assaying Effects of Genes on Anoikis in Stable Expression Experiments

Generating Stable Expressors. We have focused on the cell line MDCK because of its sensitivity to anoikis and its authentic epithelial cell behavior. It has two drawbacks, however. First, the fact that it is a canine cell line limits the choice of antibodies and nucleic acid probes, a problem that we usually solve by testing several antibodies or generating canine homologs of genes by polymerase chain reaction (PCR). The second is the low transfectability by calcium phosphate. This latter feature has prompted us to use retroviral vectors, which yield extremely efficient and stable expression.

Chiefly, we use the vector pBABE.[6] This vector drives expression of the insert through the viral long terminal repeat (LTR) enhancer and has an internal simian virus 40 (SV40) promoter to drive the expression of a puromycin-resistance gene for selection. After subcloning of the gene of interest into pBABE [usually in a FLAG-, myc-, or hemagglutinin (HA) epitope-tagged form], the retrovirus vector is transfected by the standard calcium phosphate method into the amphotropic packaging cell line ϕNX,[7] which, being based on 293 cells, is much more transfectable than its 3T3-based predecessors. Two to 3 days after transfection, the viral supernatant is removed and cleared by centrifugation (3000g for 10 min), Polybrene is added to give a 4-μg/ml final concentration, and it is applied to a subconfluent monolayer of MDCK cells. The MDCK cells can either be on tissue culture plastic, or, for higher efficiency, on permeable cell culture inserts of 25-mm diameter and 3.0-μm pore size (Falcon; Becton Dickinson Labware, Lincoln Park, NJ), which fit into 35-mm wells. For the latter, the viral supernatant is placed on the cells from which all media have been removed, and then allowed to flow through the filter by gravity, which takes about 1 min/ml of supernatant. The infected cells are washed briefly with medium and incubated for 1 day, followed by being trypsinized and replated on tissue culture dishes at various low (i.e., colony) densities. After cell attachment (6–8 hr), puromycin is then added to give 1.5 μg/ml, and colonies are selected with refeeding every 3 days, for a total elapsed time of about 2 weeks. Colonies are then ring-isolated, transferred to 10-mm-diameter wells, and expanded, and lysates are tested for expression of the transgene by Western blotting with anti-epitope antibody. The total elapsed time for subcloning the gene into pBABE through obtaining positive cell lines is about 1 month.

[6] J. Morgenstern and H. Land, *Nucleic Acids Res.* **18**, 3587 (1990).
[7] T. Kinsella and G. Nolan, *Hum. Gene Ther.* **7**, 1405 (1996).

Alternatively, Lipofection using the Qiagen (Chatsworth, CA) Effectene reagent can be used for stable expression. This method is described below for transient transfection experiments.

Materials

MDCK cells: American Type Culture Collection (Manassas, VA)

Tissue culture medium: DME-high glucose (Irvine Scientific, Santa Ana, CA) supplemented with 10% (v/v) fetal bovine serum (GIBCO-BRL, Gaithersburg, MD) and 1× glutamine–penicillin–streptomycin (from 100× stock; GIBCO-BRL)

Cell culture inserts (Falcon 3090; 2.4 cm)

Trypsin (0.25%, w/v)–EDTA (TE): GIBCO-BRL

Microcentrifuge (model 5414; Eppendorf, Hamburg, Germany)

Phosphate-buffered saline (PBS): 1× PBS (minus calcium and magnesium): GIBCO-BRL

Apoptosis lysis buffer (ALB): 15 mM Tris (pH 8), 20 mM EDTA, 0.25% (v/v) Triton X-100

Sodium acetate: 2.5 M sodium acetate, pH 7.3

Glycogen (nuclease free; Boehringer Mannheim, Indianapolis, IN)

Ethanol (100%)

TE master mix: 50 mM NaCl in TE containing 10 μl of RNase (DNase free; Boehringer Mannheim) per milliliter

Ficoll loading buffer: 10% (v/v) Ficoll (MW 400,000; Sigma, St. Louis, MO), 20 mM

EDTA, 0.05% (w/v) bromphenol blue

Agarose (1.5% w/v) gel in 1× TBE (5× TBE: 54 g of Tris, 27.5 g of boric acid, 20 ml 0.5 ml of 0.5 M EDTA per liter)

Cytomix:
  KCl, 120 mM
  Calcium chloride, 0.15 mM
  Potassium phosphate (pH 7.6), 10 mM
  HEPES (pH 7.6), 25 mM
  EGTA, 2 mM
  Magnesium chloride, 5 mM

Adjust to pH 7.6. Before use, take out a 50-ml aliquot and add ATP to 2 mM and glutathione to 5 mM, readjust to pH 7.6 on a pH meter, and filter sterilize

0.4 cm Gap electroporation cuvettes (0.4 cm; Bio-Rad, Hercules, CA)

Gene Pulser electroporator (Bio-Rad)

ImaGene Green reagent (Molecular Probes, Eugene, OR)

4′,6-Diamidino-2-phenylindole (DAPI; Sigma)

PBS containing bovine serum albumin (2 mg/ml; Sigma)
Sodium dodecyl sulfate–polyacrylamide gel electrophoresis (SDS–PAGE) sample buffer: 62.5 mM Tris (pH 6.8), 0.4% (w/v) SDS, 10% (v/v) glycerol, 0.05% (w/v) bromphenol blue
Tris–glycine minigels (14%; Novex, San Diego, CA)
Immobilon transfer membranes (Millipore, Bedford, MA)
Anti-myc epitope antibody Ab-1 (Oncogene Science/Calbiochem, La Jolla, CA)
Super-signal chemiluminescent detection kit (Pierce, Rockford, IL)
Phosphorimager (Bio-Rad)
Superfect reagent (Qiagen)

Methods

Assaying Stable Expressors for Anoikis

The following is our optimized anoikis-assay protocol for MDCK cells, which generates DNA fragmentation ladders. Note that MDCK cells must be grown to confluence (preferably, beyond) in order to be sensitive to anoikis. Also, note that the cells are extracted with Triton rather than SDS, thus leaving behind intact genomic DNA. Nonapoptotic cells will produce blank lanes on the gel as a result.

1. Plate our MDCK cells on 2.4-cm cell culture inserts (Falcon 3090), and grow to confluence. Continue growing for an additional 1 day.

2. Trypsinize the cells, spin them down, resuspend in 2.0 ml of medium, and count the cells. Transfer 5 × 10^5 cells to a 2.0-ml microcentrifuge tube and fill the tube to the top with medium. For some clones of MDCK (e.g., those that grow on plastic as "islands"), severe clumping will occur during suspension. To avoid this, use 10 ml of medium in a 14-ml snap-cap polypropylene tube.

3. Place the tube on a wheel in a 37° incubator for 2.5–3.5 hr.

4. Microcentrifuge the cells for 8 sec at 8000 rpm (in an Eppendorf model 5414 microcentrifuge); suction off the supernatant (For 10-ml cell samples, spin at 3200 rpm in a tabletop tissue culture centrifuge for 30 sec, suction off all but 0.5 ml of medium, resuspend the cell pellet in this remaining medium, and then transfer to a microcentrifuge tube).

5. Wash with 600 μl of PBS by inverting, and respin in the microcentrifuge.

6. Resuspend the pellets in 500 μl of apoptosis lysis buffer (ALB) by pipetting up and down with a P1000.

7. Vortex for 20 sec. Spin out the debris in a cold microcentrifuge for 10 min at maximum speed.

8. Transfer the supernatant to a new microcentrifuge tube. Add 80 μl of water, 25 μl of 20% (w/v) SDS, and 15 μl of proteinase K (15 mg/ml). Incubate for 1 hr at 50°.

9. Extract with 600 μl of phenol–chloroform–isoanyl alcohol; spin for 2 min in the microcentrifuge.

10. Transfer 380 μl of aqueous phase to a microcentrifuge tube and add 40 μl of 2.5 M sodium acetate (pH 7.3), 1.5 μl of glycogen (20 mg/ml), and 1 ml of cold ethanol. Precipitate overnight.

11. Spin in a cold microcentrifuge for 15 min. Remove the supernatant by pipetting with a P1000.

12. Wash with 300 μl of cold 70% (v/v) ethanol. Respin for 4 min and remove the supernatant by pipetting with a P200. Respin briefly, take off as much of the remaining supernatant as possible with a P20.

13. Redissolve the pellets in 25 μl of 50 mM NaCl in TE containing 1 μl of RNase (DNase free; Boehringer Mannheim) (make a master mix for all samples); incubate at 37° for 20 min.

14. Add 3.3 μl of Ficoll loading buffer, and load onto a 1.5% (w/v) agarose gel in TBE (can put 10 μl of ethidium bromide in a 200–ml gel before pouring). Run the blue dye half way to the end of the gel, and photograph the gel.

Assaying Effects of Genes on Anoikis in Transient Expression Experiments

Two methods are presented. In the first, a vector containing the *Escherichia coli* β-galactosidase gene under the control of a mammalian promoter is used to express the gene of interest. After transfection, cells are placed in suspension for various periods of time and then doubly stained for nuclear morphology with the fluorescent nuclear stain DAPI and for β-galactosidase activity using the viable stain ImaGene Green. The β-galactosidase-positive cells are then scored for apoptotic nuclear morphology.

The second method is technically easier to perform since it requires no microscopy. It is based on the principle that, in epithelial cells undergoing apoptosis, effector caspases specifically cleave the keratin 18 (K18) protein to yield two discrete products.[8] We have taken advantage of this observation to develop a new assay for anoikis (Fig. 1). The keratin 18 gene was N-terminally tagged with an myc epitope and substituted for the neomycin resistance gene in the *Sma*I–*Bst*BI site of pcDNA3.1-. The gene of interest (i.e., the candidate anoikis regulator) is subcloned into the polylinker downstream of the cytomegalovirus (CMV) promoter. The resulting plasmid is

---

[8] C. Caulin, G. Salvesen, and R. Oshima, *J. Cell Biol.* **138**, 1379 (1997).

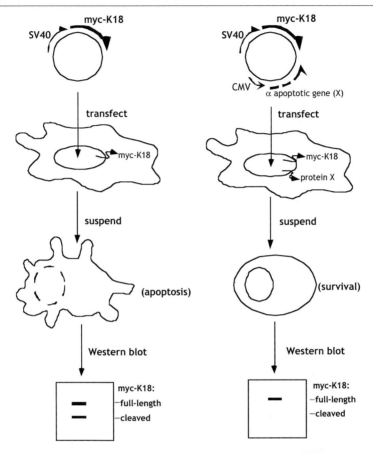

FIG. 1. Transient anoikis assay. MDCK cells are transfected with the vector depicted (*top*), into which coding sequences to be tested for anoikis effects are subcloned (CMV promoter-driven transcription unit). Thirty-six hours later, cells are suspended, lysed, and analyzed for caspase-mediated cleavage of the myc-K18 marker gene by Western blotting. Results are quantitated by phosphoimaging.

then transfected into epithelial cells. After expression, the cells are placed in suspension, and total cell lysates are Western blotted, using myc epitope antibody. The ratio of cleavage product to intact K18 is a measure of the degree of anoikis. Note that the assay scores only transfected cells, thus providing a high degree of sensitivity and selectivity. We also find it relatively simple to perform. Also, the transfection frequency may vary from one sample to the next but this does not affect the final results, as they are automatically normalized against intact myc-K18.

*β-Galactosidase/4',6-Diamidino-2-phenylindole
Double-Staining Method*

Construction of Expression/Reporter Plasmid. We have modified the
expression plasmid pHook2 (InVitrogen, San Diego, CA) by inserting a
β-galactosidase-coding sequence in the unique *Eco*RV site upstream of the
signal peptide of the Hook epitope gene, thus placing it under the control
of the Rous sarcoma virus (RSV) promoter. This leaves most of the cloning
sites in the polylinker downstream of the CMV available for inserting the
candidate anoikis-regulating gene (the complete sequence of pHook2 can
be found on the InVitrogen home page, *www.invitrogen.com*).

Electroporation of MDCK Cells. MDCK cells are electroporated as
follows.

1. Ethanol precipitate 75 μg of plasmid DNA and redissolve it in
0.5 ml of cytomix.

2. Trypsinize a 50% confluent 150-mm dish of MDCK cells and spin
down.

3. Wash the cells once with 5.0 ml of sterile PBS and respin.

4. Resuspend the cells in the DNA solution and transfer to a 0.4-cm
gap Bio-Rad cuvette; keep on ice.

5. Electroporate at 0.3 kV, 960 μF (Bio-Rad Gene Pulser), put the
cuvette back on ice, and keep it on ice for 10 min.

6. Resuspend the cells and transfer to a tube containing 5.0 ml of
complete medium; spin down the cells.

7. Resuspend the cells and plate onto one 25-mm cell culture insert
(0.2 or 1.0 μm) for transient anoikis assay.

8. The next day, wash dead cells off the top of the filter and refeed.

9. Cells should be confluent enough 1 day later (48 hr postelectropora-
tion) to perform the anoikis assay.

The cells are trypsinized and placed in suspension as described for the
stable transfection protocol above. During the final 15 min of suspension,
both stains are added: ImaGene Green is added to 33 μM final concentration, and DAPI to 1 μM. The cells are then centrifuged for 8 sec at 10,000
rpm in an Eppendorf 5414 microcentrifuge, washed twice with PBS plus
BSA (2 mg/ml), resuspended in 30 μl of the latter, and transferred to a
microscope slide. The doubly stained cells are viewed on a fluorescence
microscope, using an FITC filter to find transfected cells and a UV filter
to assess normal versus apoptotic nuclear morphology.

Transient Transfection Using Lipofection (Effectene). MDCK cells are
lipofected as follows.

1. Plate cells in 35-mm wells of six-well cluster dishes the day before the transfection, to give about 50% confluence the next day.

2. Dilute 0.4 μg of DNA and 3.2 μl of Qiagen enhancer in Qiagen buffer EC to give a 100-μl final volume. Vortex and incubate for 2–5 min at room temperature.

3. Add 4 μl of Qiagen Effectene solution. Vortex for 10 sec. Incubate for 5–10 min at room temperature.

4. Add 600 μl of complete cell growth medium, mix by pipetting, and add to a well of cells containing 1.6 ml of medium.

5. Incubate overnight. Wash twice with medium and assay the cells 24–48 hr later.

Keratin 18-Cleavage Transient Assay for Effects of Transgenes on Anoikis

We have N-terminally tagged the human K18-coding sequence with an myc epitope tag and substituted it for the neomycin resistance gene by insertion in the *Sma*I–*Bst*BI site of pcDNA3.1- (InVitrogen; the complete sequence of the parental plasmid is on the homepage, *www.invitrogen.com*). Sequences encoding candidate anoikis regulators are most conveniently blunt-end ligated into the *Pme*I site of the plasmid, resulting in their being driven by the CMV promoter.

Cells are electroporated and subjected to suspension conditions as described above. After suspension for various times (generally 0, 1, 2, and 3 hr provides a good time course for MDCK cells), cells are washed twice with PBS and lysed by boiling in 300 μl of SDS–PAGE sample buffer per time point (four time points can be obtained from one 25-mm-diameter cell culture insert of cells). Western blots are run on 14% minigels (Novex), electroblotted onto polyvinylidene difluoride (PVDF) filters, and probed with anti-myc epitope (clone 9E10) monoclonal antibody (Oncogene Science). After development by enzyme-linked chemiluminescence (Pierce) we scan the blot on a Bio-Rad Phosphorimager and quantitate the band intensities so as to reveal relative degrees of apoptosis, as judged by the percentage of myc-K18 cleaved.

[42] Transient Transfection Assay of Cell Death Genes

By MASAYUKI MIURA and JUNYING YUAN

Introduction

Gene overexpression in cultured cells is a common method used to examine the function of a gene; however, genes that have detrimental effects on cell proliferation or survival are incompatible with the establishment of stable overexpressing cell lines. Even when such cell lines are established, the expression levels of the gene in question are almost always low and may quickly diminish to nothing after just a few passages in culture. Thus, the choices for expression of genes that have a deleterious effect on cell proliferation or cell survival include inducible expression and transient expression systems. The advantage of an inducible expression system is that it allows one to analyze the function of a gene in question in a longer time frame and more controlled fashion than in a transient expression system. The establishment of an inducible system, however, is time consuming and a certain leakiness of the suppression, which is invariably present, may prevent the establishment of such an inducible cell line. On the other hand, transient transfection allows rapid analysis of the gene function, if such function can be detected within 72 hr of expression. Thus, it is almost always recommended to examine the functions of genes in a transient assay first, even if an inducible cell line is desired, because it allows screening for the most responsive cell lines to the gene in a quick and inexpensive fashion. The other advantage of a transient transfection system is that the possibility of selecting for one or two stable cell lines that may carry unrelated mutations and skew experimental results is avoided. Transient transfection assays are especially suited, and in fact may be the only choice, for analysis of genes that induce apoptosis because such genes usually induce apoptosis within a few hours of expression and it is often difficult if not impossible to establish stable cell lines expressing cell death genes.

Standard transfection methods are used for transient transfection assays. The procedure for calcium phosphate transfection can be found in Sambrook et al.[1] The procedure for LipofectAMINE transfection should follow the manufacturer protocol. Many cell lines such as Rat-1, HeLa, 293, and COS cells can be successfully transfected by calcium phosphate precipita-

[1] J. Sambrook, E. F. Fritsch, and T. Maniatis, "Molecular Cloning: A Laboratory Manual," 2nd Ed. pp. 16.33–16.36. Cold Spring Harbor Laboratory Press, Cold Spring Harbor, New York, 1989.

tion[2] and LipofectAMINE (GIBCO-BRL, Gaithersburg, MD).[3–7] In comparing these two methods of transfection, LipofectAMINE has the advantage of higher transfection efficiency than that of calcium phosphate precipitation. LipofectAMINE, however, has the following disadvantages: (1) it may be toxic to certain cells, resulting in high background cell death; (2) it may not be efficient in introducing multiple plasmids into a cell for unknown reasons; and (3) it is expensive. For these reasons, calcium phosphate precipitation is always recommended if your cells have a reasonable transfection efficiency. Calcium phosphate precipitation has also been used to transfect primary neurons.[2]

In this chapter, we review the methods and protocols that have been used successfully to examine the functions of genes involved in regulating apoptosis. We concentrate on one of the most frequently asked questions in apoptosis: How to determine if expression of the gene of interest can indeed induce apoptosis? Different methods have been developed or adapted to address this question. They can be divided into morphology-based assays, dye uptake, DNA-chelating dye staining, terminal deoxy-nucleotidyltransferase-mediated dUTP nick end labeling (TUNEL), fluorescence-activated cell sorting (FACS) analysis, and enzymatic assays. When deciding which method to use, one needs to consider whether the method is compatible with the transfection marker used.

Morphology-Based Cell Death Assays

The first functional evidence that caspases are mammalian homologs of the *Caenorhabditis elegans* cell death gene *ced-3* was obtained by transient transfection of ICE-lacZ and CED-3-lacZ into Rat-1 cells.[3] In this experiment, bacterial *Escherichia coli* gene *lacZ* encoding β-galactosidase was used as a reporter. *lacZ* was directly fused to the sequence encoding the C terminus of caspases, and then these chimeric genes were transfected into Rat-1 cells. One day after transfection, cells expressing the chimeric genes can be easily detected by 5-bromo-4-chloro-3-indolyl-β-D-galactopy-

[2] Z. Xia, H. Dudek, C. K. Miranti, and M. E. Greenberg, *J. Neurosci.* **16,** 5425 (1996).
[3] M. Miura, H. Zhu, R. Rotello, E. A. Hartwieg, and J. Yuan, *Cell* **75,** 653 (1993).
[4] L. Wang, M. Miura, L. Bergeron, H. Zhu, and J. Yuan, *Cell* **78,** 739 (1994).
[5] S. Wang, M. Miura, Y.-K. Jung, H. Zhu, V. Gagliaridini, L. Shi, A. H. Greenberg, and J. Yuan, *J. Biol. Chem.* **271,** 20580 (1996).
[6] H. Duan, K. Orth, A. M. Chinnaiyan, G. G. Poirier, C. J. Froelich, W.-W. He, and V. M. Dixit, *J. Biol. Chem.* **271,** 16720 (1996).
[7] H. Duan, A. M. Chinnaiyan, P. L. Hudson, J. P. Wing, W.-E. He, and V. M. Dixit, *J. Biol. Chem.* **271,** 1621 (1996).

ranoside (X-Gal) staining. Since Rat-1 is a fibroblast cell line, healthy cells are flat and well attached to the plates and round cells are usually dead cells. The percentage of dead cells was calculated as round blue versus total blue cells. Such a morphology-based assay is the easiest way to identify dying cells and is compatible with all transfection markers. Proapoptotic function of caspases 1, 2, and 11 was examined in this way.[3–5] One should be cautious, however, when interpreting the data, since (1) this method distinguishes only dead versus live cells and not apoptotic versus necrotic cells; (2) certain cells, such as HeLa cells, may round up without dying, especially when plated in high densities; (3) for nonadherent cells, it is not possible to use a morphology-based method to identify live versus dead cells since they are usually round already. Other methods described below can be used to identify live versus dead nonadherent cells.

Since it is somewhat time consuming to construct a fusion with *lacZ,* some of the later experiments were done by cotransfection of *lacZ.*[6–11] By increasing the ratio of the gene in question versus *lac-Z,* one can ensure the majority of cells expressing the gene in question will have the *lacZ* reporter. We must point out that although cotransfection has the convenience of no special fusion construct requirement, it may lead to higher background, i.e., cells expressing *lacZ* or the gene encoding green fluorescent protein (GFP) only, compared with the single gene transfection of a chimeric construct. Table I shows a list of reporter genes used in transient transfection assays employed in the study cell death. Green fluorescent protein from the jellyfish *Aequorea victoria* has been widely used as a reporter in cotransfection experiments (Fig. 1). The advantage of GFP as a reporter is that one can observe the morphology of transfected cells without fixation, which is a useful feature for observing apoptosis since dead cells are often floating and can be removed easily during washing in a fixation procedure.[12]

Procedure for X-Gal Assay

This procedure is taken from Ref. 3.

1. Cells are fixed with 1% (v/v) glutaraldehyde for 5 min at room temperature.

[8] H. Hsu, J. Xiong, and D. Goeddel, *Cell* **81,** 495 (1995).
[9] A. M. Chinnaiyan, K. O'Rourke, M. Tewari, and V. M. Dixit, *Cell* **81,** 505 (1995).
[10] B. Z. Stanger, P. Leder, T.-H. Lee, E. Kim, and B. Seed, *Cell* **81,** 513 (1995).
[11] H. Hisahara, H. Kanuka, S. Shoji, S. Yoshikawa, H. Okano, and M. Miura, *J. Cell Sci.* **111,** 667 (1998).
[12] S. Wang, M. Miura, Y.-K. Jung, H. Zhu, E. Li, and J. Yuan, *Cell* **92,** 501 (1998).

TABLE I
REPORTER GENES USED IN CELL DEATH STUDY

Reporter gene[a]	Detection	Comments
lacZ (3–11, 20)	X-Gal staining	Easy identification of dying cells on the basis of their morphology
	FACS Immunocytochemistry Enzyme assay	
Green fluorescent protein (12, 30)	Fluorescence microscopy	Easy identification of dying cells on the basis of their morphology, without fixation
	FACS Immunocytochemistry	
Luciferase (17, 18)	Enzyme assay	Easy quantification of the relative ratio of cell death; more sensitive than that of β-galactosidase

[a] Numbers in parentheses refer to references.

2. Cells are rinsed with phosphate-buffered saline (PBS) three times, and then stained with X-Gal buffer [X-Gal] (0.5 mg/ml), $K_3Fe(CN)_6$ and $K_4Fe(CN)_6 \cdot 3H_2O$ (3 mM each), 1 mM $MgCl_2$, 10 mM KCl, 0.1% (v/v) Triton X-100 in 0.1 M sodium phosphate buffer, pH 7.5) at 37° until a suitable color develops, usually 3–5 hr.

3. To examine the nuclear morphology, cells are rinsed with PBS and then incubated with Hoechst 33342 dye (final concentration, 5 μM) for 1 min, and washed again with distilled H_2O.

4. Samples are mounted using PermaFluor aqueous mounting medium (Shandon–Lipshaw, Pittsburgh, PA) and the X-Gal-positive cells are examined. Apoptotic cells have small, dense, or split nuclei, whereas non-apoptotic cells have large, round nuclei.

Identification of Dying Cells by Dye Uptake

A more objective method to determine if the cells expressing the gene of interest have indeed died (not just rounded up) is by differential uptake of fluorescent DNA-binding dyes. When added to unfixed cells, propidium iodide (PI) and ethidium bromide (EB) are taken up only by dying cells. Thus, the percentage of dead cells can be estimated by the percentage of cells taking up the dye. Hoechst 33342 and metachromatic dye acridine orange (AO) can be taken up by both live and dead cells, which can be distinguished by their different morphology: live cells have flat nuclei with noncondensed chromatin and dead cells have small nuclei with condensed,

GFP   **PI**

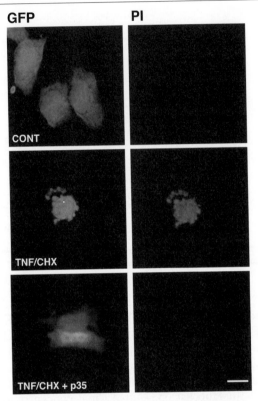

Fig. 1. Prevention of TNF-induced cell death by p35. HeLa cells were transfected with GFP expression vector alone or with GFP and p35 expression vectors, and 48 hr later cells were treated with TNF-α (TNF, 10 ng/ml) and cycloheximide (CHX, 10 μg/ml). Twelve hours later, cells were stained with PI (1 μg/ml) for 1 min. Overexpression of caspase-inhibitor p35 prevents TNF-induced HeLa cell death.

sometimes fragmented chromatin. If GFP is used as a reporter, count the dead cells among the transgene-expressing cells by PI or EB staining under nonfixed conditions (live cells are green; dead cells are green with red nuclei). Fluorescent dyes frequently used to identify dying cells are listed in Table II. When choosing which dye to use, consider whether the dye may cover the markers of transfection. For example, Hoechst dye, ethidium bromide (EB), and propidium iodide (PI) are compatible for GFP. Hoechst dye, however, can be used only with a light X-Gal staining, since a dark X-Gal staining will cover the fluorescence of Hoechst dye.

TABLE II
COMMON FLUORESCENT DYES USED TO IDENTIFY DYING CELLS

	Stock solution	Working concentration	Excitation filter (wavelength, nm)
Hoechst 33342	1 mM in distilled water	1–10 μM	350
Propidium iodide	1 mg/ml in distilled water	1–10 μg/ml	535
Acridine orange	1 mg/ml in distilled water	1–2 μg/ml	500
Ethidium bromide	1 mg/ml in distilled water	1–10 μg/ml	518

Procedure

1. Add fluorescent dyes (Hoechst 33342, AO, PI, or EB) directly into cell culture medium at the concentration described in Table II.
2. Incubate the cells for 5 min.
3. Observe the cells directly under an inverted fluorescence microscope.

If desired, fix the cells after vital staining of transfected cells for further detailed analysis of morphology:

1. Wash the cells with PBS (this step can be omitted), and then fix the cells with 4% (w/v) paraformaldehyde or 1% (v/v) glutaraldehyde for 5 min at room temperature.
2. Wash the cells with PBS.
3. Samples are mounted with PermaFluor aqueous mounting medium and cells are examined under a fluorescence microscope equipped with the appropriate filters.

Washing should be done as gently as possible to minimize the loss of dead cells.

Identification of Dying Cells by Apoptotic Nuclear Morphology

Identification of dying cells by morphology or dye uptake has the problem of not addressing whether it is apoptosis or necrosis. An easy way to determine whether apoptosis is induced by expressing the genes of interest is to examine the nuclear morphology. Apoptotic cells have characteristic condensed and fragmented nuclei that can be easily distinguished from normal cells, which have well-rounded nuclei. *Note:* Be careful about the nuclear morphology of dividing cells; the nuclei of dividing cells which can also be condensed, are distinguishable from apoptotic nuclei since only the latter is fragmented. It may be difficult, however, at times to distinguish necrotic from normal cells by their nuclear morphology since necrotic nuclei do not condense. AO/EB staining may be used to distinguish necrotic cells from normal and apoptotic cells (see below). If antibody staining is used to detect cells expressing reporter genes (e.g., anti-β-galactosidase, anti-GFP), any dyes listed in Table II can be applied to examine the nuclear morphology of transfected cells after immunocytochemistry.

Procedure

1. Wash cells with PBS (this step may not be necessary).
2. Fix the cells with 4% (w/v) paraformaldehyde or 1% (v/v) glutaraldehyde for 5 min at room temperature.
3. Stain the cells with Hoechst 33342 (1–10 μM), PI (1–10 μg/ml), AO (1–2 μg/ml) or EB (1–10 μg/ml) for 5 min at room temperature.
4. Wash the cells with PBS.
5. Samples are mounted with PermaFluor aqueous mounting medium and the cells are examined by fluorescence microscopy.

Hoechst 33342 can stain the nuclei of both living and dying cells. If fluorescent dyes are applied to live cells, the intact plasma membrane of live cells excludes charged dyes, including PI or trypan blue. Dying cells cannot exclude such charged dyes. PI can penetrate into dying cells (apoptotic and necrotic cells) and binds to DNA and fluoresces intensively. If cells are first exposed to PI, and then subsequently stained with Hoechst 33342 under unfixed conditions, Hoechst 33342 staining is suppressed and PI fluoresces intensely in dying cells by UV light excitation.

Since AO is lipophilic, it penetrates cells easily and is accumulated and trapped within the lysozomes of living cells. In contrast to the normal green fluorescence associated with the monomer form of the dye, concentrated and aggregated AO luminesces red. In dying cells, AO is taken up less efficiently and will bind to nucleic acids (DNA and RNA) as monomers. At the concentration mentioned above, only live cells accumulate AO in lysosomes and luminesce red, and dead cells stain uniformly green.[13] If a higher concentration of AO (100 μM) is used, dead cells stain red but live cells have red lysosomes and green nuclei and cytoplasm.[14] In the early stages of apoptosis, however, cells retain their ability to concentrate AO into lysosome and stain red. Caution should be exercised in interpreting data on AO staining.

---

[13] Z. Darzynkiewicz, X. Li, and J. Gong, *Methods Cell Biol.* **41**, 15 (1994).
[14] F. Traganos and Z. Darzynkiewicz, *Methods Cell Biol.* **41**, 185 (1994).

Acridine Orange/Ethidium Bromide Double Staining to Distinguish Necrotic versus Apoptotic Cells

If live and dead cells can be distinguished in the same microscopic field by different colors, the ratio of live to dead cells can be determined more accurately. A method to differentiate live and dead cells is AO and EB double staining.[15,16] It is also possible to distinguish apoptotic from necrotic cells on the basis of their nuclear morphology. A disadvantage of this staining is that AO/EB staining is not compatible with either GFP or X-Gal staining. This method may be applied only when most of the cells express transfected genes.

Procedure

1. Add both AO (final concentration, 1–10 μg/ml) and EB (final concentration, 1–10 μg/ml) directly to the culture medium.
2. Incubate the cells for 5 min.
3. Observe the cells directly under an inverted fluorescence microscope with a fluorescein filter. We obtain good results with the following filters: DBP 485/20, 578/14; FT 500/600; and BP 515-540/LP 610 (Zeiss, Thornwood, NY).

The status of cells can be identified by counting the number of each cell exhibiting staining pattern:
> Bright green chromatin with organized cell structure: VN (viable cells with normal nuclei, live cells)
> Bright green chromatin that is highly condensed or fragmented: VA (viable cells with apoptotic nuclei, early phase of apoptosis)
> Bright orange chromatin with organized cell structure: NVN (nonviable cells with normal nuclei, necrotic cells)
> Bright orange chromatin that is highly condensed or fragmented: NVA (nonviable cells with apoptotic nuclei, late phase of apoptosis)

Enzyme Reporter-Based Apoptosis Assays

In contrast to microscopic examination of apoptosis, enzymatic assays of reporter genes may be rapidly and easily performed. When cultured adherent cells die, they often detach from dishes. As a result, the number of cells expressing the reporter gene decreases as dead cells increase. Thus,

[15] R. C. Duke and J. J. Cohen, Morphological and biochemical assays of apoptosis. In "Current Protocols in Immunology" (J. E. Coligan, A. M. Kruisbeek, D. H. Margulies, E. M. Shevach, and W. Strober, eds.), pp. 3.17.1–3.17.3. John Wiley & Sons, New York, 1994.
[16] N. Inohara, L. Ding, S. Chen, and G. Nunez, *EMBO J.* **16,** 1688 (1997).

the loss of a reporter enzymatic activity as a result of cell death can be used as an indicator. Luciferase or β-galactosidase expression constructs can both be used as reporters, but luciferase is preferred since the sensitivity of the β-galactosidase assay is much lower than that of the luciferase assay. After transient transfection, cells are washed and lysed, and then aliquots are assayed for luciferase activity.[17,18] To use enzymatic assays as an indicator of cell death, the transfection efficiency of the cells must be high since such assays are not sensitive enough to detect a few cells expressing the reporter construct. In addition, one should consider the possibility that expression of the gene in question may suppress the expression of the reporter construct rather than induce apoptosis. Proper controls should be used to demonstrate that loss of reporter expression is not due to inhibition of gene expression.

Luciferase Assay

This assay is based on Ref. 19.

1. Cells are cotransfected with luciferase expression vector (e.g., pRSV-luciferase[20]) together with the gene of interest (DNA amount, 1:10) in a six-well dish.

2. After 48–72 hr of transfection, cells are collected and washed with PBS twice.

3. Cells are scraped and collected by centrifugation, and cell pellets are suspended with 100 μl of extraction buffer [100 mM potassium phosphate (pH 7.8), 1 mM dithiothreitol (DTT)], then frozen and thawed three times.

4. Lysates are clarified by centrifugation at 15,000 rpm for 3 min.

5. Lysates (10–20 μl) are mixed with 350 μl of ATP solution [25 mM glycylglycine (pH 7.8), 5 mM ATP, 15 mM MgSO$_4$].

6. Measure the background light emission of the samples by luminescence reader.

7. Add 100 μl of 1 mM D-luciferin and mix well.

8. Measure the light emission by luminescence reader.

Convenient luciferase detection kits are available from various companies (e.g., the luciferase assay system from Promega, Madison, WI).

[17] J. N. Lavoie, M. Nguyen, R. C. Marcellus, P. E. Branton, and G. C. Shore, *J. Cell Biol.* **140**, 637 (1997).
[18] J. N. Lavoie, G. L'Allemain, A. Brunet, R. Muller, and J. Pouyssegur, *J. Biol. Chem.* **271**, 20608 (1996).
[19] J. R. de Wet, K. W. Wood, M. deLuca, D. R. Helinski, and S. Subramanni, *Mol. Cell. Biol.* **7**, 725 (1987).
[20] L. del Peso, V. M. Gonzalez, and G. Nunez, *J. Biol. Chem.* **273**, 33495 (1998).

β-Galactosidase Assay

This assay is from Refs. 20, 21.

1. Cells are cotransfected with β-galactosidase expression vector (e.g., pactβGal'[2]) together with the gene of interest (DNA amount, 1:10) in a six-well dish.

2. After 48–72 hr of transfection, cells are collected and washed with PBS twice.

3. Cells are scraped and collected by centrifugation, and cell pellets are suspended with 50 μl of extraction buffer [0.25 M sucrose, 10 mM Tris-HCl (pH 7.4), 10 mM EDTA], then frozen and thawed three times.

4. Lysates are clarified by centrifugation at 15,000 rpm for 3 min.

5. Cell lysates (20 μl) are mixed with 180 μl of Z buffer (60 mM Na_2HPO_4, 40 mM NaH_2PO_4, 10 mM KCl, 1 mM $MgSO_4$, 50 mM 2-mercaptoethanol).

6. Reaction mixtures are incubated for 5 min at 28°.

7. Forty microliters of o-nitrophenyl-β-galactopyranoside (ONPG) solution (ONPG, 4 mg/ml in 60 mM Na_2HPO_4, 40 mM NaH_2PO_4) is added and incubated at 28° until a visible yellow color is achieved.

8. Stop the reaction by adding 100 μl of 1 M Na_2CO_3.

9. Measure the colorimetric change in a spectrophotometer at 420 nm.

Prepare a blank containing a nontransfected cellular lysate, because mammalian cells contain an isozyme of β-galactosidase.

Caspase Activity as an Indicator of Apoptosis

Mammalian caspases are homologs of C. elegans cell death gene ced-3 and have been shown to play an important role in mediating apoptosis.[22] Activation of caspases has been demonstrated in many types of cells undergoing apoptosis.[22] Thus, it is possible to use the activation of caspases as an indicator of apoptosis.[23] In some cell lines, such as human embryonic kidney 293 cells, African green monkey kidney COS cells, or Drosophila S2 cells, standard transfection gives 20–70% transfection efficiency. In such cases, detection of caspase activity can be used to evaluate the relative levels of apoptosis.

[21] J. H. Miller, "Experiments in Molecular Genetics," pp. 352–355. Cold Spring Harbor Laboratory Press, Cold Spring Harbor, New York, 1972.

[22] V. Cryns and J. Yuan, Genes Dev. 12, 1551 (1998).

[23] Q. L. Deveraux, R. Takahashi, G. S. Salvesen, and J. C. Reed, Nature (London) 388, 300 (1997).

Procedure for Caspase Activity Measurement

This procedure is from Ref. 24.

1. Cells are transfected in a six-well dish.

2. After 24–48 hr of transfection, cells are collected and washed with PBS twice.

3. Cells are scraped and collected by centrifugation, and cell pellets are suspended with 100 μl of extraction buffer [50 mM Tris HCl (pH 7.4), 1 mM EDTA, 10 mM EGTA], then frozen and thawed three times.

4. After addition of 10 μM digitonin, the cells are incubated at 37° for 10 min. Lysates are clarified by centrifugation at 15,000 rpm for 3 min.

5. Clear lysates (2–10 μg of protein in 10 μl of extraction buffer) are mixed with 10 μl of assay buffer [20 mM HEPES-NaOH (pH 7.4), 100 mM NaCl, 0.05% (v/v) Nonidet P-40 (NP-40), 5 mM MgCl$_2$], and then incubated at 37° for 30 min.

6. Add 1 ml of assay buffer containing 10 μM Ac-DEVD-MCA, and then incubate at 37° for 30 min.

7. The levels of released 7-amino-4-methylcoumarin (AMC) are measured with a spectrofluorometer with excitation at 380 nm and emission at 460 nm.

Terminal Deoxynucleotidyltransferase-Mediated dUTP Nick End
   Labeling to Identify Apoptotic Cells

Fragmentation of DNA into 180-bp fragments, originally identified by Wyllie,[25] is a hallmark of apoptosis.[26] Cells with fragmented DNA can be identified *in situ* by the terminal deoxynucleotidyl transferase-mediated dUTP nick end-labeling (TUNEL) method.[27] Cells transfected with the gene of interest can be fixed and stained by TUNEL to determine if expression of the gene of interest has induced DNA fragmentation. The CADD/ICADD system, which is activated by caspases, plays a role in this DNA fragmentation.[28] In this context, caspase-mediated cell death can be detected by the TUNEL method. One should be cautious, however, when using this method to distinguish apoptosis and necrosis since necrosis can

[24] H. Kanuka, S. Hisahara, K. Sawamoto, S. Shoji, H. Okano, and M. Miura, *Proc. Natl. Acad. Sci. U.S.A.* **96**, 145 (1999).

[25] A. H. Wyllie, *Nature (London)* **284**, 555 (1980).

[26] M. J. Arends, R. G. Morris, and A. H. Wyllie, *Am. J. Pathol.* **136**, 593 (1990).

[27] Y. Gavrieli, Y. Sherman, and S. A. Ben-Sasson, *J. Cell Biol.* **119**, 493 (1992).

[28] M. Enari, H. Sakahira, H. Yokoyama, A. Okawa, A. Iwamatsu, and S. Nagata, *Nature (London)* **391**, 43 (1998).

induce DNA damage as well. Details of this procedure are described in this volume.[29]

Fluorescence-Activated Cell Sorting Analysis of Apoptosis

Examination of apoptotic cells by fluorescence microscope is simple but it can be time consuming to obtain a large number of cell death data by visual inspection. The apoptosis-inducing activity of a proapoptotic gene can be measured by a reduction in the number of cells that express the reporter protein (e.g., GFP) relative to that obtained by transfection with a control empty vector. This method is useful for adherent cells because a dying cell occasionally detaches from dishes and escapes from cell counting. Direct counting of GFP-positive cells under a fluorescence microscope or by FACScan flow cytometry analysis can be performed to quantitate the number of positive cells. This method, however, should be used with caution and with proper controls to ensure that the transfection efficiency is equal among all cells. In addition, the expression of the gene may have a positive or negative effect on the cellular levels of RNA and protein expression in general, which may inhibit the expression of the reporter gene and thus lead to an erroneous conclusion that the expression of the gene induces apoptosis. Therefore, it is important to have additional evidence to support a conclusion that the gene of interest induces apoptosis. Quantification of apoptotic cells can be performed by FACS analysis.[30] To limit analysis to cells that have taken up plasmid DNA, reporter constructs (e.g., CD4 or GFP expression vector) are cotransfected with the genes of interest. First, these cells are gated by staining for reporter, and then cell deaths are analyzed by staining with PI, which indicates loss of plasma membrane integrity, and binding of annexin V, which interacts with phosphatidylserine on the surface of apoptosis cells. Double-positive cells are in the later stage of apoptosis; double-negative cells are viable. PI-negative/annexin V-positive cells are in the early stage of apoptosis; PI-positive/annexin V-negative cells are necrotic.

Summary

In conclusion, transient transfection is an efficient and powerful method to determine quickly whether a gene has a detrimental effect on cell survival. We have described a variety of assay systems from which to choose. Each

[29] *Methods Enzymol.* **322,** 2000, this volume.
[30] S. A. Marsters, J. P. Sheridan, C. J. Donahue, R. M. Pitti, C. L. Gray, A. D. Goodard, K. D. Bauer, and A. Ashkenazi, *Curr. Biol.* **6,** 1669 (1996).

system has its own advantages and disadvantages. It is important to back up any experimental conclusion with more than one type of assay if possible. In addition, one must consider the fact that transient transfection can often achieve artificially high levels of a protein product that may be unrealistic *in vivo*. High levels of certain proteins may have an adverse effect on cell survival even when they have nothing to do with apoptosis *in vivo*. Thus, we must emphasize here that a transient transfection assay is just the first test to determine the function of a gene.

Acknowledgments

We thank Dr. Honglin Li for critical reading of this manuscript. This work is supported in part by a Grant-in-Aid from the Ministry of Education (Japan), Science and Culture (to M.M.), a Grant-in-Aid from the Tokyo Biochemical Research Foundation (to M.M.), an R01 from the National Institute of Aging (United States) (to J.Y.), and an American Heart Established Investigatorship (to J.Y.).

[43] Sindbis Virus Vector System for Functional Analysis of Apoptosis Regulators

By J. Marie Hardwick and Beth Levine

Introduction

A method was developed to quantitate the antiapoptotic and proapoptotic activities of a variety of proteins, using a virus vector system. On infection of susceptible host cells, Sindbis virus induces classic apoptotic cell death, which is blocked by overexpressed Bcl-2.[1,2] To study Bcl-2 function, the Bcl-2-coding sequences were inserted into an infectious clone of Sindbis virus. The ability of Bcl-2 encoded by the virus to impair Sindbis virus-induced cell death provides a reliable measurement of the antiapoptotic function of Bcl-2 and its homologs and derivatives.[3,4] That is, the virus serves both as an expression vector and as a cell death stimulus. This rapid

---

[1] B. Levine, Q. Huang, J. T. Isaacs, J. C. Reed, D. E. Griffin, and J. M. Hardwick, *Nature* (*London*) **361,** 739 (1993).

[2] S. Ubol, P. C. Tucker, D. E. Griffin, and J. M. Hardwick, *Proc. Natl. Acad. Sci. U.S.A.* **91,** 5202 (1994).

[3] E. H.-Y. Cheng, B. Levine, L. H. Boise, C. B. Thompson, and J. M. Hardwick, *Nature* (*London*) **379,** 554 (1996).

[4] B. Levine, J. E. Goldman, H. H. Jiang, D. E. Griffin, and J. M. Hardwick, *Proc. Natl. Acad. Sci. U.S.A.* **93,** 4810 (1996).

functional assay facilitates the analysis of a large number of constructs without the cumbersome task of generating stably transfected cell lines. Proteins that induce apoptosis also can be assayed with this system as these proteins accelerate apoptosis induced by Sindbis virus.[3] Furthermore, Sindbis virus infects and kills newborn mice, permitting the functional analysis of cell death regulators *in vivo*. Thus, the use of recombinant Sindbis viruses expressing a variety of both viral and cellular genes that regulate apoptosis provides a unique agent with which to study both the cellular death pathway as well as the molecular mechanisms of viral pathogenesis.

With the exception of lymphoid-derived cells, Sindbis virus infects a wide range of cell types, including most cell lines as well as postmitotic neurons both in culture and in a newborn mouse model.[1,4,5] The most common outcome of infection with Sindbis virus is the induction of apoptosis, but in some cell types and in older animals Sindbis virus can establish a persistent infection.[1,6] Both viral and cellular determinants regulate the outcome of infection leading to apoptosis or long-term persistence.[2] Thus, it is also possible to utilize Sindbis virus as an expression vector under conditions in which the virus does not kill the cell, or in which cell death that is induced by the virus occurs late. Under such conditions, Sindbis virus can be utilized as a gene transfer vector in matured cultured neurons, which are otherwise difficult to transfect.[7]

Biology of Sindbis Virus

Biosafety Issues

Sindbis virus is an alphavirus in the Togaviridae family that is transmitted to vertebrate hosts via mosquitoes. While some alphaviruses cause devastating human disease, Sindbis virus is largely an avian virus and is rarely associated with human infections.[8] The laboratory strain AR339 isolated from mosquitoes in the 1950s is not known to cause human disease. Infection of a laboratory worker by accidental needle stick was associated with a mild local inflammatory response that was resolved within a few days. Infectious virus and derived clones and vectors, which are less virulent

[5] V. E. Nava, A. Rosen, M. A. Veliuona, R. J. Clem, B. Levine, and J. M. Hardwick, *J. Virol.* **72**, 452 (1998).

[6] B. Levine, J. M. Hardwick, and D. E. Griffin, *Trends Microbiol.* **2**, 25 (1994).

[7] D. S. Park, B. Levine, G. Ferrari, and L. A. Greene, *J. Neurosci.* **17**, 8975 (1997).

[8] R. E. Johnston and C. J. Peters, in "Field's Virology" (B. N. Fields, D. M. Knipe, and P. M. Howley, eds.), 3rd Ed., p. 843. Lippincott-Raven, Philadelphia, 1996.

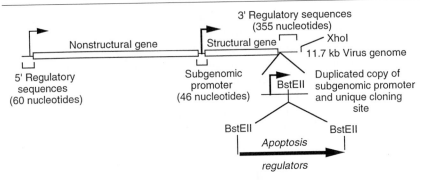

Fig. 1. Diagram of the double subgenomic Sindbis virus vector.

than AR339, are maintained in the laboratory under biosafety level 2 conditions.

Sindbis Virus Life Cycle

The Sindbis virus genome is a single-strand messenger RNA of 11.7 kb (Fig. 1).[9] The 5' two-thirds of the genome encodes a polyprotein precursor that is cleaved into four nonstructural proteins (nsP1, nsP2, nsP3, and nsP4) by a cysteine protease encoded in the C-terminal half of nsP2.[10] The viral RNA polymerase is contained in nsP4, which is produced only on read-through of a termination codon following nsP3. Other functions including helicase, ATPase, and methyltransferase activities are localized to the nsP1–3 proteins.[11] The nsP123 polyprotein together with the nsP4 polymerase make up the replication complex responsible for transcribing the negative-strand RNA (a replication intermediate complementary to the viral genome). Subsequent cleavage of the nsP123 polyprotein into its individual components (nsP1, nsP2, and nsP3) switches the template specificity, leading to production of the plus-strand genomic RNA and the subgenomic RNA derived from the 3' third of the genome.[12,13]

The subgenomic 26S RNA, initiated from an internal (subgenomic) promoter sequence, is translated into a second polyprotein that when processed forms the virus particle (Fig. 1). The N-terminal capsid protein contains a chymotrypsin-like serine proteinase that autocatalytically cleaves

---

[9] M. J. Schlesinger and S. Schlesinger, in "The Togaviridae and Flaviviridae" (S. Schlesinger and M. J. Schlesinger, eds.), p. 121. Plenum Press, New York, 1986.

[10] E. G. Strauss, R. J. De Groot, R. Levinson, and J. H. Strauss, Virology 191, 932 (1992).

[11] J. H. Strauss and E. G. Strauss, Microbiol. Rev. 58, 491 (1994).

[12] D. J. Barton, S. G. Sawicki, and D. L. Sawicki, J. Virol. 65, 1496 (1991).

[13] J. A. Lemm, A. Bergqvist, C. M. Read, and C. M. Rice, J. Virol. 72, 6546 (1998).

itself from the precursor polyprotein.[14–17] The remaining structural proteins, P62(E3 + E2), 6K, and E1, are generated by cellular proteases.[9,18] E2 and E1 heterodimerize through noncovalent interactions to form the virion transmembrane glycoprotein spikes. These proteins are responsible for binding to one of several potential cellular receptors and subsequent fusion of the viral envelope to the endocytic membrane.[19–21] The virus RNA remains associated with cytoplasmic membranes/vesicles where virus replication occurs.[22,23] Nucleocapsid assembly is facilitated by an interaction between a 68-amino acid segment of the capsid protein with nucleotides 945 to 1076 (within the nsP1-coding sequence) of the newly synthesized genomic RNA.[24] Tyr-400 and Leu-402 of the E2 cytoplasmic tail are thought to bind in the pocket domain of the nucleocapsid and facilitate assembly of virions at the plasma membrane.[25] Infectious virus is formed by budding of the nucleocapsid through the viral glycoprotein-laden plasma membrane. Thus, the lipid membrane and glycoproteins are acquired by budding and there is no infectious intracellular virus.

Induction of Apoptosis by Sindbis Virus

Sindbis virus induces apoptotic cell death as evidenced by membrane blebbing, chromatin condensation, and DNA ladder formation in culture and in mouse brains.[1,26] Although Sindbis virus infection leads to caspase activation,[5] the molecular mechanisms by which Sindbis virus activates programmed cell death are not clear and may be multifactorial. Virus replication is required to induce apoptosis when infection occurs at low multiplicities. However, addition of 100 or more infectious units per cell is sufficient to trigger apoptosis under some conditions in the absence of

[14] D. T. Simmons and J. H. Strauss, *J. Mol. Biol.* **86**, 397 (1974).
[15] E. G. Strauss and J. H. Strauss, *in* "The Togaviridae and Flaviviridae" (S. Schlesinger and M. J. Schlesinger, eds.), p. 35. Plenum Press, New York, 1986.
[16] Y. S. Hahn, A. Grakoui, C. M. Rice, E. G. Strauss, and J. H. Strauss, *J. Virol.* **63**, 1194 (1989).
[17] H.-K. Choi, L. Tong, W. Minor, P. Dumas, U. Boege, M. G. Rossmann, and G. Wengler, *Nature (London)* **354**, 37 (1991).
[18] P. Liljestrom and H. Garoff, *J. Virol.* **65**, 147 (1991).
[19] J. M. Wahlberg, R. Bron, J. Wilschut, and H. Garoff, *J. Virol.* **66**, 7309 (1992).
[20] K.-S. Wang, R. J. Kuhn, E. G. Strauss, S. Ou, and J. H. Strauss, *J. Virol.* **66**, 4992 (1992).
[21] S. Ubol and D. E. Griffin, *J. Virol.* **65**, 6913 (1991).
[22] J. Peranen and L. Kaariainen, *J. Virol.* **65**, 1623 (1991).
[23] S. Froshauer, J. Kartenbeck, and A. Helenius, *J. Cell Biol.* **107**, 2075 (1988).
[24] B. Weiss, U. Geigenmuller-Gnirke, and S. Schlesinger, *Nucleic Acids Res.* **22**, 780 (1994).
[25] S. Lee, K. E. Owen, H.-K. Choi, H. Lee, G. Lu, G. Wengler, D. T. Brown, M. G. Rossmann, and R. J. Kuhn, *Structure* **4**, 531 (1996).
[26] J. Lewis, S. L. Wesselingh, D. E. Griffin, and J. M. Hardwick, *J. Virol.* **70**, 1828 (1996).

virus replication, implying that cell surface receptor interactions/fusion events may contribute to apoptotic cell death.[27] Sindbis virus deleted of the glycoprotein-coding sequences replicates its genome efficiently and produces nonstructural proteins (but no infectious virions). However, the cytopathic effects of this virus are significantly diminished, implicating P62, 6K, and/or E1 transmembrane proteins in the induction of apoptosis.[28] In fact, expression of either P62 or 6K–E1 in the absence of other viral proteins is sufficient to induce cell death in AT-3 cells.[29] The prodeath domains of these proteins were mapped to the membrane-spanning region near the C terminus of E2 and E1. However, the cell death pathway triggered by these transmembrane domains is not universal as they fail to induce cell death in BHK (baby hamster kidney) cells, which undergo classic apoptosis after Sindbis virus infection. Similarly, Sindbis virus induces apoptosis of AT-3 cells in an NF-κB-dependent manner, although NF-κB is not required for apoptosis of BHK cells.[30,31] The downstream targets of NF-κB in the death pathway are unknown.

Methods

Sindbis Virus Vector Construction

The Sindbis virus vector is constructed by inserting a duplicated copy of the viral subgenomic promoter plus a unique cloning site (*Bst*EII) between the structural gene and the 3′ regulatory region of the virus genome (Fig. 1). The details of the cloning strategy for construction of the double subgenomic Sindbis virus vector have been reported previously[3,4] and are patterned after vectors described by Bredenbeek and Rice.[32] The coding sequences for genes of interest are inserted into the *Bst*EII site (GGTNACC), although the *Xba*I site can also serve as a cloning site (when plasmids are prepared from Dam methylase-minus bacteria). Inserts that are generated by polymerase chain reaction (PCR) in which the primers contain *Bst*EII sites will allow for unidirectional cloning (5′ primer se-

[27] J.-T. Jan and D. E. Griffin, *J. Virol.* **73,** 10269 (1999).

[28] I. Frolov and S. Schlesinger, *J. Virol.* **68,** 1721 (1994).

[29] A. K. Joe, H. H. Foo, L. Kleeman, and B. Levine, *J. Virol.* **72,** 3935 (1998).

[30] K.-I. Lin, S.-H. Lee, R. Narayanan, J. M. Baraban, J. M. Hardwick, and R. R. Ratan, *J. Cell Biol.* **131,** 1149 (1995).

[31] K.-I. Lin, J. A. DiDonato, A. Hoffman, J. M. Hardwick, and R. R. Ratan, *J. Cell Biol.* **141,** 1479 (1998).

[32] P. J. Bredenbeek and C. M. Rice, *Semin. Virol.* **3,** 297 (1992).

quence, GGT<u>C</u>ACC; 3' primer sequence, GGT<u>G</u>ACC). PCR primers that contain a *Bst*EII site immediately adjacent to the ATG start codon of the gene of interest reconstitute a consensus Kozak initiation site for translation: ACC ATG. To avoid PCR errors and the necessity of sequencing multiple inserts, blunt-end ligations of restriction fragments are preferred and permit the generation of both forward and reverse orientations of the insert. The length of untranslated sequences in the insert should be minimized to increase expression and avoid exceeding the capacity of the vector. It is recommended that two independent constructs of the forward orientations be used to generate recombinant viruses, although variability between clones is usually minimal. A plasmid map showing restriction sites frequently used for determining the presence/absence and orientation of inserts is shown in Fig. 2.

Difficulties with cloning inserts into the vector most often result from (1) incorrect ends (e.g., Klenow failed to produce blunt ends), (2) insufficient levels of insert and/or vector (check on a gel prior to ligation), or (3) incomplete digestion of the vector with *Bst*EII (performed at 60°; check by transforming bacteria with cut, unligated plasmid). Because the vector is relatively large, adjusting the ratio of vector and insert DNA concentrations during ligation may also eliminate problems. The frequent production of small plasmids that replicate to high copy numbers is presumably due to end-to-end ligation of the vector, resulting in unstable plasmids that delete large segments. Reducing the vector concentration during ligation will correct this problem.

Because the insert size will somewhat affect the replication efficiency of the virus, appropriate control viruses are always necessary. Perhaps the easiest control is to insert the gene of interest in reverse orientation. Although there are theoretical concerns that the production of complementary RNAs could potentially function as antisense inhibitors of the endogenous gene, this problem is rarely encountered. An alternative control is to insert a premature stop codon in the gene of interest. This is accomplished most simply by cutting the recombinant vector plasmid with a restriction enzyme and ligating in a commercially available unphosphorylated nonsense linker that contains a distinct restriction site for screening. A potential concern with this approach is that a truncated protein product could function by a dominant negative mechanism. However, inserting the stop codon near the 5' end will help eliminate this possibility. An alternative control is to test a recombinant virus encoding a heterologous gene of the same size but of irrelevant function. A drawback to this approach is the potential unknown effect of the irrelevant gene. A combination of two or more different controls increases the confidence level.

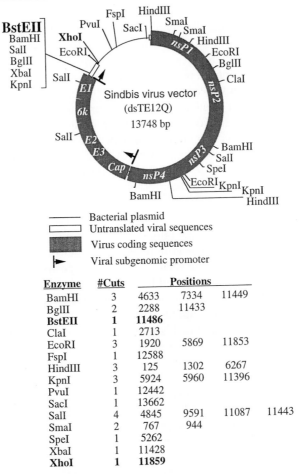

FIG. 2. Restriction enzyme map of the double subgenomic Sindbis virus vector.

Enzyme	#Cuts	Positions			
BamHI	3	4633	7334	11449	
BglII	2	2288	11433		
BstEII	**1**	**11486**			
ClaI	1	2713			
EcoRI	3	1920	5869	11853	
FspI	1	12588			
HindIII	3	125	1302	6267	
KpnI	3	5924	5960	11396	
PvuI	1	12442			
SacI	1	13662			
SalI	4	4845	9591	11087	11443
SmaI	2	767	944		
SpeI	1	5262			
XbaI	1	11428			
XhoI	**1**	**11859**			

Preparation of Recombinant Viruses

The Sindbis virus vector plasmid contains an SP6 promoter upstream of the 5′ viral sequences (Fig. 3). Infectious virus RNA is synthesized *in vitro* from a linearized plasmid and transfected into cells. At 24–48 hr posttransfection, cell supernatants containing infectious progeny virus are collected, titered, and stored at $-70°$.

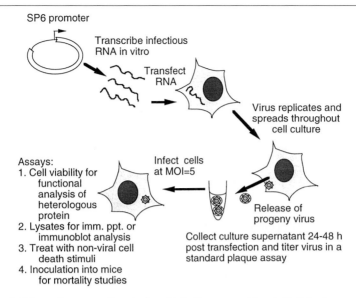

SP6 promoter

Transcribe infectious
RNA in vitro

Transfect
RNA

Virus replicates and
spreads throughout
cell culture

Assays:
1. Cell viability for
   functional
   analysis of
   heterologous
   protein
2. Lysates for imm. ppt. or
   immunoblot analysis
3. Treat with non-viral cell
   death stimuli
4. Inoculation into mice
   for mortality studies

Infect cells
at MOI=5

Release of
progeny virus

Collect culture supernatant 24-48 h
post transfection and titer virus in a
standard plaque assay

FIG. 3. Scheme for generating stocks of infectious recombinant Sindbis viruses, starting with plasmid DNA.

Linearization of Plasmid DNA

Plasmid DNA should be prepared with a Qiagen (Chatsworth, CA) prep or related product and not by minilysis protocols. Linearization at the *Xho*I site (downstream of the 3' untranslated/regulatory sequences ending at nucleotide 11850) yields the most efficient production of RNA. If the insert contains an *Xho*I site, an alternate vector in which the *Not*I site replaces the *Xho*I site at 11859 is available. Linearization at *Fsp*I (nucleotide 12588), *Pvu*I (nucleotide 12442) or *Sac*I (nucleotide 13622) is also possible but yields less RNA. Blunting the 3' overhangs of the *Pvu*I and *Sac*I sites with Klenow improves efficiency.

1. Incubate 5 μg of plasmid DNA with 3 μl of *Xho*I (or appropriate restriction enzyme) and 10 μl of 10× reaction buffer in a 100-μl volume at 37° for 3 hr.

2. Add 1 μl of proteinase K (Boehringer Mannheim, Indianapolis, IN) for 30 min at 37° (optional).

3. Phenol–chloroform extract and ethanol precipitate the linearized DNA.

4. Resuspend the DNA in 12 μl of RNase-free Tris–EDTA (TE) buffer or H$_2$O.

5. Run 1 μl on a gel to confirm linearization.

In Vitro Transcription of Infectious Viral RNA

1. Combine the following in the order listed, mix, and incubate at 37° for 50 min:

rNTP mix (10 mM rUTP, 10 mM rCTP, 10 mM rATP, 3 mM rGTP; Stratagene, La Jolla, CA)	1 μl
rGTP (10 mM; Stratagene), diluted 1:10	1 μl
CAP analog (1404S; New England BioLabs, Beverly, MA) (~10 mM)	2.5 μl
SP6 buffer (5×)	5 μl
Dithiothreitol (DTT, 10–100 mM)	1 μl
RNasin (Promega, Madison, WI)	1 μl
RNase-free water	10 μl
SP6 RNA polymerase (GIBCO-BRL, Gaithersburg, MD)	1 μl
Linearized DNA template (~1 μg)	2.5 μl

2. Use the RNA immediately for transfection (most efficient), or phenol extract and store in ethanol at −70°. In vitro-transcribed RNA can be examined on an RNase-free agarose gel. However, verification of the size requires a denaturing gel.

Transfection of Viral RNA into BHK Cells

1. Seed 1–2.5 × 10^5 BHK cells suspended in 2 ml of culture medium [Dulbecco's modified Eagle's medium (DMEM)–10% (v/v) fetal bovine serum (FBS)–glutamine–gentamicin] per well in a six-well plate (35-mm wells) and incubate at 37° with 5% CO$_2$ until the cells are 40–60% confluent (16–24 hr).

2. On the day of transfection incubate 30 μl of Lipofectin (GIBCO-BRL; preferred over LipofectAMINE) with 170 μl of serum-free DMEM at room temperature for at least 20 min in a polystyrene tube. OptiMEM, a complete serum-free medium, may be substituted for serum-free DMEM.

3. Mix 20 μl of the RNA transcription reaction with 180 μl of serum-free DMEM.

4. Combine the Lipofectin mix with the RNA mix at room temperature for 10–15 min.

5. Add 600 μl of serum-free DMEM to the transfection mix and incubate at room temperature for 15 min immediately prior to transfection. To generate a larger stock of virus, the transcription mix can be divided into two separate 1-ml transfection mixes, using 12 μl of RNA in each, and collecting virus from two 35-mm wells.

6. Wash the cells with 2 ml of serum-free DMEM, add 1 ml of transfection mix per well, and incubate for 2 hr.

7. Wash the cells with 2 ml of DMEM–10% (v/v) FBS with penicillin–streptomycin (Pen/Strep) and incubate with 2–3 ml of the same medium at 37° with 5% CO_2.

Harvesting Infectious Virus

1. Harvest supernatants from the transfected cells when cytopathic effects are maximal but before most of the cells detach (24–48 hr).

2. Briefly centrifuge the supernatant to remove cellular debris.

3. Store supernatants in 10- to 100-μl aliquots at $-70°$. Sindbis virus is stable at $-70°$, but decreases in titer when freeze–thawed multiple times. Therefore, aliquots should be discarded after thawing. Passaging recombinant viruses is not recommended as the inserts may not be genetically stable.

If cytopathic effects (rounding/blebbing of cells initially in foci) are not observed, no virus is present. The most common explanation for failure to produce recombinant viruses is degradation of the in vitro-transcribed RNA. Use of proteinase K, RNasin, and RNase-free reagents in the steps described above, and direct transfection of the transcription reaction mix immediately after synthesis, will correct this problem.

Plaque Assay for Titration of Recombinant Sindbis Viruses

1. Seed six-well plates (35-mm wells) with approximately 5×10^5 BHK cells per well in 2 ml of culture medium [DMEM–10% (v/v) FBS] per well.

2. Incubate overnight at 37° with 5% CO_2; the cells should be approximately 80–90% confluent when infected. Lower cell density may cause holes in the monolayer, which may be confused with virus plaques.

3. Perform serial 10-fold dilutions of virus stock in DMEM with 1% (v/v) FBS, beginning with 0.1 ml of virus stock into 0.9 ml of DMEM–1% (v/v) FBS (tube A); 0.1 ml of tube A with 0.9 ml of DMEM–1% (v/v) FBS (tube B), etc. The titers of recombinant virus stocks fall between 10^6 and 10^9; therefore recommended plating dilutions are 10^{-4} to 10^{-7}.

4. Suction off medium, three wells at a time, to prevent the drying of cells and pipette 0.2 ml of virus dilutions into each well in triplicate. Vigorously rock plates with a jerking motion both width and lengthwise so that virus inoculum covers the entire surface of the wells. Repeat every 5 to 10 min while plates incubate at 37° for 60 min.

5. In advance, prepare 1.2% Bacto-Agar (Difco, Detroit, MI) in purified water and autoclave in 50-ml aliquots in glass bottles. Before beginning virus infections, melt the necessary number of agar aliquots in a microwave, and then incubate the liquefied agar in a 40° water bath for at least 30 min prior to use.

6. Incubate 2× modified Eagle's medium (MEM) with 2% (v/v) FBS, glutamine, gentamicin, and Fungizone in 50-ml aliquots in a 40° water bath for at least 30 min. *Note:* It is critical that no phenol red be present, or the plaque assay will not work.

7. At the end of the infection time, combine 1 : 1 agar and 2× MEM–2% (v/v) FBS (mix well). Quickly add 2 ml of diluted agarose to each well by pipetting down the sides of wells to prevent overheating the cells; swirl each plate to mix the virus solution with the agarose. Allow the agar to solidity in a hood, approximately 10–15 min, and incubate the cells for 48 hr at 37°. Recombinant viruses generally form smaller plaques than wild-type viruses and take a full 48 hr to develop. Recombinant viruses expressing apoptosis inhibitors still induce a significant level of cell death, thereby permitting reliable determination of the titer in a standard plaque assay. However, extending the incubation period from 48 to 50–55 hr in some cases can be helpful.

8. To stain plaques 48 hr after infection, dilute 10% (w/v) neutral red in the following manner (scale up amounts as needed for plaque assay): 15 ml of 2× MEM, 15 ml of sterile distilled water, 3.3 ml of 10% (w/v) neutral red stain. (Perform this step outside of the cell culture hood.) Add 1 ml of diluted neutral red solution to each well and incubate at 37° for 2 hr. Suction off the stain, using a drawn-out pipette or a needle to avoid aspirating the monolayer. Count plaques within 2–4 hr by dotting each one with a Flair pen on the bottom of each well. Calculate the number of plaque-forming units (PFU) per milliliter:

Number of plaques at $10^{-x} = N$

PFU/ml $= N \times 10^x \times 5$ (dilution factors)

For example, 25 plaques are counted in a 10^{-2} well; therefore PFU/ml $= 25 \times 10^2 \times 5 = 1.25 \times 10^4$.

Infection of BHK Cells with Recombinant Viruses for Functional Analysis of Heterologous Genes

Cells should be infected with a multiplicity of 5 PFU per cell to ensure uniform infection. Low multiplicities of infection can delay cell death, giving the false impression that the heterologous gene is protective. Infection at multiplicities of 5–100 produces identical results. *Note:* Experiments by Park *et al.*[7] using the Sindbis vector for expressing cell cycle inhibitors in neurons were performed at an MOI of 1–2 because of the toxicity of the virus (see below).

For viability assays it is critical that cells be free of *Mycoplasma,* which is a common "silent" contaminant of cultured cells.

Dilute virus so that the number of desired plaque-forming units will be contained within a 0.2-ml inoculum for a 35-mm well. Virus should be diluted in medium used to propagate the given cell type being infected, except with only 1% (v/v) serum, as higher concentrations of serum reduce infectivity of the virus. Incubate the virus inoculum with the cells at 37° in 5% CO_2 (with frequent rocking of the dishes to cover the cells with inoculum, as described above). Alternatively, infection can be performed in a 1-ml volume. After infection change the medium to a standard concentration of serum, or simply add 1–2 ml of complete medium/35-mm well and return the cells to the incubator. Alternatively. OptiMEM can be used for both transfection of RNA and during virus infections. With cell types that will not tolerate serum reduction or medium changes, infect in a low volume of complete medium, then increase to full volume after 1–2 hr, or infection can be performed in complete medium.

Viability Assays

To detect inhibition of Sindbis virus-induced apoptosis by heterologous genes, cell viabilities of most cell types are determined approximately 48 hr after infection with recombinant viruses. Cells are detached from the plate by scraping with a rubber tip or by trypsinization and then combined with floating cells before pelleting. Some cell types (especially when stressed) are killed by one or both of these detachment techniques and must be removed from the dish by shaking or another method. Viability can be assessed by trypan blue exclusion, fluorescence-activated cell sorting (FACS) analysis, propidium iodide (PI) stain 3-(4,5-dimethylthiazol-2-yl)-2,5-diphenyltetrazolium bromide (MTT), terminal deoxynucleotidyltrans-ferase-mediated dUTP nick end labeling (TUNEL), or other assay. By trypan blue exclusion, BHK cells infected with control viruses should range between 5 and 20% viable at 48 hr postinfection. Viruses expressing Bcl-2, Bcl-x_L, CrmA, P35, XIAP/ILP, and other apoptosis inhibitors range between 40 and 60% viable at 48 hr, although earlier time points may be required when analyzing less robust apoptosis inhibitors. The presence of *Mycoplasma* in the cultures will prevent the detection of antiapoptotic function. To detect acceleration of apoptosis by Bax and other proteins, cell viability should be determined between 24 and 36 hr postinfection, before the control viruses have significantly reduced cell viability. Some assays such as FACS analyses are performed at somewhat earlier times than trypan blue exclusion.

Detection of Heterologous Proteins

For detection of heterologous proteins by immunoblot analysis, it is recommended to harvest cells at 16–24 hr postinfection with recombinant viruses, as lower protein levels are detected at later times. COS cells are preferred over BHK cells for the detection of proapoptotic proteins. Nevertheless, potently proapoptotic proteins can be difficult to detect, presumably because the cells expressing these proteins are underrepresented.

Applications of Sindbis Virus Vector System

Functional Analysis under Conditions in which Sindbis Virus Induces Apoptosis

The expression of heterologous proteins using the Sindbis virus vector has permitted rapid analysis of multiple proteins without the necessity of generating stably transfected cell lines. This system permitted the timely analysis of more than 30 mutants of Bcl-x_L and Bcl-2, which led to the conclusion that these proteins do not need to bind Bax or Bak in order to prevent cell death. Two mutants of Bcl-x_L and one of Bcl-2 were identified that fail to bind Bax or Bak but protect cells from Sindbis virus-induced apoptosis with 50–80% of the efficiency of their wild-type counterparts.[3] The search for these mutants among stably transfected cell lines that express similar levels of mutant proteins would be a difficult task. The results obtained with the Sindbis virus vector system were subsequently confirmed using other death stimuli and cell types, indicating that using Sindbis virus as a death stimulus produces the same results as at least several other death stimuli.[3,33] Furthermore, Sindbis virus infection represents a biologically relevant death stimulus.

Sindbis virus also infects mice, permitting the analysis of protein function *in vivo.* Newborn mice (1–5 days of age) are infected with 5000 PFU in 30 μl of Hanks' balanced salt solution, using a tuberculin syringe and inoculating just under the skull midway between the eye and the ear.[26] Mice are observed daily for 21 days after infection for mortality and/or clinical signs. Recombinant viruses expressing Bcl-2, CrmA, and others are impaired in their ability to kill newborn CD-1 mice.[4,5,34] Although viruses that express some apoptosis inhibitors may replicate to slightly lower titers *in vivo,* this appears to be a consequence of the apoptosis inhibitor rather than a cause for survival differences. Suppression of virus replication is not required for protection from fatal disease.[35] Sindbis virus targets to neurons of the central nervous system, including spinal cord motor neurons. Thus, Sindbis

[33] C. S. Duckett, V. E. Nava, R. W. Gedrich, R. J. Clem, J. L. Van Dongen, M. C. Gilfillan, H. Shiels, J. M. Hardwick, and C. B. Thompson, *EMBO J.* **15**, 2685 (1996).
[34] X. H. Liang, L. Kleeman, H. Jiang, G. Gordon, J. E. Goldman, G. Berry, B. Herman, and B. Levine, *J. Virol.* **72**, 8586 (1998).
[35] J. Lewis, G. A. Oyler, K. Ueno, J. Vornov, S. J. Korsmeyer, S. Zou, and J. M. Hardwick, *Nature Medicine* **5**, 832 (1999).

virus also will be a useful tool to study the role of genes involved in motor neuron disease in an animal model.

Sindbis Virus: An Efficient Gene Delivery System under Conditions in which the Virus Does Not Kill the Cell

The Sindbis virus vector system also can be used to study the effects of candidate death regulatory genes on apoptosis induced by nonviral stimuli. This application is feasible only in experimental paradigms where death induced by the nonviral stimulus occurs more rapidly than death induced by the virus. Sindbis virus-induced apoptosis generally occurs rapidly (e.g., within 12–48 hr) in most dividing cell lines. However, in nondividing terminally differentiated cells such as neurons, Sindbis virus may establish a prolonged infection (days to weeks) without inducing cell death. This property, coupled with the limited availability of alternative techniques for gene delivery into neurons, makes the Sindbis virus vector system particularly useful for gene transfer studies in experimental paradigms involving trophic factor deprivation, oxidative stress, and excitotoxic or DNA-damaging agent-induced neuronal death. It is possible that the Sindbis virus vector system may also prove to be useful in nonviral apoptotic paradigms in other differentiated cell types in which Sindbis virus replicates (e.g., muscle), but there is as yet no experience in this area.

To explore the role of cell cycle regulatory molecules in neuronal apoptosis induced by nerve growth factor (NGF) deprivation, the Sindbis virus vector was used to express several flag epitope-tagged cell cycle inhibitors, including p16, p21, and p27, and dominant-negative mutant forms of cyclin-dependent kinases 2, 3, 4, and 6, in primary rat sympathetic neuron cultures obtained from superior cervical ganglia of postnatal day 1 rats. On the third day in culture, sympathetic neurons were infected with Sindbis virus vector constructs at a multiplicity of infection of 2 and deprived of NGF. Immunofluorescence staining performed 2 days after infection revealed expression of flag epitope-tagged proteins in 95% of the neurons. In most experiments, more than 75% of sympathetic neurons were dead by day 2 and more than 95% of neurons were dead by day 3 after NGF deprivation. In contrast, in the presence of NGF, no cytotoxicity attributable to Sindbis virus infection was observed before 4–5 days after infection. Thus, the Sindbis virus vector system could be used effectively to assess the effects of heterologous gene expression on NGF deprivation-induced death on days 1, 2, and 3 after infection and NGF deprivation. These studies revealed a significant protective effect of p16, p21, p27, dominant-negative Cdk4, and dominant-negative Cdk6 on NGF deprivation-induced death of sympathetic neurons.

There are several advantages of this system compared with other gene delivery systems for the study of neuronal apoptosis. First, stable recombinant viruses that express the gene of interest can be readily generated within a short period of time. Second, Sindbis virus efficiently targets virtually all neurons in a given culture population. Third, Sindbis virus replicates efficiently in all types of rodent neurons examined to date (including rat and mouse dorsal root ganglion neurons, sympathetic neurons, cortical neurons, and hippocampal neurons). Fourth, it permits high levels of expression of the heterologous genes in neurons. Fifth, we have observed relatively long-term expression (up to at least 14 days) of heterologous proteins. These factors, and the demonstrated utility of the vector system in the study by Park et al.[7] suggest that Sindbis viruses are extremely efficient neuronal-targeting vectors for apoptosis studies.

Limitations of the Sindbis Virus Vector System

Size Limitations of the Vector

The 11.7-kb RNA genome of Sindbis virus is surprisingly tolerant of insertions, allowing the expression of heterologous genes from an internal Sindbis virus promoter. The double subgenomic vector described here has an approximate 2-kb upper limitation on the size of the insert. Other versions of the vector have been generated and are commercially available (InVitrogen, San Diego, CA), in which the structural gene function is supplied on a helper virus. This vector could theoretically accept 5- to 6-kb inserts. However, the expression of Bcl-2 in this vector leads to excessive overexpression and at these levels Bcl-2 becomes proapoptotic. Therefore, this vector in its current form may not be suitable for the study of programmed cell death.

Strain- and Age-Dependent Susceptibility of Mice to Sindbis Virus

Mice exhibit age-dependent susceptibility to apoptosis induced by Sindbis virus. Most but not all strains of mice become resistant to Sindbis virus-induced mortality within a few days after birth unless the virus is carrying an exceptionally potent proapoptotic protein, or unless the vector is modified to increase virulence (e.g., Q55H in E2 glycoprotein).[2,36] Thus, assays for apoptosis inhibitors generally are performed on 2- to 3-day-old pups, making it necessary to establish breeding colonies or to elicit the

---

[36] P. Lee, D. Kerr, T. Larson, J. M. Hardwick, D. Irani, and D. E. Griffin, unpublished data.

cooperation of the vendor of timed pregnant mice regarding the precise time of delivery. In addition, mice of different strains have considerable variability in their susceptibility to a fatal Sindbis virus infection. Therefore, it is critical always to include the appropriate control strains, which are not available for some lines of genetically deficient animals. In these cases the knockout animal of interest can be back-crossed onto an appropriate strain. Infecting the offspring of heterozygous matings and correlating mortality with genotype provides optimal control littermates.

Requirement for Transfection of RNA

While DNA versions of the Sindbis virus vector have been generated,[37,38] transfection of the DNA plasmid for the vector reported here does not yield infectious virus, necessitating the synthesis and transfection of infectious RNA. However, because the RNA is highly infectious and the virus readily spreads from cell to cell, low transfection efficiencies result in high virus titers.

Limited Dual Infections with Sindbis Virus

It is extremely useful to infect cells and animals simultaneously with two viruses, each encoding a different heterologous protein of interest, in order to study protein–protein interactions. However, Sindbis virus, like some other viruses, has an exclusion mechanism whereby a second virus cannot infect a cell that is already infected by Sindbis virus or a closely related virus. The molecular mechanism for the exclusion mechanism apparently lies with the viral nsP2 protease of the primary infecting virus. nsP2 cleaves the nsP123 polyprotein of the superinfecting virus into its individual components prior to the synthesis of negative-strand RNA replication intermediates of the incoming virus, which requires the nsP123 polyprotein precursor for transcription (see above).[39] Dual infections presumably occur when two distinct viruses enter and replicate simultaneously such that neither is subject to the exclusion mechanism. Dual infection rates up to 40% have been achieved in cultured cells infected with two different viruses, one expressing the immunoglobulin heavy chain and the other expressing

---

[37] H. Herweijer, J. S. Latendresse, P. Williams, G. Zhang, I. Danko, S. Schlesinger, and J. A. Wolff, *Hum. Gene Ther.* **6,** 1161 (1995).

[38] T. W. Dubensky, Jr., D. A. Driver, J. M. Polo, B. A. Belli, E. M. Latham, C. E. Ibanez, S. Chada, D. Brumm, T. A. Banks, S. J. Mento, D. J. Jolly, and S. M. W. Chang, *J. Virol.* **70,** 508 (1996).

[39] A. R. Karpf, E. Lenches, E. G. Strauss, J. H. Strauss, and D. T. Brown, *J. Virol.* **71,** 7119 (1997).

the immunoglobulin light chain.[40] Coimmunoprecipitation experiments performed with cells dually infected with Bcl-x_L- and Bax-expressing viruses have also permitted the analysis of Bcl-x_L mutants for their ability to heterodimerize with the Bax protein.[3] Dual infections are likely to be much less efficient in mice.

Future Directions

Although only one Sindbis virus vector is described here for use in studying programmed cell death, a growing array of Sindbis and related virus vectors has been generated to fulfill multiple purposes. Similar virus vectors are being utilized to (1) deliver antigens as vaccines,[41] (2) to target specific cell types by modification of the Sindbis virion surface,[42] and (3) to generate "transgenic" mosquitoes.[43] Combining these vectors with cell death regulatory genes poses exciting prospects.

Acknowledgments

We thank members of the Hardwick and Levine laboratories for their expertise with the Sindbis virus vector. This work was supported by National Institutes of Health Grants NS34175 and CA73581 (J.M.H.), AIO1217 and AI40246 (B.L.), the Muscular Dystrophy Association (J.M.H.), and the James S. McDonnell Foundation (B.L.).

[40] X. H. Liang, H. H. Jiang, and B. Levine, *Mol. Immunol.* **34,** 907 (1997).
[41] P. Pushko, M. Parker, G. V. Ludwig, N. L. Davis, R. E. Johnston, and J. F. Smith, *Virology* **239,** 389 (1997).
[42] T. Wickham, *Nature Biotechnol.* **15,** 717 (1997).
[43] S. Higgs, D. Traul, B. S. Davis, K. I. Kamrud, C. L. Wilcox, and B. J. Beaty, *BioTechniques* **21,** 660 (1996).

[44] Transduction of Full-Length Tat Fusion Proteins Directly into Mammalian Cells: Analysis of T Cell Receptor Activation-Induced Cell Death

By ADAMINA VOCERO-AKBANI, NATALIE A. LISSY, and STEVEN F. DOWDY

Currently, delivery of expression vectors, proteins, and/or pharmacologically important peptidyl mimetics to target cells is problematic because of the low percentage of cells targeted, overexpression, size constraints, and bioavailability. Concentration-dependent transduction of full-length proteins and domains directly into cells would serve to alleviate these problems. Previous researchers have demonstrated the ability of proteins linked to the human immunodeficiency virus (HIV) Tat transduction domain to transduce into cells; but because of inefficiencies, this methodological potential has not significantly progressed since 1988. We describe, in this chapter, a significant increase in transduction efficiency of proteins and ease of use by (1) generation of a Tat protein transduction domain in-frame bacterial expression vector, pTAT-HA, and (2) development of a purification protocol yielding denatured proteins. We have transduced full-length Tat fusion proteins ranging in size from 15 to 115 kDa into ~100% of all target cells examined, including peripheral blood lymphocytes, all cells present in whole blood, bone marrow stem cells, diploid fibroblasts, fibrosarcoma cells, and keratinocytes. Transduction occurs in a concentration-dependent manner, achieving maximum intracellular concentrations in less than 10 min. We conclude that our methodology generates highly efficient transducible proteins that are biologically active and have broad potential in the manipulation of biological experimental systems, such as apoptotic induction, cell cycle progression, and differentiation, and in the delivery of pharmacologically relevant proteins.

Introduction

Deletion of antigen-activated T cells after an immune response and during peripheral negative selection after strong T cell receptor (TCR) engagement of cycling T cells occurs by an apoptotic process termed TCR activation-induced cell death (TCR-AID).[1] Execution of TCR-AID requires both the presence of the stimulated T cell at a late G_1-phase cell cycle position, termed the death check point, and functional retinoblastoma protein (pRb).[2] To dissect the question of specific cell cycle sensitivity to AID required the ability to manipulate the biology and *in vivo* biochemistry of ~100% of T cells in a population.

Transfection or viral introduction of cDNA expression vectors and microinjection of proteins into cells present various difficulties, including massive overexpression, broad cell-to-cell intracellular concentration ranges of expressed proteins, and a low percentage of cells targeted.[3,4]

---

[1] D. R. Green and D. W. Scott, *Curr. Opin. Immunol.* **6,** 476 (1994).
[2] N. A. Lissy, L. Van Dyk, M. Becker-Hapak, J. H. Mendler, A. Vocero-Akbani, and S. F. Dowdy, *Immunity* **8,** 57 (1998).
[3] W. F. Anderson, *Nature (London)* **392,** 25 (1998).
[4] D. Bar-Sagi, *Methods Enzymol.* **255,** 436 (1995).

In addition, the use of antisense approaches to manipulate intracellular processes have both specific gene and cell type restrictions. Thus, the ability to manipulate cellular processes by introduction of full-length proteins or protein domains in a concentration-dependent fashion into 100% of cells would serve to alleviate these technological problems.

In 1988, Green and Loewenstein[5] and Frankel and Pabo[6] independently uncovered the ability of HIV Tat protein to cross cell membranes. In 1994, Fawell et al.[7] expanded on this observation by demonstrating that heterologous proteins chemically cross-linked to a 36-amino acid domain of Tat were able to transduce into cells. However, these reports did not develop into a broadly usable method to efficiently transduce proteins into cells. Subsequent to the Tat discovery, other transduction domains have been identified that reside in the Antennapedia (Antp) protein from Drosophila[8] and herpes simplex virus (HSV) VP22 protein.[9] The exact mechanism of transduction across cellular membranes remains unclear; however, small Tat peptides have been shown to transduce into cells at 4° in a receptorless fashion.[10] This observation suggests that all cell types are potentially targetable by this methodology.

Although Tat-mediated protein transduction was first discovered in 1988, no method to harness this technological potential has been devised. We describe the development of full-length protein transduction methodology by utilization of urea-denatured, shock-misfolded, genetic in-frame Tat fusion proteins that can be applied to a broad spectrum of proteins regardless of size or function. Briefly, bacterially expressed N-terminal in-frame Tat fusion proteins are isolated from bacteria by sonication in 8 M urea. The use of 8 M urea achieves two goals. First, the majority of recombinant proteins in bacteria are present in inclusion bodies as denatured insoluble proteins. Sonication in urea solubilizes this material, thus allowing for its isolation. Second, transduction of denatured Tat fusion proteins elicits biological responses more efficiently than correctly folded soluble proteins.[11] The denatured proteins are made aqueously soluble and added directly to the tissue culture medium. We have transduced Tat fusion pro-

[5] M. Green and P. M. Loewenstein, Cell 55, 1179 (1988).

[6] A. D. Frankel and C. O. Pabo, Cell 55, 1189 (1988).

[7] S. Fawell, J. Seery, Y. Daikh, C. Moore, L. L. Chen, B. Pepinsky, and J. Barsoum, Proc. Natl. Acad. Sci. U.S.A. 91, 664 (1994).

[8] D. Derossi, A. H. Joliot, G. Chassaings, and A. Prochiantz, J. Biol. Chem. 269, 10444 (1994).

[9] G. Elloit and P. O'Hare, Cell 88, 223 (1997).

[10] E. Vives, P. Brodin, and B. Leblus, J. Biol. Chem. 272, 16010 (1997).

[11] H. Nagahara, A. Vocero-Akbani, E. L. Snyder, A. Ho, D. G. Latham, N. A. Lissy, M. Becker-Hapak, S. A. Ezhevsky, and S. F. Dowdy, Nat. Med. 4, 1449 (1998).

teins into a variety of primary and transformed cell types, including peripheral blood lymphocytes (PBLs), diploid human fibroblasts, keratinocytes, bone marrow stem cells, osteoclasts, fibrosarcoma cells, leukemic T cells, osteosarcoma, glioma, hepatocellular carcinoma, renal carcinoma, NIH 3T3 cells, and all cells present in whole blood, including both nucleated and enucleated cells.[2,11–13] Most importantly, we have generated and transduced more than 50 full-length proteins and domains from 15 to 115 kDa by this method, suggesting that most proteins may be transduced into cells.[11]

Materials and Reagents

Buffers

Buffer Z: 8 M urea–100 mM NaCl–20 mM HEPES (pH 8.0)
Buffer A: 50 mM NaCl–20 mM HEPES (pH 8.0)
Buffer B: 1 M NaCl–20 mM HEPES (pH 8.0)
Phosphate-buffered saline (PBS)
Paraformaldehyde (4%, w/v)
Imidazole (5 M)

Reagents

pTAT-HA plasmid
BL21(DE3)LysS bacteria (Novagen, Madison, WI)
12CA5 anti-hemagglutinin (HA) antibodies (BabCO, Berkeley, CA)
Protein fast protein liquid chromatography (FITC) labeling kit (Pierce, Rockford, IL)
Ni-NTA resin (Qiagen, Chatsworth, CA)
Resource Q and S resin (Pharmacia, Uppsala, Sweden)
LB medium (Sigma, St. Louis, MO)

Equipment

Gravity columns (Bio-Rad, Hercules, CA)
Mono Q and S columns (5/5 or 10/10; Pharmacia)
Slide-a-Lyzer dialysis cassettes (Pierce)
Sonicator
FPLC (optional; Pharmacia)
Sodium dodecyl sulfate–polyacrylamide gel electrophoresis (SDS–PAGE) and protein transfer units

[12] S. A. Ezhevsky, H. Nagahara, A. Vocero-Akbani, D. R. Gius, M. Wei, and S. F. Dowdy, *Proc. Natl. Acad. Sci. U.S.A.* **94,** 10699 (1997).
[13] A. Vocero-Akbani, N. Vander Heyden, N. A. Lissy, L. Ratner, and S. F. Dowdy, *Nat. Med.* **5,** 29 (1999).

Fluorescence-activated cell sorter (FACS)
Fluorescence microscope

Generation of Transducible Tat Fusion Proteins

To produce genetic in-frame Tat fusion proteins, we constructed a bacterial expression vector, pTAT-HA, that contains an N-terminal 6-histidine leader followed by the 11-amino acid Tat protein transduction domain[5] flanked by glycine residues for free bond rotation of the domain, a hemagglutinin (HA) tag, and a polylinker (Fig. 1A). As examples, pTAT-HA vectors expressing full-length wild-type and mutant cDNAs from the p16^{INK4a} tumor suppressor protein (Tat–p16, 20 kDa), human papilloma virus E7 oncoprotein (Tat–E7, 18 kDa), and adenovirus 13S E1A oncoprotein (Tat–E1A, 60 kDa) were cloned into pTAT-HA.[2,12] To obtain a genetic N-terminal in-frame fusion with the Tat leader, the 5′ untranslated regions (UTRs) of each cDNA were deleted. The plasmid is transformed into a bacterial strain that yields a high copy plasmid number, such as DH5α, and the genetics are assayed. The pTAT-HA plasmid is then transformed into the high-expressing BL21(DE3)LysS bacterial strain.

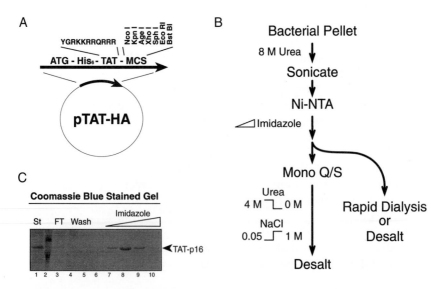

FIG. 1. Characterization of pTAT vector and purification protocol. (A) pTAT expression vector. (B) Purification and "shock" misfolding protocol. (C) Purification of Tat–p16^{WT} fusion protein over Ni-NTA resin, resolution by SDS–PAGE, and staining with Coomassie blue. St, Start; FT, flowthrough.

Overnight cultures (1 ml) from 6–10 isolates are grown in the presence of isopropyl-β-D-thiogalactopyranoside (IPTG). The bacteria are boiled in 2× SDS gel loading buffer and analyzed by SDS–PAGE. One gel is stained with Coomassie blue and expression of the recombinant protein is compared with a nonexpressing BL21(DE3)LysS control lysate. Please note that not all Tat fusion proteins are expressed at a substantial enough level to be viewed in this manner. Another gel is transferred to a filter and probed with anti-HA antibodies by immunoblot analyses.[14] The information derived from both of these analyses will allow the focus to center on a single, high-expressing isolate. Please note that the Tat leader is approximately 3.5 kDa in size; however, the addition of the leader to some proteins results in a further increase in apparent molecular weight as measured by SDS–PAGE. Glycerol stocks are made of all the high-expressing BL21(DE3)LysS isolates. For Tat fusion proteins that are cytotoxic to bacteria, such as Tat–CPP32, Tat–pRb, and Tat–Bcl-2, we routinely obtain higher yields from 1-liter overnight stationary-phase cultures compared with log-phase cultures.

Urea Denaturation of Tat Fusion Proteins

The exact mechanism of transduction across bilipid membranes is currently unknown; however, an analysis of Tat–p27^{Kip1} protein revealed that urea-denatured proteins elicit biological phenotypes more efficiently than soluble, correctly folded protein.[11] We hypothesized that, because of reduced structural constraints, higher energy (ΔG), denatured proteins may transduce more efficiently into cells than lower energy, correctly folded proteins. Once inside the cell, transduced denatured proteins would be correctly refolded by chaperones[15] such as heat shock protein 90 (HSP90).[16] Therefore, we devised a urea-denaturing protein purification protocol (Fig. 1B). The protocol was originally designed for use with an FPLC protein purification system; but because of the use of single steps to perform buffer changes, commonly available gravity flow columns can be substituted for FPLC columns.

Start a 100- to 200-ml overnight culture of high-expressing pTAT-HA BL21(DE3)LysS isolate in LB. The next morning, inoculate into 1 liter of LB plus 100 μM IPTG and shake for 5–10 hr at 37°. Isolate and wash the

[14] S. F. Dowdy, L. VanDyk, and G. S. Schreiber, in "Human Genome Methods" (K. W. Adolph, ed.), p. 121. CRC Press, Boca Raton, Florida, 1997.

[15] S. Gottesman, S. Wickmer, and M. R. Maurizi, Genes Dev. 11, 815 (1997).

[16] C. Schneider, L. Sepp-Lorenzino, E. Nimmesgern, O. Ouerfelli, S. Danishefsky, N. Rosen, and F. U. Hartl, Proc. Natl. Acad. Sci. U.S.A. 93, 14536 (1996).

cell pellet in PBS, and then resuspend in 10 ml of buffer Z and sonicate on ice three times (15-sec pulses) or until turbid. Clarify by centrifugation (5000g at 4° for 10 min) and save the supernatant. To prevent nonspecific binding of cellular proteins to the Ni-NTA resin, add imidazole to buffer Z to achieve a final concentration of 10–20 mM, then add sample at 4° or room temperature to a preequilibrated 3- to 10-ml Ni-NTA column in buffer Z plus 10–20 mM imidazole. Allow the sample to flow through the column by gravity or apply slight air pressure via a syringe as required. Save some of the Start (St) and flowthrough (FT) for SDS–PAGE analysis. Wash the column in ~50 ml of buffer Z plus 10–20 mM imidazole. Elute the Tat fusion protein by stepwise addition of 5–10 ml of buffer Z containing 100 mM, 250 mM, 500 mM, and 1 M imidazole.

Analyze the St, FT, and each column fraction by Coomassie blue (Fig. 1C) and immunoblot analysis as outlined above and pool appropriate fractions. If the Tat protein is detected primarily in the FT, decrease the imidazole concentration in loading buffer Z and if a high background of contaminating bacterial proteins is detected, increase the imidazole concentration of the loading buffer. In addition, because of the "head" binding nature of the Ni-NTA column, removing the urea (see below) on this column results in aggregation and precipitation of the protein on the column.

Solubilization of Tat Fusion Proteins into an Aqueous Buffer

To use the Tat fusion proteins in an aqueous environment, such as tissue culture media, rapid removal of the 8 M urea is required. This also achieves the goal of obtaining high energetic (ΔG), denatured Tat fusion proteins. Several choices for this procedure are described below. In our experience, the use of Mono Q or S ion-exchange chromatography yields superior results than rapid dialysis or utilization of a desalting column. However, we have generated transducible Tat fusion proteins by both of these approaches (Fig. 1B).

Ion-Exchange Chromatography

Fast Protein Liquid Chromatography

The isoelectric point (pI) of the fusion protein will in large measure determine whether to use a Mono Q (acidic proteins) or Mono S (basic proteins) column. The Tat leader is a basic entity (8 of 11 residues are basic); however, it has been our experience that ~50% of all Tat fusion proteins will bind to the Mono S resin regardless of pI predictions.

Dilute the pooled Ni-NTA fractions from above 1:1 with 20 mM HEPES, pH 8.0 for Mono Q and pH 6.5 for Mono S. This will result in 50 mM NaCl and 4 M urea final concentrations. Inject the sample into a 10/10 (preferably) Mono Q/S column attached to an FPLC equilibrated in buffer A plus 4 M urea. Wash with ~50 ml of buffer A (no urea) and elute with a single 1 M NaCl step in buffer A. Analyze St, FT, and column fractions by SDS–PAGE as outlined above and pool appropriate fractions. The sample is then desalted on a PD-10 (G-25 Sephadex) desalting column equilibrated in PBS, collected, and analyzed by SDS–PAGE.

By reducing the urea from 4 to 0 M in a single step, the denatured proteins are forced to become soluble in an aqueous environment. Because of the mixed population of protein configurations and hence, biophysical properties, use of a single 1 M NaCl step is preferred and will result in a sharper protein peak resolution compared with a linear NaCl gradient. Check St, FT, and column fractions as outlined above for SDS–PAGE. If failure to bind or weak binding of protein is observed, try the other column type regardless of predicted pI. If strong binding is observed with weak elution, decrease (Q resin) or increase (S resin) the buffer A pH by steps of 0.5 pH units until a small amount of protein is detected in the FT fraction.

Gravity Columns

Because of the design of the protocol to use single elution steps, an FPLC or HPLC is not required. Set up a 10- to 40-ml ion-exchange column, using 30-μm Resource Q or S resin (Pharmacia). The sample is diluted as described above and, because of a low back pressure, is injected into the preequilibrated column via a syringe. Wash the column with ~50 ml of buffer A plus imidazole and elute Tat fusion protein with a single step of 1 M NaCl. The sample is then desalted on a PD-10 desalting column, collected, and analyzed by SDS–PAGE.

Rapid Dialysis

Removal of urea by dialysis is generally a slow process, allowing partial or complete refolding of the protein and, therefore, counter to the goal of obtaining soluble, denatured Tat fusion proteins. Therefore, rapid dialysis is performed by utilization of dialysis cassettes with a high surface area-to-protein volume ratio. The $\leq$2 ml of Tat fusion protein present in buffer Z from above (Ni-NTA column) is placed in a 15-ml capacity "Slide-a-Lyzer" cassette and dialyzed against PBS or the buffer of choice in a large volume (~4 liter at 4°) that is changed twice. After dialysis, spin out the insoluble precipitate and check the protein concentration/integrity on a Coomassie blue-stained SDS–polyacrylamide gel in comparison with protein standards, such as bovine serum albumin (BSA). Dialysis can result in aggregation and precipitation of the Tat fusion protein that is both concentration and protein dependent.

Desalting Column

The theory behind rapid desalting is that passage of a denatured protein from 8 M urea through the interface into PBS on the other side forces the protein to rapidly hide its hydrophobic residues and become aqueously soluble. The desalting column has a 1:1.4 dilution factor, and therefore denatured proteins are separating from each other, helping to avoid aggregation of the proteins and subsequent precipitation on the column. A total of 1–1.5 ml of the Tat fusion protein in 8 M urea is applied to a disposable PD-10 (G-25 Sephadex) desalting column equilibrated in PBS. Column fractions of 1 ml are isolated and analyzed by SDS–PAGE as described above. Tat fusion proteins usually elute in fractions 6 and 7. We have had a reasonable amount of success using this rapid and inexpensive procedure.

Regardless of the method of generating denatured, aqueously soluble Tat fusion proteins, the protein concentration is determined by SDS–PAGE in comparison with a standard, such as BSA, and/or by Bio-Rad protein concentration analysis. In addition, all proteins are flash frozen in 10% (v/v) glycerol and stored at $-80°$. To check for loss of protein due to freeze–thaw-induced precipitation, thaw a test vial, spin at 12,000–100,000g at 4° for 10 min, and analyze by SDS–PAGE.

Transduction of Fluorescein-Labeled Tat Fusion Proteins into Cells

Tat fusion proteins can be labeled with fluorescein isothiocynate (FITC), which conjugates to lysine residues. Tat fusion protein (20–50 μg) is placed in a 300-μl reaction mix as per the manufacturer instructions for 2 hr at room temperature. The nonconjugated FITC is then removed from the FITC-labeled Tat protein by either gel-filtration chromatography or use of a PD-10 desalting column. We routinely use a PD-10 column equilibrated in PBS. Check the column fractions by Bio-Rad protein analysis and/or SDS–PAGE.

To analyze for transduction into cells, add 100–400 μl of purified FITC-labeled Tat fusion protein to 10^5 to 10^6 cells in 1 ml of medium. Analyze for transduction by flow cytometry (FACS) analysis at 10 and 20 min postaddition of the FITC-labeled Tat fusion protein (Fig. 2A). FITC-labeled Tat fusion proteins rapidly transduced into ~100% of cells, achieving maximum intracellular concentration in less than 10 min. In addition, we note a narrow intracellular concentration range of the transduced protein within

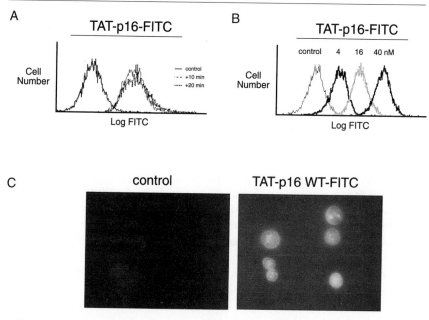

FIG. 2. Analysis of FITC-labeled Tat fusion proteins added to cells. (A) FITC-labeled Tat–p16^{WT} fusion protein was added to Jurkat T cell cultures and analyzed by FACS at 10 and 20 min postaddition. Note the rapid shift of the entire cellular population within 10 min. (B) Equilibrium FACS analysis of Jurkat cells 1 hr posttransduction with 4, 16, and 40 nM Tat–p16^{WT}–FITC protein. Note the concentration-dependent transduction and narrow intracellular concentration as measured by nearly equivocal FACS peak widths between control and transduced cells. (C) Confocal microscopy analysis of Jurkat T cells transduced with control (*left*) or Tat–p16^{WT}–FITC protein (*right*). Note cytoplasmic and nuclear fluorescence of Tat–p16^{WT}–FITC protein and not mere attachment to the cellular membrane.

the population as supported by the narrow FACS peak width between control and transduced cells. Furthermore, FACS analysis of cells 1 hr postaddition (steady state levels) of 4, 16, and 40 nM FITC-labeled Tat fusion protein demonstrates a concentration dependency for protein transduction and, hence, the ability to modulate intracellular concentrations (Fig. 2B). In addition, confocal analysis of transduced cells fixed in 4% (w/v) paraformaldehyde shows the presence of FITC-labeled Tat proteins in both the nuclear and cytoplasmic compartments, and not merely bound to the cellular membrane (Fig. 2C). We have successfully transduced Tat fusion proteins into a wide variety of cell types including peripheral blood lymphocytes (PBLs), diploid human fibroblasts, keratinocytes, bone marrow stem cells, osteoclasts, fibrosarcoma cells, leukemic T cells, osteosarcoma, glioma, hepatocellular carcinoma, renal carcinoma, NIH 3T3 cells, and all cells present in whole blood, including both nucleated and enucleated cells[2,11-13] (our unpublished observations, 1999).

Biological Activities of Transduced Tat Fusion Proteins

Previously, p16[INK4a] and E1A proteins have been shown to bind Cdk6 and the retinoblastoma protein (pRb), respectively.[17] Therefore, to analyze for *in vivo* biochemical function of transduced proteins, p16(−/−) Jurkat T cells were treated with 100 or 200 nM (final concentration) Tat–p16 protein during concomitant [^{35}S]methionine labeling of cellular proteins. Cellular lysates were prepared and Tat fusion proteins were immunoprecipitated with anti-p16 antibodies followed by reimmunoprecipitation with anti-Cdk6 antibodies (Fig. 3A). Transduced Tat–p16 wild-type protein, but not mutant protein, bound endogenous Cdk6 in a concentration-dependent fashion. In addition, by transducing Tat–p16 or Tat–E1A into the cells at various times after labeling cellular proteins with [^{35}S]methionine, we observed a rapid refolding and binding of the transduced Tat–p16 to cellular Cdk6 and Tat–E1A to cellular pRb (Fig. 3B and C). We have observed similar results with transduced Tat–E7 and Tat–p27, which rapidly bind intracellular targets.[2,11]

TCR activation-induced cell death (TCR-AID) occurs from a late G$_1$ death check point in the presence of active pRb, a transcriptional negative regulator of G$_1$-phase cell cycle progression.[2] To directly test if both cell cycle position and functional pRb were required to mediate TCR-AID, we TCR-stimulated centrifugally elutriated early G$_1$-phase Jurkat T cells and transduced Tat–E7 to inactivate pRb and Tat–p16 to effect a premature cell cycle arrest (Fig. 4). Both Tat–E7 and Tat–p16 wild-type proteins rescued T cells from TCR-AID in a concentration-dependent manner, whereas mutant versions did not. These observations demonstrate that TCR-AID requires the T cell at the death check point and functional pRb. In addition, these observations demonstrate the utility of transducing proteins into cells to dissect biological and *in vivo* biochemical processes.

Discussion

We have utilized the procedure described here to generate and transduce more than 50 urea-denatured proteins into ~100% of all cells assayed thus far, including primary cells from blood, osteoclasts, bone marrow stem cells, and peripheral blood lymphocytes.[11-13] The use of a genetic in-frame fusion combined with denaturation of the proteins achieves several goals,

---

[17] C. J. Sherr, *Science* **274**, 1672 (1996).

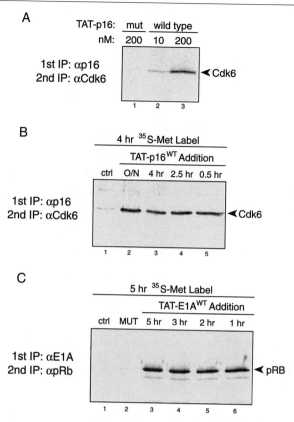

FIG. 3. Formation of *in vivo* complexes of Tat–p16 with Cdk6 and Tat–13S E1A with pRb. (A) p16(–/–) Jurkat T cells were transduced with 10 and 200 n*M* Tat–p16^(WT/MUT) fusion proteins during [35S]methionine labeling of cellular proteins for 4 hr. Cellular lysates were immunoprecipitated with anti-p16 antibodies, then reimmunoprecipitated with anti-Cdk6 antibodies and resolved by SDS–PAGE. (B and C) Jurkat T cells were transduced at various times with 100 n*M* Tat–p16^(WT) or Tat–E1A fusion proteins during [35S]methionine labeling of cellular proteins for 4 or 5 hr, respectively. Cellular lysates were immunoprecipitated with anti-p16 or anti-E1A antibodies, then reimmunoprecipitated with anti-Cdk6 or anti-pRb antibodies and resolved by SDS–PAGE. The positions of Cdk6 and pRb are as indicated.

including isolation of the bulk of recombinant proteins that are usually present in inclusion bodies, increased efficiency of biological response, and ease of use. Once inside the cell, transduced denatured proteins appear to be rapidly refolded by chaperones,[15] such as HSP90,[16] and are capable of binding their cognate intracellular targets and performing biochemical functions, such as protection from apoptosis (see Fig. 4) as well as induction of apoptosis by caspase 3 (CPP32).[13]

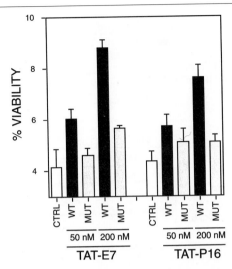

FIG. 4. Rescue of AID by direct transduction of TAT–E7 and TAT–p16 protein. Purified TAT–E7 and TAT–p16 wild-type and mutant proteins were transduced into PDBu/iono-mycin-stimulated, elutriated, G_1 Jurkat T cells at 50 or 200 nM final concentration and then assayed for cell viability at 9 hr poststimulation. Both TAT–E7 and TAT–p16 wild-type proteins rescued cells from TCR-AID in a concentration-dependent fashion. TAT–E7 mutant protein retains a low affinity for pRb and shows a weak level of protection compared with TAT–E7 wild-type protein. Cell viability was determined by trypan blue dye exclusion. All experiments were performed in triplicate and error bars indicate standard deviation.

Because of size constraints, research into therapeutically important proteins has focused on miniaturization of protein-specific functions, such as contact sites. However, large protein domains have been evolutionarily selected to yield high-affinity active sites and protein:protein and protein:DNA contact sites. As an example, in our hands, a 300 μM concentration of a 20-amino acid p16 peptidyl mimetic[18] fused to the Tat transduction domain is required to elicit a G_1-phase cell cycle arrest (our unpublished observation, 1999); however, only a 150 nM concentration of full-length Tat–p16 protein is required.[12] Thus, transduction of full-length p16 protein into cells increases the specificity of the biological response by three orders of magnitude compared with a peptidyl mimetic. In addition, the Tat–p16 peptidyl mimetic induces both apoptotic and cell adhesion side effects that are not associated with the Tat–p16 full length protein.

We conclude that transduction of denatured proteins directly into ~100% of primary or transformed cells has broad implications for regulating

[18] R. Fahraeus, J. M. Paramio, K. L. Ball, S. Lain, and D. P. Lane, *Curr. Biol.* **6,** 84 (1996).

intracellular processes such as apoptosis, cell cycle arrest, and differentiation in experimental systems and has the potential to allow development of new therapeutic strategies utilizing full-length proteins and protein domains that retain high affinities for their intracellular targets.

Author Index

Numbers in parentheses are footnote reference numbers and indicate that an author's work is referred to although the name is not cited in the text.

A

Abrams, J. M., 65, 67, 67(1), 68, 68(1), 70, 70(17), 71, 75, 76(1, 27)
Achyuthan, K. E., 455, 457
Acosta, D., Jr., 225
Adachi, S., 213, 220(8)
Adam, D, 383
Adam-Klages, S., 383
Adams, J. L., 389
Adams, V., 244
Adany, R., 440, 465(28)
Adida, C., 155
Adolphe, M., 444, 454(1)
Aebersold, R., 126, 128(8)
Aeschlimann, D., 434, 442, 443(36), 444, 444(6), 445(38)
Afanas'ev, V. N., 9
Agapite, J., 68
Aggarwal, B., 347
Ahmad, M., 145, 155, 183, 195(10), 235, 326
Ahmed, N. N., 402
Ahmed, T., 24, 33(45, 46, 48)
Aimé-Sempé, C., 162, 255, 256(13), 257(13), 265(13), 283
Akerman, K. E., 227
Albertson, D. G., 78, 79(18)
Albrecht, H., 413
Aldape, R. A., 101
Alessi, D. R., 402, 405, 407, 407(25), 408, 409, 409(24)
Alexander, C., 379
Alexander, H. R., 347
Al-Habori, M., 422
Ali, A., 101, 106(4), 177
Allen-Hoffmann, B. L., 442, 443(36)
Alnemri, E. S., 10, 53, 85, 86(35), 98, 144, 145(13), 154(13), 155, 164, 165(20), 168(20), 169(20), 170(20), 177, 195, 235, 326, 370
Alonso-Llamazares, A., 389
Altermatt, S., 444
Altieri, D., 155
Altman, A., 382(8), 383
Ambros, V., 84
Ambrose, C. M., 348, 351(16), 355(16)
Ambrosini, G., 155
Ameisen, J. C., 297
Amendola, A., 439, 457(26)
Amiri, P., 70
Anderson, D. C., 164, 168(18)
Anderson, J. W., 418
Anderson, K., 408
Anderson, W. F., 509
Andjelkovic, M., 402, 405, 407, 407(25), 409(24)
Andreoli, J., 435, 465(9), 468(9)
Andreyev, A. Y., 222, 233
Andrieu-Abadie, N., 414
Androlewicz, M. J., 348, 361
Annichhiarico-Petruzzelli, M., 439, 440(27), 459(27)
Anton, R., 184, 413, 416(3)
Antonsson, B., 184, 195(12), 242, 271, 276, 278(18)
Apgar, J. R., 14
Appel, J. R., 102
Appelbaum, E. R., 326
Arato, G., 445
Arch, R. H., 183
Arends, M. J., 20, 22(10, 16), 23(10, 16), 39(10), 490
Arias, A. M., 66
Armstrong, R., 162, 195, 283, 284(10), 370
Arsura, M., 393
Ashburner, M., 66, 69(12), 75(11, 12)
Ashkenazi, A., 325, 326, 351, 483(29), 491

523

J

Jacinto, E., 390, 391
Jacob, W. A., 218, 415
Jacobson, D. M., 77, 78(9), 79(9), 80(9)
Jacobson, M. D., 177
Jaffe, E., 255
Jakubowicz, T., 401
James, C., 162, 283
James, I. E., 326
James, S. R., 408
Jamison, W. C. L., 374, 376(8)
Janes, W., 347, 364
Jans, D. A., 127
Jans, P., 127
Jansen, G., 77
Jarrett, J., 347
Jasaitis, A. A., 232
Jayadev, S., 382(7), 383
Jemmerson, R., 157, 158(29), 159(29), 177, 183, 185(9), 195(9), 198, 205, 218, 235, 243
Jenkins, G. M., 373
Jiang, C., 65
Jiang, H. H., 492, 493(4), 496(4), 504, 504(4), 506, 508
Jiang, T., 225
Jiang, Y., 378, 388
Jia-Tsrong, J., 496
Jin, Y. S., 84
Jockel, J., 112
Joe, A. K., 496
Johansen, H., 73
Johnson, A. L., 10
Johnson, E. M., 415
Johnson, G., 472
Johnson, G. L., 391
Johnson, G. S., Jr., 445
Johnson, K. S., 391, 392(21)
Johnson, R. E., 493
Johnson, R. S., 326, 347
Johnsson, N., 164, 168(17)
Johnston, R. E., 493, 508
Joliot, A. H., 510
Jolly, D. J., 507
Jones, D. P., 73, 85, 195, 235, 413, 415(6)
Jones, E. Y., 347
Jones, P. F., 401
Jones, S., 244
Jongeneel, C. V., 413

Joris, I., 20, 39(13)
Jouavill, L. S., 243
Juan, G., 9, 10(38), 19, 20(3), 28(3)
Jung, S.-W., 457
Jung, Y.-K., 481, 482, 482(5), 483(5, 12)
Jürgensmeier, J. M., 159, 162, 184, 195(13), 232, 242, 283, 284(10), 301

K

Kaariainen, L., 495
Kadam, S., 164, 168(15)
Kagan, B. L., 271, 276
Kagi, D., 126
Kaipia, A., 187
Kakimoto, S.-I., 286, 289(15)
Kakizuka, A., 483(19), 488
Kalashnikova, G., 21, 29(31), 38(31)
Kam, C. M., 126, 128, 128(8)
Kamada, S., 113, 144
Kamen, R., 101
Kamentsky, L. A., 19, 20, 20(4, 5), 30(4, 5)
Kamentsky, L. D., 19, 20(4, 5), 30(4, 5)
Kameyama, K., 401
Kamiike, W., 21, 29(30), 212
Kamrud, K. I., 508
Kane, D. J., 184, 413, 416(3), 421(8)
Kang, J. J., 164, 165(20), 168(20), 169(20), 170(20)
Kantardjieff, K. A., 275
Kantrow, S. P., 243
Kanuka, H., 482, 483(11)
Kaplan, D. R., 401, 402, 403, 403(22), 404(22), 405(22), 406, 406(22), 407(22, 23), 408(23), 481
Kapuscinski, J., 38
Karin, M., 389, 390, 391, 394, 395, 399(11)
Karn, J., 77
Karp, J. E., 7
Karpf, A. R., 507
Kartasova, T., 437, 440, 461(31)
Kartenbeck, J., 495
Katz, D., 393
Katz, L., 164, 168(15)
Kaufmann, S. H., 3, 4, 7, 13, 183, 185, 198
Kavanaugh, W. M., 408
Kawata, M., 458
Kazlauskas, A., 402, 403(22), 404(22), 405(22), 406(22), 407(22)

M

MacCross, M., 101, 106(5)
MacDonald, H. R., 4, 12(18)
MacFarlane, M., 326
Machleidt, T., 382, 385, 385(5), 387(19)
Macho, A., 21, 29(27), 200, 209, 240, 415
MacIvor, D. M., 126
Mackey, F., 358
Macleod, H. A., 267
Madeo, F., 302
Madi, A., 446
Magnani, M., 420
Mahajan, N., 275, 283, 284(8), 285(8), 292(8)
Mahboubi, A., 360, 361(35), 382(8), 383
Maher, P., 413
Mai, S., 127
Majno, G., 20, 39(13)
Malstrom, S., 256
Mandle, R. J., 128
Mandon, E. C., 379
Maniatis, T., 10, 50, 460, 480
Mankovich, J. A., 93, 94(5), 101
Mann, M., 367, 370, 370(8), 372(8), 394
Mannella, C. A., 243
Mannherz, H. G., 4, 48
Manning, A., 394
Manon, S., 195, 284
Mantsygin, Y. A., 9
Mao, C., 373
Mao, X., 255
Marcellus, R. C., 483(17), 488
March, C. J., 330, 347
Marchetti, P., 21, 29(27), 198, 205, 209, 244, 251(15), 415
Marekov, L. N., 435, 437, 440, 461(30, 31), 465(9, 17), 466(17), 467(17, 30, 32), 468(9, 17)
Margolin, N., 12
Marhefka, G., 23, 31(40)
Mari, M., 415
Markkov, L. N., 450, 468(54)
Markova, N. G., 469
Maroni, G., 73
Marsh, J. B., 374
Marshall, C. J., 388
Marsters, S. A., 325, 326, 483(29), 491
Martin, S. J., 15(3), 16, 157, 157(32), 158, 158(28), 159(28, 32), 183, 185, 195(30, 41), 198, 360, 361(35), 382(10), 383

Martinetto, H., 391
Martinou, I., 255, 276, 278(18)
Martinou, J. C., 242, 255, 276, 278(18)
Martins, L. M., 13
Marzatico, F., 223
Marzku, S., 364
Marzo, I., 208, 243, 244(5), 251, 251(5), 275, 302, 379
Mashima, T., 164, 168(19)
Masters, S., 409
Masui, Y., 192
Masuzumi, M., 224
Mathias, S., 185, 198, 380
Mat Suda, H., 242
Matsuda, H., 21, 29(30)
Matsuno-Yagi, A., 195, 413
Matsuyama, S., 184, 268, 271, 272(7), 273(7), 274(7), 276, 283, 285, 293, 296, 302, 303(17)
Matsuzaki, H., 401, 402
Mattaliano, M. D., 402
Mattmann, C., 343, 349, 351(19)
Mauel, J., 41, 43(11)
Maundrell, K., 276, 278(18), 300
Maurer, F., 401
Mauri, D. N., 351, 361(21), 362
Maurizi, M. R., 513, 519(15)
Maybaum, J., 3
Mayer, B. J., 185
Mazat, J.-P., 243
Mazzei, G., 242, 276, 278(18)
McBride, O. W., 442, 443(35), 469
McCall, J. O., 164, 168(15)
McCall, K., 65
McConnell, M., 256, 257(18), 263, 264(22), 265(22)
McCormick, F., 408
McCulloch, E. A., 21, 22(28), 29(28)
McDevitt, H., 351
McDonnell, J. M., 266, 275, 275(15)
McDonnell, P. C., 326, 389
McElhaney, J. E., 41
McGahon, A. J., 15(3), 16
McGarry, T. J., 113, 125(13)
McGuiness, L., 101
McGuire, M. J., 127
McHugh, T., 415
McKinney, J. S., 225
McLaughlin, M., 379, 389
McLoughlin, M., 382(10), 383

Subject Index

A

ISBN 0-12-182223-0

90038